D1272570

Asymptotic Structure of Space-Time

Asymptotic Structure of Space-Time

Edited by

F. Paul Esposito and Louis Witten

University of Cincinnati
Cincinnati, Ohio

Plenum Press · New York and London

Library of Congress Cataloging in Publication Data

Symposium on Asymptotic Structure of Space-Time,
University of Cincinnati, 1976.
Asymptotic structure of space-time.

Includes index.
1. Space and time–Congresses. 2. Mathematical physics–Congresses. I. Esposito, F. Paul. II. Witten, Louis. III. Title.
QC173.59.S65S98 1976 530.1'1 77-487
ISBN 0-306-31022-8

Proceedings of a Symposium on Asymptotic Structure of Space-Time,
held at the University of Cincinnati, Ohio, June 14–18, 1976

A Division of Plenum Publishing Corporation
227 West 17th Street, New York, N.Y. 10011

Printed in the United States of America

Preface

The Symposium on Asymptotic Structure of Space-Time (SOASST) was held at the University of Cincinnati, June 14-18, 1976. We had been thinking of organizing a symposium on the properties of "infinity" for several years. The subject had reached a stage of maturity and had also formed a basis for important current investigations. It was felt that a symposium, together with a publication of the proceedings, would review, summarize, and consolidate, the more mature aspects of the field and serve as an appropriate introduction to an expanding body of research. We had from the first the enthusiastic support and encouragement of many colleagues; with their cooperation and advice, the Symposium acquired its final form. These proceedings will attest to the value of the Symposium. The Symposium consisted of thirty lectures and had an attendance of approximately one hundred and thirty.

The final impetus to our decision to go forward was the Bicentennial Anniversary of the independence of our country. A most appropriate celebration on a University Campus surely is an intellectual Symposium which pays honor to the histories and traditional purposes of a University. The Symposium was supported financially by the University of Cincinnati Bicentennial Committee, the National Science Foundation, the Gravity Research Foundation, and by Armand Knoblaugh, Professor Emeritus of Physics of the University of Cincinnati.

We also have the pleasant task of recognizing the people who, in various ways, contributed to the smooth operation of the Symposium. During the days preceding the Symposium and during the week in which it was held, much of the actual leg work was done by Gillian Mayers and Richard Hatfield, degree candidates in the Physics Department of the University of Cincinnati. Shari Dillmore deserves a special mention for the patience and skill she displayed with manuscript typing and other secretarial tasks. Of course, the hidden variable who made everything work from the beginning was our organizing secretary, Elaine Marmel. We are grateful to all of them and to our colleagues in the Physics Department who assisted so graciously.

F. Paul Esposito and Louis Witten Cincinnati, 1976

Contents

Asymptotic Structure of Space-Time

ASYMPTOTIC STRUCTURE OF SPACE-TIME

Robert Geroch*

Department of Physics, University of Chicago

Chicago, Illinois 60637

I. INTRODUCTION

Many physical theories have the feature that one can distinguish within the theory a certain class of models which one regards as representing "isolated systems". In Newtonian gravitation, to take one example, one might define a solution as representing an isolated system if i) the mass density vanishes outside some compact set in the Euclidean 3-space, and ii) the Newtonian gravitational potential approaches zero in the limit far from that compact set. Normally, one would not expect that the models so distinguished will actually be realized in our World. Thus, with respect to the example above, one might expect that no matter how far one recedes from a given system in our own Universe one will encounter additional galaxies, whence i) will fail in our Universe. Nonetheless less, it turns out that the solutions so distinguished within a given theory can be of considerable physical interest, for one often encounters in the physical World systems to which these solutions are a good approximation, e.g., in the Newtonian example, our solar system. Indeed, one could perhaps argue for a much stronger statement: It is in a sense only through a suitable notion of an isolated system that one acquires any ability at all to deal individually with various subsystems in the Universe - in particular, to assign to subsystems such physical attributes as mass, angular momentum, character of emitted radiation, etc. With no ability to isolate, one would presumably be restricted to consideration only of solutions within each theory which purport to represent our Uni-

* Supported in part by the NSF, contract number MPS 74-17456, and by the Sloan Foundation.

verse in every detail. In effect, one would be unable to make much progress at all in physics. In any case, the standard procedure within a given theory is first to obtain a more or less precisely defined class of solutions "representing isolated systems" within the mathematical framework of the theory, and then to attribute to solutions within that class various properties of possible physical interest, relegating the approximate character of these solutions and associated properties to the point of their application to actual physical systems.

General relativity is a physical theory: One would like to carry out such a program within it. This turns out to be a comparatively difficult task, despite the fact that the same program is rather simple in other theories (so much so that it often goes without mention). Why should this be so? Apparently, the answer is the standard reason why almost everything is more difficult in general relativity: The fields present in other theories are normally separated into one nondynamical background field and other fields which describe the physics (e.g., in Newtonian gravitation, the Euclidean metric and the gravitational potential); in general relativity, by contrast, the metric is both. The extraction of physical meaning from dynamical fields is more difficult without a nondynamical reference point. Thus, for example, given a solution in general relativity which represents "two systems, rather far apart", there is no algorithm for writing down a new solution representing "the same two systems, but now somewhat farther apart", nor is there a simple way, e.g., to split the Riemann tensor at a point into contributions from the two systems individually. It is perhaps not surprising, then, that it should turn out to be a rather delicate matter even to provide a definition of "isolated system" in general relativity. We shall here survey what has been done in formulating such a notion, and extracting its implications, in general relativity.

One begins with the following observation: Minkowski space (R^4 with a flat complete Lorentz metric and zero stress-energy tensor) is a solution of Einstein's equation. One might therefore expect on physical grounds that "as one recedes from an isolated system (whatever that means), the space-time geometry should approach this one". The first step, then, is to translate "isolated system" to mean "asymptotically flat". Recognition of this step is, however, far from solving the problem, for what is "asymptotically flat" to mean? Presumably, something such as "far from the matter, the metric of space-time approaches some flat metric", or "far from the matter, the Riemann tensor of space-time approaches zero". Unfortunately, the notion of one tensor field's approaching another on a manifold does not in general make sense. Comparison of components with respect to some chart, for example, will not do, for the result of the comparison will in general depend crucially on the choice of chart. [E.g.: Given two nowhere equal vector fields on a manifold, and a curve on the manifold which goes to infinity in the sense that $\gamma: R \rightarrow M$ has $\gamma^{-1}[C]$ compact for all compact C, then there exists a chart with respect to which the components of the vector fields approach each other in the limit along the curve - and also a chart with respect to which they do not.] Furthermore, even if one were able somehow to eliminate the ambiguity of chart-choices, obtaining a geometrical notion of asymptotic flatness, one would still face a description of asymptotic structure which involves the taking of limits "as one goes to infinity", a procedure which, to say the least, is mathematically awkward.

There is an idea - perhaps the key one in this subject - which at one stroke takes care of both of these difficulties. In some sense, "infinity is metrically far away". Consider, then, the replacement of the physical metric $\tilde{g}_{ab}$ of space-time by a new metric, $g_{ab} = \Omega^2 \tilde{g}_{ab}$, where Ω is some smooth positive scalar field, i.e., consider a conformal transformation. One might hope that, if the conformal factor Ω were chosen to approach zero asymptotically at an appropriate rate, then the resulting "scaling down of the metric" might "bring infinity in (with respect to g_{ab}) to a finite region". Perhaps one could then attach to the original space-time manifold $\tilde{M}$ additional points which, according to the physical metric $\tilde{g}_{ab}$, are "at infinity". Of course, these additional points are not to be at infinity with respect to the "scaled-down" metric g_{ab}, a condition one expresses by requiring that this metric admit a smooth extension from $\tilde{M}$ to the space consisting of $\tilde{M}$ together with these additional points. Finally, the appropriate asymptotic behavior of the conformal factor Ω is expressed in terms of the local behavior of Ω near these additional points. That is to say, one might think of <u>defining</u> asymptotic flatness of a space-time to mean the possibility of attaching to the manifold certain additional points such that the physical metric $\tilde{g}_{ab}$, scaled by a conformal factor Ω having certain local behavior at these points, yields a metric g_{ab} with smooth extension to these points. This of course is precisely what one does. There results a definition which, on the one hand, avoids completely the inspection of components and the corresponding introduction of unnecessary gauge, and, on the other, replaces limits by local differential geometry at the points at infinity.

Appropriate asymptotic behavior of the metric g_{ab} is expressed in terms of the behavior of Ω and g_{ab} near the points at infinity; that of other physical fields on the space-time in terms of their behavior near these same points. The precise conditions to be imposed on all these fields turns out to be a rather delicate issue. Conditions too strong will have the effect of eliminating solutions which would seem clearly to represent isolated systems; conditions too weak may have the effect of admitting too many solutions or, what is worse, may result in a structure which is so weak that potentially useful aspects of the asymptotic behavior of one's fields are lost in a sea of bad behavior. In Newtonian gravitation, for example, it is certainly a mistake to require anything other than that the gravitational potential go to zero as 1/r. Whereas the decision as to what asymptotic conditions are to be imposed is made with ease in Newtonian gravitation (perhaps in part because one has there a non-dynamical "r" to work with), the decisions are not so easy in general relativity. It may be that on this issue general relativity is more the rule and Newtonian gravitation the exception - that one would have expected the exercise of judgement in the formulation of a suitable definition. After all, a given physical theory - and general relativity in particular - has no need whatever of any notion of "isolated system": The theory is, at least in principle, as viable, as self-contained, as predictive without such a notion as with it. A definition is sought, rather, for its convenience, for we choose to understand the Universe through analysis of smaller, simpler systems, one at a time. There are no "correct" or "incorrect" definitions, only more or less useful ones. It is perfectly possible that there turn out to be a number of competing definitions, applicable to differing physical systems, or a single definition as in Newtonian gravitation, or none at all. What happens in general relativity is that one is able to limit the range of possibilities to a considerable extent by means of various external criteria, but that some ambiguity remains nontheless.

The next step in the treatment of asymptotic structure in general relativity is a comparatively minor one. One "detaches" from the original physical space-time the additional points at infinity, and forms them into an abstract manifold. The various fields on the physical space-time, considered locally near the points at infinity (i.e., considered "asymptotically"), are then found to yield corresponding fields on this abstract manifold. This step is for convenience: It permits one to isolate a complete summary of what is available characteristic of the asymptotic structure of the original space-time - excluding everything non-asymptotic - within the simple framework of a manifold with fields. That is, an "abstract asymptotic structure"becomes an object of study in its own right.

It is conventional not to treat all the fields which arise on this abstract manifold equally. Rather, the fields are divided into two classes: the universal, or geometrical fields; and the physical fields. The former are thought of as merely providing a background geometry on this manifold while having no direct physical significance; the latter, as providing asymptotic information about the various physical processes which may be occurring in the physical space-time. How these two classes of fields are regarded is a reflection of the mathematical distinction between the two classes: The former includes those fields which are, at least locally, always the same, no matter what the original physical space-time; the latter, those which are not. That one makes such a distinction the moment it becomes possible comes as no surprise. The root of many difficulties in general relativity, including that of defining "isolated system", is that the metric is both a geometrical and a physical object. In the asymptotic limit, it becomes possible to disentangle these two roles - and one does.

The universal fields give rise to what is called the asymptotic symmetry group: the group of all diffeomorphisms on the abstract manifold of points at infinity which leave invariant the universal fields. This group is analogous to, for example, the Poincaré group for electrodynamics in special relativity. There, the universal field is the flat metric, the physical field is the electromagnetic field, and the Poincare (symmetry) group is the group of all diffeomorphisms on the Minkowski manifold which leave invariant the universal field. One expects that the asymptotic symmetries of an asymptotically flat space-time will be closely related to the symmetries of flat space-time, i.e., that the asymptotic symmetry group will be closely related to the Poincaré group. The two groups are in fact closely related, but, as it turns out, not identical. It is not hard to guess where the differences will lie. Killing fields in Minkowski space which represent rotations and boosts "become large at infinity", while those representing translations "remain finite". In particular, "near infinity, rotation-boost Killing fields dominate translations". What happens, then, is that rotations and boosts, i.e., the Lorentz-part of the Poincaré group, carry over to the asymptotic symmetry group without change, while the translations suffer distortion (Perhaps "lack of proper recognition" is a better description.) in the asymptotic limit. The final asymptotic symmetry group, then, becomes the Poincaré group with "a distorted translation subgroup".

Consider again electrodynamics in special relativity. The electromagnetic field provides the detailed description of the various physical processes occurring in the space-time. However, one also constructs certain additional fields by combining the physical field with the symmetries, for example, $T^a{}_m \xi^m$, from the stress-energy of the electromagnetic field and a Killing field. One thus obtains local quantities of physical interest (in this case, flux of energy, momentum, angular momentum), and, by performing integrals, quantities characteristic of the system as a whole (in this case, total energy, momentum, angular momentum). In general relativity, by contrast, one has no symmetries in general (or, if you prefer, every diffeomorphism is a symmetry), whence such quantities cannot normally be defined. But in the asymptotic limit, one recovers asymptotic symmetries, and so the situation reverts to that of special relativity. That is to say, one combines the physical fields on the abstract manifold of points at infinity with the asymptotic symmetries to obtain local fields and global quantities which are then to be interpreted physically in terms of what they say about the original physical space-time. There are essentially two ways - about equal in importance - of arriving at these physical interpretations: by evaluating the quantities in special cases in which one "knows" what everything means, and by using the resemblance of the asymptotic symmetry group to the Poincaré group. Here, again, one can see why the original definitions must be formulated with care. Asymptotic conditions too strong may force some of the physical fields on the abstract manifold to be zero, while conditions too weak may result in an asymptotic symmetry group having little resemblance to the Poincare group. In the former case, one will not be able to obtain enough interesting physical quantities, while in the latter one will not know what the quantities mean.

The discussion above was intended to be an overview of this subject. There was omitted, however, one important feature, namely that there are actually two distinct regimes in which asymptotic structure has been studied. The two correspond, roughly speaking, to passage from the system to infinity in null directions, or in spacelike directions. As regards both mathematical structure and physical meaning, these two regimes have a great deal in common. Indeed, all of the remarks above are applicable to each. However, there is also an important sense in which the two differ from each other. In mathematical terms, it is that "moving in null directions, one meets at infinity a null surface; in spacelike directions, a timelike surface". In physical terms, it is that information can travel from the physical space-time (say, from a small compact region) to reach null infinity, but not spatial infinity. This difference, in turn, is reflected at several points in the details of the treatments of the two regimes.

At null infinity, one's description involves "what the system is doing through time", i.e., the dynamics of the system. At spatial infinity, "what the system does" is never even recorded. Rather, one sees only "the state of the system", once and for all. For example, one obtains in each regime a quantity which is called the total energy of the system. In the null case, one further derives a formula which expresses "the time rate of change of this energy in terms of energy lost by radiation to null infinity". In the spatial case, the energy is essentially just a number: There is no corresponding formula. As a general statement, the study of null infinity tends to be "differential"; that of spatial infinity somewhat more "algebraic".

Spatial infinity being a "timelike surface", one expects to find, among the universal fields on the corresponding abstract manifold, something resembling an invertible metric. In the null case, one does not. This expectation is confirmed, and has several consequences.

The asymptotic symmetry group at spatial infinity turns out to be finite-dimensional, as one might have expected. [The group of isometries on a manifold with invertible metric is always finite-dimensional.] What happens in the spatial case is that the "distortion" of the translation subgroup becomes its elimination, i.e., the asymptotic symmetry group is just the Lorentz group. This is perhaps not surprising, for at least the time-translations, which in Minkowski space generate dynamical evolution, have nothing to generate at spatial infinity. In the null case, on the other hand, the "distortion" takes the form of an enormous enlargement of the translation subgroup. What happens is that in the asymptotic limit, and in the presence of a rotation or boost, it becomes difficult to distinguish a "true translation" from a large class of impostors. When no rotations or boosts interfere, the true translations can again be distinguished. The asymptotic symmetry group in the null case is infinite-dimensional, with all the extra dimensions in a large abelian subgroup of generalized translations. In some sense, the same enlargement could have occurred in the spatial case, but one there essentially takes a quotient by the generalized translations, since they play no physical role anyway.

The invertible metric at spatial infinity yields a unique derivative operator on the corresponding abstract manifold. Thus, in this case, one works in the familiar context of (pseudo-) Riemannian geometry, easily writing one's differential equations, etc. In the null case, one has no unique universal derivative operator, and so one works more with Lie and exterior derivatives, and with other differential concomitants. As a general rule, it is considerably more difficult in the null case to write down formulae which say what one wants to say.

One would expect a priori that this general approach to asymptotic structure should lead to gauge freedom, since one obtains the abstract manifold and its fields via a conformal transformation while, presumably, the particular choice of conformal factor will not be uniquely determined by the physical space-time. That is, this freedom of conformal factor should produce different mathematical descriptions of the same physical situation. Gauge freedom does appear, but it is dealt with quite differently in the two regimes. Because of the presence of an invertible metric at spatial infinity, one is able there to impose what is in essence a gauge condition - a condition which actually eliminates all the gauge freedom. In particular, everything one can write down at spatial infinity makes sense. At null infinity, although quantities one deals with are ultimately gauge-invariant, and although one could at least in principle deal only with such quantities, it turns out to be inconvenient to do so. Instead, one contends with the intrusions of a large gauge group.

What the discussion above suggests is, on the whole, true: Asymptotic structure at null infinity is somewhat more complicated than that at spatial infinity. It seems to me that the former is also of slightly more physical interest.

Our purpose is to state the main definitions and results, and the central issues, concerning asymptotic structure in general relativity. The emphasis will be on the broad geometrical framework of the two regimes. We shall not on the whole be concerned with the many generalizations and extensions of the basic ideas. We consider first null infinity (Sect. II), and then spatial (Sect. III). Conformal transformations of one sort or another will be used constantly and without further discussion: A few definitions and formulae are collected in the appendix. There appears at the end a brief list of representative papers. A few things are, as far as I am aware, new, primarily some of the discussion of the appropriateness of definitions for null infinity, the treatment of Bondi mass, the introduction of a surface for spatial infinity, and the use of the Weyl tensor for the gravitational field at spatial infinity. The basic ideas, however, are taken from the original papers in this subject - by Arnowitt, Deser, Misner, Geroch, Bondi, Van Der Berg, Metzner, Sachs, and Penrose; the first five papers in the list of references - and from the public domain.

II. ASYMPTOTIC STRUCTURE AT NULL INFINITY

1. Asymptotes

Let $\tilde{M}$, $\tilde{g}_{ab}$ be a space-time (i.e., a 4-manifold with smooth ($= C^\infty$) metric of Lorentz signature). By an <u>asymptote</u> of $\tilde{M}$, $\tilde{g}_{ab}$ we mean a manifold M with boundary I, together with a smooth Lorentz metric g_{ab} on M, a smooth function Ω on M, and a diffeomorphism ψ from $\tilde{M}$ to M - I (by means of which we shall identify $\tilde{M}$ and M - I), satisfying the following conditions:

1. On $\tilde{M}$, $g_{ab} = \Omega^2\, \tilde{g}_{ab}$.
2. At I, $\Omega = 0$, $\nabla_a \Omega \neq 0$, and $g^{ab}(\nabla_a\Omega)(\nabla_b\Omega) = 0$, where "$\nabla_a$" denotes the gradient on M.

This g_{ab} is called the <u>unphysical metric</u> (to distinguish it from the physical metric $\tilde{g}_{ab}$), while I is called the <u>boundary</u> (at null infinity). Note that the definition requires that the unphysical metric be defined and have Lorentz signature also at points of the boundary. By contrast, the physical metric is not even defined at I (and, indeed, according to the first condition and the first part of the second, could not be given sensible meaning there). We also note that it follows from the definition that Ω is nonzero on $\tilde{M}$ (By convention, we choose it positive there.), and that I is a null surface (since $\nabla_a\Omega$ at I is normal to I, nonzero, and null).

It is intended that the definition represent the intuitive idea of "the attachment to the space-time manifold $\tilde{M}$ of additional ideal points at null infinity". The additional points are of course those of I, while the diffeomorphism ψ inserts $\tilde{M}$ in M; thus, M itself represents the physical space-time manifold with points at infinity attached. The first condition states that the conformal factor rescales the physical metric to the unphysical. The first part of the second condition, together with the requirement that the unphysical metric be well-behaved on I, states that "infinity is $\tilde{g}_{ab}$-far away". The second part of the second condition fixes the asymptotic behavior of Ω. In effect, it states that Ω fall to zero "as $1/r$". The third part of the second condition states essentially that we are dealing with null infinity.

Perhaps these remarks at least make it seem reasonable that some definition along the lines above might reflect the intuitive idea of "asymptotic flatness at null infinity". We must, however, still make a case that the actual definition as given is appropriate to this intuitive idea. There are several pieces of evidence on this issue, which we now discuss briefly.

First, one would expect that space-times which are "obviously asymptotically flat" would admit corresponding asymptotes. We give one example in detail, and remark on some others.

Let $\tilde{M}$, $\tilde{g}_{ab}$ be the Schwarzschild space-time, so $\tilde{M}$ is $S^2 \times R^2$, and $\tilde{g}_{ab}$ is the metric which, in the usual Schwarzschild coordinate system, takes the form

$$-(1-2m/r)\,dt^2 + (1-2m/r)^{-1}\,dr^2 + r^2\,(d\theta^2 + \sin^2\theta\,d\varphi^2). \qquad (1)$$

Now replace t and r by two new coordinates, u and x, according to the formulae

$$u = t - r - 2m \log(r - 2m), \qquad x = r^{-1}. \qquad (2)$$

Rewritten in terms of these coordinates, the Schwarzschild metric is

$$x^{-2}\{2\,du\,dx - x^2\,(1-2mx)\,du^2 + (d\theta^2 + \sin^2\theta\,d\varphi^2)\}. \qquad (3)$$

Let M be the manifold with boundary consisting of the points of the Schwarzschild manifold $\tilde{M}$, together with those boundary points labeled by "$x = 0$" in the above chart. Since "$x = 0$" corresponds to "$r = \infty$", the boundary I was not included in the original Schwarzschild manifold. Let the differentiable structure of M be that induced by the original Schwarzschild manifold, and by this chart. Let the diffeomorphism ψ be the identity. Let $\Omega = x$, a smooth function on M, and let the metric g_{ab} on M be that given by the expression in curly brackets in (3) (and, outside of the range of this chart, by $\Omega^2\,\tilde{g}_{ab}$).

This set-up, we claim, is an asymptote. Indeed, the first condition is immediate from our choice of g_{ab}. For the second condition, we first note that $\Omega = 0$ at I, and that, since Ω is in a

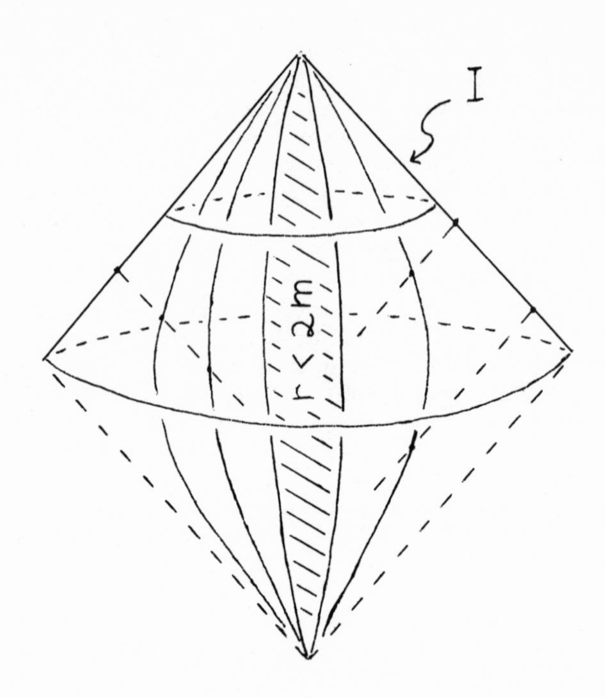

Fig. 1. The exterior $(r > 2m)$ Schwarzschild space-time with a boundary I attached at null infinity. (One dimension of spherical symmetry is suppressed.) The solid lines represent world-lines of constant r, θ, and φ. The dashed lines represent future-directed, outgoing, radial null geodesics.

neighborhood of I one of the coordinates in a chart including I, we must have $\nabla_a \Omega \neq 0$ there. Finally, it is easily checked directly from (3) that $\nabla_a \Omega$ is null at I.

A picture of this asymptote of the Schwarzschild space-time is shown in Fig. 1. Whereas some null geodesics reach I, the timelike geodesics (with respect to $\tilde{g}_{ab}$) do not. There are, however, some timelike curves which reach I, e.g., certain ones with acceleration of constant magnitude, directed outward and radially. The manifold I is $S^2 \times R$. There are at least three other asymptotes of the Schwarzschild space-time. One is obtained just as above, but with the first two minus signs in (2) reversed. The corresponding picture would again be Fig. 1, but now I would be the "lower cone" in the figure. Two others are found, similarly, on "the other side of the bridge connecting the two halves of the extended Schwarzschild space-time". The special case m = 0 was not excluded in the calculation above. Thus, Minkowski space-time admits asymptotes. The corresponding picture would look essentially the same as Fig. 1, except that the interior region $r \leq 2m$, excluded in Fig. 1, may in this case be restored.

The technique of the example above can be applied to certain other space-times to show the existence of a non-trivial asymptote. One introduces two "angular coordinates", an appropriate retarded null coordinate u, and another coordinate, x, whose inverse measures affine length along the integral curves of the gradient of u. Reexpressing the metric in terms of these coordinates, one often obtains a chart which is suitable for the direct attachment of a boundary at null infinity ($x = 0$). One selects one's conformal factor, and explicitly checks the conditions of the definition. In this way, for example, one shows that the Kerr space-time and the asymptotically flat Weyl space-times admit appropriate asymptotes. (The pictures, again, are similar to Fig. 1.) One can in fact do a bit better than this: One can write down various general criteria satisfaction of which ensures existence of a non-trivial asymptote. Such criteria may be constructed as follows. Let $\tilde{M}$, $\tilde{g}_{ab}$ be a space-time, and let M, g_{ab}, Ω be as in the definition. Fix a point p of I, and choose coordinates x^1, x^2, x^3, Ω in a neighborhood of p. (There are such admissible charts on M, since $\nabla_a\Omega \neq 0$ at I, and hence in some neighborhood of p.) The components of the metric g_{ab} in this chart yield ten smooth functions of the coordinates. The second condition in the definition is now expressed as conditions on these ten functions. Now consider the chart x^1, x^2, x^3, $r = \Omega^{-1}$ on $\tilde{M}$ (i.e., I is now excluded). Using the first condition, one writes down the components of the physical metric $\tilde{g}_{ab}$ with respect to this chart in terms of the ten functions previously obtained. One then translates the second condition in the definition, through the ten functions, co conditions on these components. One thus obtains a statement of the following form: If $\tilde{M}$, $\tilde{g}_{ab}$ admits a non-trivial asymptote, then there exists a chart, x^1, x^2, x^3, r, in terms of which the components of $\tilde{g}_{ab}$ satisfy ... One can further refine such statements by imposing additional conditions on the original x^1, x^2, x^3. The converses of these statements will of course also be true. Although the statements which result from this program do lend some support to the assertion that the definition is an appropriate one, these statements in practice are perhaps on the whole less useful than simply guessing the asymptote. All of the above applies only to space-times presented in a coordinate system in which they "look asymptotically flat". As far as I am aware, we are not even close to having a practical algorithm for deciding whether or not a general space-time admits a non-trivial asymptote.

We are, recall, concerned with the issue of whether or not the definition of an asymptote captures the intuitive notion of "asymptotically flat". A second aspect of this issue is the question of whether or not the existence of a boundary at infinity is persistent in any sense. Suppose for example that one lived in a space-time which does admit an asymptote, and suppose that, at some event, one decided to cause some disturbance in the space-time, e.g., by emitting gravitational radiation. One would wish it to be true that such a disturbance could not result in destruction of the boundary, for otherwise, since such disturbances are presumably commonplace in our World, the physically realistic space-times would not admit a non-trivial asymptote. Does the present definition satisfy such a criterion?

To simplify matters at this point, let us restrict attention to gravitational disturbances only. (Other fields, it turns out, are generally much easier to deal with; we shall return to them later.) Let, then, $\tilde{M}$, $\tilde{g}_{ab}$ be a space-time which satisfies Einstein's equation with zero stress-energy, and which admits a non-trivial ($I \neq \emptyset$) asymptote, M, g_{ab}, Ω. Let us interpret "disturbance" to mean a linearized field, i.e., a symmetric tensor field $\tilde{\gamma}_{ab}$ on $\tilde{M}$. Let the individual creating the disturbance operate within a compact region of space-time, i.e., let $\tilde{\gamma}_{ab}$ satisfy the linearized Einstein equation,

$$\tilde{\nabla}^2 \tilde{\gamma}_{ab} + \tilde{\nabla}_a \tilde{\nabla}_b \tilde{\gamma}_m{}^m - 2 \tilde{\nabla}^m \tilde{\nabla}_{(a} \tilde{\gamma}_{b)m} = 0, \tag{4}$$

outside some compact subset C of $\tilde{M}$, where indices are raised and lowered with the physical metric, and where $\tilde{\nabla}_a$ is the derivative operator associated with that metric. We are interested in the effect of this field on the boundary. But implicit in the word "effect" is the assumption that our space-time is deterministic: So, let us work only in the domain of dependence of some surface T. We may in effect ensure that all of $\tilde{\gamma}_{ab}$ was created by the disturbance in C by requiring that the support of $\tilde{\gamma}_{ab}$ intersect T in a compact set.

Finally, we must express the idea that our disturbance does not destroy the boundary. We say that $\tilde{\gamma}_{ab}$ preserves the asymptote if the field $\gamma_{ab} = \Omega^2 \tilde{\gamma}_{ab}$ on $\tilde{M}$ has smooth extension to M, and if, under that extension, $g^{am} g^{bn} (\nabla_a \Omega)(\nabla_b \Omega) \gamma_{mn} = 0$ at I. This will be recognized as the "linearization" of the definition of an asymptote. [The first condition in the definition is reflected in the definition of γ_{ab}; the first two parts of the second condition, being metric-independent, need not be reflected; the third part of the second condition is reflected in the equation.] Note that, since we have imposed no gauge-conditions on $\tilde{\gamma}_{ab}$, we need not permit perturbation of the conformal factor Ω, since such a perturbation could always be absorbed through gauge into γ_{ab}. Preservation of the asymptote is not, of course, gauge-invariant, i.e., it is possible that one of $\tilde{\gamma}_{ab}$ and $\tilde{\gamma}_{ab} + \tilde{\nabla}_{(a} \tilde{\eta}_{b)}$ (with $\tilde{\eta}_b$ some field) preserves the asymptote while the other does not.

These remarks motivate the following persistence criterion on the definition:

Conjecture 1. Let $\tilde{M}$, $\tilde{g}_{ab}$ be a space-time with zero Ricci tensor, M, g_{ab}, Ω an asymptote, and T a spacelike submanifold, with boundary in I, of M. Let $\tilde{\gamma}_{ab}$ on $\tilde{M}$ satisfy (4) outside a compact set, and have support intersecting T compactly. Then, for some field $\tilde{\eta}_b$ on $\tilde{M}$, $\tilde{\gamma}_{ab} + \tilde{\nabla}_{(a}\tilde{\eta}_{b)}$ preserves that part of the asymptote in the domain of dependence of T.
This conjecture is open. Its resolution one way or the other would, it seems to me, be of some interest: If in particular it should turn out to be false in some essential way, then I should think that this subject would be in serious jeopardy. The conjecture is likely to be true. We shall later suggest a possible approach to a proof.

Assuming that Conjecture 1 were settled in the positive, one might try to generalize it to the case in which one does not linearize the disturbance. It appears to be difficult to state the corresponding question. For example, the new metric would in general change the domain of dependence of T; considerably more care would be necessary to define preservation of the asymptote. Furthermore, even should one succeed in stating a conjecture, one expects the proof to be difficult.

In any case, the evidence in favor of the present definition consists of some examples, some general criteria, Conjecture 1 and some related theorems (for other fields; later), and the fact that the definition leads to quantities of apparent physical interest.

We next consider a somewhat different aspect of the definition. The ultimate goal is to describe the asymptotic structure of a physical space-time $\tilde{M}$, $\tilde{g}_{ab}$ in terms of the local behavior at I of various fields. One would like to have some guarantee that this description will have genuine physical significance, i.e., that statements about fields near I will actually say something about the physical space-time. What is needed essentially is a proof that, given the physical space-time, its asymptote is in some sense unique, for otherwise statements about I may refer only to the choice of asymptote. There are two distinct senses in which the asymptote is certainly not unique. Let $\tilde{M}$, $\tilde{g}_{ab}$ be a space-time, M, g_{ab}, Ω an asymptote, and ω a smooth positive scalar field on M. Then M, $\omega^2 g_{ab}$, $\omega\Omega$ is clearly also an asymptote. [Note: ω-factors so chosen that $\Omega^{-2} g_{ab}$, the physical metric, remains the same.] We call two asymptotes so related (The sense is easy to make precise.) <u>equivalent</u>. Thus, one always has the freedom of additional conformal transformations. The second sense of non-uniqueness is the following: For C any closed subset of I, M - C, g_{ab}, Ω (the latter two fields now restricted to M - C) is also an asymptote. We call M, g_{ab}, Ω an <u>extension</u> of this one. That is, one can remove part of the boundary.

It turns out that these two are the only ambiguities, at least for sufficiently well-behaved asymptotes. One suitable behavioral criterion is this: We call an asymptote <u>regular</u> if, given any point p of I, and any non-zero null vector l^a at point x of $\tilde{M}$ such that the null geodesic in M generated by l^a does not meet p, then there are neighborhoods U of p in M and V of l^a in the space of null vectors in $\tilde{M}$ such that no null geodesic in M generated by an element of V enters U. The definition requires essentially that null geodesics near a geodesic which does not meet p all stay away from p. The point of the definition is that regular asymptotes do not interfere with each other, in the following sense. Fix space-time $\tilde{M}$, $\tilde{g}_{ab}$, let M_1 and M_2 be regular asymptotes, and let p_1 and p_2 be points of the respective boundaries, I_1 and I_2. Then, we claim, either i) the null geodesics in $\tilde{M}$ which, in M_1, meet p_1 are precisely the same as the null geodesics in $\tilde{M}$ which, in M_2, meet p_2 (i.e., p_1 and p_2 should be thought of as the "same" boundary point), or ii) there are neighborhoods of p_1 in M_1 and p_2 in M_2 such that that these neighborhoods do not intersect in $\tilde{M}$ (i.e., p_1 and p_2 are "separated"). Indeed, suppose that i) fails: Let γ be a null geodesic in $\tilde{M}$ which meets p_1 but not p_2, let x be a point of γ, and let l^a be its tangent at x. Then by regularity of M_2 there are neighborhoods U_2 of p_2 and V of l^a as in the definition. Since I_1 is null, the geodesic γ cannot meet p_1 tangentially to I_1. Hence, the union U_1 of all null geodesics in M_1 generated by elements of V is a neighborhood of p_1. By construction of U_2 and V, this U_1 and U_2 are the neighborhoods required for ii).

The sense in which equivalence and extension are the only ambiguities in selecting an asymptote is the following.

Theorem 2. Let $\tilde{M}$, $\tilde{g}_{ab}$ be a space-time. Then there exists a regular asymptote, M, g_{ab}, Ω, unique up to equivalence, which is maximal: Any other regular asymptote of $\tilde{M}$, $\tilde{g}_{ab}$ is equivalent to one of which M is an extension.

Sketch of proof: For p_1 and p_2 boundary points of regular asymptotes of $\tilde{M}$, $\tilde{g}_{ab}$, write $p_1 \sim p_2$ if case i) of the previous paragraph obtains. This is an equivalence relation: Denote by I the set of equivalence classes, and set $M = \tilde{M} \cup I$. Each regular asymptote of $\tilde{M}$, $\tilde{g}_{ab}$ is naturally represented as a subset of M (associating with a boundary point of a regular asymptote the element of I which is the corresponding equivalence class). Let Γ be the manifold with boundary consisting of all triples (x, l^a, c), where l^a is a nonzero null vector at point x of $\tilde{M}$ and c is a number which either i) is positive, and is such that $\exp(c^{-1} l^a)$ exists, or ii) is zero, and the geodesic generated by l^a meets the boundary of some regular asymptote. If $c \neq 0$, set $\varphi(x, l^a, c) = \exp(c^{-1} l^a)$; if $c = 0$, set $\varphi(x, l^a, 0)$ the corresponding element of I. Then φ is a mapping from Γ to M. The requirement that φ be smooth induces on M a differentiable structure, compatible with that of each embedded regular asymptote. Since the embedded regular asymptotes have conformal structures compatible on $\tilde{M}$, and differentiable structures compatible on M, they have conformal structures compatible on M. Since the embedded regular asymptotes cover M, this M inherits a conformal structure. This

M with this conformal structure is the required maximal asymptote.

This theorem is not to be interpreted as saying that every space-time is "asymptotically flat": The boundary can be empty. One can weaken regularity somewhat, and still obtain the theorem. Furthermore, one can prove a result analogous to Theorem 2, but which ignores regularity and abandons uniqueness, asserting that up to equivalence every asymptote possesses an extension having no proper extension.

In any case, this discussion suggests how one deals with the various ambiguities. For that of extension, one can invoke Theorem 2, and pass to the maximal regular asymptote. Alternatively, one could use the result suggested in the previous paragraph. From the way the proof of Theorem 2 works, one expects that the resulting ambiguity will not be too severe. The final, and perhaps best, alternative would be to just consider space-times on a case by case basis, for none of these problems of extension seem normally to arise in practice. Equivalence, on the other hand, is dealt with in quite a different way. One retains the conformal transformations as a gauge group, working ultimately only with expressions which are gauge-invariant.

2. Einstein's Equation

Let $\tilde{M}$, $\tilde{g}_{ab}$ be a space-time, and M, g_{ab}, Ω an asymptote. Einstein's equation on the physical space-time equates the Ricci tensor of $\tilde{g}_{ab}$ to a certain expression involving the matter fields on the manifold $\tilde{M}$. We wish to write this equation and its consequences in terms of fields appropriate to the asymptote, g_{ab} and Ω. As a first step, we consider the following program: Working only at points of $\tilde{M}$ (i.e., where $\tilde{g}_{ab}$ is defined), write what equations one can the left sides of which involve only g_{ab} and Ω, and the right sides of which involve the physical metric only through its Ricci tensor. Ultimately, of course, we shall take the limits at I of these formulae.

Indices of tensor fields with tildes are raised and lowered with the physical metric; indices of fields without, with the unphysical metric. The derivative operators associated with the two metrics are denoted $\tilde{\nabla}_a$ and ∇_a, respectively. The formulae are simplified somewhat by the following choices of variables. Replace the unphysical Ricci tensor* by the combination $S_{ab} = R_{ab} - 1/6\, R\, g_{ab}$. Thus, for example, we have $S^m{}_m = 1/3\, R$, and also

$$R_{abcd} = C_{abcd} + g_{a[c}\, S_{d]b} - g_{b[c}\, S_{d]a}, \tag{5}$$

* Our conventions for the Riemann tensor, Ricci tensor, and scalar curvature are: For any k_c, $\nabla_{[a}\nabla_{b]}\, k_c = 1/2\, R_{abc}{}^d\, k_d$; $R_{ab} = R^m{}_{amb}$; $R = R^m{}_m$.

where C_{abcd} is the unphysical Weyl tensor. We further write n_a for $\nabla_a \Omega$, and f for the combination $\Omega^{-1} n^m n_m$. Finally, we replace the physical Ricci tensor by the combination $\tilde{R}_a{}^b - 1/6 \, \tilde{R} \, \delta_a{}^b$. This is a tensor field on $\tilde{M}$, fixed by the physical metric once and for all. We write it $L_a{}^b$, and raise and lower L's indices with the unphysical metric.

From the formula (Appendix) for the behavior of the Ricci tensor under conformal transformations, we have

$$\Omega \, S_{ab} + 2 \nabla_a n_b - f \, g_{ab} = \Omega^{-1} L_{ab}. \tag{6}$$

Clearly, this is essentially the only formula of the desired type (unphysical on the left, L on the right) at our disposal. From it we must derive all else. Contracting (6) with n^b and using the definition of f, we obtain

$$S_{ab} n^b + \nabla_a f = \Omega^{-2} L_{ab} n^b. \tag{7}$$

Taking the curl of (6), we obtain

$$\begin{aligned} &\Omega \nabla_{[a} S_{b]c} + n_{[a} S_{b]c} + 2 \nabla_{[a} \nabla_{b]} n_c - \nabla_{[a} f \, g_{b]c} \\ &= \nabla_{[a} (\Omega^{-1} L_{b]c}). \end{aligned} \tag{8}$$

But the third term on the left in (8) equals $R_{abcd} n^d$. Replacing R_{abcd} via (5), and using (7) to eliminate the resulting $S_{ab} n^b$ -term, we obtain

$$\begin{aligned} &\Omega \nabla_{[a} S_{b]c} + C_{abcd} n^d \\ &= \nabla_{[a} (\Omega^{-1} L_{b]c}) - \Omega^{-2} g_{c[a} L_{b]d} n^d. \end{aligned} \tag{9}$$

Contracting the Bianchi identity, $\nabla_{[a} R_{bc]de} = 0$, twice, and eliminating the Ricci tensor in favor of S_{ab}, we obtain

$$\nabla^m S_{am} - \nabla_a S^m{}_m = 0. \tag{10}$$

Contracting the Bianchi identity once, and substituting (5), we have

$$\nabla^m C_{abcm} + \nabla_{[a} S_{b]c} = 0. \tag{11}$$

Eliminating the second term via (9), this can be written

$$\begin{aligned} &\nabla^m (\Omega^{-1} C_{abcm}) \\ &= - \Omega^{-2} \nabla_{[a} (\Omega^{-1} L_{b]c}) + \Omega^{-4} g_{c[a} L_{b]d} n^d. \end{aligned} \tag{12}$$

Finally, taking the double dual of (12), or, substituting (5) into the Bianchi identity without contractions, we obtain

$$\nabla_{[a}(\Omega^{-1} C_{bc]}{}^{de}) = 2\,\Omega^{-2}\,\delta_{[a}{}^{[d}\nabla_b(\Omega^{-1} L_{c]}{}^{e]}) - 2\,\Omega^{-4}\delta_{[a}{}^{[d}\delta_b{}^{e]} L_{c]m}\, n^m. \tag{13}$$

These are the desired equations.

The basic fields are g_{ab} and Ω, the two derived from these, n^a and f, and the curvature fields, S_{ab} and C_{abcd}. These fields are related to each other by their definitions and by (10). These fields are also related to the physical Ricci tensor, represented by L_{ab}, by Eqns. (6), (7), (9), (12), and (13). We shall of course need these equations later.

We observed in the previous section that there is available the gauge freedom of passing from a given asymptote to an equivalent one. That is, one may replace the metric g_{ab} by $g'_{ab} = \omega^2\, g_{ab}$ and Ω by $\Omega' = \omega\Omega$, where ω is a positive scalar field on M. Under this replacement, all the other basic fields also change. We want the formulae expressing those fields as obtained from g'_{ab} and Ω' in terms of those from g_{ab} and Ω. We first note (Appendix) that the derivative operator, ∇'_a, associated with g'_{ab} acts as follows: For any k_b,

$$\nabla'_a k_b = \nabla_a k_b - 2\omega^{-1} k_{(a}\nabla_{b)}\omega + g_{ab}\omega^{-1} g^{mn} k_m \nabla_n \omega. \tag{14}$$

Now, again using the technique of the Appendix, the formulae for the primed versions of the other basic fields follow:

$$n'^a = \omega^{-1} n^a + \omega^{-2}\Omega\,\nabla^a\omega, \tag{15}$$

$$f' = \omega^{-1} f + 2\,\omega^{-2} n^m \nabla_m\omega + \omega^{-3}\Omega(\nabla^m\omega)(\nabla_m\omega), \tag{16}$$

$$S'_{ab} = S_{ab} - 2\,\omega^{-1}\nabla_a\nabla_b\omega + 4\,\omega^{-2}(\nabla_a\omega)(\nabla_b\omega) - \omega^{-2} g_{ab}(\nabla^m\omega)(\nabla_m\omega), \tag{17}$$

$$C'_{abcd} = \omega^2 C_{abcd}. \tag{18}$$

Of course, Eqns. (6), (7), (9), (10), (12), and (13) are also true for the primed fields, i.e., are invariant under (14) - (18), provided one assigns to L_{ab} that dimension (See Appendix.) suggested by its definition: zero.

We now wish to impose Einstein's equation. The immediate problem we face is that, in the presence of matter, Einstein's equation is not a purely geometrical condition on $L_a{}^b$: Rather, it merely equates $L_a{}^b$ to some expression involving the matter fields. We, however, are principally concerned with fields near infinity, and one expects on physical grounds that there will be very little matter there. Suppose, for example, that the matter is represented by a particle traveling on a timelike geodesic in the physical space-time. Then, at least in this example, there would normally be no matter near infinity, for, as it is not difficult to show, no timelike geodesic in the physical space-time can have endpoint on I. On such physical grounds, then, one might be tempted to impose Einstein's equation in the following geometrical form: $L_{ab} = 0$ in some neighborhood of I. Unfortunately, this formulation would be too restrictive. First, although it is indeed true, in our example, that no single particle can by itself reach infinity, it is possible to have a gas of particles which fills space-time. There would then in particular be a nonzero contribution to the stress-energy in every neighborhood of I. Secondly, and perhaps even more seriously, massless fields such as the Maxwell field are able to propagate to infinity, e.g., because the characteristics of the equations are null surfaces. In short, the requirement of vanishing stress-energy in a neighborhood of the boundary would rule out numerous space-times of potential physical interest. On the other hand, one would like to take advantage of the observation that the stress-energy would normally be "small" near infinity. It is clear that what is needed is a treatment whose criterion is, not "vanishing or non-vanishing" of the stress-energy near infinity, but rather "rate of approach to zero". We seek, therefore, a treatment in which this rate is given expression.

Let $\tilde{M}$, $\tilde{g}_{ab}$ be a space-time, M, g_{ab}, Ω an asymptote, and s a real number (shortly, a positive integer). We say that with respect to this asymptote the stress-energy of this space-time <u>vanishes</u> asymptotically <u>to order</u> s if $\Omega^{-s} L_{ab}$ on $\tilde{M}$ admits a smooth extension to I. Although this, the most naive such notion, will suffice for our purposes, several refinements are possible. For example, one could impose a lower differentiability class on $\Omega^{-s} L_{ab}$. More generally, one could require only that $\Omega^{-s+k} \nabla_{a_1} \ldots \nabla_{a_k} L_{ab}$ be locally bounded on I for $k = 0, 1, \ldots, k_o$. A further refinement would involve introducing separate orders for the various pieces of L_{ab}, e.g., $L_{ab} n^b$, $L_{ab} n^a n^b$, $n_{[a} L_{b]c}$, etc. We note that vanishing of the stress-energy to order s is gauge-invariant, and that order s implies order s-1. The question of to what order one should require the stress-energy to vanish is essentially a physical one: It will depend on what types of matter are admitted into the space-time, and on what further assumptions one wishes to make as to how much matter there is to be far from the system of interest. As a general rule of thumb, order four is normally not unreasonable.

Now, depending on what asymptotic order is imposed, the right sides of the equations we have written down earlier will vanish or remain finite at I. Thus, the right side of (6) will vanish for order two, the right sides of (7) and (9) for order three, and the right sides of (12) and (13) for order five. In this way one obtains, for sufficiently large order, equations on just the unphysical fields. We shall later make extensive use of this idea. For the moment, we just give one example of the interaction of asymptotic order with the equations of this section. Consider Eqn. (6). Since g_{ab} and Ω are smooth on M, the first two terms on the left are smooth. Since $f = \Omega^{-1} n_a n^a$, $n^a n_a$ is smooth on M, and since, by the definition of an asymptote, $n_a n^a$ vanishes on I, f is also smooth on M*. Thus, the left side of (6) is smooth, whence the stress-energy always vanishes asymptotically to order one. Physically, "too much matter at infinity destroys asymptotic flatness", as one might expect. One could of course have proceeded in a different way, requiring vanishing of the stress-energy to order one, and then using (6) to prove the third part of the second condition in the definition of an asymptote.

3. The Local Geometry of Infinity

Let $\tilde{M}$, $\tilde{g}_{ab}$ be a space-time, and let M, g_{ab}, Ω be an asymptote. Denote by $\mathcal{J}$ a diffeomorphic copy of the three-dimensional manifold I, and let $\zeta : \mathcal{J} \rightarrow M$ be the corresponding smooth mapping, so ζ sends $\mathcal{J}$ to I diffeomorphically. This manifold $\mathcal{J}$, then, represents "I, detached from M"; It will be convenient to describe asymptotic structure in terms of it. In this section, we first introduce the technique which yields, from tensor fields on M, fields on $\mathcal{J}$. By means of this technique, we then induce on $\mathcal{J}$ (at least, whenever the stress-energy vanishes asymptotically to an appropriate order) those fields which characterize its universal geometry. Finally, we study this geometry.

We denote by ζ^* the pullback - the operator, now to be defined, which carries certain fields from M to $\mathcal{J}$. We shall define its action in two stages. We first demand that, for any covariant field $\mu_{a..c}$ on M, $\zeta^*(\mu_{a..c})$ be a covariant field on $\mathcal{J}$ of the same rank, and that this mapping from fields on M to fields on $\mathcal{J}$ satisfy the following three conditions:

1. For $\mu : M \rightarrow R$ a scalar field, $\zeta^*(\mu) = \mu \circ \zeta$, i.e., is μ evaluated on I and regarded as a function on $\mathcal{J}$.

2. The pullback of the gradient of a scalar field is the gradient of its pullback.

3. The pullback of the sum (resp., outer product) of two covariant tensor fields on M is the sum (resp., outer product) of their pullbacks.

---

* Here, and on several occasions hereafter, we make use of the following fact: If $\alpha^{a..c}{}_{b..d}$ is a smooth tensor field on M which vanishes on I, then $\Omega^{-1} \alpha^{a..c}{}_{b..d}$ has smooth extension from $\tilde{M}$ to M. Proof: Passing to components and suppressing dimensions, it suffices to show that if u(x,y) is smooth with u(0,y) = 0 then x^{-1} u(x,y) is smooth. But this is true, for $u(x,y) = \int_0^1 (d/dt\, u(tx,y))\, dt = x \int_0^1 u'(tx,y)\, dt$, where prime denotes derivative with respect to the first variable.

There exists, we claim, one and only one γ^* satisfying these conditions. Indeed, the first condition specifies the action of γ^* on scalar fields, while, since every covariant tensor field can be written as a sum of outer products of scalars and gradients of scalars, the other two conditions specify the action of γ^* on all other covariant fields. Thus, for example, we have $\gamma^*(n_a) = 0$, since $n_a = \nabla_a \Omega$, while, by the second condition for an asymptote, $\Omega \circ \gamma = 0$. Clearly, $\gamma^*(\mu_a) = 0$ if and only if μ_a is, at each point of I, a multiple of n_a; more generally, $\gamma^*(\nu_{a..c}) = 0$ if and only if $\nu_{a..c}$ can be written, at each point of I, as a sum of outer products of vectors in such a way that at least one vector in each term in the sum is n_a.

We now extend the action of this γ^* to certain other fields as follows. Denote by $\mathcal{C}$ the collection of all smooth tensor fields, e.g., $\alpha^{a..c}{}_{b..d}$, on M having the following property: Whenever $\gamma^*(\nu_{a..c}) = 0$, then also $\gamma^*(\alpha^{a..c}{}_{b..d}\nu_{a..c}) = 0$. It is on the fields in the collection $\mathcal{C}$ that we may define γ^*. The action of γ^* on $\alpha^{a..c}{}_{b..d}$ in $\mathcal{C}$ is specified by imposing the following requirement: For any field $\mu_{a..c}$,

$$\gamma^*(\alpha^{a..c}{}_{b..d}\mu_{a..c}) = \gamma^*(\alpha^{a..c}{}_{b..d})\ \gamma^*(\mu_{a..c}). \tag{19}$$

We note that the field on the left, and the second field on the right, were defined already in the previous paragraph, so (19) serves as a definition of $\gamma^*(\alpha^{a..c}{}_{b..d})$. Clearly, membership of $\alpha^{a..c}{}_{b..d}$ in $\mathcal{C}$ is precisely what is needed to ensure that (19) yield one and only one field $\gamma^*(\alpha^{a..c}{}_{b..d})$. The covariant fields are in $\mathcal{C}$, and the action of γ^* here defined on these reduces to that of the previous paragraph.

As examples, μ^a is in $\mathcal{C}$ if and only if μ^a is tangent to I at each point of I; $\tau_a{}^b$ is in $\mathcal{C}$ if and only if, at each point of I, $\tau_a{}^b n_b$ is a multiple of n_a, or, what is the same thing, if and only if $\tau_a{}^b\mu^a$ is tangent to I for each μ^a tangent to I. Raising and lowering of indices may of course take tensors in or out of $\mathcal{C}$. The sum (resp., outer product) of two fields in $\mathcal{C}$ is again in $\mathcal{C}$, and the pullback of the sum (resp., outer product) is just the sum (resp., outer product) of the pullbacks. Unfortunately, the pullback does not in general commute with contraction. For example, for $\delta_a{}^b$ the unit tensor field in M, $\gamma^*(\delta_a{}^b) = \underline{\delta}_a{}^b$, the unit field in $\mathcal{I}$. But $\delta_a{}^a = 4$, while $\underline{\delta}_a{}^a = 3$. It is not difficult to write down the formula for the contraction of the pullback of a field on M in terms of the pullbacks of certain fields on M. [Caution: It is false in general that the contraction of a field in $\mathcal{C}$ is even in $\mathcal{C}$.] We shall, however, make do with the following observation: If $\alpha^{..p..}{}_{..q..}n_p = 0$ at points of I, then the right side of $\underline{\delta}_p{}^q\ \gamma^*(\alpha^{..p..}{}_{..q..}) = \gamma^*(\alpha^{..p..}{}_{..p..})$ makes sense, and this equation holds. The pullback commutes with Lie derivatives, e.g., since the latter can be expressed in terms of gradients. That is, for k^a and $\alpha^{\cdots}{}_{\cdots}$ in $\mathcal{C}$,

$\mathcal{L}_k(\alpha^{\cdots}{}_{\cdots})$ is also in $\mathcal{C}$, and $\mathfrak{f}^*(\mathcal{L}_k(\alpha^{\cdots}{}_{\cdots}))$ $= \mathcal{L}_{\mathfrak{f}^*(k)}\ \mathfrak{f}^*(\alpha^{\cdots}{}_{\cdots})$. Similarly, the pullback commutes with exterior differentiation.

Thus, we now have our 3-manifold $\mathcal{J}$ representing abstract points at null infinity, together with the techniques for carrying tensor fields and their algebraic and differential relations from M to $\mathcal{J}$. Of course, we shall ultimately apply $\mathfrak{f}^*$ to every field and every equation we can get our hands on. However, as discussed in the Introduction, the resulting fields are divided into two classes: the universal and the physical. We are in the present section concerned with the former.

The two tensor fields n^a and g_{ab} on M are in the collection $\mathcal{C}$: Set $\underline{n}^a = \mathfrak{f}^*(n^a)$ and $\underline{g}_{ab} = \mathfrak{f}^*(g_{ab})$. These two fields on $\mathcal{J}$ essentially describe the universal geometry of this manifold. (We shall shortly motivate this choice, and explain why only "essentially".)

We wish to write down all statements we can about $\underline{n}^a$ and $\underline{g}_{ab}$ on $\mathcal{J}$. First note that $\underline{n}^a$ vanishes nowhere, as follows from the second condition in the definition of an asymptote. Furthermore, applying $\mathfrak{f}^*$ to $g_{ab}\, n^b = n_a$, we obtain $\underline{g}_{ab}\, \underline{n}^b = 0$. Thus, $\underline{g}_{ab}$ is not invertible; indeed, it is clear from the geometry (since I is a null surface and $\underline{g}_{ab}$ is essentially the induced metric thereon) that $\underline{g}_{ab}$ has signature (0, +, +). Hence, up to a factor at each point, $\underline{n}^b$ is the only vector annihilated by $\underline{g}_{ab}$, so in particular $\underline{g}_{ab}$ determines $\underline{n}^b$ up to a factor. The final property follows from Eqn. (6), which we may rewrite in the form

$$\Omega\, S_{ab} + \mathcal{L}_n\, g_{ab} - f\, g_{ab} = \Omega^{-1}\, L_{ab}. \tag{20}$$

Let us now suppose that the stress-energy vanishes asymptotically to order two, a very weak supposition, which we keep in force for the remainder of this section. Then, applying $\mathfrak{f}^*$ to (20), we obtain $\mathcal{L}_{\underline{n}}\, \underline{g}_{ab} = \underline{f}\, \underline{g}_{ab}$, where we have set $\underline{f} = \mathfrak{f}^*(f)$. That is, $\underline{n}^a$ is a "conformal Killing field" for $\underline{g}_{ab}$. One convinces oneself, by perusal of the equations of the previous section, that there are no other local properties of $\underline{n}^a$ and $\underline{g}_{ab}$, even, e.g., on requiring vanishing of the stress-energy to higher order.

We turn next to the gauge-freedom, described by Eqns. (14) - (18). The gauge-function ω on M is represented, in terms of $\mathcal{J}$, by the (positive) function $\underline{\omega} = \mathfrak{f}^*(\omega)$ on this manifold. Applying $\mathfrak{f}^*$ to $g'_{ab} = \omega^2\, g_{ab}$ and to (15), we obtain

$$\underline{g}'_{ab} = \underline{\omega}^2\, \underline{g}_{ab}, \qquad \underline{n}'^a = \underline{\omega}^{-1}\, \underline{n}^a. \tag{21}$$

That is to say, $\underline{g}_{ab}$, $\underline{n}^a$ and $\underline{g}'_{ab}$, $\underline{n}'^a$, related by (21) for some positive $\underline{\omega}$, are to represent the same geometrical situation. Applying

$\mathscr{I}^*$ to (16), we have

$$\underline{f}' = \underline{\omega}^{-1}\, \underline{f} + 2\, \underline{\omega}^{-2}\, \underline{n}^a\, D_a \underline{\omega}, \tag{22}$$

where D_a denotes the gradient on $\mathscr{I}$.

Two observations follow from the gauge-formulae of the previous paragraph. For the first, set $\Gamma^{ab}{}_{cd} = \underline{n}^a\, \underline{n}^b\, \underline{g}_{cd}$. Then this tensor field $\Gamma^{ab}{}_{cd}$ on $\mathscr{I}$ is gauge-invariant, i.e., under (21) we have $\underline{n}'^a\, \underline{n}'^a\, \underline{g}'_{cd} = \underline{n}^a\, \underline{n}^b\, \underline{g}_{cd}$. Furthermore, all the properties of $\underline{g}_{cd}$ and $\underline{n}^a$ above can be written directly, i.e., without mention of gauge, in terms of this $\Gamma^{ab}{}_{cd}$. One suitable, although not very illuminating, set of properties is the following: i) $\Gamma^{ab}{}_{cd} = \Gamma^{(ab)}{}_{(cd)} \neq 0$ and $\Gamma^{a[b}{}_{cd}\, \Gamma^{e]f}{}_{gh} = 0$ (ensures that $\Gamma^{ab}{}_{cd} = \underline{n}^a\, \underline{n}^b\, \underline{g}_{cd}$ for <u>some</u> $\underline{n}^a$ and $\underline{g}_{cd}$), ii) $\Gamma^{am}{}_{cm} = 0$ (ensures that $\underline{n}^a\, \underline{g}_{ab} = 0$), iii) whenever $w_c\, v^{[a}\, \Gamma^{b]c}{}_{de} \neq 0$, $w_a\, w_b\, v^c\, v^d\, \Gamma^{ab}{}_{cd}$ is positive (signature of $\underline{g}_{ab}$), and iv) whenever $v^{[a}\, \Gamma^{b]c}{}_{de} = 0$, $\mathscr{L}_v\, \Gamma^{ab}{}_{cd}$ is a multiple of $\Gamma^{ab}{}_{cd}$ (ensures that $\underline{n}^a$ is a $\underline{g}_{ab}$ - conformal Killing field). Thus, we may regard $\Gamma^{ab}{}_{cd}$ as representing the complete universal structure of $\mathscr{I}$ in a gauge-invariant way. By an <u>asymptotic</u> <u>geometry</u> we shall mean a three-dimensional manifold $\mathscr{I}$ with a tensor field $\Gamma^{ab}{}_{cd}$ satisfying the four properties above.

For the second observation, set $\underline{f}' = 0$ in Eqn. (22) to obtain $\underline{n}^a\, D_a\, (\log \underline{\omega}) = -\, 1/2\; \underline{f}$. Clearly, there always exists, at least locally, a positive $\underline{\omega}$ satisfying this equation. That is to say, by a gauge transformation we can always arrange locally to have $\underline{f}' = 0$, i.e., to have $\mathscr{L}_{\underline{n}'}\, \underline{g}'_{ab} = 0$. Furthermore, $\underline{f} = 0$ is preserved by (22) when and only when $\underline{n}^a\, D_a\, \underline{\omega} = 0$, i.e., when and only when $\underline{\omega}$ is constant along the $\underline{n}$ - integral curves. For $\mathscr{I}$, $\Gamma^{ab}{}_{cd}$ an asymptotic geometry, by a <u>decomposition</u> of $\Gamma^{ab}{}_{cd}$ we mean fields $\underline{n}^a$ and $\underline{g}_{ab}$ such that $\Gamma^{ab}{}_{cd} = \underline{n}^a\, \underline{n}^b\, \underline{g}_{cd}$ and $\mathscr{L}_{\underline{n}}\, \underline{g}_{ab} = 0$. What we have shown, then, is that every asymptotic geometry possesses, locally, a decomposition, and that it is unique up to (21) with $\underline{\omega}$ constant along the $\underline{n}$ - integral curves.

We may now explain why $\Gamma^{ab}{}_{cd}$ is called "universal". We claim: For $\mathscr{I}$, $\Gamma^{ab}{}_{cd}$ and $\hat{\mathscr{I}}$, $\hat{\Gamma}^{ab}{}_{cd}$ asymptotic geometries, and p and $\hat{p}$ points of $\mathscr{I}$ and $\hat{\mathscr{I}}$, respectively, there exist neighborhoods U of p and $\hat{U}$ of $\hat{p}$, and a diffeomorphism from U to $\hat{U}$ which sends p to $\hat{p}$ and $\Gamma^{ab}{}_{cd}$ to $\hat{\Gamma}^{ab}{}_{cd}$. The proof will be immediate after the following section. One could however construct a direct proof at this point, e.g., by showing the existence of a local chart in terms of which the components of $\Gamma^{ab}{}_{cd}$ are: $\Gamma^{11}{}_{22} = \Gamma^{11}{}_{33} = 1$, all others zero. In any case, the claim asserts that "all Γ's are locally the same", so, in particular, this field does not tell one anything whatever about the structure, asymptotic or otherwise, of the physical space-time from which this field was obtained. The physics of the underlying space-time will be described by other fields.

The metric of general relativity gives rise to the raising and lowering of indices, to an alternating tensor field unique up to sign (i.e., to duals), to a unique derivative operator, and finally to the curvature tensor. We ask what the corresponding apparatus is for $\Gamma^{ab}{}_{cd}$. Let $\mathscr{I}$, $\Gamma^{ab}{}_{cd}$ be an asymptotic geometry, and let $\underline{n}^a$ and $\underline{g}_{ab}$ be a decomposition. We shall obtain what fields, operations, etc. we can from $\underline{n}^a$, $\underline{g}_{ab}$, and then obtain their behavior under the remaining gauge freedom.

We first note that the indices of various tensor fields on can be contracted with those of $\underline{n}^a$ and $\underline{g}_{ab}$. In particular, we may "lower" indices with $\underline{g}_{ab}$. One cannot of course "raise" indices, since $\underline{g}_{ab}$ is not invertible. It is convenient to have available a field $\underline{g}^{ab}$, defined by the property $\underline{g}_{am}\, \underline{g}^{mn}\, \underline{g}_{nb} = \underline{g}_{ab}$. This $\underline{g}^{ab}$ is the "closest thing" to an inverse of $\underline{g}_{ab}$. There certainly exists such a $\underline{g}^{ab}$, and it is clearly unique up to addition of a term of the form $v^{(a}\, \underline{n}^{b)}$, for some v^a. We may use this $\underline{g}^{ab}$ to "raise" indices under those circumstances under which the lack of uniqueness does not lead to an ambiguous result. For example, for α_a such that $\alpha_a\, \underline{n}^a = 0$, the fields $\underline{g}^{ab}\, \alpha_a\, \alpha_b$ and $\alpha_m\, \underline{g}^{m[a}\, \underline{n}^{b]}$ are unchanged under addition to $\underline{g}^{ab}$ of $v^{(a}\, \underline{n}^{b)}$. The former, in particular, shows that we may take "norms" of covariant vectors orthogonal to $\underline{n}^a$. We define an alternating tensor field ϵ^{abc}, up to sign, on $\mathscr{I}$ by the formula

$$\epsilon^{amn}\, \epsilon^{bpq}\, \underline{g}_{mp}\, \underline{g}_{nq} = 2\, \underline{n}^a\, \underline{n}^b \tag{23}$$

and antisymmetry. This sign having been chosen, ϵ_{abc} is fixed uniquely by its antisymmetry and $\epsilon^{abc}\, \epsilon_{abc} = 6$. [In fact, $\epsilon^{abc} = \zeta^*(\epsilon^{abcm}\, n_m)$, but $\zeta^*(\epsilon_{abcm}\, n^m) = 0 \neq \epsilon_{abc}$.] The usual identities for alternating tensors of course hold, and so we may take duals, up to the standard sign ambiguity, cross products, etc.

We next obtain what remnant we can of a derivative operator on $\mathscr{I}$. Let α_a be orthogonal to $\underline{n}^a$, and set $\alpha^a = \underline{g}^{ab}\, \alpha_b$, so α^a is only defined up to a multiple of $\underline{n}^a$. Consider now the right side of

$$D_a\, \alpha_b = D_{[a}\, \alpha_{b]} + 1/2\ \mathcal{L}_{\alpha}\, \underline{g}_{ab}, \tag{24}$$

where the first term on the right is the exterior derivative. [This will be recognized as the formula which, in the invertible case, defines the derivative operator.] The second term on the right is the Lie derivative with respect to (the ambiguous) α^a. But for any function β we have $\mathcal{L}_{\beta \underline{n}}\, \underline{g}_{ab} = \beta \mathcal{L}_{\underline{n}}\, \underline{g}_{ab} + 2\, \underline{n}^m\, \underline{g}_{m(a}\, D_{b)}\beta = 0$, and so this ambiguity disappears from (24). Thus, Eqn. (24) defines an operator D_a which, however, is applicable only to covariant vectors orthogonal to $\underline{n}^a$. By taking sums of outer products, we may extend the action of this D_a to arbitrary covariant tensors, $\alpha_{a..c}$, having the property that any one contraction with $\underline{n}^a$ yields zero. The resulting D_a deserves to be called the derivative: It is additive on sums, and obeys the Leibnitz rule on outer products. As in the invertible case, the derivatives of the fields which give rise to the derivative vanish: $D_a\, g_{bc} = 0$ and $D_a\,(\epsilon_{bcm}\, \underline{n}^m) = 0$. [Note that neither "$D_a\, \underline{n}^b$" nor "$D_a\, \epsilon_{bcd}$" make sense.] Finally, the derivative is connected to $\underline{n}^a$ by the following two properties: For any $\alpha_{a..c}$ orthogonal to $\underline{n}^a$, $\underline{n}^m D_m\, \alpha_{a..c} = \mathcal{L}_{\underline{n}}\, \alpha_{a..c}$, and $\underline{n}^p\, D_m\, \alpha_{a..p..c} = 0$. As an example, we verify these two properties for the case of one index. For α_a and α^a as in (24), we have

$$\underline{n}^a\, D_{[a}\, \alpha_{b]} = 1/2\, \mathcal{L}_{\underline{n}}\, \alpha_b - 1/2\, D_b\,(\alpha_a\, \underline{n}^a) = 1/2\, \mathcal{L}_{\underline{n}}\, \alpha_b, \tag{25}$$

where the first step is an identity, and also

$$\begin{aligned} \underline{n}^a\, \mathcal{L}_\alpha\, g_{ab} &= \mathcal{L}_\alpha (\underline{n}^a\, g_{ab}) - g_{ab}\, \mathcal{L}_\alpha\, \underline{n}^a = -\, g_{ab}\, \mathcal{L}_\alpha\, \underline{n}^a \\ &= g_{ab}\, \mathcal{L}_{\underline{n}}\, \alpha^a = \mathcal{L}_{\underline{n}}\, (g_{ab}\, \alpha^a) - \alpha^a\, \mathcal{L}_{\underline{n}}\, g_{ab} = \mathcal{L}_{\underline{n}}\, \alpha_b. \end{aligned} \tag{26}$$

Contracting (24) successively with $\underline{n}^a$ and $\underline{n}^b$, using (25) and (26), we immediately obtain the two properties in this case.

Now let there be carried out a gauge transformation (21), with $\underline{n}^a\, D_a\, \underline{\omega} = 0$. We ask for what then happens to the objects introduced above. The conformal behavior of the algebraic fields follows in each case directly from the definition:

$$\epsilon'^{abc} = \underline{\omega}^{-3}\, \epsilon^{abc}, \quad \epsilon'_{abc} = \underline{\omega}^3\, \epsilon_{abc}, \quad g'^{ab} = \underline{\omega}^{-2}\, g^{ab}. \tag{27}$$

For the derivative, one makes the appropriate replacements ($g'_{ab} = \underline{\omega}^2\, g_{ab}$, $\alpha'^a = \underline{\omega}^{-2}\, \alpha^a$) in (24):

$$D'_a\, \alpha_b = D_a\, \alpha_b - 2\, \underline{\omega}^{-1}\, \alpha_{(a}\, D_{b)}\underline{\omega} + \underline{\omega}^{-1}\, g_{ab}\, g^{mn}\, \alpha_m\, D_n \underline{\omega}. \tag{28}$$

Note that the last term on the right in (28) is unambiguous.

To summarize, the geometrical structure of $\mathcal{J}$ permits the taking of duals, the lowering of indices at will, the raising of indices subject to an ambiguity, and the taking of derivatives of covariant fields orthogonal to $\underline{n}^a$; with gauge-behavior (27) and (28). The situation is certainly more complicated and less satisfactory than, e.g., that of an invertible metric.

We conclude this section with the introduction of the "curvature tensor" of an asymptotic geometry. Let α_a satisfy $\alpha_a\, \underline{n}^a$ and $\mathcal{L}_{\underline{n}}\alpha_a = 0$. By the first, $D_a\, \alpha_b$ makes sense; by the second, $\underline{n}^a D_a\, \alpha_b$ and $\underline{n}^b\, D_a\, \alpha_b$ vanish. Hence, $D_a\, D_b\, \alpha_c$ makes sense. Consider, then, the left side of

$$D_{[a}\, D_{b]}\, \alpha_c = 1/2\, \mathcal{R}_{abc}{}^d\, \alpha_d. \tag{29}$$

This left side is clearly additive in α_c, and, as is seen by using the Leibnitz rule twice, $D_{[a} D_{b]} (\beta \alpha_c) = \beta D_{[a} D_{b]} \alpha_c$ for any scalar field β. Hence, there certainly exists some tensor field $\mathcal{R}_{abc}{}^d$ on $\mathcal{I}$ such that (29) holds for all allowed α_c. Since, however, we can only impose (29) for α_c satisfying $\alpha_d \underline{n}^d$, this equation only defines $\mathcal{R}_{abc}{}^d$ up to a multiple of $\underline{n}^d$. Hence, $\mathcal{R}_{abcd} = \mathcal{R}_{abc}{}^m \underline{g}_{md}$, the curvature tensor of $\mathcal{I}$, $\Gamma^{ab}{}_{cd}$ (with respect to this particular decomposition), is uniquely determined.

The symmetries of $\mathcal{R}_{abcd}$ are obtained by restating the arguments for the invertible case. Evaluating the left side of $D_{[a} D_{b]} \underline{g}_{cd} = 0$ using (29), we obtain that $\mathcal{R}_{abcd}$ is also antisymmetric in its last two indices. Evaluating the left side of $D_{[a} D_b \alpha_{c]} = 0$ using (29), we obtain that $\mathcal{R}_{[abc]d} = 0$. Thus, $\mathcal{R}_{abcd}$ has all the symmetries of a curvature tensor. Any contraction of $\mathcal{R}_{abcd}$ with $\underline{n}$ must yield zero, since contraction with $\underline{n}^d$ certainly does. Thus, the corresponding Ricci tensor, $\mathcal{R}_{ab} = \underline{g}^{mn} \mathcal{R}_{ambn}$, and scalar curvature, $\mathcal{R} = \underline{g}^{mn} \mathcal{R}_{mn}$, are unambiguous. Since $\mathcal{R}_{abcd}$ lives in the two dimensions orthogonal to $\underline{n}^a$, it may be reconstructed from its scalar curvature alone: $\mathcal{R}_{abcd} = \mathcal{R}\, \underline{g}_{a[c} \underline{g}_{d]b}$. Expanding $D_{[a} D_b D_{c]} \alpha_d$ in two ways using (29) (first replacing "ab" with a Riemann tensor, then "bc"), we obtain the Bianchi identity, $D_{[a} \mathcal{R}_{bc]de} = 0$, noting that this makes sense. Finally, applying $\mathcal{L}_{\underline{n}}$ to (29), we obtain $\mathcal{L}_{\underline{n}} \mathcal{R}_{abcd} = 0$. The behavior of $\mathcal{R}_{abcd}$ under a new choice, (21), of decomposition is obtained by substituting (28) into (29):

$$\mathcal{R}' = \underline{\omega}^{-2} \mathcal{R} - 2\, \underline{\omega}^{-3}\, \underline{g}^{mn} D_m D_n \underline{\omega} + 2\, \underline{\omega}^{-4}\, \underline{g}^{mn} D_m \underline{\omega}\, D_n \underline{\omega}. \quad (30)$$

We shall see in the next section that these properties can be understood in a more natural way.

4. The Global Structure of Infinity

We first introduce a number of possible global conditions on an asymptotic geometry, and, where feasible, describe these conditions in terms of the physical space-time. We then summarize an alternative framework for much of the discussion of the previous section.

Is every asymptotic geometry possible, i.e., does every one arise from some physical space-time? Locally, the answer is of course yes, since all asymptotic geometries are locally the same and since, from the examples of Sect. 1, at least one asymptotic geometry arises from a physical space-time. The answer is also yes globally. Let $\mathcal{I}$, $\Gamma^{ab}{}_{cd}$ be an asymptotic geometry, and set $\Gamma^{ab}{}_{cd} = \underline{n}^a \underline{n}^b \underline{g}_{cd}$ (not necessarily with $\mathcal{L}_{\underline{n}} \underline{g}_{ab} = 0$). Set $M = \mathcal{I} \times [0, 1)$, a four-dimensional manifold with boundary $I = \mathcal{I} \times \{0\}$. Let Ω be the smooth scalar field on M whose value at (p, r) (p in $\mathcal{I}$, r in [0, 1)) is r. Denote by l^a the field of tangents in M to the curves given by p in $\mathcal{I}$ fixed as r varies in [0, 1). Then, in M, $l^a \nabla_a \Omega = 1$. Via the slicing of M by $\mathcal{I}$ and $[0, 1)$, the fields $\underline{n}^a$ and $\underline{g}_{ab}$ on $\mathcal{I}$ give rise naturally to corresponding fields, $\underline{n}^a$ and $\underline{g}_{ab}$, on M. These satisfy, e.g., $\underline{n}^a \nabla_a \Omega = 0$ and $1^a \underline{g}_{ab} = 0$. Now let 1_a be any covariant vector field on M satisfying $1_a 1^a = 0$ and $1_a \underline{n}^a = 1$. The tensor field $g_{ab} = \underline{g}_{ab} + 2\, 1_{(a} \nabla_{b)}\Omega$ on M is then a metric of the proper signature, while, with respect to this metric, we have $\underline{n}^a g_{ab} = \nabla_b \Omega$ and $\underline{n}^a \underline{n}^b g_{ab} = 0$ on M. Thus, setting $\tilde{g}_{ab} = \Omega^{-2} g_{ab}$ on $\tilde{M} = M - I$, we have an asymptote of the physical space-time $\tilde{M}$, $\tilde{g}_{ab}$. The asymptotic geometry of this asymptote is clearly the one with which we began. This example will in general have stress-energy vanishing asymptotically only to the minimum order necessary: one. Although it seems rather unlikely that imposition of vanishing of the stress-energy to higher order will lead to further global restrictions on $\mathcal{I}$, $\Gamma^{ab}{}_{cd}$, this question is, as far as I know, open. (One would presumably approach this question through the argument above and Eqn. (20).) We shall see in a moment that there do exist "nonstandard" asymptotic geometries which arise nontrivially from flat physical space-times.

We next make a few observations regarding orientability. Let $\tilde{M}$, $\tilde{g}_{ab}$ be a space-time, and M, g_{ab}, Ω an asymptote. Then $\tilde{M}$ is necessarily time-orientable in (its intersection with) a neighborhood of I, for $n^a = \nabla^a \Omega$ is null on I, and hence defines a preferred half of the full light-cone in a neighborhood of I. On the other hand, $\tilde{M}$ is orientable (equivalently, by the above, space-orientable) in sufficiently small neighborhoods of I if and only if the manifold I is orientable. Indeed, in the parenthetical remark following Eqn. (23) is a formula for an alternating tensor field on I in terms of one on M. Suppose, finally, that our physical space-time has been endowed with a global time-orientation. Let p be a point of I. Then, since I is null at p, the timelike curves in $\tilde{M}$ which, in M, have endpoint p are either all future-directed or all past-directed. We say that p is, respectively, a <u>future</u> or <u>past</u> boundary point. Since by continuity all points of any connected component of I must fall in the same class, each such component is either a future boundary or past boundary. These correspond, of course, to passage to infinity in future or past null directions. For example, endowing the Schwarzschild space-time with the usual time-orientation associated with the chart (1), the boundary indicated in Fig. 1 is a future boundary; the other, past.

Let $\mathcal{I}$, $\Gamma^{ab}{}_{cd}$ be an asymptotic geometry. A maximally extended integral curve γ of $\underline{n}^a$ is said to be <u>almost closed</u> if, for some point p of γ, γ reenters every sufficiently small neighborhood of p. Thus, a closed integral curve is almost closed. Let γ and p be as above, and let γ' and p' be the corresponding curve and point in I. Then γ' is a null geodesic in M. Hence, given any sufficiently small neighborhood of p' in M, some null geodesic in $\tilde{M}$ passes through this neighborhood more than once. Unfortunately, even this last criterion is not entirely in terms of the physical space-time, for it requires the point p' of I and its neighborhoods. It is by no means clear, for example, that a causality condition on the physical space-time would rule out almost-closed integral curves of $\underline{n}^a$ in the corresponding asymptotic geometry.

We now turn to the more interesting conditions. Let $\mathcal{J}$, $\Gamma^{ab}{}_{cd}$ be an asymptotic geometry having no almost-closed integral curves of $\underline{n}^a$. Denote by B the set of all maximally extended integral curves of $\underline{n}^a$, and let $\pi : \mathcal{J} \rightarrow B$ be the mapping which sends each point of $\mathcal{J}$ to the integral curve on which it lies. We obtain some charts on the set B. Consider a chart in $\mathcal{J}$ based on open set U, such that no $\underline{n}$ - integral curve passes through U more than once, and such that two of the coordinate functions are constant along the $\underline{n}$ - integral curves in U. Projecting these two coordinate functions down to B via π, we obtain a chart in B based on $\pi[U]$. These charts are all compatible with each other, and, by absence of almost-closed integral curves, cover B. Thus, B becomes a two-dimensional manifold - although it may not be Hausdorff. [E.g., remove one point from I in the example of Sect. 1.] When B is Hausdorff, we call it the base space of the asymptotic geometry. In this case, the mapping π is smooth, and the manifold $\mathcal{J}$ is just $B \times R$. By a cross section of $\mathcal{J}$ in this case, we mean a smooth mapping κ from B to $\mathcal{J}$ such that $\pi \circ \kappa$ is the identity on B. That is to say, a cross section represents a "lifting" of B back into $\mathcal{J}$ such that each point p of B is sent to a point of the integral curve in $\mathcal{J}$ which defines p.

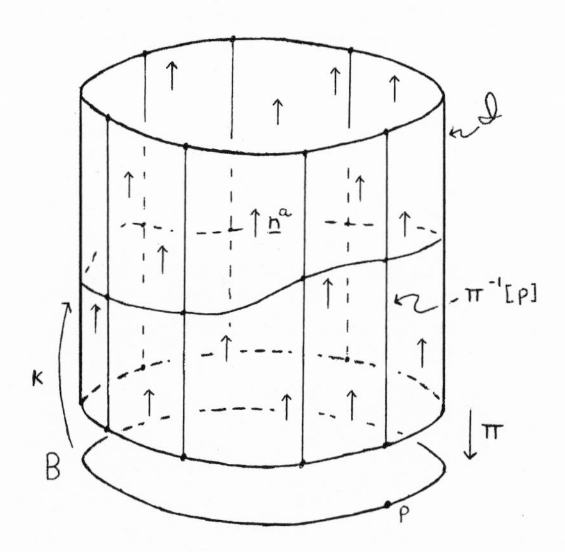

Fig. 2. The asymptotic geometry of the Schwarzschild space-time. (One dimension of spherical symmetry is suppressed.) The base space is a two-sphere (circle in the figure). The mapping π acts vertically downward; the vertical lines in $\mathcal{J}$ are integral curves of $\underline{n}^a$.

Consider, as an example, the asymptotic geometry of the Schwarzschild space-time obtained in Sect. 1. (See Fig. 2.) A suitable chart on I, and hence on $\mathscr{I}$, is that given by the u, θ, and φ of Eqn. (3), for I was there given by x = 0. The integral curves of $\underline{n}^a$ are the curves of constant θ and φ. Hence, we may take θ and φ as coordinates on B. The base space B is therefore a two-sphere, while the mapping π sends the point of $\mathscr{I}$ labeled by u_0, θ_0, φ_0 to the point of B labeled by θ_0, φ_0. The manifold $\mathscr{I}$ is $B \times R$. In terms of these charts, a cross section is described by a function f of the two spherical variables, θ and φ: The mapping κ sends the point of B labeled by θ_0, φ_0 to the point of $\mathscr{I}$ labeled by $u_0 = f(\theta_0, \varphi_0)$, θ_0, φ_0. Any two cross sections, then, are related by a "sliding up or down the $\underline{n}$ - integral curves".

Let $\mathscr{I}$, $\Gamma^{ab}{}_{cd}$ be an asymptotic geometry, with base space B. Then there necessarily exists a global decomposition of $\Gamma^{ab}{}_{cd}$. [Indeed, (See the second paragraph after Eqn. (22).) one can choose $\underline{\omega} = 1$ on the image of a cross section, and define $\underline{\omega}$ elsewhere on $\mathscr{I}$ by the equation $\underline{n}^a \nabla_a (\log \underline{\omega}) = -1/2\, \underline{f}$. Since every $\underline{n}$ - integral curve meets the cross section once and only once, this yields a unique $\underline{\omega}$.] Let such a decomposition be chosen. Let $\nu_{a..c}$ be a covariant tensor field on the manifold B, and let $\mu_{a..c} = \pi^*(\nu_{a..c})$, defined via the same three conditions as at the beginning of the previous section, be its pullback to $\mathscr{I}$. Then the tensor field $\mu_{a..c}$ on $\mathscr{I}$ satisfies $\mathcal{L}_{\underline{n}}\, \mu_{a..c} = 0$ and also the condition that contraction of any of its indices with $\underline{n}^a$ yields zero. Conversely, any $\mu_{a..c}$ on $\mathscr{I}$ satisfying these two conditions is of the form $\pi^*(\nu_{a..c})$ for a unique field $\nu_{a..c}$ on B. In particular, the metric, $\underline{g}_{ab}$, of our decomposition satisfies these two conditions: hence, $\underline{g}_{ab} = \pi^*(h_{ab})$ for some (positive-definite) metric h_{ab} on B. That is to say, any decomposition of $\Gamma^{ab}{}_{cd}$ yields a positive-definite metric on the base space. Next, let α_a be any field on $\mathscr{I}$ satisfying $\alpha_a \underline{n}^a = 0$ and $\mathcal{L}_{\underline{n}}\, \alpha_a = 0$, so $\alpha_a = \pi^*(\beta_a)$ for some β_a on B. Taking the derivative of this β_a (using the natural derivative operator on B, h_{ab}), and applying π^* to the result, we obtain a second-rank, covariant field on $\mathscr{I}$. This field is just the "$D_a \alpha_b$" defined by Eqn. (24) (although the "D_a" there defined was applicable to a wider class of covariant fields). That is, the derivative we obtained in the previous section was essentially the pullback, via π^*, of the natural derivative operator on B, h_{ab}. Similarly, the curvature tensor, $\mathcal{R}_{abcd}$, of Sect. 3 is now seen to be the pullback of the natural curvature tensor on B, h_{ab}. Here, then, is the "reason" this $\mathcal{R}_{abcd}$ turned out to have all the properties of a curvature tensor. Finally, we express the gauge freedom in terms of the base space. Let two decompositions be related by (21), with $\underline{n}^a \nabla_a \underline{\omega} = 0$. This last equation implies that $\underline{\omega} = \pi^*(\sigma)$ for some scalar field σ on B. Then, from (21), we have $h'_{ab} = \sigma^2 h_{ab}$. Gauge transformations between two decompositions thus yield conformal transformations on the metric of the base space: It is only the conformal geometry of the base space which is gauge-invariant. To summarize, the "$\underline{n}^a$ - part" of $\Gamma^{ab}{}_{cd}$ is reflected in the construction of the base space B, and the "$\underline{g}_{ab}$ - part" in a conformal metric on B. In this way, much of the geometry inherent in $\Gamma^{ab}{}_{cd}$ is carried by the base space.

As examples of applications of these remarks, we make three observations. First, since every asymptotic geometry locally admits a base space, since every conformal metric in two dimensions can locally, by a choice of conformal factor, be made flat, and since all flat Riemannian 2-manifolds are locally isometric, it follows immediately that all asymptotic geometries are locally the same. For the second observation, consider the case in which B is the torus, $S^1 \times S^1$. Then one can choose the conformal factor so that the metric of B is flat, and, repeating the construction at the beginning of this section, can arrange that this asymptotic geometry arises from a flat physical space-time. Thus, choices of $\mathscr{I}$ other than $S^2 \times R$ are certainly compatible with vanishing of the stress-energy to higher order. For the third observation, we note that one can impose global conditions on an asymptotic geometry beyond the mere existence of a base space. In particular, one could require that B be compact, or that the vector field $\underline{n}^a$ of a decomposition be complete (noting that the latter condition is independent of decomposition). These two taken together imply that the asymptotic geometry is inextendible, i.e., that no "holes" have been cut in it. Thus, one has available simple, intrinsic versions (as opposed to the more complicated one of Theorem 2) of the idea that "all of the boundary at infinity has been included".

By far the most important case is that in which there exists a base space, it is the two-sphere, and $\underline{n}^a$ is complete, i.e., that of Minkowski space, the Schwarzschild space-time, etc. We call such an asymptotic geometry Minkowskian. These correspond to physical space-times which are not only "asymptotically flat", but also "have the same global asymptotic structure as Minkowski space". That is to say, these correspond to physical space-times which genuinely represent "isolated systems", in the stronger sense that not only does no geometrical information permeate from the outside world, but also no "topological information". From a practical viewpoint, not a great deal would be lost by restricting consideration throughout to Minkowskian asymptotic geometries. [Nor would a great deal be gained: We shall normally deal with more general cases only to make it more explicit what follows from what.] It might be of interest to look for properties which i) are simple and are intrinsic to the physical space-time, ii) have the intuitive connotation of "asymptotically Minkowskian", and which iii) imply that the corresponding asymptotic geometry is Minkowskian. One result [5] along these lines - not completely intrinsic, but probably the best one can do in this regard without destroying simplicity - states that if every maximally extended null geodesic in the physical space-time has two endpoints on I, then the asymptotic geometry consists of one future part and one past part, each with base space S^2. It is apparently not known whether or not completeness of $\underline{n}^a$ also follows.

5. Asymptotic Symmetries

Let $\mathcal{I}$, $\Gamma^{ab}{}_{cd}$ be an asymptotic geometry, with base space B. A symmetry on $\mathcal{I}$, $\Gamma^{ab}{}_{cd}$ is a diffeomorphism from $\mathcal{I}$ to $\mathcal{I}$ which sends $\Gamma^{ab}{}_{cd}$ to itself. The set of symmetries, under composition, clearly has the structure of a group. An infinitesimal symmetry on $\mathcal{I}$, $\Gamma^{ab}{}_{cd}$ is a vector field ξ^a on $\mathcal{I}$ satisfying $\mathcal{L}_\xi \Gamma^{ab}{}_{cd} = 0$. Under the bracket of vector fields, the infinitesimal symmetries have the structure of a Lie algebra. Of course, infinitesimal symmetries represent in some sense "symmetries infinitesimally near the identity". As discussed in the Introduction, these are analogous to the Poincaré group and Killing fields in Minkowski space, respectively. In this section, we consider these algebraic objects.

We begin with the following observation.

Theorem 3. Let $\tilde{M}$, $\tilde{g}_{ab}$ be a space-time with Killing field $\tilde{\eta}^a$, and let M, g_{ab}, Ω be an asymptote, with corresponding $\mathcal{I}$ and $\Gamma^{ab}{}_{cd}$. Then $\tilde{\eta}^a$ has a unique extension to a smooth vector field on M, the resulting field η^a is in the class $\mathcal{C}$, and $\zeta^*(\eta^a)$ is an infinitesimal symmetry.

Proof: In $\tilde{M}$, we have $\mathcal{L}_{\tilde{\eta}}\, g_{ab} = \mathcal{L}_{\tilde{\eta}}(\Omega^2\, \tilde{g}_{ab}) = 2\,\Omega\, \tilde{g}_{ab}\, \mathcal{L}_{\tilde{\eta}}\Omega = 2\,\Omega^{-1}\, g_{ab}\, \mathcal{L}_{\tilde{\eta}}\Omega$. Thus, $\tilde{\eta}^a$ is a g_{ab} - conformal Killing field in $\tilde{M}$, and so has a smooth extension, η^a, to M. From this and the formula above, $\alpha = \Omega^{-1}\mathcal{L}_\eta\Omega$ must be smooth in M. Hence, $\eta^a\ \nabla_a\Omega = \Omega\alpha$ vanishes on I, whence η^a is in the class $\mathcal{C}$. By direct computation, $\mathcal{L}_\eta\, n^a = \mathcal{L}_\eta\,(g^{ab}\ \nabla_b\Omega) = -\,\alpha\, n^a + \Omega\ \nabla^a\alpha$ whence, expanding, we see that $\mathcal{L}_\eta\ (n^a\, n^b\, g_{cd}) = 0$ at points of I. Applying ζ^*, we obtain the last assertion.

Thus, infinitesimal symmetries of the physical space-time always lead to infinitesimal symmetries at infinity. The converse, as we shall see, is not even close to being true. One might suspect that, since "asymptote" is a conformal notion, Theorem 3 would also be true with "Killing field" replaced by "conformal Killing field". Although the argument does show that, under this change, $\tilde{\eta}^a$ has a smooth extension to M, the rest of the conclusion fails in general. In Minkowski space, for example, the situation is this. The dilations vanish on I, and thus result in the zero infinitesimal symmetry, while the remaining conformal Killing fields turn out not to be in the class $\mathcal{C}$, and so define no vector fields on $\mathcal{I}$.

We discuss first the Lie algebra $\mathcal{L}$ of infinitesimal symmetries. It will be convenient to have an expression of the statement that ξ^a is an infinitesimal symmetry in terms of a decomposition of $\Gamma^{ab}{}_{cd}$ $(= \underline{n}^a\ \underline{n}^b\ \underline{g}_{cd}$, with $\mathcal{L}_{\underline{n}}\ \underline{g}_{cd} = 0)$. We have $\mathcal{L}_\xi\ \Gamma^{ab}{}_{cd} = \underline{n}^a\ \underline{n}^b\ \mathcal{L}_\xi\ \underline{g}_{cd} + 2\ \underline{g}_{cd}\ (\mathcal{L}_\xi\ \underline{n}^{(a})\ \underline{n}^{b)}$. The right side vanishes (contracting with an arbitrary α^{cd}, then with an arbitrary β_{ab}) if and only if

$$\mathcal{L}_\xi \underline{g}_{ab} = 2\,\kappa\,\underline{g}_{ab}, \quad \mathcal{L}_\xi \underline{n}^a = -\,\kappa\,\underline{n}^a \tag{31}$$

for some scalar field κ on $\mathcal{J}$. Eqn. (31), then, is an alternative statement that ξ^a is in $\mathcal{L}$.

An infinitesimal symmetry ξ^a is called an <u>infinitesimal supertranslation</u> if ξ^a is a multiple of $\underline{n}^a$. We denote the set of infinitesimal supertranslations by $\mathcal{S}$. The terminology is motivated by the fact that the translations in Minkowski space give rise to elements of $\mathcal{S}$.

<u>Theorem</u> 4. Let $\tilde{M}$, $\tilde{g}_{ab}$ be a space-time with Killing field $\tilde{\eta}^a$, and let M, g_{ab}, Ω be an asymptote. Then $\tilde{\eta}^a$ yields, via Theorem 3, an infinitesimal supertranslation if and only if $\Omega^2\ \tilde{\eta}^a\ \tilde{\eta}^b\ \tilde{g}_{ab}$ vanishes on I (and, in particular, whenever $\tilde{\eta}^a\ \tilde{\eta}^b\ \tilde{g}_{ab}$ is bounded).
Proof: We have $\eta^a\ \eta^b\ g_{ab} = \Omega^2\ \tilde{\eta}^a\ \tilde{\eta}^b\ \tilde{g}_{ab}$. Thus, the right side vanishes if and only if η^a is g_{ab}-null on I, i.e., since I is null and η^a is tangent to I, if and only if η^a is a multiple of n^a on I.
We also note from Theorem 4 that rotations and boosts in Minkowski space do not give rise to infinitesimal supertranslations.

Let $\xi^a = \alpha\,\underline{n}^a$. Then $\mathcal{L}_\xi \underline{g}_{ab} = \alpha\,\mathcal{L}_{\underline{n}}\,\underline{g}_{ab} + 2\,\underline{n}^m\,\underline{g}_{m(a}\,D_{b)}\alpha = 0$, and $\mathcal{L}_\xi\,\underline{n}^a = -\,(\underline{n}^m\,D_m\alpha)\,\underline{n}^a$. Thus, by (31), $\alpha\,\underline{n}^a$ is in $\mathcal{S}$ if and only if $\underline{n}^m\,D_m\alpha = 0$, i.e., if and only if α is constant along the $\underline{n}$-integral curves. But such α's on $\mathcal{J}$ are precisely those of the form $\alpha = \pi^*(\beta)$ for some scalar field β on the base space B. Thus, $\mathcal{S}$ is essentially the same as the set of scalar fields on B. This representation of an element of $\mathcal{S}$ by a field α depends of course, on our particular choice of decomposition of $\Gamma^{ab}{}_{cd}$. In order to have $\alpha'\,\underline{n}'^a = \alpha\,\underline{n}^a$ under a new choice, (21), we must have $\alpha' = \underline{\omega}\,\alpha$. Thus, the set $\mathcal{S}$ is in fact the same as the set of scalar fields on B of dimension +1.

Any linear combination (with constant coefficients!) of infinitesimal supertranslations is again an infinitesimal supertranslation. Thus, $\mathcal{S}$ is a vector subspace of the vector space $\mathcal{L}$ - and, by the paragraph above, an infinite-dimensional subspace. For $\xi^a = \alpha\,\underline{n}^a$ and $\eta^a = \beta\,\underline{n}^a$ in $\mathcal{S}$, we have $\mathcal{L}_\xi\,\eta^a = \mathcal{L}_\xi\,(\beta\,\underline{n}^a) = (\mathcal{L}_\xi\,\beta)\,\underline{n}^a + \beta\,\mathcal{L}_\xi\,\underline{n}^a = 0$. That is, any two elements of $\mathcal{S}$ have vanishing Lie bracket. We conclude that $\mathcal{S}$ is even a subalgebra of the Lie algebra $\mathcal{L}$, and in fact an abelian one. Finally, for ξ^a in $\mathcal{L}$ and $\beta\,\underline{n}^a$ in $\mathcal{S}$, we have $\mathcal{L}_\xi\,(\beta\,\underline{n}^a) = (\mathcal{L}_\xi\,\beta)\,\underline{n}^a + \beta\,\mathcal{L}_\xi\,\underline{n}^a = (\mathcal{L}_\xi\,\beta - \beta\kappa)\,\underline{n}^a$, with κ given by (31). But the last expression, an element of $\mathcal{L}$, is a multiple of $\underline{n}^a$, and so is an element of $\mathcal{S}$. That is, the bracket of an element of $\mathcal{L}$ with an element of $\mathcal{S}$ is again in $\mathcal{S}$, i.e., the infinitesimal supertranslations form an ideal in the Lie algebra $\mathcal{L}$.

To summarize, we have so far found an infinite-dimensional, commutative subalgebra $\mathcal{S}$, the infinitesimal supertranslations, of the Lie algebra $\mathcal{L}$ of infinitesimal symmetries. This $\mathcal{S}$ turns out also to be an ideal, i.e., the commutator of any element of $\mathcal{S}$ with any element of $\mathcal{L}$ is back in $\mathcal{S}$. Further, we have found that the elements of $\mathcal{S}$ can be represented explicitly in terms of the base space B - as scalar fields on this manifold of dimension +1. In this sense, then, we "understand the $\mathcal{S}$-part of $\mathcal{L}$". What remains is to understand the rest of $\mathcal{L}$. This is accomplished as follows. Since $\mathcal{S}$ is an ideal in $\mathcal{L}$, one can form the quotient algebra, $\mathcal{L}/\mathcal{S}$. [That is, an element of the quotient is an element of $\mathcal{L}$, specified only up to addition of an arbitrary element of $\mathcal{S}$ (more precisely, an equivalence class under the relation "differ by an element of $\mathcal{S}$"). Define a bracket on this quotient as follows. For ξ^a and η^a in $\mathcal{L}$ and μ^a and ν^a in $\mathcal{S}$, we have $[\xi+\mu, \eta+\nu] = [\xi,\eta] + [\xi,\nu] + [\mu,\eta] + [\mu,\nu]$. By the ideal property, each of the last three terms on the right is in $\mathcal{S}$. Hence, $[\xi+\mu, \eta+\nu]$ is equal to $[\xi,\eta]$, up to addition of an arbitrary element of $\mathcal{S}$. That is to say, the bracket of two elements of the quotient yields a well-defined element of the quotient. The quotient $\mathcal{L}/\mathcal{S}$, then, becomes a Lie algebra.] This quotient algebra just represents the "rest of $\mathcal{L}$"; we wish, therefore, to understand its structure.

It turns out that $\mathcal{L}/\mathcal{S}$ can be represented explicitly within $\mathcal{I}$. Fix a decomposition of $\Gamma^{ab}{}_{cd}$ (which exists, since we are assuming a base space), let ξ^a be any infinitesimal symmetry, and set $\xi_a = \underline{g}_{ab}\,\xi^b$. Then, we claim, this ξ_a satisfies

$$\underline{n}^a\,\xi_a = 0, \quad D_{(a}\,\xi_{b)} = \kappa\,\underline{g}_{ab}, \quad \mathcal{L}_{\underline{n}}\,\xi_a = 0, \tag{32}$$

for some κ. Indeed, the first is immediate from the definition, the second from (24) and the first equation in (31), and the third by applying $\underline{g}_{ab}$ to the second equation in (31). We next claim, conversely, that any ξ_a satisfying (32) is of the form $\underline{g}_{ab}\,\xi^b$ for some infinitesimal symmetry ξ^b. Let such a ξ_a be given. Then the first equation in (32) implies that $\xi_a = \underline{g}_{ab}\,\eta^b$ for <u>some</u> η^b; set $\xi^a = \eta^a + \alpha\,\underline{n}^a$. Then the second equation in (32) yields $\mathcal{L}_\xi\,\underline{g}_{ab} = 2\kappa\,\underline{g}_{ab}$, while the third yields $\mathcal{L}_\xi\,\underline{n}^a = -(\underline{n}^m D_m\alpha)\,\underline{n}^a$. By (31), this ξ^a will therefore be an infinitesimal symmetry if and only if α satisfies $\underline{n}^m D_m\alpha = \kappa$. But, since we have a base space, we can always find some α satisfying this equation, i.e., we can always find some infinitesimal symmetry ξ^a such that $\underline{g}_{ab}\,\xi^b$ is our original solution of (32). Finally, we note that infinitesimal symmetries ξ^a and τ^a differ by an infinitesimal supertranslation if and only if $\underline{g}_{ab}\,\xi^b = \underline{g}_{ab}\,\tau^b$, i.e., if and only if ξ^a and τ^a define the same solution of (32). Thus, a solution of (32) determines an element of $\mathcal{L}$ up to addition of an arbitrary element of $\mathcal{S}$, i.e., the solutions of (32) realize the quotient algebra, $\mathcal{L}/\mathcal{S}$.

As one might expect, the solutions of (32) have a simple geometrical interpretation in terms of the base space B. The first and third equations in (32) are necessary and sufficient for the existence of a field μ_a on B with $\xi_a = \pi^*(\mu_a)$. The second equation in (32) is then just π^* applied to the conformal Killing equation for μ_a on B, h_{ab}. Thus, the solutions of (32) are precisely the pullbacks, using π^*, of conformal Killing fields (covariantly represented) on B, h_{ab}. The Lie algebra $\mathcal{L}/\mathcal{S}$, then, is naturally isomorphic with the Lie algebra of conformal Killing fields on the base space B with its conformal geometry.

This, then, is the structure of the Lie algebra $\mathcal{L}$. It possesses an ideal $\mathcal{S}$ isomorphic with the abelian Lie algebra of scalar fields on B of dimension +1. The quotient, $\mathcal{L}/\mathcal{S}$, is isomorphic with the Lie algebra of solutions of (32), or, what is the same thing, with the Lie algebra of conformal Killing fields on B, h_{ab}. In terms of algebraic structure, the infinitesimal symmetries on $\mathcal{I}$, $\Gamma^{ab}{}_{cd}$ are analogous to the infinitesimal symmetries on Minkowski space. Our $\mathcal{L}$ corresponds to the Lie algebra of the Poincaré group, $\mathcal{S}$ to the infinitesimal translations on Minkowski space, and $\mathcal{L}/\mathcal{S}$ to the Lie algebra of the Lorentz group. In each case, the second is an abelian ideal in the first, with quotient the third.

This analogy goes even deeper. For example, there is no single, natural Lorentz subalgebra of the Poincarè Lie algebra. Similarly, there is no single, natural $\mathcal{L}/\mathcal{S}$ subalgebra of $\mathcal{L}$. However, it is possible to realize the Lorentz Lie algebra as a subalgebra of the Poincaré Lie algebra. For example, fix a point of Minkowski space, and consider the collection of all Killing fields which vanish at that point. These form a subalgebra of Poincarè, isomorphic with Lorentz. Of course, this subalgebra is not "natural", because its determination requires the choice of a point of Minkowski space. We may ask the corresponding question in the asymptotic case. Does there exist a subalgebra of $\mathcal{L}$ which is isomorphic with $\mathcal{L}/\mathcal{S}$, and which is such that every element of $\mathcal{L}$ can be written in one and only one way as the sum of one element from this subalgebra and one element of $\mathcal{S}$? The answer is yes. Fix a cross section of $\mathcal{I}$, and denote its image by C. Denote by $\mathcal{L}_C$ the collection of all infinitesimal symmetries which, at points of C, are tangent to C. Then, since the Lie bracket of two vector fields tangent to a submanifold is again tangent to that submanifold, $\mathcal{L}_C$ is a subalgebra of $\mathcal{L}$. Now let ξ^a be any infinitesimal symmetry. Then, at each point of C, there is one and only one number α such that $\xi^a - \alpha \underline{n}^a$ is tangent to C at that point. Define scalar field α everywhere on $\mathcal{I}$ by the conditions that at points of C α be this number, and that $\underline{n}^a D_a \alpha = 0$ (noting that these conditions determine one and only one scalar field α). But now we have $\xi^a = (\xi^a - \alpha \underline{n}^a) + \alpha \underline{n}^a$. Since the second term on the right is an infinitesimal supertranslation, the first is in $\mathcal{L}$; by construction, then, the first is in $\mathcal{L}_C$. Thus, every infinitesimal symmetry can be written in one and only one way as the sum of one element of $\mathcal{L}_C$ and one of $\mathcal{S}$. We conclude, therefore, that this $\mathcal{L}_C$ is the required realization of $\mathcal{L}/\mathcal{S}$ in $\mathcal{L}$. Of course, this $\mathcal{L}_C$ is not "natural", because its determination requires a cross section. Thus, at least via the present analogy of Lie algebras, cross sections of $\mathcal{I}$ correspond to points of Minkowski space.

All of the remarks above apply, of course, to the special case in which the asymptotic geometry $\mathcal{I}$, $\Gamma^{ab}{}_{cd}$ is Minkowskian. The Lie algebra of infinitesimal symmetries in this case is called the BMS (Bondi-Metzner-Sachs) Lie algebra. In this case, B is the two-sphere, and so $\mathcal{L}/\mathcal{S}$ is isomorphic with the Lie algebra of conformal Killing fields on the conformal two-sphere. But the latter is isomorphic to the Lie algebra of the Lorentz group, and, in particular, is finite-dimensional. Thus, for the BMS Lie algebra, not only is $\mathcal{L}/\mathcal{S}$ analogous to the Lorentz Lie algebra - it is isomorphic with it.

By Theorems 3 and 4, every translational Killing field in Minkowski space leads to an infinitesimal supertranslation on the corresponding asymptotic geometry. But the vector space of the former is four-dimensional; of the latter, infinite-dimensional. Thus, not every element of $\mathcal{S}$ arises in this way. Is it possible to distinguish, within the geometrical structure of the Minkowskian asymptotic geometry $\mathcal{I}$, $\Gamma^{ab}{}_{cd}$, those infinitesimal supertranslations which do arise from translational Killing fields in Minkowski space? It turns out, as we now show, that one can.

It will be convenient, for both now and later, to have available an auxiliary geometrical field on $\mathcal{I}$.

Theorem 5. Let $\mathcal{I}$, $\Gamma^{ab}{}_{cd}$ be a Minkowskian asymptotic geometry, and $\underline{g}_{ab}$, $\underline{n}^a$ a decomposition. Then there exists one and only one symmetric field ρ_{ab} on $\mathcal{I}$ satisfying

$$\rho_{ab}\,\underline{n}^b = 0, \qquad \rho_{ab}\,\underline{g}^{ab} = \mathcal{R}, \qquad D_{[a}\,\rho_{b]c} = 0. \tag{33}$$

Proof: We first note, from (28) and (30), that ρ_{ab} satisfies (33) for decomposition $\underline{g}_{ab}$, $\underline{n}^a$ if and only if

$$\begin{aligned}\rho'_{ab} = \rho_{ab} - 2\,\underline{\omega}^{-1}\,D_a D_b\underline{\omega} + 4\,\underline{\omega}^{-2}\,(D_a\underline{\omega})\,(D_b\underline{\omega})\\ - \underline{\omega}^{-2}\,\underline{g}_{ab}\,(\underline{g}^{mn}\,D_m\underline{\omega}\,D_n\underline{\omega})\end{aligned} \tag{34}$$

satisfies the primed version of (33) for the decomposition $\underline{g}'_{ab}$, $\underline{n}'^a$ given by (21). For existence, it suffices by this observation to show existence in any one decomposition. Consider that with $\mathcal{R}$ constant; then $\rho_{ab} = 1/2\,\mathcal{R}\,\underline{g}_{ab}$ is a solution.

For uniqueness, we first note that the difference, σ_{ab}, of two solutions of (33) again satisfies (33), but with the right side of the second equation zero. Fix any infinitesimal symmetry ξ^a. Then, from (32) and the third equation (33), we have $D_{[a}(\sigma_{b]c}\, \xi_d) = k\, \rho_{c[b}\, \underline{g}_{a]d}$. Applying $\underline{g}^{cd}$ to each side, using symmetry of σ_{ab} on the right, we obtain $D_{[a}(\sigma_{b]c}\, \xi^c) = 0$. Hence, since $\mathscr{I}$ is simply connected, we have $\sigma_{bc}\, \xi^c = D_b \alpha$ for some scalar field α ; and, by the first equation (33), $\underline{n}^a D_a \alpha = 0$. Set $\epsilon_{ab} = \epsilon_{abc}\, \underline{n}^c$. Then, we claim, $\epsilon_{am}\, \underline{g}^{mn}\, \sigma_{nb}$ is symmetric in "ab", as one easily checks by contracting this expression with ϵ^{sab}, removing the two ϵ's with the epsilon-identity, and using the first two equations (33), obtaining zero. Hence, $\epsilon_{am}\, \underline{g}^{mn}\, \sigma_{nb}\, \xi^b = \xi^b \epsilon_{bm}\, \underline{g}^{mn}\, \sigma_{na}$. Taking the curl of both sides of this equation, the right side vanishes by an argument just like that above; hence, replacing $\sigma_{nb}\, \xi^b$ by $D_n \alpha$ we obtain $D_{[c}(\epsilon_{a]m}\, \underline{g}^{mn}\, D_n \alpha) = 0$. Now, $\underline{n}^a D_a \alpha = 0$ implies that $\alpha = \pi^*(\beta)$ for some β on B, while this last equation is precisely π^* of the equation which asserts that β is a harmonic function on B. But B is compact, whence the only harmonic functions are constants, whence α is constant on $\mathscr{I}$, whence $\sigma_{ab}\, \xi^b = 0$ for every infinitesimal symmetry ξ^b. Since the infinitesimal symmetries span the tangent space at every point of $\mathscr{I}$, we must therefore have $\sigma_{ab} = 0$.

What makes the auxiliary field ρ_{ab} useful is its very complicated conformal behavior, (34). This field, which is "practically all gauge", will be used to simplify the conformal behavior of other things. Note where the Minkowskian character of the asymptotic geometry was used in the proof: in existence, to obtain a decomposition with $\mathcal{R}$ constant (presumably possible much more generally); in uniqueness, to conclude that $\mathscr{I}$ is simply connected, that the only harmonic functions are constants, and that the infinitesimal symmetries span the tangent space. Generally speaking, it is Theorem 5 which endows the Minkowskian case with many of its special features. Insofar as one can, in other cases, find a solution of (33), one can repeat many of the same constructions available in the Minkowskian case. As an example of Theorem 5, we have: For any infinitesimal symmetry ξ^a,

$$\mathcal{L}_\xi\, \rho_{ab} = -\, 2\, D_a\, D_b\, k\, , \tag{35}$$

where k is given by (31). One proves this by verifying that the difference of the two sides satisfies (33) with zero on the right in the second equation. Hence, this difference is zero.

We now establish:

Theorem 6. Let $\mathscr{I}$, $\Gamma^{ab}{}_{cd}$ be a Minkowskian asymptotic geometry, and $\underline{g}_{ab}$, $\underline{n}^a$ a decomposition. Then the vector space of scalar fields α on $\mathscr{I}$ satisfying

$$\underline{n}^a\, D_a \alpha = 0, \qquad D_a\, D_b\, \alpha + 1/2\, \alpha\, \rho_{ab} = \mu\, \underline{g}_{ab} \tag{36}$$

for some μ (possibly different for each solution) is four-dimensional.

Proof: We first note, from (28) and (34), that α satisfies (36) if and only if $\alpha' = \underline{\omega}\alpha$ satisfies the primed version of (36), with $\underline{g}'_{ab}$ and $\underline{n}'^a$ given by (21). [Note the role of $\mathcal{P}_{ab}$ here: To make (36) conformally invariant.] It suffices, therefore, to prove the theorem for a particular gauge-choice, e.g., that with $\mathcal{R}$ constant. But for this choice, $\mathcal{P}_{ab} = 1/2\, \mathcal{R}\, \underline{g}_{ab}$, whence, setting $\alpha = \pi^*(\beta)$ by the first equation (36), the second equation just requires that the gradient of β on B there be a conformal Killing field. But the only functions on a metric two-sphere whose gradients are conformal Killing fields are the constants and the functions whose gradients correspond to the boosts - a four-dimensional vector space. [In other words, β must be a sum of "$\ell = 0$" and "$\ell = 1$" spherical harmonics.]

We postpone until the following section, in which we shall return to the relation between Killing fields and infinitesimal symmetries, the proof of the result which ties all of this together:

Theorem 7. The infinitesimal supertranslations obtained via Theorem 3 from the infinitesimal translations in Minkowski space are precisely those of the form $\alpha\, \underline{n}^a$ for α satisfying (36).

Let $\mathcal{I}$, $\Gamma^{ab}{}_{cd}$ be a Minkowskian asymptotic geometry. The infinitesimal supertranslations characterized in Theorem 7 are called the infinitesimal translations (on $\mathcal{I}$, $\Gamma^{ab}{}_{cd}$), noting that this characterization is gauge-invariant by the first line of the proof. The infinitesimal translations clearly form a four-dimensional, abelian subalgebra of $\mathcal{S}$, and hence of $\mathcal{L}$. Since $\mathcal{S}$ is abelian, the infinitesimal translations form an ideal in $\mathcal{S}$. It turns out that they also form an ideal in $\mathcal{L}$. Indeed, let ξ^a be any infinitesimal symmetry, and let α satisfy (36). We have $\mathcal{L}_\xi(\alpha\, \underline{n}^a) = (\mathcal{L}_\xi \alpha - K\alpha)\, \underline{n}^a$. But, applying $\mathcal{L}_\xi$ to the second equation (36), using (35), we find that $(\mathcal{L}_\xi \alpha - K\alpha)$ also satisfies (36). Hence, the infinitesimal translations, $\mathcal{T}$, form an ideal in $\mathcal{L}$. The quotient algebra, $\mathcal{L}/\mathcal{T}$ therefore makes sense: It is very big, and apparently not very illuminating.

In Minkowski space, the norm of any translational Killing field is constant, i.e., is just a number. Within a Minkowskian asymptotic geometry, is there a number associated with each infinitesimal translation? There is. Taking the curl of the second equation (36) using (29) for the first term and (33) for the second, we obtain $\mathcal{R}\, \underline{g}_{b[c}\, D_{a]}\alpha + \mathcal{P}_{b[a}\, D_{c]}\alpha = 2\, D_{[c}\mu\, \underline{g}_{a]b}$. Contracting with $\underline{g}^{bc}$, using (33), we have $\mathcal{P}_{am}\, \underline{g}^{mn}\, D_n\alpha = -2\, D_a\mu$. Next, contract (36) with $\underline{g}^{bm}\, D_m\alpha$, and use this last equation to eliminate the resulting $\mathcal{P}_{ab}\, \underline{g}^{bm}\, D_m\alpha$ - term to obtain $D_a(\underline{g}^{mn}\, D_m\alpha\, D_n\alpha - 2\mu\alpha) = 0$. Finally, replacing μ in this last equation by the expression in terms of α which results from contracting (36) with $\underline{g}^{ab}$, we obtain:

$$\underline{g}^{mn}\, D_m\alpha\, D_n\alpha - \alpha\, \underline{g}^{mn}\, D_m D_n\alpha - 1/2\, \mathcal{R}\, \alpha^2 = \text{const.} \qquad (37)$$

What we have shown, then, is that every solution of (36) is also a solution of (37). [I suspect that the converse is also true.] Now let a new decomposition of $\Gamma^{ab}{}_{cd}$, via (21), be selected. Setting $\alpha' = \underline{\omega}\,\alpha$ and using (28), one verifies that the constant which appears on the right in (37) remains the same. Thus, Eqn. (37) assigns to every infinitesimal translation a number, quadratically. This number is precisely the (constant) norm of the corresponding translational Killing field in Minkowski space. [Proof: The norm is, up to a factor, the only Poincaré-invariant quadratic form on the translational Killing fields in Minkowski space.]

Given an arbitrary Killing field in Minkowski space, its derivative, covariantly represented, is a constant, antisymmetric tensor field. We obtain the corresponding asymptotic statement. Let ξ^a be an arbitrary infinitesimal symmetry on a Minkowskian asymptotic geometry, and let α represent an infinitesimal translation. Then, as we saw on the previous page, $\mathcal{L}_\xi \alpha - K\alpha$ also represents an infinitesimal translation. Thus, we obtain a linear mapping, L_ξ, from the four-dimensional vector space $\mathcal{T}$ of infinitesimal translations to itself. But Eqn. (37) defines a metric - by the remarks above, of Lorentz signature - on $\mathcal{T}$: We may write the constant on the right $\langle \alpha, \alpha \rangle$. Applying $\mathcal{L}_\xi$ to (37), one finds that $\langle \alpha, \mathcal{L}_\xi \alpha - K\alpha \rangle = 0$, i.e., that L_ξ is anti-self-adjoint with respect to this metric, i.e., that when the metric on $\mathcal{T}$ is used to lower the index of L_ξ, the result is an antisymmetric, covariant tensor over $\mathcal{T}$. In this way, each infinitesimal symmetry leads to an antisymmetric, second-rank, covariant tensor over the four-dimensional vector space $\mathcal{T}$ of infinitesimal translations. Every infinitesimal supertranslation, for example, yields the zero tensor. More abstractly, we have a homomorphism from the Lie algebra $\mathcal{L}$ to the Lie algebra of infinitesimal Lorentz transformations on the vector space $\mathcal{T}$ with its Lorentz metric.

Nothing surprising happens in the transition from the infinitesimal symmetries to the full symmetries: One simply "de-infinitesimalizes" all the definitions, observations, and proofs above. Since furthermore the full symmetries are of considerably less interest than the infinitesimal, we here merely summarize the situation. For $\Upsilon: \mathcal{I} \to \mathcal{I}$ a symmetry, Υ must send $\underline{n}^a$ to a multiple of $\underline{n}^a$ and $\underline{g}_{ab}$ to a multiple of $\underline{g}_{ab}$, with these multiples related as in (21). By the first, Υ sends integral curves of $\underline{n}^a$ to integral curves. Hence, since B is the manifold of such integral curves, we must have $\pi \circ \Upsilon = \varphi \circ \pi$ for some diffeomorphism φ on B. By the second property, this φ must send the metric h_{ab} of B to a multiple of itself, i.e., must be a conformal mapping on B, h_{ab}. The supertranslations are the symmetries for which φ is the identity on B, i.e., those which send each $\underline{n}$ - integral curve to itself. The supertranslations form an abelian normal subgroup of the group of symmetries. Every conformal mapping φ on B gives rise to some symmetry (at least, when $\underline{n}^a$ is complete), and two symmetries differ by a factor of a supertranslation if and only if they produce the same "φ". Hence, the quotient group of the symmetries by the supertranslations is isomorphic with the group of conformal mappings on B. Although there is no single, natural realization of the quotient in the full group of symmetries, there are (again, when $\underline{n}^a$ is complete) many "unnatural" realizations. For example, one could consider the subgroup of the group of symmetries consisting of those which send the image of a fixed cross section to itself. In the Minkowskian case, one further distinguishes a four-dimensional, abelian, normal subgroup, the translations, of the symmetry group. This group acquires a left- and right-invariant Lorentz metric, and one further obtains a homomorphism from the symmetry group to the group of all metric-preserving isomorphisms on the translation subgroup.

The list of structural properties of these groups and algebras, and their relations with the asymptotic geometry, is clearly almost infinite in length. We have here just summarized the main features and the techniques by which one can settle questions in this area.

6. The Physical Fields at Infinity

What we have done so far is introduce and study two notions: that of an asymptote and of an asymptotic geometry. These two, as discussed in the Introduction, merely provide a geometrical framework: The physics itself is to be characterized in terms of certain other fields which arise on $\mathcal{I}$ from the various physical fields in the physical space-time. In this section, we introduce such fields. There are of course numerous possibilities, for there are numerous physical fields in general relativity. Rather than attempt to give an exhaustive list, we shall largely restrict consideration to the main one - the gravitational - and to two other examples - Klein-Gordon and Maxwell.

Let $\widetilde{M}$, $\widetilde{g}_{ab}$ be a space-time. By a (zero-mass, conformally invariant) Klein-Gordon field on $\widetilde{M}$, $\widetilde{g}_{ab}$, we mean a real scalar field $\widetilde{\varphi}$ on $\widetilde{M}$ satisfying

$$(\widetilde{\nabla}^2 - 1/6\,\widetilde{R})\,\widetilde{\varphi} = 0. \tag{38}$$

[Other possibilities, to which we shall occasionally refer: Omit the "$- 1/6\,\widetilde{R}$", or replace it by "$- m^2$", where m is a constant.] By a Maxwell field on $\widetilde{M}$, $\widetilde{g}_{ab}$, we mean an antisymmetric tensor field $\widetilde{F}_{ab}$ satisfying

$$\widetilde{\nabla}_{[a}\,\widetilde{F}_{bc]} = 0, \qquad \widetilde{\nabla}_{[a}\,{}^*\widetilde{F}_{bc]} = 0, \tag{39}$$

where "*" denotes the dual: ${}^*\widetilde{F}_{ab} = 1/2\ \widetilde{\epsilon}_{abcd}\,\widetilde{g}^{cm}\,\widetilde{g}^{dn}\,\widetilde{F}_{mn}$. Since we only work near the boundary, we may permit sources on the right in (38) and (39), provided they vanish in a neighborhood of the boundary (or, still more generally, vanish to an appropriate asymptotic order).

Let M, g_{ab}, Ω be an asymptote of our physical space-time. We say that Klein-Gordon field $\tilde{\varphi}$ (resp., Maxwell field $\tilde{F}_{ab}$) is asymptotically regular, with respect to this asymptote, if the fields

$$\varphi = \Omega^{-1}\tilde{\varphi}, \qquad F_{ab} = \tilde{F}_{ab}, \tag{40}$$

respectively, on $\tilde{M}$ permit smooth extensions to I in M. We refer to $\tilde{\varphi}$ and $\tilde{F}_{ab}$ as the physical fields; to φ and F_{ab} as the unphysical. That is, Klein-Gordon fields are assigned dimension -1, Maxwell -2: More generally, dimension = - (spin) - 1.

It is intended that regularity be the correct asymptotic boundary conditions to be imposed on these fields. We are thus in a situation similar to that of Sect. 1: We must make a case that the definitions are the appropriate ones. In particular, one would like some result analogous to Conjecture 1. It turns out, fortunately, that what was a conjecture in the gravitational case in Sect. 1 is now a theorem.

Theorem 8. Let $\tilde{M}$, $\tilde{g}_{ab}$ be a space-time, and M, g_{ab}, Ω an asymptote. Let T be an achronal, spacelike submanifold, with boundary in I, of M. Let $\tilde{\varphi}$ (resp., $\tilde{F}_{ab}$) on $\tilde{M}$ satisfy (38) (resp., (39)) outside a compact set, and have support in $\tilde{M}$ intersecting T compactly. Then $\tilde{\varphi}$ (resp., $\tilde{F}_{ab}$) is regular on that part of I consisting of points from which every past-directed timelike and null curve in M meets T.

Proof: First note that Eqns. (38) and (39), expressed in terms of the unphysical variables via (40), become

$$(\nabla^2 - 1/6\ R)\varphi = 0; \quad \nabla_{[a}F_{bc]} = 0, \quad \nabla_{[a}{}^*F_{bc]} = 0, \tag{41}$$

respectively. Let p be a point of I as in the theorem, and let γ be the integral curve of $-n^a$ from p. Then γ, as a past-directed null geodesic from p, meets T, say at point q. Now let $\hat{M}$, $\hat{g}_{ab}$ be an extension of M, g_{ab} such that the closed segment of γ between p and q is in its interior, and extend T in $\hat{M}$ through q to obtain a spacelike $\hat{T}$ in $\hat{M}$. Then p is in the interior of $D^+(\hat{T})$, the future domain of dependence of $\hat{T}$ in $\hat{M}$. Denote by $\hat{\varphi}$ (resp., $\hat{F}_{ab}$) the solution of (41) in int $D^+(\hat{T})$ which arises from the initial data on $\hat{T}$ and sources induced by the original unphysical field. These fields satisfy linear hyperbolic differential equations with smooth sources and smooth coefficients, and so are smooth in int $D^+(\hat{T})$ - in particular, in a neighborhood of p. But $\hat{\varphi}$ (resp., $\hat{F}_{ab}$) coincide with φ (resp., F_{ab}) in $\tilde{M}$. Hence, the original unphysical fields possess an extension to I in a neighborhood of p.

In simpler situations, such as those of the asymptotes of Minkowski space and the Schwarzschild space-time of Sect. 1, one can obtain as a corollary of Theorem 8 a version which, requiring fewer global safeguards, is somewhat simpler to state. One can let T be without boundary, entirely within $\tilde{M}$, and a Cauchy surface for the physical space-time; regularity is obtained with respect to the entire asymptote. Theorem 8 also holds for $\tilde{\varphi}$ satisfying (38) without the "$- 1/6\ R$", provided one requires vanishing of the stress-energy asymptotically to order two. This requirement is just what is needed in this case to ensure that the corresponding first equation in (41) have coefficients smooth on I. The present proof of Theorem 8 fails with "$- 1/6\ \tilde{R}$" in (38) replaced by "$- m^2$", for the corresponding first equation in (41) then acquires an extra term, "$- m^2 \Omega^{-2} \varphi$", which will not in general be smooth on I. This is particularly surprising, since one would have expected in this case not only that Theorem 8 would hold, but also that the corresponding φ would vanish on I. Is Theorem 8 true for a massive Klein-Gordon field? One would like to prove Conjecture 1 along the same lines as above. The problem is that, whereas Eqns. (38) and (39) lead to equations on the unphysical fields with coefficients smooth on I, this is not the case for (4). It seems likely that, if one imposes a suitable gauge-condition on $\tilde{\gamma}_{ab}$ in the physical space-time, then breaks $\tilde{\gamma}_{ab}$ up into pieces using n^a, and finally imposes on these pieces possibly different dimensions, one can, from (4) and the gauge equation, obtain a deterministic system of coupled hyperbolic equations with coefficients smooth on I.

Having now introduced two fields which contribute to the stress-energy, we may return to the question of to what asymptotic order that stress-energy should vanish. The stress-energies of Klein-Gordon and Maxwell fields are given, respectively, by

$$\tilde{T}_{ab} = \tilde{\nabla}_a\tilde{\varphi}\ \tilde{\nabla}_b\tilde{\varphi} - 1/2\ \tilde{\varphi}\ \tilde{\nabla}_a\ \tilde{\nabla}_b\tilde{\varphi} + 1/4\ \tilde{\varphi}^2\ \tilde{R}_{ab} - 1/4\ \tilde{g}_{ab}\ (\tilde{\nabla}^m\tilde{\varphi}\ \tilde{\nabla}_m\tilde{\varphi} + 1/6\ \tilde{\varphi}^2\ \tilde{R}), \tag{42}$$

$$\tilde{T}_{ab} = \tilde{F}_a{}^m\ \tilde{F}_{bm} + {*\tilde{F}_a}{}^m\ {*\tilde{F}_{bm}}, \tag{43}$$

where, by our previous convention, indices are raised and lowered with the physical metric. Replacing $\tilde{T}_{ab}$ by the L_{ab} of Sect. 2, and replacing physical fields everywhere by unphysical, we have

$$L_{ab} = \Omega^4 \left\{ \nabla_a\varphi\ \nabla_b\varphi - 1/2\ \varphi\ \nabla_a\ \nabla_b\varphi + 1/4\ \varphi^2\ R_{ab} - 1/4\ g_{ab}\ (\nabla^m\varphi\ \nabla_m\varphi + 1/6\ \varphi^2\ R)\right\}, \tag{44}$$

$$L_{ab} = \Omega^4 \left\{ F_a{}^m\ F_{bm} + {*F_a}{}^m\ {*F_{bm}} \right\}, \tag{45}$$

respectively, with indices now raised and lowered with the unphysical metric. We conclude, therefore, that regular Klein-Gordon and Maxwell fields produce stress-energies vanishing asymptotically to order four. A strange thing happens when "$- 1/6\ \tilde{R}$" is omitted from

(38). Eqns. (42) and (44) are then replaced by

$$\tilde{T}_{ab} = \tilde{\nabla}_a \tilde{\varphi} \; \tilde{\nabla}_b \tilde{\varphi} - 1/2 \; \tilde{g}_{ab} \; (\tilde{\nabla}^m \tilde{\varphi} \; \tilde{\nabla}_m \tilde{\varphi}), \tag{46}$$

$$\begin{aligned} L_{ab} = \Omega^2 \{ &\varphi^2 n_a n_b + 2\Omega \varphi \, n_{(a} \nabla_{b)} \varphi + \Omega^2 \nabla_a \varphi \; \nabla_b \varphi \\ &- 1/6 \, \Omega \, g_{ab} \, (\varphi^2 f + 2 \varphi \, n^m \nabla_m \varphi + \Omega \nabla^m \varphi \; \nabla_m \varphi) \end{aligned} \tag{47}$$,

respectively. Thus, in this case regularity only implies vanishing of the stress-energy asymptotically to order two. Thus, the right side of (6) vanishes. It turns out, however, because of the form of (47), that the right side of (7), and even of (9), also vanishes. Furthermore, the right sides of (12) and (13) turn out, on substituting (47), to remain finite at I. Thus, although the stress-energy only vanishes asymptotically to order two, it acts, as far as our equations are concerned, as though it vanished to order four.

Let $\tilde{\varphi}$ be an asymptotically regular Klein-Gordon field. Then $\underline{\varphi} = \gamma^*(\varphi)$ is a smooth scalar field on the manifold $\mathcal{J}$. This represents, in terms of $\mathcal{J}$, the asymptotic structure of the physical field $\tilde{\varphi}$.

Let $\tilde{F}_{ab}$ be an asymptotically regular Maxwell field. Then, since $\tilde{*F}_{ab} = 1/2 \; \tilde{\epsilon}_{abcd} \; \tilde{g}^{cm} \; \tilde{g}^{dn} \; \tilde{F}_{mn} = 1/2 \; \epsilon_{abcd} \; g^{cm} \; g^{dn} \; F_{mn}$, the field $*F_{ab}$ on M also has smooth extension to M. The corresponding fields on $\mathcal{J}$ are given by $\underline{F}_{ab} = \gamma^*(F_{ab})$ and $*\underline{F}_{ab} = \gamma^*(*F_{ab})$. [Note that the two stars on the right in the second equation have different meanings.] In the Klein-Gordon case, $\underline{\varphi}$ on $\mathcal{J}$ is essentially arbitrary; in the Maxwell case, by contrast, $\underline{F}_{ab}$ and $*\underline{F}_{ab}$ satisfy a number of equations. We now derive these. We first note that, by definition of the dual, $*F_{ab} \, n^b = 1/2 \; g_{am} \, (\epsilon^{mcdp} \, n_p) \, F_{cd}$. Applying γ^* to this equation, and to the analogous one obtained by interchange of F_{ab} and $*F_{ab}$, we obtain

$$*\underline{F}_{ab} \, n^b = 1/2 \; \underline{g}_{am} \, \epsilon^{mcd} \, \underline{F}_{cd}, \quad \underline{F}_{ab} \, \underline{n}^b = - 1/2 \; \underline{g}_{am} \, \epsilon^{mcd} \, *\underline{F}_{cd}. \tag{48}$$

Eqns. (48), then, reflect in $\mathcal{J}$ the fact that $\underline{F}_{ab}$ and $*\underline{F}_{ab}$ began as mutual duals. The other equations on these fields are obtained by applying γ^* to Maxwell's equations in terms of the unphysical fields, the last two equations in (41):

$$D_{[a} \underline{F}_{bc]} = 0, \qquad D_{[a} *\underline{F}_{bc]} = 0, \tag{49}$$

where the left sides are the exterior derivatives. Thus, a Maxwell field is described asymptotically by two fields, $\underline{F}_{ab}$ and $*\underline{F}_{ab}$, on $\mathcal{J}$, satisfying (48) and (49).

By (48), $\underline{F}_{ab}$ and $*\underline{F}_{ab}$ together have, at each point of $\mathcal{J}$, four algebraically independent components. But the original F_{ab} and $*F_{ab}$ have, at each point of I, six. Thus, at least locally, information has been lost by application of ζ^*. However, it turns out that, in the Minkowskian case, this "loss of information" is only apparent.

Theorem 9. Let $\tilde{M}$, $\tilde{g}_{ab}$ be a space-time, and M, g_{ab}, Ω an asymptote, with Minkowskian asymptotic geometry. Then, for any asymptotically regular Maxwell field with $\underline{F}_{ab} = *\underline{F}_{ab} = 0$, we have also $F_{ab} = 0$ at points of I.

Proof: Since $\zeta^*(F_{ab}) = \zeta^*(*F_{ab}) = 0$, we have $n_b F^{ab} = n_b *F^{ab} = 0$ at points of I. Hence, F^{ab} and $*F^{ab}$ are in the class $\mathcal{C}$: Set $K_a = \epsilon_{amn}\ \zeta^*(F^{mn})$ and $*K_a = \epsilon_{amn}\ \zeta^*(*F^{mn})$. Then each of K_a and $*K_a$ on $\mathcal{J}$ is orthogonal to $\underline{n}^a$, while mutual duality gives $*K_a = \epsilon_{amn}\ \underline{n}^n\ \underline{g}^{mp}\ K_p$. Maxwell's equations, expressed in terms of the K's, give $D_{[a}\ K_{b]} = D_{[a}\ *K_{b]} = 0$. By the first and simple connectivity of $\mathcal{J}$, $K_a = D_a \alpha$. By $K_a\ \underline{n}^a = 0$, we have $\alpha = \pi^*(\beta)$ for some scalar field β on the base space. But the second, together with the dual relationship between K_a and $*K_a$, implies that β is harmonic on the base space, and so, by compactness, constant. Hence, $K_a = *K_a = 0$, whence $\zeta^*(F^{ab}) = 0$, whence $F^{ab} = 0$ at points of I.

To what extent do the asymptotic fields determine the original fields in the physical space-time? If I were a Cauchy surface for a neighborhood of I, then the fields on I would determine the physical fields in that neighborhood. Unfortunately, I, being null, can never be a Cauchy surface. It seems reasonably likely, however, that the asymptotic fields may still determine the physical fields whenever I is "as nearly Cauchy as possible, consistent with its being null".

Conjecture 10. Let $\tilde{M}$, $\tilde{g}_{ab}$ be a space-time, and M, g_{ab}, Ω an asymptote such that every maximally extended, past-directed null geodesic in $\tilde{M}$ has past endpoint on I. Let $\tilde{\varphi}$ be a Klein-Gordon field (resp., $\tilde{F}_{ab}$ a Maxwell field) on $\tilde{M}$ which is asymptotically regular, and which is such that φ (resp., F_{ab}) vanishes on I. Then $\tilde{\varphi}$ (resp., $\tilde{F}_{ab}$) vanishes in $\tilde{M}$.

Consider, for example, Minkowski space with the past boundary given in Sect. 1. This arrangement satisfies the hypothesis of Conjecture 10. In this case, however, the conjecture is true, for the propagators for the fields in Minkowski space have support on the light cone, whence the field at any interior point in Minkowski space can be expressed as an integral over I of the field there. In the presence of curvature, the support of the propagator includes also the interior of the light cone. The statement of the conjecture, then, is that nonetheless, no influences, via such fields, can make their way into the space-time without registering on I. A similar question can be asked in the gravitational case. It would be of interest to resolve these.

One might expect on physical grounds that, for an asymptotically regular Maxwell field, "the 1/r part of the field, as one approaches the boundary at infinity, will be an outgoing radiation field, i.e., that it will be null, with repeated principal null direction the direction of approach to I". Statements of this general type (suitably made precise, and applied to the Maxwell, as well as other, fields) go under the name of the peeling-off property. In general terms, this behavior arises from the following geometrical circumstance. The asymptotic Maxwell field is of course the value of F_{ab}, the unphysical field, at points of I: It need have no special algebraic properties (as, e.g., the proof of Theorem 8 makes clear). However, the "natural" frame with respect to which one views the Maxwell field from the physical space-time suffers an "infinite boost", as one approaches I, relative to the natural frame from the unphysical. Furthermore, it is to the behavior of the Maxwell field in the unphysical space that regularity refers. Thus, what one "sees" in the physical space-time, in the asymptotic limit, is a regular but otherwise essentially arbitrary field to which an "infinite boost" has been applied. In this way, there arises the peeling-off property. We now carry out the construction in more detail.

Let $\tilde{M}$, $\tilde{g}_{ab}$ be a space-time, and M, g_{ab}, Ω an asymptote. Consider first the following geometrical arrangement. Let $\gamma: [0,1] \rightarrow M$ be a smooth curve such that $\gamma(0)$ is in I and $\gamma(r)$ is otherwise in $\tilde{M}$, and such that, writing $-v^a$ for its tangent vector, $v^a n_a = -1$ at $\gamma(0)$. Set $p_o = \gamma(0)$ and $q = \gamma(1)$. [See Fig. 3.] Now fix r in (0,1], and consider the following sequence of operations on a covariant vector at p_o: First parallel transport from p_o to $\gamma(r)$ using the unphysical derivative operator, then from $\gamma(r)$ to q using the physical, and finally from q back to p_o using the unphysical again. [Note that we avoid parallel transport up to p_o using the physical derivative operator: It is not well-behaved there.] The result of all this is again a covariant vector at p_o. Thus, we obtain a certain linear mapping from covariant vectors at p_o to covar-

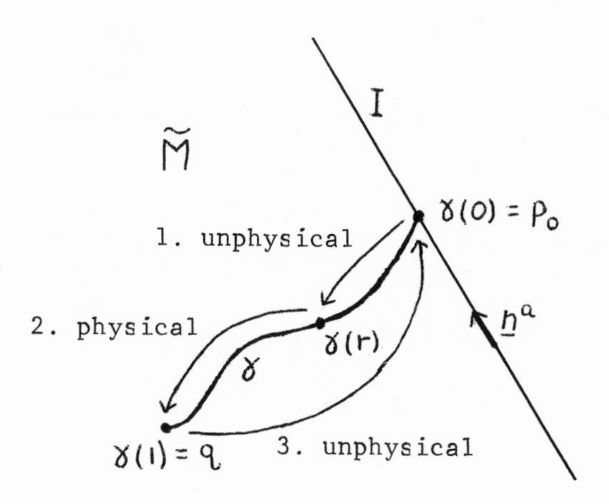

Fig. 3. The geometrical arrangement for the peeling-off property.

iant vectors at p_o, i.e., we obtain a tensor, $L^a{}_b(r)$, at p_o. Since the physical and unphysical metrics are conformally related, our mapping is a Lorentz transformation and dilation, i.e., we have that $g_{mn} L^m{}_a(r) L^n{}_b(r)$ is a multiple of g_{ab}, for all r.

The meaning of all this is the following. From the standpoint of the physical space-time, the curve γ "goes off to null infinity". The condition that $v^a n_a = -1$ at p_o ensures that $\tilde{r} = 1/r$ is, asymptotically, a "typical radial parameter" along the curve. The idea is to describe the asymptotic behavior of the physical Maxwell field in terms of its limiting behavior along this curve. Implicit in this idea, however, is some ability to compare the value of the Maxwell field at one point of the curve with its value at another - just the sort of comparison which is not normally available in curved spaces. In the present case, however, a natural mode of comparison is available: Parallel transport back along the curve γ, using the physical derivative operator, to q. That is to say, one would think of studying the asymptotic behavior of the physical Maxwell field by studying the limit, as $r \to 0$, of the antisymmetric tensor at q obtained, for each r, by parallel transport of the Maxwell field at $\gamma(r)$ back to q. This is fine in the physical space-time; but it is only in the unphysical space that we know anything about the limiting behavior of the Maxwell field - namely, that it is regular. In the unphysical space, on the other hand, the natural method of comparing tensors at different points is via unphysical parallel transport along the curve, and the natural point of comparison is p_o. What we do, therefore, is use unphysical parallel transport to carry the entire construction above back to p_o. The result is that "physical transport from $\gamma(r)$ to q" is replaced by "unphysical from p_o to $\gamma(r)$, physical from $\gamma(r)$ to q, and unphysical from q back to p_o". That is, we obtain $L^a{}_b(r)$. Since unphysical transport does not seriously distort anything (all unphysical fields being smooth over all of γ), nothing essential is lost in the transition from the construction in the physical space-time to $L^a{}_b(r)$. In short, if we can determine the behavior of $L^a{}_b(r)$ for r near zero, then we shall also know the asymptotic behavior along γ of the physical fields in the physical space-time.

We wish, then, to study the behavior of $L^a{}_b(r)$ as $r \to 0$. What distinguishes $L^a{}_b(r)$ from $L^a{}_b(r + \Delta r)$ is whether physical or unphysical transport, respectively, was used from $\gamma(r)$ to $\gamma(r + \Delta r)$. For comparison of transports, on the other hand, we have, since the two metrics are conformally related, that

$$v^a \tilde{\nabla}_a k_b - v^a \nabla_a k_b = \Omega^{-1} (v^n n_n \delta^m{}_b + v^m n_b - n^m v_b) k_m \tag{50}$$

for all k_b. [Note that the quantity in parenthesis on the right is an infinitesimal dilation plus an infinitesimal Lorentz transformation.] Write $v^a(r)$ (resp., $n^a(r)$) for the vector at p_o resulting from parallel transport of v^a (resp., n^a) from $\gamma(r)$ to p_o, and set $\Omega(r) = \Omega(\gamma(r))$. Then, from Eqn. (50) and the remark preceding it, we have

$$d/dr\, L^a{}_b(r) = -\Omega(r)^{-1} L^a{}_m(r) \left[v^n(r)\, n_n(r)\, \delta^m{}_b + v^m(r)\, n_b(r) - n^m(r)\, v_b(r) \right]. \qquad (51)$$

This is the differential equation which gives $L^a{}_b(r)$: The initial condition is $L^a{}_b(1) = \delta^a{}_b$. To see the asymptotic behavior of $L^a{}_b(r)$, set $v_o{}^a = v^a(0)$, $n_o{}^a = n^a(0)$, and

$$L_o{}^a{}_b(r) = r\, \delta^a{}_b + r(1-r)\, v_o{}^a\, n_{ob} - (1-r)\, n_o{}^a\, v_{ob} - 1/2\, (1-r)^2\, v_o{}^m\, v_{om}\, n_o{}^a\, n_{ob}. \qquad (52)$$

Eqn. (52) is just the solution of (51) with $v^a(r)$, $n^a(r)$, and $\Omega(r)$ on the right replaced by $v_o{}^a$, $n_o{}^a$, and r, respectively. Now set $L^a{}_b(r) = K^a{}_m(r)\, L_o{}^m{}_b(r)$. Whenever $K^a{}_b(r)$, so defined, remains finite and invertible up to and including $r = 0$, i.e., whenever the asymptotic behavior of $L^a{}_b(r)$ is dominated by $L_o{}^a{}_b(r)$, we say that our curve <u>generates the peeling-off property</u>. Whenever this is the case, all the irrelevant details - of the behavior of the curve γ away from p_o, of the physical metric away from p_o, of the choice of the point q - are carried by $K^a{}_b(r)$, for (52) involves only the values of v^a and n^a at p_o. We can find at least a sufficient condition for this by substituting $L^a{}_b(r) = K^a{}_m(r)\, L_o{}^a{}_b(r)$ into (51), to obtain the differential equation for $K^a{}_b(r)$:

$$d/dr\, K^a{}_b(r) = K^a{}_m(r) \left\{ L_o{}^m{}_n(r)\, F^n{}_p(r)\, L_o^{-1}{}^p{}_b(r) \right\}, \qquad (53)$$

where "-1" denotes the inverse, and where $F^n{}_p(r)$, the difference between the coefficient of $L^a{}_m(r)$ on the right in (51) and that coefficient with $v^a(r)$, $n^a(r)$, and $\Omega(r)$ replaced, respectively, by $v_o{}^a$, $n_o{}^a$, and r, is finite at $r = 0$. A sufficient condition that our curve generate the peeling-off property, then, is that the expression in curly brackets on the right in (53) remain finite at $r = 0$. Unfortunately, this will not always be the case, for $L_o^{-1}{}^a{}_b(r)$ blows up as $1/r^2$. It is, however, the case when γ is a null geodesic, and so, by the remarks above, whenever γ is asymptotically a null geodesic, to second order. Presumably, there are many other curves which generate the peeling-off property. Is there a simple, geometrical characterization of this class?

Finally, we return to our Maxwell field. Let γ generate the peeling-off property. Writing $F_{ab}(r)$ for the result of parallel transport of F_{ab} at $\gamma(r)$ to p_o, the asymptotic behavior of the Maxwell field is described by $F_{mn}(r)\, L^m{}_a(r)\, L^n{}_b(r)$. Set $l_o{}^a = v_o{}^a + 1/2\, v_o{}^m\, v_{om}\, n_o{}^a$, the other null vector in the v-n tangent plane, and $P^a{}_b = \delta^a{}_b + l_o{}^a n_{ob} + n_o{}^a l_{ob}$, the projection operator into the plane orthogonal to the l-n plane. Then (52) takes the simple form

$$L_o{}^a{}_b(r) = - n_o{}^a l_{ob} + r P^a{}_b - r^2 l_o{}^a n_{ob}. \tag{54}$$

Hence,

$$\begin{aligned} &\tilde{F}_{mn}(r) L^m{}_a(r) L^n{}_b(r) = [F_{mn}(r) K^m{}_p(r) K^n{}_q(r)] \\ &\{- 2r (n_o{}^p l_{o[a} P^q{}_{b]}) + r^2 (P^p{}_a P^q{}_b + 2 n_o{}^p l_o{}^q l_{o[a} n_{ob]}) \\ &- 2 r^3 (l_o{}^p n_{o[a} P^q{}_{b]})\}. \end{aligned} \tag{55}$$

This is the desired formula. The quantity in square brackets on the right is some antisymmetric tensor depending on r, which is well-behaved as r goes to zero. It includes all the irrelevant things: the details of the curve, the choice of endpoint q - and the actual Maxwell tensor. The quantity in curley brackets on the right consists of three terms. The first term goes to zero as r $(= 1/\tilde{r})$, while its tensor in round brackets is precisely the projection operator from antisymmetric tensors at p_o to antisymmetric tensors which are null with repeated principal null direction $l_o{}^a$. The second term goes to zero as r^2, and projects to antisymmetric tensors with principal null directions $l_o{}^a$ and $n_o{}^a$. The third term goes to zero as r^3, and projects to antisymmetric tensors which are null with repeated principal null direction $n_o{}^a$. That is to say, $F_{mn}(r) L^m{}_a(r) L^n{}_b(r)$ has $1/\tilde{r}$ - term with principal null directions $l_o{}^a$, $l_o{}^a$, $1/\tilde{r}^2$ - term with principal null directions $l_o{}^a$, $n_o{}^a$, and $1/\tilde{r}^3$ - term with principal null directions $n_o{}^a$, $n_o{}^a$. This is the peeling-behavior for the Maxwell field. Of course, one has to go through all this work only once: To find the peeling-behavior of anything else, one just uses (54). [As an exercise, one might let the stress-energy vanish asymptotically to order four, and find its peeling behavior.]

This completes the discussion of our two examples: Klein-Gordon fields and Maxwell fields. We turn now to the gravitational field. It turns out in this case that one obtains four objects on $\mathcal{I}$: a derivative operator, its curvature tensor, and two other fields. One thinks of the derivative operator as the "potential" for the curvature tensor, and of the curvature tensor as the potential for the remaining fields.

Let $\tilde{M}$, $\tilde{g}_{ab}$ be a space-time, and M, g_{ab}, Ω an asymptote. We begin with the following observation. Let μ_b be a covariant vector field on $\mathcal{I}$. Then $\mu_b = \zeta^*(\nu_b)$ for some ν_b on M, and this ν_b is uniquely determined up to addition of terms of the form $\alpha n_b + \Omega \tau_b$. But, in M, we have $\nabla_a (\alpha n_b + \Omega \tau_b) = (\nabla_a \alpha) n_b + \alpha \nabla_a n_b + n_a \tau_b + \Omega \nabla_a \tau_b$. Now choose the conformal factor [See (16).] such that f = 0 on I, i.e., such that n^a and g_{ab} lead to a decomposition of $\Gamma^{ab}{}_{cd}$. Then $\zeta^*(\nabla_a (\alpha n_b + \Omega \tau_b)) = 0$, as one sees by applying ζ^* to the equation above, noting that the first, third, and fourth terms vanish immediately, while the second vanishes by (6), assuming vanishing of the stress-energy to order two. Thus, $\zeta^*(\nabla_a \nu_b)$ on $\mathcal{J}$ depends only on the original field μ_b on $\mathcal{J}$: We write this field $D_a \mu_b$. In this way, we obtain a derivative operator (on covariant vector fields, and hence on all tensor fields) on $\mathcal{J}$. We note further that, for μ_b satisfying $\mu_b \underline{n}^b = 0$, this $D_a \mu_b$ coincides with the field, also denoted $D_a \mu_b$, defined in Sect. 3. Thus, the present derivative operator, applicable to all tensor fields, is an extension of the derivative of Sect. 3, applicable only to covariant fields orthogonal to $\underline{n}^a$. We have immediately from (6) that $D_a \underline{n}^b = 0$. Thus, for example, we have $D_p \Gamma^{ab}{}_{cd} = 0$, and, for any tensor field $\alpha^{\cdot\cdot}{}_{\cdot\cdot}$, $\underline{n}^m D_m \alpha^{\cdot\cdot}{}_{\cdot\cdot} = \mathcal{L}_{\underline{n}} \alpha^{\cdot\cdot}{}_{\cdot\cdot}$. This derivative operator is the first of our gravitational fields.

The second field is obtained from (7). Let the stress-energy vanish asymptotically to order three (or, at least, let the right side of (7) vanish at I), and keep in force our gauge-choice f = 0 on I. Then, at points of I, $\nabla_a f$ is a multiple of n_a, whence (7) gives that $S_a{}^b n_b$ is a multiple of n_a there. But this is precisely the condition that $S_a{}^b$ be in the class $\mathcal{C}$: Set $\underline{S}_a{}^b = \zeta^*(S_a{}^b)$. This is the second gravitational field: essentially the pullback of the unphysical Ricci tensor. Since $n^a S_a{}^b$ is a multiple of n^b, we have, applying ζ^*, that $\underline{n}^a \underline{S}_a{}^b = \sigma \underline{n}^b$ for some σ on $\mathcal{J}$. Set $\underline{S} = \underline{S}_m{}^m$, and $\underline{S}_{ab} = \underline{S}_a{}^m \underline{g}_{mb}$. Then $\underline{S}_{ab} \underline{n}^b = 0$, and $\underline{g}^{ab} \underline{S}_{ab} = \underline{S} - \sigma$. These are the algebraic properties of $\underline{S}_a{}^b$. A differential property follows from (7) and (10), assuming vanishing of the stress-energy to order four. We have $\nabla_{[a} (S_{b]}{}^c n_c) = (\nabla_{[a} S_{b]}{}^c) n_c + S_{[b}{}^c \nabla_{a]} n_c$. Evaluate on I. Then the left side vanishes by (7) and vanishing of the stress-energy to order four, while the second term on the right vanishes by (6). Hence, $(\nabla_{[a} S_{b]}{}^c) n_c = 0$. Hence (Section 3) ζ^* of the contraction of $\nabla_{[a} S_{b]}{}^c$ equals the contraction of $\zeta^*(\nabla_{[a} S_{b]}{}^c)$. But the former vanishes, by (10). Hence,

$$D_b (\underline{S}_a{}^b - \underline{S}\, \underline{\delta}_a{}^b) = 0. \tag{56}$$

In particular, contracting this equation with $\underline{n}^a$ and using $D_b \underline{n}^a = 0$ and $\underline{n}^a \underline{S}_a{}^b = \sigma \underline{n}^b$, we obtain $\underline{n}^a D_a(\underline{S} - \sigma) = 0$. That is, $\underline{S} - \sigma = \underline{S}_{ab} \underline{g}^{ab}$ is constant along the $\underline{n}$-integral curves.

The last two gravitational fields, as expected, come from the unphysical Weyl tensor. We first obtain

Theorem 11. Let $\tilde{M}$, $\tilde{g}_{ab}$ be a space-time, and M, g_{ab}, Ω an asymptote, such that the stress-energy vanishes asymptotically to order four, and such that the asymptotic geometry is Minkowskian. Then the unphysical Weyl tensor, C_{abcd}, vanishes at I.

Proof: By (9), $C^{abcd} n_d$ vanishes at I, whence C^{abcd} is in the class $\mathcal{C}$: Set $\underline{C}^{abcd} = \zeta^*(C^{abcd})$. Now let w^a be any vector field in a neighborhood of I satisfying $w^a n_a = 1$. We then have, at points of I,

$\nabla_m (C^{abcd} - \Omega w^d \nabla_p C^{abcp}) n_d = \nabla_m (C^{abcd} n_d - \Omega w^d n_d \nabla_p C^{abcp})$
$= \nabla_m (C^{abcd} n_d - \Omega \nabla_p C^{abcp}) = \nabla_m (\Omega^2 \nabla_p (\Omega^{-1} C^{abcp})) = 0$,
where the first equality follows from the fact that $\nabla_m n_d = 0$ on I, the second from $w^d n_d = 1$, and the fourth from (12), using the fact that, since the stress-energy vanishes asymptotically to order four, the right side of (12) is finite. Hence, the contraction of $\mathfrak{I}^*$ of $\nabla_m (C^{abcd} - \Omega w^d \nabla_p C^{abcp})$ is $\mathfrak{I}^*$ of its contraction. But, again using (12), the latter vanishes on I, while the former is $D_d \underline{C}^{abcd}$. Hence, $D_d \underline{C}^{abcd} = 0$. Now set $C_{ab} = \epsilon_{amn} \epsilon_{bpq} \underline{C}^{mnpq}$. Then, from what we have just shown, we have $D_{[a} C_{b]c} = 0$. From the symmetries of the Weyl tensor and $C_{abcd} n^d = 0$ at I, we also have $C_{ab} \underline{n}^b = 0$ and $C_{ab} \underline{g}^{ab} = 0$. Consider now the ρ_{ab} whose existence is guaranteed by Theorem 5. From the properties of C_{ab} above, $\rho_{ab} + C_{ab}$ also satisfies the conditions of Theorem 5, whence, by uniqueness, $C_{ab} = 0$. Hence, $\underline{C}^{abcd} = 0$, and so $C^{abcd} = 0$ at points of I.

In order to obtain the remaining two gravitational fields, we must restrict consideration to those cases in which the unphysical Weyl tensor vanishes at I, e.g., by Theorem 11, to the Minkowskian case. Then, by the footnote of Sect. 3, $\Omega^{-1} C_{abcd}$ is smooth up to and including I. Set $K^{ab} = \epsilon^{amn} \epsilon^{bpq} \mathfrak{I}^*(\Omega^{-1} C_{mnpq})$ and $*K^{ab} = \epsilon^{amn} \epsilon^{bpq} \mathfrak{I}^*(\Omega^{-1} *C_{mnpq})$. [The ϵ's here merely serve to replace four indices by two for easier writing. We have, for example, $K^{mn} \epsilon_{mab} \epsilon_{ncd} = 4 \mathfrak{I}^*(\Omega^{-1} C_{abcd})$.] These are the remaining two gravitational fields.

We derive the properties of K^{ab} and $*K^{ab}$. Since the Weyl tensor and its dual are trace-free, we have $K^{ab} \underline{g}_{ab} = *K^{ab} \underline{g}_{ab} = 0$, i.e., that K^{ab} and $*K^{ab}$ are also trace-free. Since the Weyl tensor and its dual are mutual duals, we have $1/2\, g_{ap} \epsilon^{pqmn} n_q C_{mncd} = *C_{amcd} n^m$. Multiplying by Ω^{-1}, applying $\mathfrak{I}^*$, and expressing the result in terms of the K's, we obtain the first of

$$\underline{g}_{am} K^{mb} = - \epsilon_{amp} \underline{n}^p *K^{mb}, \quad \underline{g}_{am} *K^{mb} = \epsilon_{amp} \underline{n}^p K^{mb}. \tag{57}$$

These equations are analogous to (48) in the electromagnetic case: They reflect the fact that K^{ab} and $*K^{ab}$ came from fields which were mutual duals. These (together, of course, with the symmetry of K^{ab} and $*K^{ab}$) are the only algebraic properties. Next, the differential. Multiplying the left side of (9) by Ω^{-1} and raising an index, that left side can be written $\nabla_{[a} S_{b]}{}^c - 1/2\, \Omega^{-1} *C_{abmn} \epsilon^{mncd} n_d$. The right side of the resulting equation vanishes at I, provided the stress-energy vanishes asymptotically to order four. Applying $\mathfrak{I}^*$, we obtain

$$D_{[a} \underline{S}_{b]}{}^c = 1/4\, \epsilon_{abm} *K^{mc}. \tag{58}$$

The trace of this equation again gives (56). The final two differential equations come from (12) and (13). Again, we suppose that the stress-energy vanishes asymptotically to order four, and let

$L_{ab} = \Omega^4 L_{o\,ab}$, with $L_{o\,ab}$ finite on I. Then (13) can be written

$$\nabla_{[a} (\Omega^{-1} C_{bc]de}) = 2\,\Omega\, g_{d[a} \nabla_b L_{o\,c]e} + 6\, g_{d[a} n_b L_{oc]e} - 2\, g_{d[a} g_{b|e|} L_{o\,c]m} n^m, \tag{59}$$

where antisymmetrization over "de" is to be applied on the right. Now apply ζ^* to this equation. The first two terms on the right give zero. Eliminating $\zeta^*(\Omega^{-1} C_{bcde})$ on the left in favor of K^{ab}, and setting $\underline{L}_{ab} = \zeta^*(L_{o\,ab})$, we then obtain

$$D_m K^{am} = -4\, \underline{n}^a (\underline{L}_{mn}\, \underline{n}^m\, \underline{n}^n). \tag{60}$$

Proceeding in the same way with (12), first taking a dual to bring the left side into the form $\nabla_{[a} (\Omega^{-1}\, {}^*C_{bc]de})$, we obtain

$$D_m\, {}^*K^{am} = 0. \tag{61}$$

Thus, the asymptotic stress-energy, expressed as the field $\underline{L}_{ab}$ on $\mathcal{J}$, acts as a source for the K's. We shall interpret the right side of (60) in the following section. There is one further differential equation in this system - not actually on K^{ab} and $^*K^{ab}$, but rather a consequence of the assumption of the vanishing of the unphysical Weyl tensor on I. In the unphysical space, we have, for any k_c, $\nabla_{[a} \nabla_{b]} k_c = 1/2\, R_{abc}{}^d k_d$. Substituting (5), applying ζ^* (with care: There is a contraction.), and using the vanishing of C_{abcd} on I, we obtain

$$D_{[a} D_{b]} \underline{k}_c = 1/2\, (\underline{g}_{c[a} \underline{S}_{b]}{}^d + \underline{S}_{c[a} \underline{\delta}_{b]}{}^d)\, \underline{k}_d, \tag{62}$$

where $\underline{k}_c = \zeta^*(k_c)$. This equation holds for all fields $\underline{k}_c$ on $\mathcal{J}$. That is to say, the tensor field in parentheses on the right is the curvature tensor, $\mathcal{R}_{abc}{}^d$, of the derivative operator D_a on $\mathcal{J}$. Note the distinction between this equation and (29). The latter was applicable only to covariant vectors which, among other things, are orthogonal to $\underline{n}^a$. As a consequence, (29) defined $\mathcal{R}_{abc}{}^d$ only up to a multiple of $\underline{n}^d$, and hence defined only $\mathcal{R}_{abcd}$ uniquely. By contrast, (62) defines $\mathcal{R}_{abc}{}^d$ uniquely. But the present derivative operator and the derivative of Sect. 3 coincide when both are applicable. Thus, lowering of the index of the present $\mathcal{R}_{abc}{}^d$ with $\underline{g}_{de}$ must yield the $\mathcal{R}_{abce}$ of Sect. 3, i.e., we must have $\underline{g}_{c[a} \underline{S}_{b]d} + \underline{S}_{c[a} \underline{g}_{b]d} = \mathcal{R}\, \underline{g}_{c[a} \underline{g}_{b]d}$. But all indices of this equation are orthogonal to $\underline{n}^{\cdot}$, and so this equation is equivalent to its contraction with $\underline{g}^{ac} \underline{g}^{bd}$, i.e., to

$$\underline{S}_{mn}\, \underline{g}^{mn} = \mathcal{R}. \tag{63}$$

This equation ties together the observation of Sect. 3 that $\mathcal{R}$ is constant along the $\underline{n}$ - integral curves with that, following Eqn. (56), that $\underline{S}_{mn}\, \underline{g}^{mn}$ is also constant along the $\underline{n}$ - integral curves.

To summarize, the asymptotic gravitational field is characterized by i) a derivative operator D_a, which extends the action of the derivative of Sect. 3, and which annihilates $\underline{n}^a$ and $\underline{g}_{ab}$, ii) a tensor field $\underline{S}_a{}^b$, which has $\underline{n}^a$ as an eigenvector and which is such that $\underline{S}_{ab} = \underline{S}_a{}^m \underline{g}_{mb}$ is symmetric, and iii) two symmetric, trace-free tensor fields, K^{ab} and $*K^{ab}$, which are "mutual duals", in the sense of (57). These fields further satisfy the differential equations (56), (58), (60), (61), and (62). These equations require two assumptions: vanishing of the stress-energy asymptotically to order four, and vanishing of the unphysical Weyl tensor on I. [The former could of course be weakened somewhat, e.g., to include the massive Klein-Gordon field.] One thinks of the asymptotic stress-energy as a source for the K's (Eqns. (60) and (61)), of $\underline{S}_a{}^b$ as a potential for the K's (Eqn. (58)), and of D_a as a potential for $\underline{S}_a{}^b$ (Eqn. (62)). One convinces oneself, by thinking of other things to do to (6) and applying $\mathfrak{J}^*$ to the results, that one will obtain no further equations on these fields. One further convinces oneself that the taking of curls, divergences, duals, contractions, contractions with $\underline{n}^a$, etc. yield either identities or more complicated derived equations. [For example, imposing the Bianchi identity on the curvature tensor of (62), and using (56), one obtains an identity, by symmetry of $*K^{ab}$; taking the divergence of (58), switching the order of derivatives by (62), and using (56), one obtains just (61). There are at least ten or fifteen more possibilities.] The situation is clearly considerably more complicated than that of Klein-Gordon or Maxwell fields.

All of this was carried out with respect to a particular asymptote, so chosen that f = 0 on I, i.e., that the corresponding $\underline{n}^a$, $\underline{g}_{ab}$ be a decomposition of $\Gamma^{ab}{}_{cd}$. We of course have the freedom to pass to an equivalent asymptote, again satisfying this condition. We ask for the behavior of the gravitational fields under such a gauge transformation. Let ω be the additional conformal factor, as in Sect. 2. Setting f = 0 in (16), this ω will result in f' = 0 at I if and only if $\nabla^m \omega\, n_m = 0$ at I, i.e., if and only if $\nabla^m \omega$ is in the class $\mathcal{C}$. That is, we wish to deal only with ω's with $\nabla^m\omega$ in $\mathcal{C}$. Let ω be such, and set $\underline{\omega}^a = \mathfrak{J}^*(\nabla^a\omega)$. Then, setting $\underline{\omega} = \mathfrak{J}^*(\omega)$ as usual, application of $\mathfrak{J}^*$ to $g_{am}\nabla^m\omega = \nabla_a\omega$ yields $\underline{g}_{am}\,\underline{\omega}^m = D_a\underline{\omega}$. Thus, the complete specification, in terms of $\mathscr{I}$, of a gauge transformation requires not only a positive scalar field $\underline{\omega}$ on $\mathscr{I}$, but also a vector field $\underline{\omega}^a$ whose covariant version is the gradient of $\underline{\omega}$. The gauge behavior of the gravitational fields is now obtained by applying $\mathfrak{J}^*$ to the equations of Sect. 2. Applying $\mathfrak{J}^*$ to (14), to (17) after raising one index, and to (18) after multiplication by Ω'^{-1}, we obtain

$$D'_a \alpha_b = D_a \alpha_b - 2\underline{\omega}^{-1}\alpha_{(a}\underline{\omega}_{b)} + \underline{\omega}^{-1}\underline{g}_{ab}\underline{\omega}^m\alpha_m, \tag{64}$$

$$\underline{S}_a'{}^b = \underline{\omega}^{-2}\underline{S}_a{}^b - 2\underline{\omega}^{-3}D_a\underline{\omega}^b + 4\underline{\omega}^{-4}\underline{\omega}_a\underline{\omega}^b \tag{65}$$

$$- \underline{\omega}^{-4} \, \delta_a{}^b \, \underline{\omega}^m \, \underline{\omega}_m,$$

$$K'^{ab} = \underline{\omega}^{-5} K^{ab}, \qquad *K'^{ab} = \underline{\omega}^{-5} *K^{ab}, \tag{66}$$

for arbitrary α_b in (64), and where, of course, $\underline{\omega}_a = D_a \underline{\omega}$. Thus, the derivative operator has rather complicated gauge-behavior, as one might expect. The field $\underline{S}_a{}^b$ has even more complicated behavior. The K's, on the other hand, behave simply: They are fields of dimension -3. Assigning the source $\underline{L}_{ab}$ dimension -4, all our gravitational field equations are of course unchanged under (64), (65), and (66).

One can, in part, "correct" the awkward gauge-behavior of $\underline{S}_a{}^b$. First, lower the index of (65), to obtain $\underline{S}'_{ab} = S_{ab} - 2\,\underline{\omega}^{-1} D_a D_b \underline{\omega} + 4\,\underline{\omega}^{-2} D_a \underline{\omega} D_b \underline{\omega} - \underline{\omega}^{-2} \underline{g}_{ab} \underline{g}^{mn} D_m \underline{\omega} D_n \underline{\omega}$. Next, compare with (34): The gauge-terms here are precisely the same as the gauge-terms there. [This is not exactly a coincidence, for the last equation in (33) and (58) must both be conformally invariant.] Thus, the combination

$$N_{ab} = \underline{S}_{ab} - \rho_{ab} \tag{67}$$

has gauge-behavior $N'_{ab} = N_{ab}$, i.e., dimension -2. This N_{ab} is called the news tensor field. Note that N_{ab} does not carry all the information in $\underline{S}_a{}^b$. The rest, however, is "pure gauge". The properties of N_{ab} are obtained directly from the properties (33) of ρ_{ab}, the algebraic properties of $\underline{S}_{ab}$ and (58) and (63): N_{ab} is symmetric, $\underline{n}^b N_{ab} = 0$, $\underline{g}^{mn} N_{mn} = 0$, and

$$D_{[a} N_{b]c} = 1/4 \; \epsilon_{abm} *K^{mn} \underline{g}_{nc}. \tag{68}$$

That is, $*K^{ab}$ acts as a source for the news. The news, then, is the unphysical Ricci tensor, pulled-back to infinity, and corrected for its complicated gauge-behavior.

Since we now have gravitational fields, one can ask how things look from the standpoint of the physical space-time, i.e., one can ask for the peeling behavior of various fields. The interesting field is the Weyl tensor. For the physical Weyl tensor, $\tilde{C}_{abcd}$, we have $\tilde{C}_{abcd} = \Omega^{-1} (\Omega^{-1} C_{abcd})$. Hence, in the notation of Eqn. (55),

$$\begin{aligned} &\tilde{C}_{mnpq} \, L^m{}_a(r) \, L^n{}_b(r) \, L^p{}_c(r) \, L^q{}_d(r) \\ &= \Omega^{-1} [\Omega^{-1} C_{uvwx}(r) \, K^u{}_m(r) \, K^v{}_n(r) \, K^w{}_p(r) \, K^x{}_q(r)] \\ &\quad \times L_o{}^m{}_a(r) \, L_o{}^n{}_b(r) \, L_o{}^p{}_c(r) \, L_o{}^q{}_d(r). \end{aligned} \tag{69}$$

Choosing a curve which generates the peeling-off property, and keeping in force the assumption that the unphysical Weyl tensor vanishes at I, we have that the quantity in square brackets on the right in (69) remains finite at $r = 0$. Now substitute (54) for $L_o{}^a{}_b(r)$ on the right. The result is that the quantity in square brackets will be multiplied by a sum of five terms, with respective r-dependences r, r^2, r^3, r^4, r^5, and with tensors which project Weyl tensors to Weyl tensors with respective principal null directions $(l_o{}^a, l_o{}^a, l_o{}^a, l_o{}^a)$, $(l_o{}^a, l_o{}^a, l_o{}^a, n_o{}^a)$, $(l_o{}^a, l_o{}^a, n_o{}^a, n_o{}^a)$, $(l_o{}^a, n_o{}^a, n_o{}^a, n_o{}^a)$, $(n_o{}^a, n_o{}^a, n_o{}^a, n_o{}^a)$. This, then, is the peeling behavior of the Weyl tensor.

We shall conclude this section with an example: that of the Schwarzschild space-time of Sect. 1. It is convenient, however, to first return to the question, discussed at the beginning of Sect. 5, of the relationship between Killing fields in the physical space-time and asymptotic symmetries.

We first obtain a version of Theorem 3 which includes the gravitational fields.

Theorem 12. Let $\tilde{M}$, $\tilde{g}_{ab}$ be a space-time with Killing field $\tilde{\eta}^a$. Let M, g_{ab}, Ω be an asymptote with, on the corresponding $\mathcal{I}$, universal fields $\underline{n}^a$, $\underline{g}_{ab}$ (a decomposition), and gravitational fields D_a, $\underline{S}_a{}^b$, K^{ab}, and $*K^{ab}$. Finally, let $\underline{\eta}^a$ be the infinitesimal symmetry on $\mathcal{I}$ from Theorem 3. Then there is a vector field κ^a on $\mathcal{I}$ and a scalar field κ, satisfying $\kappa^b \underline{g}_{ab} = D_a \kappa$, and such that

$$\mathcal{L}_{\underline{\eta}}\, \underline{g}_{ab} = 2\kappa\, \underline{g}_{ab}, \quad \mathcal{L}_{\underline{\eta}}\, \underline{n}^a = -\kappa\, \underline{n}^a, \tag{70}$$

$$(\mathcal{L}_{\underline{\eta}} D_a - D_a \mathcal{L}_{\underline{\eta}})\, \mu_b = -2\mu_{(a} D_{b)}\kappa + \underline{g}_{ab}\, \kappa^m \mu_m, \tag{71}$$

$$\mathcal{L}_{\underline{\eta}}\, \underline{S}_a{}^b = -2\kappa\, \underline{S}_a{}^b - 2 D_a \kappa^b, \tag{72}$$

$$\mathcal{L}_{\underline{\eta}}\, K^{ab} = -5\kappa\, K^{ab}, \quad \mathcal{L}_{\underline{\eta}}\, *K^{ab} = -5\kappa\, *K^{ab}, \tag{73}$$

for all μ_b in (71).

Proof: Let η^a be the extension of $\tilde{\eta}^a$ to M, as in Theorem 3. Then, from the proof of Theorem 3, there is a scalar field α on M such that $\mathcal{L}_\eta g_{ab} = 2\alpha g_{ab}$ and $\mathcal{L}_\eta \Omega = \alpha\Omega$. From these, we compute $\mathcal{L}_\eta f = -\alpha f + 2 n^m \nabla_m \alpha$. But $f = 0$ on I, whence, from this last equation, $n_m \nabla^m \alpha = 0$ on I. Thus, $\nabla^m \alpha$ is in the class $\mathcal{C}$: Set $\kappa^a = \varsigma^*(\nabla^a \alpha)$ and $\kappa = \varsigma^*(\alpha)$. Now, the unphysical fields n^a, g_{ab}, ∇_a, $S_a{}^b$, and $\Omega^{-1} C_{abcd}$ are all constructed from Ω and g_{ab}. Since one knows the η-Lie derivatives of Ω and g_{ab} in terms of Ω, g_{ab}, and α, one can compute the η-Lie derivatives of all of these fields, all in the unphysical space. Applying ς^* to these equations, we obtain (70)-(73). [It's easier than it looks: Eqns. (70)-(73) are just (14), (15), (17), and (18) for "ω differing infinitesimally from one".]

Thus, a Killing field in the physical space-time gives rise not only to an infinitesimal symmetry, (70), but even to one which is a symmetry of the gravitational fields, in the sense of (71)-(73). These are just "infinitesimal versions" of the gauge-transformations, (64)-(66). Eqn. (71) gives "the $\underline{\eta}$-Lie derivative of the derivative operator D_a". It can be written in an equivalent, and more convenient, form. First expand the left side, replacing Lie derivatives by derivative operators - for example, $\mathcal{L}_{\underline{\eta}} D_a \mu_b = \underline{\eta}^m D_m D_a \mu_b + (D_m \mu_b) D_a \underline{\eta}^m + (D_a \mu_m) D_b \underline{\eta}^m$. There results, for the left side, $\underline{\eta}^m D_m D_a \mu_b - \underline{\eta}^m D_a D_m \mu_b - \mu_m D_a D_b \underline{\eta}^m$. Replacing the first two terms by a term involving the curvature tensor, and substituting the resulting "left side" back into (71), we obtain an equation linear in μ_b. Finally, since μ_b is arbitrary, it may be removed from the resulting equation. Thus, (71) is equivalent to

$$\underline{\eta}^m \mathcal{R}_{mab}{}^c - D_a D_b \underline{\eta}^c = -2\,\underline{\delta}_{(a}{}^c D_{b)}\kappa + \underline{g}_{ab}\,\kappa^c. \tag{74}$$

We now specialize to the case in which $\underline{\eta}^a$ is an infinitesimal supertranslation: $\underline{\eta}^a = \alpha\,\underline{n}^a$. Then, from (70), we have $\kappa = 0$, whence κ^a must be a multiple of $\underline{n}^a$. Next, eliminate the curvature tensor on the left in (74) in favor of $\underline{S}_a{}^b$ using (62). The result is an equation each term of which is some tensor times $\underline{n}^c$. Hence, the corresponding equation for the "some tensors" must hold, yielding

$$D_a D_b \alpha + 1/2\, \alpha\, \underline{S}_{ab} = \mu\, \underline{g}_{ab}, \tag{75}$$

where μ is some combination of scalar fields involving, among other things, the proportionality scalar between κ^a and $\underline{n}^a$. Lowering the index of Eqn. (72), the right side vanishes, and so this equation, with $\underline{\eta}^a = \alpha\, \underline{n}^a$, gives $\alpha\, \underline{n}^m D_m \underline{S}_{ab} = 0$. Hence, lowering the index of (58) and contracting with $\alpha\, \underline{n}^b$, the left side vanishes: We thus obtain $\alpha\, \epsilon_{amn}\, \underline{n}^m\, {}^*K^n{}_c = 0$. But this is equivalent to the second of

$$\alpha K^{ab} = 2\, \underline{n}^{(a} x^{b)}, \quad \alpha\, {}^*K^{ab} = 2\, \underline{n}^{(a}\, {}^*x^{b)}, \tag{76}$$

for some ${}^*x^b$. The first, for some x^b satisfying $\underline{g}_{ab}\, {}^*x^b = \epsilon_{amn} x^m \underline{n}^n$, follows from the second and (57). Substituting the second equation (76) into (61), we obtain $\alpha\, \underline{n}^m D_m\, {}^*x^a = \underline{n}^a ({}^*x^m D_m \alpha - \alpha D_m\, {}^*x^m)$. But the second equation (73) gives $\mathcal{L}_{\eta}\, {}^*K^{ab} = 0$. Expanding, using the second equation (76), we obtain $\alpha\, \underline{n}^m D_m\, {}^*x^a = \underline{n}^a\, {}^*x^m D_m \alpha$. From these two formulae there follows the first equation of

$$D_{[a} (\alpha\, \underline{g}_{b]m}\, x^m) = 0, \qquad D_{[a} (\alpha\, \underline{g}_{b]m}\, {}^*x^m) = 0. \tag{77}$$

The second is obtained, similarly, from (60), setting the source-term zero. This, then, is the structure of the asymptotic gravitational fields in the presence of a supertranslation arising from a physical Killing field.

Finally, we specialize further to the Minkowskian case, making the additional assumption that support α is $\mathcal{I}$ (probably a consequence of (74) in this case). Then, by the first equation (77) and $\underline{n}^a\ \underline{g}_{am}\ x^m = 0$, we have that $\alpha\ \underline{g}_{am}\ x^m = D_a(\pi^*(\beta))$ for some scalar field β on the base space B. The second of (77) and $\underline{g}_{am}\ {}^*x^m = \epsilon_{amn}\ x^m\ \underline{n}^n$ then imply that this β must be a harmonic function on B. Thus, β = constant, whence $\underline{g}_{am}\ x^m = \underline{g}_{am}\ {}^*x^m = 0$. That is to say, each of x^a and ${}^*x^a$ must be a multiple of $\underline{n}^a$, whence, by (76), each of K^{ab} and ${}^*K^{ab}$ must be a multiple of $\underline{n}^a\ \underline{n}^b$. But now, substituting this into (58), and lowering the index, the right side vanishes, and so we obtain $D_{[a}\ \underline{S}_{b]c} = 0$. That is, $\underline{S}_{ab}$ now has all the properties of the ρ_{ab} of Theorem 5, whence, by uniqueness, $\underline{S}_{ab} = \rho_{ab}$. We conclude, in other words, that the news vanishes. Furthermore, replacing $\underline{S}_{ab}$ by ρ_{ab} in (75) and comparing the result with (36), we see that $\eta^a = \alpha\ \underline{n}^a$ is in this case an infinitesimal translation. To summarize, we have shown that, in the Minkowskian case with source in (60) vanishing, the presence of a physical Killing field which gives rise to an infinitesimal supertranslation $\alpha\ \underline{n}^a$ (with support $\alpha = \mathcal{J}$) implies: The news vanishes, each of K^{ab} and ${}^*K^{ab}$ is a multiple of $\underline{n}^a\ \underline{n}^b$, and $\alpha\ \underline{n}^a$ is actually an infinitesimal translation. In particular, this completes the proof of Theorem 7.

Ignoring the issue of support $\mathcal{J}$, it is a consequence of the remarks above that, for any physical space-time with Minkowskian asymptotic geometry and stress-energy vanishing asymptotically to order five, the Lie algebra of Killing fields must be a subalgebra of the Poincarè Lie algebra. [If there are more than six independent Killing fields, then some linear combination of the corresponding infinitesimal symmetries on $\mathcal{J}$ must be a supertranslation, hence an infinitesimal translation. But there cannot be more than four independent of these.] Is every subalgebra possible?

We return, finally, to the example of the asymptotic geometry of the Schwarzschild space-time of Sect. 1. Of course, most of the work has already been done, above: The time-translation Killing field in the Schwarzschild solution gives rise to an infinitesimal supertranslation, $\alpha\ \underline{n}^a$, by Theorem 4, and so all the properties above are here applicable. The three rotational Killing fields in the Schwarzschild space-time give rise to additional infinitesimal symmetries, which we denote $\xi_1{}^a$, $\xi_2{}^a$, and $\xi_3{}^a$. The Lie bracket structure of the Killing fields in the Schwarzschild space-time goes over, on application of γ^*, to the corresponding Lie bracket structure on the infinitesimal symmetries. That is, we have $[\xi_1, \xi_2] = \xi_3$ and its cyclic permutations, and also $[\xi_i, \alpha\ \underline{n}] = 0$ for each i. Similarly, we have, from the corresponding property in the Schwarzschild space-time, that $\xi_1{}^a$, $\xi_2{}^a$, and $\xi_3{}^a$ span, at each point of $\mathcal{J}$, a 2-plane not containing $\underline{n}^a$. We have, at each point, $0 = \mathcal{L}_{\xi_i}(\alpha\ \underline{n}^a) = \underline{n}^a\ (\mathcal{L}_{\xi_i}\alpha - \kappa\alpha)$, by (70), and so $\mathcal{L}_{\xi_i}\alpha = \kappa\alpha$. Since this holds for all i, it follows that, if α vanished at any point, then it would vanish everywhere on $\mathcal{J}$. Thus, α is either everywhere positive or everywhere negative: Say positive. Since $\alpha' = \underline{\omega}\ \alpha$, we can, by means of a gauge transformation, set $\alpha = 1$, so the translational Killing gives the infinitesimal translation $\underline{n}^a$. For this choice, the "K's" in (70), for $\xi_1{}^a$, $\xi_2{}^a$, and $\xi_3{}^a$, must all vanish, by $\mathcal{L}_{\xi_i}\alpha = \kappa\alpha$. Hence, $\mathcal{L}_{\xi_i}\underline{g}_{ab} = 0$ for each i, by (70), and so $\mathcal{L}_{\xi_i}\mathcal{R} = 0$ for each i. That is, $\mathcal{R}$ must be constant on $\mathcal{I}$. We have with this choice a metric two-sphere for the base space, so $\mathcal{R} = 2$. Since the news vanishes, we have $\underline{S}_{ab} = \rho_{ab}$, while, since $\mathcal{R}$ is constant, we have $\rho_{ab} = 1/2\, \mathcal{R}\ \underline{g}_{ab}$; hence, $\underline{S}_{ab} = \underline{g}_{ab}$. This determines $\underline{S}_a{}^b$ uniquely up to a multiple of $\underline{n}^b$, which is as unique as it can be determined within the remaining gauge freedom ((65), with $\underline{\omega}^a$ a multiple of $\underline{n}^a$). We next compute the derivative operator. Since the $\xi_i{}^a$ span, at each point of $\mathcal{I}$, a 2-plane, there is at each point a covariant vector, unique up to a factor, orthogonal to all the $\xi_i{}^a$. Choose the factor such that the resulting vector, l_a, satisfies $l_a\,\underline{n}^a = 1$. Then, at each point, $\mathcal{L}_{\xi_i} l_a$ is orthogonal to all the $\xi_j{}^a$ (contracting, and bringing $\xi_j{}^a$ inside the Lie derivative using the commutation relations), and so must be a multiple of l_a. But $\mathcal{L}_{\xi_i}\ l_a$ is also orthogonal to $\underline{n}^a$ (contracting, and bringing $\underline{n}^a$ inside the Lie derivative using the commutation relations). Hence, $\mathcal{L}_{\xi_i}\ l_a = 0$, and, similarly, $\mathcal{L}_{\underline{n}}\ l_a = 0$. Now consider $D_a\ l_b$. Taking the Lie derivative of this field by $\xi_i{}^a$, using (71) and $\mathcal{L}_{\xi_i}\ l_b = 0$, we obtain zero. Contracting this field with $\underline{n}^a$ or $\underline{n}^b$, using $\mathcal{L}_{\underline{n}}\ l_a = 0$, we also obtain zero. That is, $D_a\ l_b$ must be orthogonal to $\underline{n}$ and invariant under the rotational symmetry, and hence must be some multiple of $\underline{g}_{ab}$. Thus, we have $D_a\ l_b \propto \underline{g}_{ab}$ and $D_a\ \underline{g}_{bc} = 0$, which together determine the derivative operator as uniquely as it can be determined within the remaining gauge freedom ((64), with $\underline{\omega}^a$ a multiple of $\underline{n}^a$). Finally, we determine the K's. In the Schwarzschild space-time, we have $*\tilde{C}_{ambn}\ \tilde{t}^m\ \tilde{t}^n = 0$, by the reflection symmetry of that space-time, where $\tilde{t}^a$ is the translational Killing field. Rewrite this as $\Omega^{-1}\ \epsilon^{aumn}\ t_u\ \epsilon^{bvpq}\ t_v\ *C_{mnpq} = 0$, recall that t_u approaches a multiple of n_u on I, and apply ζ^*, to obtain: $*K^{ab} = 0$. There remains only K^{ab}, which must be some multiple of $\underline{n}^a\ \underline{n}^b$. Since $\mathcal{L}_{\xi_i}\ K^{ab} = \mathcal{L}_{\underline{n}}\ K^{ab} = 0$, by (73), this multiple must be a constant. To compute this multiple, first note that, since the Ricci tensor vanishes in the physical space-time, and since $\tilde{t}^a$ is a Killing field, we have $\tilde{C}_{mabc}\ \tilde{t}^m = \tilde{\nabla}_a\ \tilde{\nabla}_b\ \tilde{t}_c$. Now apply $\tilde{t}^b$ and expand on the right: $\tilde{C}_{ambn}\ \tilde{t}^m\ \tilde{t}^n = \tilde{\nabla}_a(\tilde{t}^m\tilde{\nabla}_m\ \tilde{t}_b) - (\tilde{\nabla}_a\tilde{t}^m)(\tilde{\nabla}_m\ \tilde{t}_b)$. Setting $\lambda = -(1 - 2m/r)$, the squared norm of $\tilde{t}^a$, we have, since $\tilde{t}^a$ is surface-orthogonal, that $\lambda^{-1}\ \tilde{t}_a$ is a gradient. Hence, $\tilde{\nabla}_a\ \tilde{t}_b = -\lambda^{-1}\ \tilde{t}_{[a}\tilde{\nabla}_{b]}\lambda$. Substituting in the above, multiplying by Ω^{-3}, rewriting the left side as we did for $*C_{abcd}$, and applying ζ^*, we obtain the multiple: $K^{ab} = -8\ m\ \underline{n}^a\ \underline{n}^b$.

7. Physical Quantities at Infinity

In this section, we interpret the various fields on $\mathcal{I}$, and also certain other quantities, normally constructed from the fields together with infinitesimal symmetries. The purpose of these interpretations, we emphasize, is only to allow one to use one's physical intuition more effectively in guessing what will be useful for what. The physical theory itself, of course, has no need for any additional "interpretations".

We begin with a general remark concerning fields other than gravitation. Let $\tilde{M}$, $\tilde{g}_{ab}$ be a space-time, and let $\tilde{T}_{ab}$ be its stress-energy. Let S be a three-dimensional submanifold of $\tilde{M}$, and let $\tilde{k}^a$ be its normal. Then $\tilde{T}_{ab}\,\tilde{k}^b$ is interpreted as the local energy-momentum transfer across S, i.e., as the flux of energy-momentum through S. Since this is a vector at each point of S (reflecting the vector character of energy-momentum flux), there is no natural way to integrate over S to obtain the "total energy-momentum flux through S". Suppose now that our space-time admits Killing field $\tilde{\xi}^a$. Then we can form $\tilde{T}_{ab}\,\tilde{k}^a\,\tilde{\xi}^b$, a scalar field on S which, when integrated over S, yields the total flux of that energy-momentum combination associated with $\tilde{\xi}^a$. [E.g., if $\tilde{\xi}^a$ is interpreted as a time-translation, of energy; if as a rotation, of angular momentum.] Now let M, g_{ab}, Ω be an asymptote for our space-time. Replace $\tilde{T}_{ab}$ by the combination L_{ab} (so $L_a{}^b = \tilde{T}_a{}^b - 1/3\,\tilde{T}\,\tilde{\delta}_a{}^b$). Let us assume that the stress-energy vanishes asymptotically to order four: Set $\underline{L}_{ab} = \zeta^*(\Omega^{-4}\,L_{ab}$ Then, since n^a is normal to I, it is natural to interpret $\underline{L}_{ab}\,\underline{n}^b$ on $\mathscr{I}$ as the local flux of energy-momentum through I, noting that the trace-difference between $\underline{L}_{ab}$ and $\tilde{T}_{ab}$ does not matter, since $\underline{g}_{ab}\,\underline{n}^b$ vanishes. As above, the construction of integrated, total quantities requires an additional vector field which describes how to weight the contributions from different regions. Let, then, η^a be an infinitesimal symmetry, S a compact (so integrals will converge three-dimensional subset of $\mathscr{I}$, and consider

$$J = \int_S (\underline{L}_{mn}\,\underline{n}^m\,\eta^n)\,\epsilon_{abc}\,dS^{abc}. \tag{78}$$

We first note that the value of the integral is gauge-invariant: $\underline{L}_{ab}$ has dimension -4, $\underline{n}^m$ and ϵ_{abc} dimension 0, η^n dimension +1, and dS^{abc} dimension +3, for a total dimension of zero. We may interpret this J, up to a factor, as the flux through S of that energy-momentum combination associated with η^a. Thus, if η^a is an infinitesimal supertranslation, we would interpret J as "an energy-momentum, with a weighting determined by η^a"; if η^a is not a supertranslation, as "an angular momentum - boost momentum". When η^a is not a supertranslation, it will retain this character under addition of any supertranslation. This corresponds to the ambiguity in the angular momentum associated with the choice of origin. Of course, if η^a arises from a Killing field in the physical space-time, then the J of (78) is just the limit at I of the corresponding quantity in the physical space-time. We may also regard J as a linear mapping from the Lie algebra $\mathscr{L}$ of infinitesimal symmetries to the reals. Restricting this mapping, in the Minkowskian case, to the subalgebra $\mathscr{T}$ of infinitesimal translations, we obtain a linear mapping from the four-dimensional vector space $\mathscr{T}$ to the reals. This linear mapping could be interpreted as the "true" energy-momentum flux through S, obtained by restricting oneself to "reasonable" weightings of the contributions from different parts of S. In Minkowski space, e.g., the total energy-momentum is a linear mapping from the four-dimensional vector space of translational Killing fields to the reals. Let us consider the case of "just energy and momentum combinations", i.e., of $\eta^a = \alpha \underline{n}^a$ an infinitesimal supertranslation, in more detail. In this case, the part of the integrand of (78) in parentheses becomes $\alpha \underline{L}_{mn} \underline{n}^m \underline{n}^n$. Thus, J is just the integral of the single scalar field $\underline{L}_{mn} \underline{n}^m \underline{n}^n$ over S, weighted by α. That a single scalar field suffices to carry all the relevant information about the incident stress-energy is a reflection of the fact that "the incident energy-momentum flux is really null", as it would have to be to get out to I. If the original T_{ab} satisfied an energy condition, then we would have that $\underline{L}_{mn} \underline{n}^m \underline{n}^n$ is non-negative. Consider now the case when α is positive, which corresponds to an infinitesimal supertranslation which is "a (possibly) distorted time-translation". Then J would be interpreted as "a total energy, possibly with a somewhat distorted weighting". In this case, of course, J would be non-negative, since the integrand is. In physical terms, we have that a stress-energy tensor satisfying an energy condition can only radiate to infinity non-negative total non-gravitational energy. Finally, consider again the four-dimensional vector space $\mathcal{T}$, with its Lorentz metric from Sect. 5. From the remarks above we have that, for the physical stress-energy satisfying an energy condition, our linear mapping applied to any timelike η^a in $\mathcal{T}$ yields a positive number. Thus, this linear mapping, which represents the "true" total energy-momentum flux through S, is represented by a covariant vector over $\mathcal{T}$ which is timelike. Again, we conclude that an energy condition implies timelike energy-momentum flux at infinity.

Consider first the Klein-Gordon field. The asymptotic field, $\underline{\varphi}$, represents the "radiation field", i.e., the "1/r part" of the physical field. Thus, for example, we would regard there as being no flux of Klein-Gordon particles through an open region of $\mathcal{I}$ in which $\underline{\varphi} = 0$. Multiplying (44) by Ω^{-4} and applying ς^*, we obtain, in this case,

$$\underline{L}_{am} \underline{n}^m = D_a \underline{\varphi} \; (\underline{n}^m D_m \underline{\varphi}) - 1/2 \; \underline{\varphi} \; D_a \; (\underline{n}^m D_m \underline{\varphi}). \qquad (79)$$

This, then, is the local energy-momentum flux for a Klein-Gordon field. We note that $\underline{L}_{mn} \underline{n}^m \underline{n}^n$ is not in general positive, i.e., that one can have negative energy radiated away. This is not surprising, for the stress-energy of the physical Klein-Gordon field does not satisfy an energy condition. [In the Klein-Gordon case without the "- 1/6 R", the stress-energy does not vanish asymptotically to order four. However, one can compute, from (47), $\Omega^{-4} n^m (L_{am} - L g_{am})$, which corresponds to the energy-momentum flux. The resulting field is finite on I. Applying ς^*, we obtain $D_a \underline{\varphi} \; (\underline{n}^m D_m \underline{\varphi})$ for the energy-momentum flux. Now, the energy radiated is positive, as one expects, since a physical energy condition is now satisfied.]

Consider next the Maxwell field. In this case, $\underline{F}_{ab}$ and $*\underline{F}_{ab}$ have a somewhat more complex structure. As suggested by the peeling behavior, $\underline{F}_{ab}$ and $*\underline{F}_{ab}$ contain information about both the "radiation field" (characterized by $\underline{F}_{am}\, \underline{n}^m$ and $*\underline{F}_{am}\, \underline{n}^m$ - the contribution to the r-term in the peeling formula), and the "Coulomb field" (characterized by the "rest" of $\underline{F}_{ab}$ and $*\underline{F}_{ab}$). Things work out, as illustrated by the peeling behavior, so that these two, physically distinct, aspects of the Maxwell field become combined into the $\underline{F}$'s. It makes sense, for example, to say that the asymptotic Maxwell fields represent a situation with no incident electromagnetic radiation at infinity (namely, when $\underline{F}_{am}\, \underline{n}^m = *\underline{F}_{am}\, \underline{n}^m = 0$). On the other hand, "no Coulomb part" does not make sense, for there is no natural 2-plane transverse to $\underline{n}^a$. The field L_{ab} in the unphysical space for the Maxwell field is given by (45). It appears offhand that, were we to multiply this equation by $\Omega^{-4} n^b$ and apply Υ^*, the right side would not be expressible on $\mathcal{I}$ in terms of only $\underline{F}_{ab}$ and $*\underline{F}_{ab}$, for that right side involves F's with contravariant indices. However, we have $F_a{}^m F_{bm} n^b = 1/2\, F_{am}\, \epsilon^{muvw}\, n_w\, *F_{uv}$, and similarly for the dual: The right side involves only the covariant F's. Thus, we obtain

$$\underline{L}_{am}\, \underline{n}^m = 1/2\, (\underline{F}_{am}\, \epsilon^{muv}\, *\underline{F}_{uv} - *F_{am}\, \epsilon^{muv}\, F_{uv}). \tag{80}$$

This, then, is the local energy-momentum flux. Applying $\underline{n}^a$, we can simplify the resulting formula using (48):

$$\underline{L}_{mn}\, \underline{n}^m\, \underline{n}^n = \underline{g}^{ab}\, (\underline{F}_{am}\, \underline{n}^m\, \underline{F}_{bn}\, \underline{n}^n + *\underline{F}_{am}\, \underline{n}^m\, *\underline{F}_{bn}\, \underline{n}^n). \tag{81}$$

We first note that the right side involves only the "radiation parts" of the fields, $\underline{F}_{am}\, \underline{n}^m$ and $*\underline{F}_{am}\, \underline{n}^m$, as we might expect. Furthermore, the right side is non-negative, and vanishes when and only when the radiation parts. vanish.

Let there be a compact base space B, fix image C of a cross-section, and consider

$$Q = \int_C \underline{F}_{ab}\, dS^{ab}, \qquad *Q = \int_C *\underline{F}_{ab}\, dS^{ab}. \tag{82}$$

We note that Q and *Q are independent of the choice of gauge, for $\underline{F}_{ab}$ and $*\underline{F}_{ab}$ have dimension -2, and dS^{ab} dimension +2. Clearly, these should be interpreted, respectively, as the total magnetic and total electric charge of the system, for, e.g., they are just the limits to I of the corresponding integrals in the physical space-time We note furthermore that Q and *Q are independent of the choice of cross-section, for the difference between these integrals over C and over C' is in each case the integral of the curl of the integrand over the region between C and C'. But, by (49), the curls vanish. [If one permitted a charge-current density near I, then one would instead obtain a formula for the difference between the charges defined by C and C' in terms of an integral representing the total flux of electric or magnetic charge through the region between C and C'.] Finally, we note that, since cross-sections are transverse to $\underline{n}^a$, the integrals in (82) sample only "the parts of $\underline{F}_{ab}$ and $*\underline{F}_{ab}$ transverse to $\underline{n}^a$". This, again, is what we expect: The charges depend only on what we regard as the "Coulomb part" of the asymptotic fields.

We consider, finally, the gravitational fields. The fields K^{ab} and $*K^{ab}$ are analogous to the electromagnetic fields (specifically, to the combinations $\epsilon^{amn} \underline{F}_{mn}$ and $\epsilon^{amn} *\underline{F}_{mn}$). That is to say, these are the "true" gravitational fields. The leading term in the peeling formula is essentially $K^{mn} \underline{g}_{ma} \underline{g}_{nb}$ $(= - K^{mn} \epsilon_{mpa} \underline{n}^p \epsilon_{nqb} \underline{n}^q)$, and similarly for the dual. These combinations, then, represent the asymptotic gravitational radiation. From either the peeling behavior or the Schwarzschild example, one concludes that the "$\underline{n}^a$ $\underline{n}^b$ - parts" of K^{ab} and $*K^{ab}$ represent the "Coulomb gravitational field". Thus, what we showed earlier was that the existence of a Killing field in the Minkowskian case which leads to an infinitesimal supertranslation at infinity implies that the asymptotic gravitational fields are "pure Coulomb". Intermediate between these two are the "$x^{(a}$ $\underline{n}^{b)}$ - parts", represented by K^{am} $\underline{g}_{mb}$ and $*K^{am}$ $\underline{g}_{mb}$. By the peeling behavior, these parts represent radiation fields having three principal null directions coincident and transverse to I, and the fourth tangent; and falling to zero as $1/r^2$. These parts, then, represent "higher order radiation contributions". Eqn. (57) is now seen to be a connection between the "radiation parts" of K^{ab} and $*K^{ab}$. The "Coulomb parts" are essentially independent. As in the electromagnetic case, the failure of $\underline{g}_{ab}$ to be invertible means that we cannot naturally decompose a given K^{ab} as the sum of three terms, one of each type. Thus, a K^{ab}'s being "pure Coulomb" (K^{am} $\underline{g}_{mb} = 0$), or "pure Coulomb and higher order radiation" (K^{mn} $\underline{g}_{ma}$ $\underline{g}_{nb} = 0$), makes sense; there is no such thing as a "pure radiation" K^{ab}.

The interpretation of the news comes from (68). On the right, there appears the "radiation part" of $*K^{ab}$ (or of K^{ab} since, by (57), that right side equals $1/4$ ϵ_{abm} K^{mn} ϵ_{ncp} $\underline{n}^p$). Thus, the news is a potential for the asymptotic radiation field. Contracting (68) with $\underline{n}^a$, we obtain $\underline{n}^m$ D_m $N_{ab} = - 1/2$ K^{mn} $\underline{g}_{ma}$ $\underline{g}_{nb}$. Thinking of $\underline{n}^m$ as the "time-direction", this equation states that the time rate of change of the news gives the pure-radiation part of K^{ab}. We shall see in a moment that the news can also be interpreted as an "amplitude for the energy-momentum flux of gravitational radiation at infinity". This, too, is something one might have expected. Symbolically, the energy-momentum flux for the Klein-Gordon field (spin zero) is "$(D \underline{\varphi})^2$", and for the Maxwell field (spin one), "$(\underline{F})^2$"; One might therefore expect that for the gravitational field (spin two) it should be "$(D^{-1}$ $K)^2$". But "D^{-1} $K = N$".

The derivative operator D_a, in the rather weak sense of (62), is a potential for the news. The derivative operator at infinity seems to have somewhat less physical significance than does, say, the derivative operator in the physical space-time. The reason, perhaps, is that D at infinity is "two derivatives removed" from the gravitational field (K), while $\tilde{\nabla}$ in the physical space-time is "one derivative removed" from the physical gravitational field ($\tilde{C}$). The derivative operator D, then, is perhaps more analogous to the metric in the physical space-time, and in particular perhaps has more of the attributes of a universal field. Finally, the true universal fields, $\underline{n}^a$ and $\underline{g}^{ab}$, merely provide the underlying geometry of $\mathscr{I}$.

We consider next energy and momentum for the asymptotic gravitational field. We restrict consideration to the Minkowskian case. Let c be any real number, l_a any covariant vector field on $\mathscr{I}$ satisfying $\underline{n}^a l_a = c$; and let $\alpha \underline{n}^a$ be any infinitesimal supertranslation (so $\underline{n}^m D_m \alpha = 0$). Consider the vector field given by the following formula:

$$P^a = + 1/4 \, \alpha K^{am} l_m + (\alpha D_m l_n + l_m D_n \alpha) \, \underline{g}^{np} N_{pq} \, \underline{g}^{q[m} \underline{n}^{a]}. \qquad (83)$$

We first note that this formula makes sense: The "$\underline{g}^{np}$" on the right does not lead to ambiguities, since $(\alpha D_m l_n + l_m D_n \alpha) \underline{n}^n = 0$ (by $l_a \underline{n}^a$ = const.) and $\underline{n}^p N_{pq} = 0$; the "$\underline{g}^{qn}$" does not lead to ambiguities, since $\underline{n}^q N_{pq} = 0$, while the other index is antisymmetrized with an index of an $\underline{n}$. We next consider the gauge-behavior of this formula. In order to preserve $l_a \underline{n}^a = c$, we assign l_a dimension zero. Then, since α has dimension +1, we have from (64)

$$D'_m l'_n + l'_m D_n \alpha' = \underline{\omega}^2 (\alpha D_m l_n + l_m D_n \alpha) + \underline{\omega} \alpha \underline{g}_{mn} (\underline{\omega}^p l_p). \qquad (84)$$

On substituting into (83), the second term on the right in (84), involving $\underline{g}_{mn}$, will give zero, since $\underline{g}_{mn} \underline{g}^{np} N_{pq} \underline{g}^{qm} = 0$ and $\underline{g}_{mn} \underline{n}^m = 0$. Furthermore, K^{ab} has dimension -3 and N_{pq} dimension -2. Thus, we have $P'^a = \underline{\omega}^{-3} P^a$: The vector field P^a has dimension -2. Finally we consider the dependence of P^a on l_a. To this end, let us consider the special case c = 0: Then we have $\underline{n}^a l_a = 0$, and hence $l_a = \underline{g}_{ab} l^b$ for some vector field l^b. We claim that, in this case, P^a can be written

$$P^a = D_m (\alpha l^p N_{pq} \, \underline{g}^{q[m} \underline{n}^{a]}). \qquad (85)$$

Indeed, expanding the right side of (85), we obtain the expression

$$(\alpha D_m l^p + l^p D_m \alpha) N_{pq} \, \underline{g}^{q[m} \underline{n}^{a]} + \alpha l^p D_m N_{pq} \, \underline{g}^{q[m} \underline{n}^{a]}. \qquad (86)$$

Using $\underline{n}^m l_m = 0$, one checks that the last term on the right in (83) is the same as the first term in (86). As for the last term in (86), one first writes it as $+ 2 \alpha l^p D_{[m} N_{p]q} \, \underline{g}^{q[m} \underline{n}^{a]}$, noting that the extra term thus added is zero. Next, eliminate the news in this expression using (68), and then $*K^{ab}$ in favor of K^{ab} using (57). The result is precisely the first term on the right in (83). We conclude, then, that (85) holds in the case $c = 0$, i.e., that in this case P^a is the divergence of a skew tensor field. Now fix nonzero c, and consider (83) for two different l's, l_a and $\hat{l}_a$, with $\underline{n}^a l_a = \underline{n}^a \hat{l}_a = c$. Then we have, by linearity, $P^a(\hat{l}) = P^a(l) + P^a(\hat{l} - l)$. But the last field must be of the form (85). We conclude: For fixed c, changes in the choice of l_a change P^a only by the addition of a term involving the divergence of an antisymmetric tensor field.

Set $c = 1$, let P^a be given by (83) for some l_a satisfying $\underline{n}^a l_a = 1$, let C be the image of a cross-section, and consider

$$M = \frac{1}{8\pi}\int_C P^m \epsilon_{mab}\, dS^{ab}. \tag{87}$$

The number M is independent of the choice of gauge, since P^m has dimension -2, ϵ_{mab} dimension 0, and dS^{ab} dimension +2. Furthermore, the number M is independent of the choice of l_a, for a new choice changes P^m by the divergence of an antisymmetric tensor field, and hence $P^m \epsilon_{mab}$ by the curl of some covariant vector field. But the integral of a curl over the compact 2-manifold C yields zero. Thus, M depends only on the choice of infinitesimal supertranslation $\alpha\, \underline{n}^a$ and on the choice of cross-section.

This number M, we claim, is to be interpreted as "the total energy-momentum of the system (with a weighting determined by $\alpha\, \underline{n}^a$) as measured by the asymptotic gravitational field in the region of null infinity described by C". There are essentially three pieces of evidence for this interpretation. For the first, consider again (83). The important term is the first one: The second is essentially just a correction term which gives P^a all the right properties. When the first term is inserted into (87), we obtain the integral of $l_m K^{mn} \epsilon_{nab}$ over C. Since $\underline{n}$ is transverse to C, this integral involves essentially the "$\underline{n}^a \underline{n}^b$ - part" of K^{ab}, i.e., its "Coulomb part". Thus, (87) is more or less an α-weighted integral of the Coulomb part of the asymptotic Weyl tensor. It is hard to see how the result, which must be something, could be anything other than the α-weighted energy-momentum. The second piece of evidence comes from examples. Let us consider, e.g., the Schwarzschild space-time. Then the news vanishes, and $K^{ab} = -8\, m\, \underline{n}^a \underline{n}^b$, and so (83) gives $P^a = -2m\alpha\, \underline{n}^a$. Substituting into (87), we obtain $M = m/4\pi \int_C \alpha\, dS$. In the special case $\alpha = 1$, corresponding to the weighting prescribed by the timelike Killing field in the Schwarzschild space-time, we obtain $M = m$. Thus, at least in the Schwarzschild case, (87) gives the "right" answer. In a similar way, one obtains the expected answer for, e.g., the Kerr and Weyl metrics. More generally, one can verify that for stationary space-times, for which a total energy can be defined in terms of the timelike Killing field, that number agrees with (87), using of course the "α" appropriate to the Killing field. Still more generally, one can show that, for a space-time whose metric "approaches the Schwarzschild metric" (for some m) to first non-trivial order, M = m. Again, one obtains the intuitively expected answer.

The final piece of evidence is perhaps the strongest. Fix α, and consider two images of cross sections, C' and C, with the former to the future of the latter. Then, denoting by M' and M the respective integrals (83) over C' and C, we have $M' - M = 1/8\pi \int_A (D_m P^m) \epsilon_{abc}\, dS^{abc}$, where A is the region of $\mathcal{I}$ between C and C'. Thus, the change in the value of M for two cross sections and fixed α is just the volume integral over the region A of the divergence of P^m. Let us compute this divergence. From (83),

$$\begin{aligned} D_a P^a &= 1/4 \, \alpha \, l_m D_a K^{am} + 1/4 \, K^{am} D_a (\alpha \, l_m) \\ &+ D_a (\alpha \, D_m l_n + l_m D_n \alpha) \, \underline{g}^{np} N_{pq} \, \underline{g}^{q[m} \underline{n}^{a]} \\ &+ (\alpha \, D_m l_n + l_m D_n \alpha) \, \underline{g}^{np} D_a N_{pq} \, \underline{g}^{q[m} \underline{n}^{a]} . \end{aligned} \tag{88}$$

We simplify. In the last term on the right in (88), we may replace "$D_a N_{pq}$" by "$2 D_{[a} N_{p]q}$", for $D_p N_{aq} \underline{g}^{qm} \underline{n}^a = 0$. But now we can eliminate the news in favor of $^*K^{ab}$ using (68), and then $^*K^{ab}$ in favor of K^{ab} using (57). Making these substitutions, we obtain for this term $(\alpha \, D_m l_n + l_m D_n \alpha)(-1/4 \, K^{mn})$. Thus, the last term on the right in (88) precisely cancels the second term. For the third term on the right in (88), we have, expanding,

$$\begin{aligned} \underline{n}^a D_{[a} (\alpha \, D_{m]} l_n + l_{m]} D_n \alpha) &= \alpha \, \underline{n}^a D_{[a} D_{m]} l_n \\ &+ \underline{n}^a (D_{[a} \alpha) D_{m]} l_n + \underline{n}^a D_{[a} l_{m]} D_n \alpha + \underline{n}^a l_{[m} D_{a]} D_n \alpha . \end{aligned} \tag{89}$$

Since $\underline{n}^a D_a \alpha = 0$ and $\underline{n}^a D_m l_a = 0$, the sum of the second and third terms on the right in (89) is just $\underline{n}^a (D_a l_{[m}) D_{n]} \alpha$. But this is antisymmetric in "mn", and hence will give zero when substituted back into (88) and thereby contracted with symmetric $\underline{g}^{np} N_{pq} \underline{g}^{qm}$. For the first term on the right in (89), we use (62) to obtain for this term $- 1/4 \, \underline{g}_{mn} \sigma - 1/4 \, \underline{S}_{mn}$. When this is substituted back into (88), the part "$- 1/4 \, \underline{g}_{mn} \sigma$" will not contribute, since $\underline{g}^{pq} N_{pq} = 0$. Finally, using $\underline{n}^a D_a D_n \alpha = 0$, the last term on the right in (89) is $- 1/2 \, D_m D_n \alpha$. Substituting all this back into (88), using (60) for the first term, we obtain:

$$\begin{aligned} D_a P^a &= - \alpha (\underline{L}_{mn} \underline{n}^m \underline{n}^n) - 1/2 \, (D_m D_n \alpha + 1/2 \, \alpha \, \underline{S}_{mn}) \\ &\quad \times \underline{g}^{np} N_{pq} \underline{g}^{qm} . \end{aligned} \tag{90}$$

This is our final expression for the divergence of P^a. It represents "the flux through I of whatever M is the value of". Consider the first term on the right in (90). It is just minus the α-weighted energy-momentum flux attributable to the stress-energy of the matter fields. For example, for α positive, corresponding to an infinitesimal symmetry representing a "time supertranslation", this term contributes to M' - M just minus the α-weighted energy flux of matter. This is what one would expect if M' and M are to be interpreted as the respective α-weighted energies at C' and C. Thus, on the basis of the appearance of this term, we are led to the present interpretation of M.

Given our interpretation of M, the second term on the right in (90) must be thought of as the α-weighted flux of energy-momentum carried away to null infinity by the gravitational radiation. We first note that this term has simple gauge behavior, for $D'_m D'_n \alpha' + 1/2\ \alpha' \underline{S}'_{mn} = \underline{\omega}(D_m D_n \alpha + 1/2\ \alpha\ \underline{S}_{mn})$ plus a multiple of $\underline{g}_{mn}$, by (64) and (65). But the multiple of $\underline{g}_{mn}$, substituted back into (90), gives zero. What is perhaps strange about this term is that it involves derivatives of α: It is as though a quickly varying α, all by itself, represents a "flux of gravitational energy-momentum to infinity". Two special cases of (90) are of particular interest. Suppose first that $\alpha\ \underline{n}^a$ arises from a Killing field in the physical space-time. Then we have (75), and hence that the second term on the right in (90) vanishes. Thus, ignoring the matter, M' = M in this case, and we obtain what we expect to obtain: Symmetries generate constants of the motion. A second special case is that in which $\alpha \underline{n}^a$ is actually an infinitesimal translation on $\mathcal{J}$, i.e., that in which we have (36). Substituting for $D_m D_n \alpha$ in (90) using (36), we obtain

$$D_a\ P^a = -\ \alpha\ (\underline{L}_{mn}\ \underline{n}^m\ \underline{n}^n)\ -\ 1/4\ \alpha\ N_{mn}\ \underline{g}^{np}\ N_{pq}\ \underline{g}^{qm}. \qquad (91)$$

Thus, in this case the derivatives of α disappear, and all we obtain for the second term is the square of the news, weighted by α. In particular, the expression multiplying α in the second term is non-negative, and vanishes when and only when N_{ab} vanishes. We interpret this as meaning that, at least for "reasonable" weightings (infinitesimal translations), the gravitational radiation always carries away to infinity positive energy (or timelike energy-momentum); or zero energy, precisely when the news vanishes. The news, then, may be interpreted as an amplitude for the flux of energy-momentum through gravitational radiation.

Fix the cross section C. Then we may regard (90) as defining a linear mapping from the vector space $\mathcal{S}$ of infinitesimal supertranslations to the reals. Restricting to the four-dimensional subspace of infinitesimal translations, we obtain a linear mapping from $\mathcal{T}$ to the reals, i.e., a covariant vector over $\mathcal{T}$. This vector is called the <u>Bondi energy-momentum vector</u>. What we showed by (91), in these terms, is that the Bondi energy-momentum vector for a later cross-section differs from that for an earlier by a timelike vector over $\mathcal{T}$.

Are there any other physically interesting "quantities" which can be constructed from these gravitational fields and infinitesimal symmetries? It seems likely that there are not. [A promising-looking candidate: For ξ^a any infinitesimal symmetry, $K^{am} \underset{\sim}{g}_{mn} \xi^n$ and $*K^{am} \underset{\sim}{g}_{mn} \xi^n$ are divergence-free, by (60) and (61). Hence, the integral over a cross-section is independent of cross-section. But, as one can show from (57) and (58), these give zero.] There are [12] however, some further "quantities" which are constructed from the derivative of the unphysical Weyl tensor in the direction off the boundary I. That is, these involve the next-higher-order term in the asymptotic behavior of the Weyl tensor - information not carried by the K's. This is the signal that the physical interpretation involves, among other things, back-scattered radiation. What one expects to obtain is a symmetric linear mapping from the quotient algebra, $\mathcal{L}/\mathcal{S}$, to the reals, more specifically, a tensor over the four-dimensional vector space $\mathcal{T}$ having all the symmetries of a Weyl tensor. It might be of interest to explicitly display this mapping in terms of tensor fields on the abstract manifold $\mathcal{I}$.

III. ASYMPTOTIC STRUCTURE AT SPATIAL INFINITY

1. Preliminaries

In this section, we collect various formulae which will be needed later: for the initial-value formulation of certain fields including the gravitational, and for the results of applying conformal transformations to these.

Let $\tilde{M}$, $\tilde{g}_{ab}$ be a space-time, and let $\tilde{T}$ be a three-dimensional, spacelike submanifold. Denote by $\tilde{t}^a$ the unit (timelike) normal to $\tilde{T}$ at points of $\tilde{T}$. A tensor field $\tilde{\alpha}^{a..b}{}_{c..d}$ defined at points of $\tilde{T}$ will be said to be in $\tilde{T}$ if each of its indices is orthogonal to $\tilde{t}^a$, i.e., if $\tilde{\alpha}^{a..b}{}_{c..d}\, \tilde{t}_a = 0, \ldots, \tilde{\alpha}^{a..b}{}_{c..d}\, \tilde{t}^d = 0$. Such fields are in one-to-one correspondence with tensor fields on $\tilde{T}$ regarded as a three-dimensional manifold in its own right. In particular, the field $\tilde{q}_{ab} = \tilde{g}_{ab} + \tilde{t}_a \tilde{t}_b$ is in $\tilde{T}$, and yields a positive-definite metric in $\tilde{T}$. Its inverse is $\tilde{q}^{ab} = \tilde{g}^{ab} + \tilde{t}^a \tilde{t}^b$, i.e., we have $\tilde{q}_{am} \tilde{q}^{mb} = \delta_a{}^b + \tilde{t}_a \tilde{t}^b$, the unit tensor in $\tilde{T}$. This unit tensor in $\tilde{T}$ is also the projection operator into $\tilde{T}$, e.g., for $\tilde{\alpha}^a{}_b$ any field defined at points of $\tilde{T}$, $(\delta^a{}_m + \tilde{t}^a \tilde{t}_m)(\delta^n{}_b + \tilde{t}^n \tilde{t}_b)\, \tilde{\alpha}^m{}_n$ is in $\tilde{T}$, and, for $\tilde{\alpha}^a{}_b$ already in $\tilde{T}$, is just $\tilde{\alpha}^a{}_b$. For tensor fields in $\tilde{T}$, raising and lowering of indices with $\tilde{g}_{ab}$ or $\tilde{q}_{ab}$ yields the same result. The tensor field $\tilde{\epsilon}_{abc} = \tilde{\epsilon}_{abcm} \tilde{t}^m$ is in $\tilde{T}$, and is the alternating tensor of $\tilde{T}$; that is, we have $\tilde{\epsilon}_{abc}\, \tilde{\epsilon}^{abc} = 6$.

Let $\tilde{\alpha}^{a..b}{}_{c..d}$ be any tensor field in $\tilde{T}$. Then the right side of

$$\tilde{D}_r \tilde{\alpha}^{a..b}{}_{c..d} = \tilde{q}_r{}^s\, \tilde{q}_m{}^a \ldots \tilde{q}_n{}^b\, \tilde{q}_c{}^u \ldots \tilde{q}_d{}^v\, \tilde{\nabla}_s\, \tilde{\alpha}^{m..n}{}_{u..v} \qquad (92)$$

makes sense (since the presence of $\tilde{q}_r{}^s = \delta_r{}^s + \tilde{t}_r \tilde{t}^s$ ensures that the derivative is only taken tangentially to $\tilde{T}$), and is also in $\tilde{T}$. This equation defines the derivative in $\tilde{T}$, $\tilde{q}_{ab}$: The $\tilde{D}_a$ so defined is indeed a torsion-free derivative operator annihilating the metric $\tilde{q}_{ab}$. We define the curvature tensor, Ricci tensor, and scalar curvature of $\tilde{T}$, $\tilde{q}_{ab}$ as usual: For any k_c in $\tilde{T}$, $\tilde{D}_{[a} \tilde{D}_{b]} k_c = 1/2\, \tilde{\mathcal{R}}_{abc}{}^m k_m$; $\tilde{\mathcal{R}}_{ab} = \tilde{\mathcal{R}}^m{}_{amb}$, $\tilde{\mathcal{R}} = \tilde{\mathcal{R}}^m{}_m$.

The field

$$\tilde{p}_{ab} = \tilde{q}_a{}^m \tilde{q}_b{}^n \tilde{\nabla}_m \tilde{t}_n = \tilde{p}_{(ab)} \tag{93}$$

called the <u>extrinsic</u> <u>curvature</u>, is in $\tilde{T}$. Set $\tilde{p} = \tilde{p}_{mn} \tilde{q}^{mn}$. Certain components of the stress-energy, which we again represent by $\tilde{L}_{ab} = \tilde{R}_{ab} - 1/6\,\tilde{R}\,\tilde{g}_{ab}$, can be expressed, at points of $\tilde{T}$, in terms of $\tilde{q}_{ab}$ and $\tilde{p}_{ab}$. The first such equation is obtained by taking the D-divergence of (93), replacing the $\tilde{D}$ by $\tilde{\nabla}$ on the right according to (92), and using the definition of the curvature tensor of the space-time:

$$\tilde{D}_m (\tilde{p}^{am} - \tilde{p}\,\tilde{q}^{am}) = \tilde{q}^{am}\,\tilde{t}^n\,\tilde{L}_{mn}. \tag{94}$$

The second is obtained by replacing $\tilde{D}$ by $\tilde{\nabla}$, via (92), in the definition of the curvature tensor of $\tilde{T}$, and contracting twice:

$$\tilde{\mathcal{R}} - \tilde{p}^{mn}\,\tilde{p}_{mn} + \tilde{p}^2 = 2\,(\tilde{L}_{mn} - \tilde{L}p_p\,\tilde{g}_{mn})\,\tilde{t}^m\,\tilde{t}^n. \tag{95}$$

This geometrical framework out of the way, we turn now to the description, in terms of $\tilde{T}$, of various physical fields which may arise on our space-time. As in the null case, there are numerous possibilities; as before, we shall restrict consideration to a few examples.

A Klein-Gordon field, a solution of (38), is represented in terms of $\tilde{T}$ by two scalar fields on this manifold: $\tilde{\varphi}$ (i.e., $\tilde{\varphi}$ on $\tilde{M}$ restricted to $\tilde{T}$) and $\tilde{\psi} = \tilde{t}^m \tilde{\nabla}_m \tilde{\varphi}$, the normal derivative of $\tilde{\varphi}$ at points of $\tilde{T}$. [The same two fields would be used also for the two other possibilities, referred to following (38).] A Maxwell field, a solution of (39), is represented by its electric and magnetic parts,

$$\tilde{E}_a = \tilde{F}_{am}\,\tilde{t}^m, \qquad \tilde{B}_a = {}^*\tilde{F}_{am}\,\tilde{t}^m, \tag{96}$$

each a vector field in $\tilde{T}$. These parts exhaust the algebraic information in $\tilde{F}_{ab}$ at points of $\tilde{T}$, for we have, at points of $\tilde{T}$, $\tilde{F}_{ab} = 2\,\tilde{t}_{[a}\tilde{E}_{b]} - \tilde{\epsilon}_{abmn}\,\tilde{t}^m\,\tilde{B}^n$. In the Klein-Gordon case, the two fields $\tilde{\varphi}$ and $\tilde{\psi}$ are essentially arbitrary; in the Maxwell case, the electric and magnetic fields are not. Indeed, contracting Eqns. (39) with $\tilde{\epsilon}^{abc}$, we obtain

$$\tilde{D}_a\,\tilde{E}^a = 0, \qquad \tilde{D}_a\,\tilde{B}^a = 0. \tag{97}$$

Eqns. (97) will be recognized from electrodynamics in special relativity. To summarize, a Klein-Gordon field is represented in terms of $\tilde{T}$ by two scalar fields; a Maxwell field by two divergence-free vector fields.

The final physical field we shall consider is the gravitational, represented initially by the Weyl tensor, $\tilde{C}_{abcd}$, of our space-time. We wish, then, to represent this Weyl tensor in terms of tensor fields in $\tilde{T}$. This is accomplished, in analogy with (96), by a decomposition into electric and magnetic parts:

$$\tilde{E}_{ab} = \tilde{C}_{ambn}\, \tilde{t}^m\, \tilde{t}^n, \qquad \tilde{B}_{ab} = *\tilde{C}_{ambn}\, \tilde{t}^m\, \tilde{t}^n. \tag{98}$$

Each is a symmetric tensor field in $\tilde{T}$, and, since $\tilde{C}_{abcd}$ has vanishing traces, each is trace-free. As in the electromagnetic case, these parts exhaust the algebraic information in the Weyl tensor, for $\tilde{C}_{abcd}$ can be reconstructed, at each point of $\tilde{T}$, from $\tilde{E}_{ab}$ and $\tilde{B}_{ab}$. The Weyl tensor satisfies an equation analogous to Maxwell's equations. To obtain it, we first again replace the Ricci tensor by $\tilde{L}_{ab}$, and then substitute $\tilde{R}_{abcd} = \tilde{C}_{abcd} + \tilde{g}_{a[c}\, \tilde{L}_{d]b} - \tilde{g}_{b[c}\, \tilde{L}_{d]a}$ into the Bianchi identity, $\tilde{\nabla}_{[a}\, \tilde{R}_{bc]de} = 0$:

$$\tilde{\nabla}_{[a}\, \tilde{C}_{bc]}{}^{de} = 2\, \delta_{[a}{}^{[d}\, \tilde{\nabla}_b\, \tilde{L}_{c]}{}^{e]}. \tag{99}$$

Thus, the stress-energy acts as a "source" for the Weyl field. Again as in the Maxwell case, this equation yields equations on the electric and magnetic parts of the Weyl tensor. For that on the electric, first contract (99) with $\tilde{\epsilon}^{abc}\, \tilde{\epsilon}_{def}$. Rewrite the left side as $\tilde{\nabla}_a\, (\tilde{\epsilon}^{abc}\, \tilde{\epsilon}_{def}\, \tilde{C}_{bc}{}^{de}) - (\tilde{\nabla}_a\, \tilde{\epsilon}^{abc})\, \tilde{\epsilon}_{def}\, \tilde{C}_{bc}{}^{de} - \tilde{\epsilon}^{abc}\, (\tilde{\nabla}_a\, \tilde{\epsilon}_{def})\, \tilde{C}_{bc}{}^{de}$. The first term, by (92) and (98), is $-4\, \tilde{D}_m\, \tilde{E}^{fm}$; for the other two terms, expand the derivative and replace the derivative of $\tilde{t}$ by $\tilde{p}_{ab}$ using (93). On the right, eliminate the two $\tilde{\epsilon}$'s using the ϵ-identity. The equation on the magnetic part is obtained in a similar way, contracting (99) with $\tilde{\epsilon}^{abc}\, \tilde{t}_d$. We thus obtain:

$$\tilde{D}_b\, \tilde{E}^{ab} = \tilde{\epsilon}^{amn}\, \tilde{p}_m{}^u\, \tilde{B}_{un} + \tilde{t}_m\, \tilde{t}_n\, \tilde{\nabla}^{[a}\, \tilde{L}^{m]n}, \tag{100}$$

$$\tilde{D}_b\, \tilde{B}^{ab} = -\tilde{\epsilon}^{amn}\, \tilde{p}_m{}^u\, \tilde{E}_{un} + 1/2\, \tilde{\epsilon}^{amn}\, \tilde{t}_u\, \tilde{\nabla}_m\, \tilde{L}_n{}^u. \tag{101}$$

These equations have structure similar to that of (97), except that one obtains correction terms on the right involving the $\tilde{p}$'s, and one is here including sources, which were not included in the Maxwell case. [It is interesting to note that the absence of "correction terms" in the Maxwell case is essentially dictated by index-counting: Even using $\tilde{\epsilon}$'s, one can construct algebraically no scalar linear in $\tilde{p}_{ab}$ and $\tilde{E}_a$, or in $\tilde{p}_{ab}$ and $\tilde{B}_a$.] Finally, there is in the gravitational case one additional set of equations having no direct Maxwell analogs. These equations assert, essentially, that $\tilde{q}_{ab}$ and $\tilde{p}_{ab}$ are potentials for $\tilde{E}_{ab}$ and $\tilde{B}_{ab}$. For the first, we replace $\tilde{D}$ by $\tilde{\nabla}$, using (92), in the definition of the curvature tensor of $\tilde{T}$, $\tilde{q}_{ab}$, and contract once. For the second, compute $\tilde{D}_{[a}\, \tilde{p}_{b]\,c}$ using (92) and (93), and dualize. There results:

$$\tilde{E}_{ab} = \tilde{\mathcal{R}}_{ab} - \tilde{p}_a{}^m\, \tilde{p}_{bm} + \tilde{p}\, \tilde{p}_{ab} - 1/2\, (\tilde{q}_a{}^m\, \tilde{q}_b{}^n\, \tilde{L}_{mn} + \tilde{q}_{ab}\, \tilde{L}_{mn}\, \tilde{q}^{mn}), \tag{102}$$

$$\tilde{B}_{ab} = \tilde{\epsilon}_{mn(a}\, \tilde{D}^m\, \tilde{p}^n{}_{b)}. \tag{103}$$

To summarize, the gravitational field is represented by the combinations (102) and (103), symmetric, trace-free tensor fields on $\tilde{T}$ whose divergences are restricted by (100) and (101).

In each of the cases above, we began with a tensor field on the space-time $\tilde{M}$, $\tilde{g}_{ab}$ satisfying some equation. We then represented that field in terms of certain fields in $\tilde{T}$, and rewrote certain components of the equation on that field as equations on the fields in $\tilde{T}$. This general procedure represents the first half of the initial-value formulation of fields in general relativity. For the second half, one introduces "nearby surfaces" to $\tilde{T}$, and asks how the fields in $\tilde{T}$ must change from surface to surface.

Let, then, t be a smooth scalar field on $\tilde{M}$ with everywhere timelike gradient, so the family of surfaces $\tilde{T}_t$, each given by t = const, are spacelike submanifolds. We denote by $\tilde{t}^a$ the field of unit (timelike) normals to this family of surfaces, so $\tilde{t}_a = \tilde{\lambda}\, \tilde{\nabla}_a\, t$, where $\tilde{\lambda} = [-(\tilde{\nabla}_a\, t)(\tilde{\nabla}^a\, t)]^{-1/2}$. Now, in each of the submanifolds $\tilde{T}_t$ in our family we obtain, as above, certain tensor fields representing the various physical fields in the space-time. We are now interested in the "rate of change" of these fields from surface to surface. We shall denote by a dot the action of the operator $\mathcal{L}_{\tilde{\lambda}\tilde{t}}$. Thus, $\dot{t} = 1$, and so one can think of the dot as "d/dt". The field $\tilde{\lambda}$ plays the role of giving the "distance between successive surfaces, infinitesimally displaced from each other". It is sometimes called the lapse field. We note that, since $\dot{\tilde{t}}^a$ and $\dot{\tilde{t}}_a$ are multiples, respectively, of $\tilde{t}^a$ and $\tilde{t}_a$, the "dot" of any tensor field in the $\tilde{T}_t$ is again in the $\tilde{T}_t$. Since $\tilde{\lambda}^{-1}\, \tilde{t}_a$ is a gradient, we have from (93) the following useful formula:

$$\tilde{\nabla}_a\, \tilde{t}_b = \tilde{p}_{ab} - \tilde{\lambda}^{-1}\, \tilde{t}_a\, D_b \tilde{\lambda}. \tag{104}$$

The idea, then, is to apply the operator "dot" to the various fields in the $\tilde{T}_t$ obtained above.

We begin with the geometrical fields. Replacing $\tilde{q}_{ab}$ on the left in

$$\dot{\tilde{q}}_{ab} = 2\, \tilde{\lambda}\, \tilde{p}_{ab}, \tag{105}$$

$$\dot{\tilde{p}}_{ab} = \tilde{D}_a\, \tilde{D}_b \tilde{\lambda} - \tilde{\lambda}\, \tilde{E}_{ab} + \tilde{\lambda}\, \tilde{p}_a{}^m\, \tilde{p}_{bm} - 1/2\, \tilde{\lambda}\, \tilde{q}_{ab}\, \tilde{L}_{mn}\, \tilde{g}^{mn} \tag{106}$$

by $\tilde{g}_{ab} + \tilde{t}_a \tilde{t}_b$, and $\tilde{p}_{ab}$ on the left by (93), expanding the Lie derivatives on the left, and using (102) and (104) in the second equation, we obtain these equations. From (105), there follows the action of "dot" on the geometrical fields obtained from $\tilde{q}_{ab}$:

$$\dot{\tilde{q}}^{ab} = -2 \tilde{\lambda} \tilde{p}^{ab} \tag{107}$$

$$\dot{\tilde{\epsilon}}_{abc} = \tilde{\lambda} \tilde{p} \tilde{\epsilon}_{abc}, \quad \dot{\tilde{\epsilon}}^{abc} = -\tilde{\lambda} \tilde{p} \tilde{\epsilon}^{abc}. \tag{108}$$

Note, from (105), (107), and (108), that the raising and lowering of indices and the taking of duals of fields in the $\tilde{T}_t$ do not commute with application of "dot".

The equations for the rate of change of the Klein-Gordon fields are:

$$\dot{\tilde{\varphi}} = \tilde{\lambda} \tilde{\gamma} \tag{109}$$

$$\dot{\tilde{\gamma}} = \tilde{D}^m(\tilde{\lambda} \tilde{D}_m \tilde{\varphi}) - \tilde{\lambda} \tilde{\gamma} \tilde{p} - 1/2 \tilde{\lambda} \tilde{\varphi} \tilde{L}_{mn} \tilde{g}^{mn}. \tag{110}$$

The first equation is immediate from the definition of $\tilde{\gamma}$. For the second, expand $\tilde{D}^m \tilde{D}_m \tilde{\varphi} = \tilde{q}^{mn} \tilde{\nabla}_m(\tilde{q}_n{}^c \tilde{\nabla}_c \tilde{\varphi}) = \tilde{q}^{mn} \tilde{\nabla}_m (\tilde{\nabla}_n \tilde{\varphi} + \tilde{t}_n \tilde{t}^c \tilde{\nabla}_c \tilde{\varphi})$ using the Klein-Gordon equation, (38), and (104). The two alternative versions of the Klein-Gordon equation result in leaving out the last term on the right in (110), or replacing it by "$- m^2 \tilde{\lambda} \tilde{\varphi}$", respectively. The corresponding equations for the Maxwell fields are:

$$\dot{\tilde{E}}^a = \tilde{\epsilon}^{amn} \tilde{D}_m(\tilde{\lambda} \tilde{B}_n) - \tilde{\lambda} \tilde{p} \tilde{E}^a, \quad \dot{\tilde{B}}^a = -\tilde{\epsilon}^{amn} \tilde{D}_m(\tilde{\lambda} \tilde{E}_n) - \tilde{\lambda} \tilde{p} \tilde{B}^a. \tag{111}$$

These are obtained, respectively, by contracting Eqns. (39) with $\tilde{t}^a \tilde{\epsilon}^{bcd}$ and using (92), (96), and (104). The second terms on the right in (111) are again correction terms arising from the extrinsic curvatures of the $\tilde{T}_t$. Ignoring these terms, Eqns. (111) are again familiar from special relativity. For example, for $\tilde{\lambda} = 1$, corresponding to equally spaced surfaces, Eqns. (111) are just two of Maxwell's equations as usually written in terms of $\tilde{E}$ and $\tilde{B}$. Note that, applying "dot" to Eqns. (97), using (105) and (111), we obtain identities, as one expects.

Finally, for the rate of change of the gravitational fields, we take components of (99). For the electric part, contract (99) with $\tilde{t}^a \tilde{\epsilon}^{bcm} \tilde{\epsilon}_{den}$; for the magnetic, with $\tilde{t}^a \tilde{\epsilon}^{bcm} \tilde{t}_d$. In each case, one eliminates $\tilde{t}^m \tilde{\nabla}_m$ in favor of $\mathcal{L}_{\tilde{\lambda}\tilde{t}}$, $\tilde{q}_a{}^m \tilde{\nabla}_m$ in favor of $\tilde{D}_a$ using (92), $\tilde{\nabla}_a \tilde{t}_b$ in favor of $\tilde{p}_{ab}$ using (104), and $\tilde{C}_{abcd}$ in favor of $\tilde{E}_{ab}$ and $\tilde{B}_{ab}$ using (98). The results of these substitutions are:

$$\dot{\tilde{E}}^{ab} = \tilde{\epsilon}^{mn(a} (\tilde{D}_m(\tilde{\lambda} \tilde{B}_n{}^{b)}) + \tilde{B}_n{}^{b)} \tilde{D}_m \tilde{\lambda}) + \tilde{\lambda} \tilde{p}^{(a}{}_m \tilde{E}^{b)m} - 2 \tilde{p} \tilde{\lambda} \tilde{E}^{ab}$$

$$- \tilde{\lambda}\, \tilde{q}^{ab}\, \tilde{E}_{mn}\, \tilde{p}^{mn} + \tilde{\lambda}\, \tilde{q}_m{}^{(a}\, \tilde{q}_n{}^{b)}\, \tilde{t}_u\, \tilde{\nabla}^{[u}\, \tilde{L}^{m]n}, \qquad (112)$$

$$\dot{\tilde{B}}^{ab} = - \tilde{\epsilon}^{mn(a}\, (\tilde{D}_m(\tilde{\lambda}\, \tilde{E}_n{}^{b)}) + \tilde{E}_n{}^{b)}\, \tilde{D}_m \tilde{\lambda}) + \tilde{\lambda}\, \tilde{p}^{(a}{}_m\, \tilde{B}^{b)m} - 2\, \tilde{p}\, \tilde{\lambda}\, \tilde{B}^{ab}$$

$$- \tilde{\lambda}\, \tilde{q}^{ab}\, \tilde{B}_{mn}\, \tilde{p}^{mn} + 1/2\, \tilde{\lambda}\, \tilde{\epsilon}^{mn(a}\, \tilde{q}^{b)u}\, \tilde{\nabla}_m\, \tilde{L}_{nu}. \qquad (113)$$

Under (112) and (113), there is preserved both the trace-free character of $\tilde{E}^{ab}$ and $\tilde{B}^{ab}$ and the divergence-conditions (100) and (101). Eqns. (112) and (113) are analogous to and similar in structure to (111).

Thus, one begins in each case with a physical field in the space-time $\tilde{M}$, $\tilde{g}_{ab}$. In each case, one obtains a description in terms of certain tensor fields in the $\tilde{T}_t$: for a Klein-Gordon field, $\tilde{\varphi}$ and $\tilde{\mathcal{F}}$, for Maxwell, $\tilde{E}^a$ and $\tilde{B}^a$, and for the gravitational field, $\tilde{E}^{ab}$ and $\tilde{B}^{ab}$. In the case of the last two, these fields in the $\tilde{T}_t$ are subject to certain constraint equations: for Maxwell, (97); for the gravitational, (100) and (101). In each case, one then obtains a further set of equations which give the "rate of change with t" of each of the fields in the $\tilde{T}_t$ in terms of the values of those fields and the additional field $\tilde{\lambda}$ which describes the displacements required to pass from one $\tilde{T}_t$ to the next. These are called the evolution equations. One thinks of one of the $\tilde{T}_t$ as representing an "initial time", and of a set of fields in that $\tilde{T}_t$ satisfying the appropriate constraint equations as initial data for the physical field at that initial time, i.e., as that information which must be known about the physical field initially to predict its future behavior. The "prediction of future behavior", in turn, is to be made using the evolution equations. Our original construction of the $\tilde{T}_t$ required that $\tilde{\lambda}$ be positive, i.e., that the successive surfaces be "positively displaced" (e.g., in the future time-direction) from each other. This restriction, however, is not at all necessary. One could allow $\tilde{\lambda}$ also to assume negative values. Thus, for a fixed $\tilde{T}_t$ and a $\tilde{\lambda}$ thereon which is positive in some regions and negative in others, there would result "infinitesimally nearby surfaces" which, say, are displaced into the future time-direction in those regions in which $\tilde{\lambda}$ is positive, and into the past where $\tilde{\lambda}$ is negative. All the equations remain the same. [It is not difficult, but is slightly more awkward, to carry out the construction from the beginning in such a way that this possibility is allowed.]

This completes our brief summary of initial-value formulations in general relativity.

We shall require in the following sections the equations which result from those above by application of a conformal transformation. The derivation of these equations is entirely straightforward, e.g., using the techniques of the Appendix. Let Ω be a positive scalar field in each of the $\tilde{T}_t$. Set $q_{ab} = \Omega^2\, \tilde{q}_{ab}$, a positive-definite metric in the $\tilde{T}_t$. We now merely assign, to every field we have introduced in the $\tilde{T}_t$, some dimension, and then make the appropriate substitutions in our equations. Logically, the assignments of dimensions are completely arbitrary at this point: Our particular assignments, of course, will be based on later convenience.

Our assignment of dimension zero to $\tilde{q}_{ab}$ dictates natural assignments, all zero, to the geometrical fields constructed from $\tilde{q}_{ab}$, i.e., $\tilde{q}^{ab}$, $\tilde{\epsilon}_{abc}$ and $\tilde{\epsilon}^{abc}$. Thus, for example, we have $\epsilon^{abc} = \Omega^{-3}\tilde{\epsilon}^{abc}$. It is convenient, as it was in the null case, to introduce a new symbol for what is essentially the derivative of the conformal factor. Set $n_a = 1/2\, \Omega^{-1/2} D_a \Omega$. The conformal behavior of $\tilde{q}_{ab}$ dictates, according to the Appendix, that of the derivative operator. We have, for any field k_b in the $\tilde{T}_t$,

$$\tilde{D}_a k_b = D_a k_b + 4\Omega^{-1/2} k_{(a} n_{b)} - 2\Omega^{-1/2} k_m n^m q_{ab}. \qquad (114)$$

We assign to the extrinsic curvature $\tilde{p}_{ab}$ dimension -1, and to the "infinitesimal displacement" field $\tilde{\lambda}$ dimension + 1/2.

To the two Klein-Gordon fields, $\tilde{\varphi}$ and $\tilde{\Upsilon}$, assign dimensions - 1/2 and -1, respectively. There are no constraint equations in this case. The evolution equations, (109) and (110), become, on removing the tildes,

$$\dot{\varphi} = -1/2\, \Omega^{-1}\dot{\Omega}\varphi + \lambda\Upsilon, \qquad (115)$$

$$\dot{\Upsilon} = -\Omega^{-1}\dot{\Omega}\Upsilon + \Omega^{1/2} D^m(\Omega^{1/2}\lambda D_m\varphi) + \lambda\varphi \times$$
$$(\Omega^{1/2} D_m n^m - 3 n^m n_m) + \Omega^{1/2}\varphi n^m D_m\lambda - 2\Omega^{1/2}\lambda n^m D_m\varphi \qquad (116)$$
$$- \Omega^{1/2}\lambda\Upsilon p - 1/2\, \Omega^{-1}\lambda\varphi(\tilde{L}_{mn}\tilde{g}^{mn}).$$

To the two Maxwell fields, $\tilde{E}^a$ and $\tilde{B}^a$, assign dimension -1. The two constraint equations, (97), then give

$$D_a E^a = 2\Omega^{-1/2} E^a n_a, \qquad D_a B^a = 2\Omega^{-1/2} B^a n_a. \qquad (117)$$

The evolution equations, (111), become

$$\dot{E}^a = -2\Omega^{-1}\dot{\Omega} E^a + \Omega^{1/2}\epsilon^{amn} D_m(\lambda B_n)$$
$$- \lambda\epsilon^{amn} n_m B_n - \Omega^{1/2}\lambda p E^a, \qquad (118)$$

$$\dot{B}^a = -2\Omega^{-1}\dot{\Omega} B^a - \Omega^{1/2}\epsilon^{amn} D_m(\lambda E_n)$$
$$+ \lambda\epsilon^{amn} n_m E_n - \Omega^{1/2}\lambda p B^a. \qquad (119)$$

To the two gravitational fields, $\tilde{E}^{ab}$ and $\tilde{B}^{ab}$, assign dimension - 3/2. [Note that the present pattern of dimensions is dimension

= - 1/2 (1 + spin), in contrast to the null case, in which the "1/2" is missing. We shall see the reason for this in the following sections.] In order to rewrite the constraint and evolution equations for this case, we must adopt some convention for the source-terms, those involving $\tilde{L}_{ab}$, in (100), (101), (112), and (113). The "proper" procedure would be to decompose $\tilde{L}_{ab}$ into tensor fields in the $\tilde{T}_t$, assign to each of these fields a dimension, and then write the source-terms out explicitly in terms of these fields, replacing $\tilde{\nabla}$ first by $\tilde{D}$ and then by D. However, the resulting expressions are rather complicated, and will anyway all give zero in the asymptotic limit. We shall therefore, in order to avoid writing lengthy expressions for "zero", proceed as follows. First note that the source-terms in (100), (101), (112), and (113) all depend only on the combination $\tilde{L}_{abc} = \tilde{\nabla}_{[a} \tilde{L}_{b]c}$. We simply assign to this field $\tilde{L}_{abc}$, and also to $\tilde{t}^a$, dimension zero. The constraint equations, (100) and (101), then become

$$D_b E^{ab} = 3 \Omega^{-1/2} E^{ab} n_b + \epsilon^{amn} p_m{}^u B_{un} + \Omega^{-5/2} L^{amn} t_m t_n, \tag{120}$$

$$D_b B^{ab} = 3 \Omega^{-1/2} B^{ab} n_b - \epsilon^{amn} p_m{}^u E_{un} + 1/2\, \Omega^{-5/2} \epsilon^{amn} t^u L_{mnu}. \tag{121}$$

For the evolution equations, (112) and (113),

$$\dot{E}^{ab} = -7/2\, \Omega^{-1} \dot{\Omega} E^{ab} + \Omega^{1/2} \epsilon^{mn(a} (D_m (\lambda B_n{}^{b)}) + B_n{}^{b)} D_m \lambda - \Omega^{-1/2} \lambda B_n{}^{b)} n_m) + \Omega^{1/2} \lambda p^{(a}{}_m E^{b)m} - 2 \Omega^{1/2} \lambda p E^{ab} - \Omega^{1/2} \lambda q^{ab} E_{mn} p^{mn} + \Omega^{-2} \lambda t^u q^{m(a} q^{b)n} L_{umn}, \tag{122}$$

$$\dot{B}^{ab} = -7/2\, \Omega^{-1} \dot{\Omega} B^{ab} - \Omega^{1/2} \epsilon^{mn(a} (D_m (\lambda E_n{}^{b)}) + E_n{}^{b)} D_m \lambda - \Omega^{-1/2} \lambda E_n{}^{b)} n_m) + \Omega^{1/2} \lambda p^{(a}{}_m B^{b)m} - 2 \Omega^{1/2} \lambda p B^{ab} - \Omega^{1/2} \lambda q^{ab} B_{mn} p^{mn} + 1/2\, \Omega^{-2} \lambda \epsilon^{mn(a} q^{b)u} L_{mnu}. \tag{123}$$

The basic equations we shall need are (115) - (123). These equations are, admittedly, rather long, and not very illuminating. It will turn out however that, in the asymptotic limit we shall take at spatial infinity, almost all the terms in these equations will vanish, resulting in very simple equations. Nothing of this sort seems to happen in the null case. There, one has right from the beginning relatively simple equations, almost every term of which ultimately has some meaning. Is there not some better choice of variables for the spatial case?

2. Asymptotes

Let $\tilde{T}$ be a 3-manifold with smooth positive-definite metric $\tilde{q}_{ab}$ and smooth symmetric tensor field $\tilde{p}_{ab}$. By an asymptote of $\tilde{T}$, $\tilde{q}_{ab}$, $\tilde{p}_{ab}$, we mean a 3-manifold T with preferred point S, a positive-definite metric q_{ab} on T (with q_{ab} C^0 at S and C^∞ elsewhere), a scalar field Ω on T (with Ω C^2 at S and C^∞ elsewhere), and a diffeomorphism ψ from $\tilde{T}$ to T - S (by means of which we shall identify $\tilde{T}$ and T - S), satisfying the following conditions:

1. On $\tilde{T}$, $q_{ab} = \Omega^2 \tilde{q}_{ab}$.
2. At the point S, $\Omega = 0$, $D_a \Omega = 0$, and $D_a D_b \Omega = 2 q_{ab}$.
3. Assigning $\tilde{p}_{ab}$ dimension -1, the field p_{ab} is bounded in a neighborhood of S.

We first note that the conditions in the definition make sense. In the second condition, in particular, we take derivatives at S no higher than the second of the C^2 field Ω. Furthermore, the second condition is independent of the choice of derivative operator used in its formulae, for $D_a \Omega = 0$ at S implies that $D_a D_b \Omega$ at S is independent of derivative operator. [This is a necessary observation, for the derivative operator defined by q_{ab} may not exist at S, since q_{ab} need only be C^0 there.] Note also that the definition requires that q_{ab} be defined and be positive-definite even at S. The original metric $\tilde{q}_{ab}$ is not defined at S, and, since $\Omega = 0$ at S, could not be given sensible meaning there. Finally, note that the definition requires that Ω be nonzero on $\tilde{T}$: By convention, we choose it positive there.

It is intended that an asymptote represent the "attachment to $\tilde{T}$ of a single point (S) at spatial infinity". The situation is very much like that of the null case: The diffeomorphism ψ performs the attachment; the conformal factor Ω serves to rescale the original metric $\tilde{q}_{ab}$ to obtain a new metric, q_{ab}, which is well-behaved at S. The second condition requires essentially that, up to second order near S, Ω behaves as (q-distance from S)^2. This, we shall maintain, is the appropriate asymptotic behavior for the conformal factor. The third condition, finally, requires that the extrinsic curvature fall to zero asymptotically at a certain rate.

Again as in the null case, we must make what case we can that this definition is an appropriate one. Again, there are several pieces of evidence on this issue.

One would expect that space-times which are "obviously asymptotically flat" would admit corresponding submanifolds $\tilde{T}$ which, with $\tilde{q}_{ab}$ the induced metric and $\tilde{p}_{ab}$ the extrinsic curvature, admit asymptotes. Let us consider again the Schwarzschild space-time, with metric (1). Let $\tilde{T}$ be the three-dimensional, spacelike submanifold given, in these coordinates, by $t = 0$. Then r, θ, φ are coordinates on this 3-manifold, in terms of which the metric $\tilde{q}_{ab}$ is represented by:

$$(1 - 2m/r)^{-1}\, dr^2 + r^2\, (d\theta^2 + \sin^2\theta\, d\varphi^2). \qquad (124)$$

Furthermore, e.g., because of the time-reflection symmetry, we have $\tilde{p}_{ab} = 0$. Choose new coordinate $R = 2\,(r + (r^2 - 2mr)^{1/2})^{-1}$. Then, in terms of R, θ, φ, (124) becomes

$$(R^2 - 1/2\, m\, R^3)^{-2}\, (dR^2 + R^2\, (d\theta^2 + \sin^2\theta\, d\varphi^2)). \qquad (125)$$

Now let T be the 3-manifold whose points are those of $\tilde{T}$, together with one additional point S, labeled by "R = 0" in terms of the spherical coordinates of the chart above. Let the differentiable structure of T be that induced by the structure of $\tilde{T}$ and this chart. [Note that "R = 0" corresponds to "$r = \infty$", i.e., that S cannot be regarded as a point of $\tilde{T}$.] Let Ω be a positive scalar field on T such that, in some neighborhood of S, $\Omega = R^2 - 1/2\, m\, R^3$. Set $q_{ab} = \Omega^2\, \tilde{q}_{ab}$ on $\tilde{T}$, whence, near S, q_{ab} is given by the expression in parentheses in (125): Define q_{ab} at S by that expression. Let Υ be the identity mapping from $\tilde{T}$ to T - S. This set-up, we claim, is an asymptote for $\tilde{T}$, $\tilde{q}_{ab}$, $\tilde{p}_{ab}$ (= 0). Indeed, the first and third conditions are immediate, while the second follows from the fact that R is q-distance from S while, to second order near S, $\Omega = R^2$. Thus, at least this spacelike submanifold of the Schwarzschild space-time admits an asymptote. The action of the mapping Υ is illustrated in the figure below.

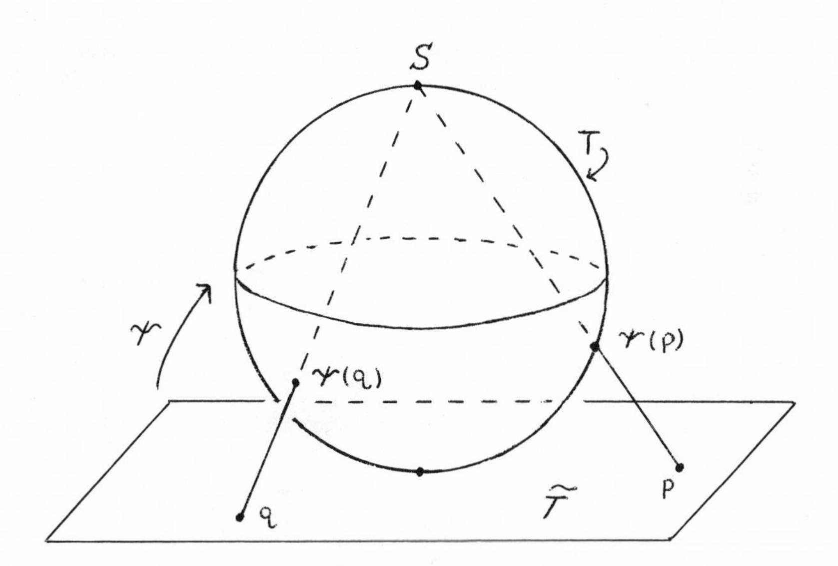

Fig. 4. Schematic diagram of the action of the mapping $\Upsilon: \tilde{T} \rightarrow T$. For p in $\tilde{T}$, $\Upsilon(p)$ is the point of T intersected by the straight line joining p and S.

The special case m = 0 is not excluded above. In this case, $\tilde{T}$ is a spacelike 3-plane in Minkowski space-time, and we again have an asymptote. Furthermore, one can verify directly that various other explicit examples - e.g., the Kerr space-time and the asymptotically flat Weyl space-times - admit asymptotes. One can also obtain, as in the null case, general criteria on $\tilde{T}$, $\tilde{q}_{ab}$, $\tilde{p}_{ab}$ for the existence of an asymptote. Such criteria and such examples do lend some support to the thesis that space-times which "look asymptotically flat" (in terms of some chart in which they are presented) do indeed admit a spacelike submanifold such that the resulting $\tilde{T}$, $\tilde{q}_{ab}$, $\tilde{p}_{ab}$ has an asymptote. As in the null case, we seem to be far from obtaining a practical algorithm for deciding whether or not a space-time, generally presented, admits a spacelike submanifold with asymptote.

In the null case, there was available a rather strong "persistence criterion" on the definition. One asks whether a perturbation, generated by sources restricted to a compact region of a space-time admitting an asymptote, destroys the existence of the asymptote. In the null case, the answer was that such perturbations will on the whole not destroy the existence of the asymptote, a result which was taken as evidence in favor of the definition. One would like a similar criterion in the spatial case. Unfortunately, such criteria, at least in this form, do not seem to be available in the spatial case, for, since $\tilde{T}$ is spacelike and since the equations governing the perturbations will, presumably, be hyperbolic, the effects of the perturbation would not be expected to propagate at all to the asymptotic region of $\tilde{T}$. This, of course, is just an expression of the idea that dynamical effects should not propagate to spatial infinity. However, one could imagine introducing rather weaker versions of these criteria, based on the constraint rather than the evolution equations. As an example, let us consider the electric field $\tilde{E}^a$ in Euclidean space, so the constraint equation is $\tilde{D}_a \tilde{E}^a = \rho$, where ρ is the charge density. We regard this ρ as the source freely specifiable by the individual responsible for the perturbation. For this example, let us suppose that we wished to test the asymptotic condition "the Euclidean components of the electric field fall to zero as $1/r^2$". Thus, our criterion would ask: Is it true that, given ρ of compact support in Euclidean space, there exists a vector field $\tilde{E}^a$ which satisfies $\tilde{D}_a \tilde{E}^a = \rho$, and whose Euclidean components fall to zero as $1/r^2$? The answer in this example is of course yes: Let $\tilde{\varphi}$ satisfy $\tilde{D}^2 \tilde{\varphi} = \rho$, with $\tilde{\varphi}$ vanishing asymptotically, and set $\tilde{E}^a = \tilde{D}^a \tilde{\varphi}$. In a similar way one could ask, for the various fields above, whether the introduction of additional, small source-terms in the constraint equations will destroy the asymptotic conditions on the fields. Unfortunately, criteria along these lines do not seem to be very telling in the spatial case, for the constraint equations are normally "too weak". Let us return to our example of the electric field. It turns out that one can impose much stronger asymptotic conditions on the electric field - conditions clearly unreasonably strong - and still obtain a positive answer for the criterion. Consider, for example, the asymptotic condition: "The vector field $\tilde{E}^a$ is strictly a Coulomb field outside some compact set". This asymptotic condition is certainly much too strong, and yet we have: Given ρ of compact support, there exists a vector field $\tilde{E}^a$ which satisfies $\tilde{D}_a \tilde{E}^a = \rho$, and which is strictly a Coulomb field outside a compact set. Indeed, let $\tilde{E}_o{}^a$ be the field found above, let $\tilde{E}_c{}^a$ be a Coulomb field manifesting the same total charge as $\tilde{E}_o{}^a$, and consider antisymmetric tensor fields $\tilde{A}^{ab}$ defined on our Euclidean space and satisfying $\tilde{D}_b \tilde{A}^{ab} = \tilde{E}_o{}^a - \tilde{E}_c{}^a$ outside a compact set large enough to include the sources and the origin. The integrability condition for such an $\tilde{A}^{ab}$ is precisely the condition that our two electric fields manifest the same total charge: So, we may find such an $\tilde{A}^{ab}$. But now $\tilde{E}_o{}^a - \tilde{D}_b \tilde{A}^{ab}$ is the desired electric field.

Thus, the strong persistence criteria of the null case do not seem to be available in the spatial. Fortunately, there arises in the spatial case a new class of closely related criteria. The idea is that one would not perhaps wish to regard a space-time as "asymptotically flat at spatial infinity" if it admits, say, just one submanifold such that $\tilde{T}$, $\tilde{q}_{ab}$, $\tilde{p}_{ab}$ has an asymptote. Rather, one would think of demanding that one's space-time admit "about as many" such submanifolds as does Minkowski space-time. However, the $\tilde{q}_{ab}$, $\tilde{p}_{ab}$ on a given $\tilde{T}$ are related to those fields on a nearby submanifold by the evolution equations, (105) and (106). Thus, the admission of a "sufficient number" of submanifolds with asymptotes is related back to the $\tilde{q}_{ab}$ and $\tilde{p}_{ab}$ on a given $\tilde{T}$. Our asymptotic conditions should be in some way related to this relationship. Specifically, one might ask: Is the existence of an asymptote preserved under (105) and (106)? With the question stated in this generality, the answer is no, for we have imposed no conditions on the scalar field $\tilde{\lambda}$ in (105) and (106). What we must do, then, is find conditions on $\tilde{\lambda}$ which reflect "nearby spacelike 3-planes in Minkowski space-time", impose those conditions, and ask our question again.

Let $\tilde{T}$ be a spacelike 3-plane in Minkowski space-time, so $\tilde{T} = R^3$, the corresponding $\tilde{q}_{ab}$ is flat, and $\tilde{p}_{ab} = 0$. Let the corresponding family $\tilde{T}_t$ of submanifolds of which this $\tilde{T}$ is one all be 3-planes. We wish to determine what this says about $\tilde{\lambda}$. To this end, consider (106). We must have $\tilde{p}_{ab} = 0$, while the last three terms on the right in this equation now vanish. Thus, we require that $\tilde{D}_a \tilde{D}_b \tilde{\lambda} = 0$, i.e., that $\tilde{\lambda} = \tilde{c} + \tilde{s}_a \tilde{x}^a$, where $\tilde{c}$ is a constant scalar field, $\tilde{s}_a$ is a constant vector field, and $\tilde{x}^a$ is the position vector field relative to some origin, i.e., a solution of $\tilde{D}_a \tilde{x}^b = \delta_a{}^b$. Geometrically, the constant generates infinitesimal time-translations in Minkowski space-time orthogonal to $\tilde{T}$, while the term $\tilde{s}_a \tilde{x}^a$ generates infinitesimal boosts about the origin in $\tilde{T}$. [Infinitesimal rotations within $\tilde{T}$ about the origin are not reflected in terms of $\tilde{\lambda}$, since they send $\tilde{T}$ to the same submanifold.] Next, we apply a conformal transformation as in the definition of an asymptote (i.e., that following Eqn. (125), with $m = 0$), and assign $\tilde{\lambda}$ dimension $+ 1/2$. The result is that $\Omega^{1/2}\lambda$ is C^1 at S, and vanishes there. [The term "$\tilde{s}_a \tilde{x}^a$" in $\tilde{\lambda}$ determines the derivative of $\Omega^{1/2}\lambda$ at S; the constant term in $\tilde{\lambda}$ yields a higher-order term in $\Omega^{1/2}\lambda$.] Thus, to first non-trivial order, this is the asymptotic behavior of λ which is appropriate to passage from a 3-plane in Minkowski space-time to a nearby one.

We now return to the general case: Let $\tilde{T}$, $\tilde{q}_{ab}$, $\tilde{p}_{ab}$ admit an asymptote. We consider (105) and (106), imposing on $\tilde{\lambda}$ the conditions we found above: that $\Omega^{1/2}\lambda$ be C^1 and vanishing at S. Imposing our conformal transformation on (105), ignoring the term in $\dot{\Omega}$, to which we shall return later, we obtain: $\dot{q}_{ab} = 2\,\Omega^{1/2}\,\lambda\, p_{ab}$. Thus, the behavior of q_{ab} under evolution depends on p_{ab}. It would be unreasonable, we suggest, to impose asymptotic conditions on q_{ab} and p_{ab} if those conditions were destroyed by this evolution, with $\Omega^{1/2}\lambda$ C^1 and vanishing at S. Thus, if we required that p_{ab} be bounded at S, it would be reasonable to require that q_{ab} be C^0 (since, for p_{ab} bounded, $\Omega^{1/2}\lambda\, p_{ab}$ is C^0); if we required that p_{ab} be C^1, it would be reasonable to require that q_{ab} be C^1; and (analyzing in more detail the differentiability class of λ) so on. What these remarks suggest, then, is a relationship between the asymptotic conditions on q_{ab} and p_{ab}. To obtain the actual asymptotic conditions, we turn to (106). Applying our conformal transformation to (106), ignoring all terms on the right but the second (the dominant one for these considerations), we obtain $\dot{p}_{ab} = -\lambda\, E_{ab}$. Thus, we now obtain a relationship between the asymptotic conditions on p_{ab} and E_{ab}. The field λ is only bounded at S; it is not in general continuous there. Thus, if we assume that E_{ab} is either bounded or continuous at S, the most we could reasonably assume on p_{ab} is that it be bounded. A stronger assumption, such as the vanishing of E_{ab} at S, would be needed to make reasonable stronger conditions on p_{ab}. Thus, the whole matter turns on what can reasonably be assumed about the behavior of E_{ab} near S. But in the Schwarzschild space-time, as we shall see later, E_{ab} does not vanish at S, and, indeed, $\lambda\, E_{ab}$ is in general at best only bounded near S. Thus, we are led to the conditions of the definition: q_{ab} C^0 and p_{ab} bounded. The second condition of the definition is also "preserved under this evolution", e.g., on setting $\dot{\Omega} = 0$. We remark that this discussion is merely a rather intuitive justification for the definition. For a proper treatment, one would perhaps wish to prove that, under reasonable asymptotic conditions on E_{ab} and the sources $\tilde{L}_{ab}$, and possibly under a slightly stronger differentiability requirement on λ, evolution of a $\tilde{T}$, $\tilde{q}_{ab}$, $\tilde{p}_{ab}$ admitting an asymptote yields another admitting an asymptote. Such a result would presumably require a delicate treatment of the partial differential equations (105) and (106). It might be of interest to prove a theorem along these lines.

In any case, the evidence in favor of the present definition consists of some examples, some general criteria, the remarks above, and the fact that the definition leads to quantities of apparent physical interest. It is my opinion that this evidence on the whole is satisfactory, although is perhaps marginally weaker than that in the null case.

We turn next, as we did in the null case, to the question of the uniqueness of an asymptote, given $\tilde{T}$, $\tilde{q}_{ab}$, $\tilde{p}_{ab}$. Again, there are at least two senses in which the asymptote need not be unique. On the one hand, one might have two asymptotes for $\tilde{T}$, $\tilde{q}_{ab}$, $\tilde{p}_{ab}$ whose "points at spatial infinity" represent two entirely different asymptotic regimes. For $\tilde{T}$ the submanifold $t = 0$ in the extended Schwarzschild space-time, for example, there are two asymptotes, corresponding to the asymptotically flat regimes on either side of the wormhole. This ambiguity is analogous to the possibility of extension in the null case, except that it is far simpler here. Indeed, we can deal with this ambiguity as follows. Let T, q_{ab} and T', q'_{ab} be two asymptotes for $\tilde{T}$, $\tilde{q}_{ab}$, $\tilde{p}_{ab}$. We may call these two <u>compatible</u> if the mapping from T to T' which is the identity on $\tilde{T}$ (regarded as a subset of T and of T') and which sends S to S' is, together with its inverse, continuous. Compatibility, then, corresponds to addition of a point at spatial infinity "in the same asymptotic regime". The second type of ambiguity is analogous to equivalence in the null case. Let T, q_{ab}, Ω be an asymptote for $\tilde{T}$, $\tilde{q}_{ab}$, $\tilde{p}_{ab}$, let ω be a positive C^2 function on T with $\omega = 1$ at S. Then, we claim, T, $\omega^2 q_{ab}$, $\omega\Omega$ is also an asymptote. [The only condition in the definition which needs checking is the second, and that is almost immediate.] Two asymptotes so related will be said to be <u>equivalent</u>. Thus, equivalent asymptotes are necessarily compatible. The statement that these are essentially the only ambiguities is:

<u>Conjecture</u> 13. Compatible asymptotes are equivalent.

Although the statement is perhaps somewhat simpler in the spatial case, the proof seems to be more difficult. The difference is that whereas in the null case one can essentially construct the boundary from the physical space-time using (conformally invariant) null geodesics, there are no analogous conformally invariant objects in the spatial case. Conjecture 13 is very likely true; one would like a proof.

In the spatial case, one will deal with the ambiguities by regarding equivalence as representing gauge transformations. It will turn out however that, whereas the gauge transformations in the null case come into play at every step, the corresponding transformations in the spatial case have no effect on quantities of interest.

3. The Geometry of Infinity

This section contains the material for the spatial case analogous to that in Sects. II.3, II.4, and II.5 for the null case. Essentially everything will be much simpler here.

Let T, q_{ab}, Ω be an asymptote for $\tilde{T}$, $\tilde{q}_{ab}$, $\tilde{p}_{ab}$. The discussion of the previous section suggests (and later examples will confirm) that we must consider fields which are bounded in a neighborhood of S, but which are not even continuous at S. We require some mechanism for dealing with such fields. To this end, let us denote by V_S the tangent space at S, so V_S has the structure of a three-dimensional vector space. Evaluating the metric q_{ab} on T at S, we obtain a positive-definite metric, which we denote $\underline{q}_{ab}$, over V_S. Let us further denote by U_S the subset of V_S consisting of q-unit tangent vectors. Thus, U_S is the unit two-sphere, centered at the origin, in V_S. By a <u>direction-dependent tensor</u> (at S) we mean a mapping from U_S to the vector space of tensors of a given index structure over V_S. That is, for $\underline{\alpha}^{a..c}{}_{b..d}$ a direction-dependent tensor, we have, for each unit tangent vector $\underline{n}^a$ at S, a tensor $\underline{\alpha}^{a..c}{}_{b..d}(\underline{n})$, with this index structure, at S. Two examples: the mapping which assigns, to each $\underline{n}^a$ in U_S, that same $\underline{n}^a$ regarded as an element of V_S (identity mapping); the mapping which assigns, to each $\underline{n}^a$, the metric $\underline{q}_{ab}$ (constant mapping). Clearly, the algebraic operations on tensors over V_S - addition, outer product, contraction, and raising and lowering of indices with $\underline{q}_{ab}$ and its inverse $\underline{q}^{ab}$ - extend immediately to similar algebraic operations on direction-dependent tensors. A direction-dependent tensor $\underline{\alpha}^{a..c}{}_{b..d}$ will be said to be <u>tangent</u> to U_S if, for each $\underline{n}^a$ in U_S, we have $\underline{\alpha}^{a..c}{}_{b..d}(\underline{n})\,\underline{n}_a = 0, \ldots, \underline{\alpha}^{a..c}{}_{b..d}(\underline{n})\,\underline{n}^d = 0$, i.e., if, for each $\underline{n}^a$, all the indices of $\underline{\alpha}^{a..c}{}_{b..d}(\underline{n})$ are tangent to U_S at the point $\underline{n}^a$ of U_S. Since V_S is a vector space and since U_S can be regarded as a two-dimensional submanifold, vectors tangent to U_S at a point of U_S can be identified with elements of V_S. Thus, we have a natural correspondence between direction-dependent tensors tangent to U_S and tensor fields in U_S regarded as a two-dimensional manifold. For example, $\underline{q}_{ab} - \underline{n}_a\,\underline{n}_b$ (more precisely, the mapping with sends $\underline{n}^a$ in U_S to this tensor over V_S) is a direction-dependent tensor tangent to U_S: The corresponding tensor field in U_S is just the induced metric on this two-dimensional manifold. The direction-dependent tensor $\delta_a{}^b - \underline{n}_a\,\underline{n}^b$ is tangent to U_S: It corresponds to the unit tensor field in U_S. This latter is also the projection operator tangent to U_S. For example, for $\underline{\alpha}_b$ any direction-dependent tensor, $\underline{\alpha}_b\,(\delta_a{}^b - \underline{n}_a\,\underline{n}^b)$ is tangent to U_S. Any direction-dependent tensor can be expressed in terms of tensors tangent to U_S. The various algebraic operations, applied to direction-dependent tensors tangent to U_S, again yield tensors tangent to U_S.

If the mapping which defines a direction-dependent tensor is C^1, then we may take its derivative. The result, evaluated at $\underline{n}^a$, of taking the derivative of direction-dependent tensor $\underline{\alpha}^{\cdot\cdot}{}_{\cdot\cdot}$ will be written $\partial_a\,\underline{\alpha}^{\cdot\cdot}{}_{\cdot\cdot}(\underline{n})$. [What this means, more precisely, is the following. The derivative of the mapping $\underline{\alpha}^{\cdot\cdot}{}_{\cdot\cdot}$, evaluated at $\underline{n}^a$, is a linear mapping from tangent vectors to U_S at $\underline{n}^a$ to tensors over V_S with index structure "$^{\cdot\cdot}{}_{\cdot\cdot}$". Let w^a be such a tangent vector, so we may regard w^a as an element of V_S satisfying $w^a\ \underline{n}_a = 0$. Then the action of this linear mapping on w^a is to be the tensor $w^a\ \partial_a\ \underline{\alpha}^{\cdot\cdot}{}_{\cdot\cdot}(\underline{n})$ over V_S. This defines $w^a\ \partial_a\ \underline{\alpha}^{\cdot\cdot}{}_{\cdot\cdot}(\underline{n})$ only for $w^a\ \underline{n}_a = 0$, and hence defines the tensor $\partial_a\ \underline{\alpha}^{\cdot\cdot}{}_{\cdot\cdot}(\underline{n})$ only up to a multiple of $\underline{n}_a$. We complete the definition by requiring that $\underline{n}^a\ \partial_a\ \underline{\alpha}^{\cdot\cdot}{}_{\cdot\cdot}(\underline{n}) = 0$.] In other words, "$\partial_a = d/d\ \underline{n}^a$", in the sense that, for w^a orthogonal to $\underline{n}^a$ (so that $\underline{n}^a + \epsilon\, w^a$ is unit to first order in ϵ), we have $\underline{\alpha}^{\cdot\cdot}{}_{\cdot\cdot}(\underline{n} + \epsilon\, w) - \underline{\alpha}^{\cdot\cdot}{}_{\cdot\cdot}(\underline{n}) = \epsilon\, w^a\, \partial_a \underline{\alpha}^{\cdot\cdot}{}_{\cdot\cdot}(\underline{n})$, to first order in ϵ. In any case, the derivative of a direction-dependent tensor is again a direction-dependent tensor, with one more index. For example, we have $\partial_a\ \underline{g}_{bc} = 0$, and $\partial_a\ \underline{n}_b = \underline{g}_{ab} - \underline{n}_a\ \underline{n}_b$. Our derivative ∂_a interacts with the algebraic operations as one would expect of a derivative: The derivative of the sum of two direction-dependent tensors is the sum of the derivatives; the derivative of the outer product satisfies the Leibnitz rule; the derivative commutes with contraction and raising and lowering of indices. Thus, for example, $\partial_a\ (\underline{g}^{bc}\ \underline{n}_b\ \underline{n}_c) = (\partial_a\ \underline{g}^{bc})\ \underline{n}_b\ \underline{n}_c + \underline{g}^{bc}\ (\partial_a\ \underline{n}_b)\ \underline{n}_c + \underline{g}^{bc}\ \underline{n}_b\ (\partial_a\ \underline{n}_c) = 0$, as one would expect, since $\underline{g}^{bc}\ \underline{n}_b\ \underline{n}_c = 1$. It is false in general that the derivative of a direction-dependent tensor tangent to U_S is again tangent to U_S. For example, for $\underline{\alpha}_b$ tangent to U_S, i.e., for $\underline{\alpha}_b(\underline{n})\ \underline{n}^b = 0$ for each $\underline{n}^b$, we have $\underline{n}^a\ \partial_a\ \underline{\alpha}_b(\underline{n}) = 0$; however $\underline{n}^b\ \partial_a\ \underline{\alpha}_b(\underline{n}) = \partial_a(\underline{n}^b\ \underline{\alpha}_b(\underline{n})) - \underline{\alpha}_b(\underline{n})\ \partial_a\ \underline{n}^b = -\underline{\alpha}_a(\underline{n}) \neq 0$ in general. We may, however, always obtain a tensor tangent to U_S by projecting the derivative: In this example, we could consider $(\delta_a{}^u - \underline{n}_a\ \underline{n}^u)\ (\delta_b{}^v - \underline{n}_b\ \underline{n}^v)\ \partial_u\ \underline{\alpha}_v(\underline{n})$. The result is a derivative on direction-dependent tensors tangent to U_S which again yields tensors tangent to U_S. Such tensors, however, correspond precisely to tensor fields on the manifold U_S. Thus, the result is a derivative operator on the two-dimensional manifold U_S. Expanding, we have $(\delta_a{}^u - \underline{n}_a\ \underline{n}^u)\ (\delta_b{}^v - \underline{n}_b\ \underline{n}^v)\ (\delta_c{}^w - \underline{n}_c\ \underline{n}^w)\ \partial_u\ (\underline{g}_{vw} - \underline{n}_v\ \underline{n}_w) = 0$. In other words, we have that this derivative operator, applied to the metric of U_S, $\underline{g}_{bc} - \underline{n}_b\ \underline{n}_c$, yields zero. That is to say, our derivative operator on the manifold U_S is precisely that associated with the metric $\underline{g}_{bc} - \underline{n}_b\ \underline{n}_c$.

To summarize, we have the notion of a direction-dependent tensor at S, algebraic operations on such tensors, and the ability to take derivatives of direction-dependent tensors. Every direction-dependent tensor can be expressed in terms of direction-dependent tensors tangent to U_S, i.e., in terms of tensor fields on the two-dimensional manifold U_S. The algebraic operations then reduce to the usual algebraic operations on tensor fields on this manifold. Furthermore, the derivative reduces essentially to the natural derivative operator on this two-sphere. In short, all this is just a roundabout way of describing the differential geometry of the two-sphere U_S. It is these direction-dependent tensors (equivalently, tensor fields on U_S) which will characterize the asymptotic behavior at spatial infinity of various fields on our original space-time.

What we now require, therefore, is some mechanism by which tensor fields on T, possibly badly-behaved at S, can give rise to direction-dependent tensors at S.

Let $\alpha^{\cdot\cdot}{}_{\cdot\cdot}$ be a tensor field on T - S, let γ be a curve in T with endpoint S, and let $\underline{n}^a$ be the unit tangent vector to γ at S. If the result of evaluating $\alpha^{\cdot\cdot}{}_{\cdot\cdot}$ at each point of γ, and then taking the limit along γ at S, exists, it will be a tensor at S. If this limit exists for every γ and $\underline{n}^a$ at S, and if this limit at S depends only on $\underline{n}^a$, being otherwise independent of γ, the resulting limit, depending only on $\underline{n}^a$, will of course be a direction-dependent tensor at S. Under these circumstances, we say that "lim $\alpha^{\cdot\cdot}{}_{\cdot\cdot}$" exists, and write $\lim \alpha^{\cdot\cdot}{}_{\cdot\cdot} = \underline{\alpha}^{\cdot\cdot}{}_{\cdot\cdot}(\underline{n})$, where the right side is the resulting direction-dependent tensor. The tensor field $\alpha^{\cdot\cdot}{}_{\cdot\cdot}$ will be said to be $C_S{}^0$ if $\lim \alpha^{\cdot\cdot}{}_{\cdot\cdot}$ exists, and is a C^0 direction-dependent tensor at S. For example, q_{ab} on T is $C_S{}^0$, and $\lim q_{ab} = \underline{q}_{ab}$; $n_a = 1/2\ \Omega^{-1/2} D_a\Omega$ is $C_S{}^0$, and $\lim n_a = \underline{n}_a$. As the second example shows, $C_S{}^0$ fields need not be continuous at S. Thus, the operation "lim" takes tensor fields on T - S to direction-dependent tensors. The algebraic operations on tensor fields of course commute with this operation. We require, finally, the interaction of "lim" with derivatives. A $C_S{}^0$ field $\alpha^{\cdot\cdot}{}_{\cdot\cdot}$ will be said to be $C_S{}^1$ if both sides of $\lim \Omega^{1/2} D_a \alpha^{\cdot\cdot}{}_{\cdot\cdot} = \partial_a (\lim \alpha^{\cdot\cdot}{}_{\cdot\cdot})$ exist and are continuous, and if this equation holds. Note the factor "$\Omega^{1/2}$" in this definition: It is essentially a conversion factor from rectangular to angular variables. To see its role, we consider an example: Let T, q_{ab} be flat in a neighborhood of S, and let $\Omega = x^a x_a$, where x^a is a position vector field with respect to the origin S. Let s_a be a constant vector field in this neighborhood, and set $\alpha = \Omega^{-1/2} s_a x^a$. Then α is $C_S{}^0$, and $\lim \alpha = \underline{\alpha}(\underline{n}) = s_a \underline{n}^a$. We now compute $\partial_a \underline{\alpha}(\underline{n}) = s_a - \underline{n}_a \underline{\alpha}$, and $D_a \alpha = \Omega^{-1/2} s_a - \Omega^{-1} x_a \alpha$. Thus, in this example, we have $\lim \Omega^{1/2} D_a \alpha = \partial_a \underline{\alpha}(\underline{n})$, i.e., we have that α is $C_S{}^1$. Returning now to the general case, we show, as a second example, that n_a is $C_S{}^1$. We have $D_a n_b = D_a (1/2\ \Omega^{-1/2} D_b\Omega) = 1/2\ \Omega^{-1/2} D_a D_b \Omega - \Omega^{-1/2} n_a n_b$. From the second condition in the definition of an asymptote, we have that $\lim \Omega^{1/2} D_a n_b = \underline{q}_{ab} - \underline{n}_a \underline{n}_b$. But $\lim n_a = \underline{n}_a$, and $\partial_a \underline{n}_b = \underline{q}_{ab} - \underline{n}_a \underline{n}_b$. Thus, n_a is $C_S{}^1$. In a similar way, one defines higher order "direction-dependent differentiability" at S, $C_S{}^2$, $C_S{}^3$, ...

The idea, then, is this. Tensor fields on the original space-time become tensor fields on $\tilde{T}$, then tensor fields on T - S (via Sect. III.1), then direction-dependent tensors at S (via "lim"), and finally tensor fields on the two-sphere U_S. This sequence of transitions is analogous to, but somewhat longer than, the corresponding sequence in the null case: from $\tilde{M}$ to M to $\mathcal{I}$. The crucial step is the passage to the asymptotic limit at spatial infinity: the application of "lim". Indeed, this operation is analogous to the application of $\mathcal{I}^*$ in the null case. As in the null case, we shall wish to apply "lim" to every field and every equation available.

We must, as it turns out, introduce one further step in the sequence of transitions above. This step arises from the fact that, given a space-time $\tilde{M}$, $\tilde{g}_{ab}$, there may be many spacelike submanifolds $\tilde{T}$ such that $\tilde{T}$, $\tilde{q}_{ab}$, $\tilde{p}_{ab}$ admits an asymptote. Our present treatment of asymptotic fields, however, is in terms of a single $\tilde{T}$ and its asymptote, and so this treatment would be expected to yield information about both the physical fields themselves and the particular choice of $\tilde{T}$. Since only information of the first type is interesting, one would like to disentangle the two types of information. A formal way of doing this would be to consider the collection of all $\tilde{T}$ in the physical space-time for which $\tilde{T}$, $\tilde{q}_{ab}$, $\tilde{p}_{ab}$ admits an asymptote, obtain for each a point S and a two-sphere U_S, obtain on each U_S the various asymptotic fields, and then simply take the disjoint union of all these two-spheres. This, however, appears to be an unnatural procedure, for the fields on the different two-spheres would not be expected to be independent of each other: Rather, they would be related by the evolution equations. This procedure, then, would result in a highly redundant description of the asymptotic fields. The question arises, then, whether one can obtain a treatment which, on the one hand, is $\tilde{T}$-independent, and, on the other, minimizes redundancy. One might think of somehow "identifying" the U_S's - specifically, embedding them in some larger manifold $\mathcal{S}$. The embeddings would have to be arranged so as to reflect the redundancies in the fields on the U_S. For example, one might demand that it be possible to introduce tensor fields on the larger manifold $\mathcal{S}$ in such a way that the induced fields on each embedded U_S be precisely the asymptotic fields, obtained via the asymptote, on that U_S. Then, this manifold $\mathcal{S}$ with its tensor fields would describe the asymptotic structure in a "$\tilde{T}$-independent way", while the choice of $\tilde{T}$ would be represented by the choice of embedded U_S. This is what we shall do - in fact, with $\mathcal{S}$ a three-dimensional manifold. The key to the construction must be the evolution equations. We first observe that the evolution equations in each case satisfy a rather special property. Let $\tilde{\lambda}_1$ and $\tilde{\lambda}_2$ be two scalar fields on $\tilde{T}$, and let us write "dot$_1$" and "dot$_2$" for the corresponding time-derivatives. Then, we claim, we have in every case the equation dot$_1$ dot$_2$ - dot$_2$ dot$_1$ = $\mathcal{L}_{\tilde{w}}$, where $\tilde{w}^a = \tilde{\lambda}_2 \tilde{D}^a \tilde{\lambda}_1 - \tilde{\lambda}_1 \tilde{D}^a \tilde{\lambda}_2$, a vector field in $\tilde{T}$, and where, in the calculation, we set dot$_1$ $\tilde{\lambda}_2$ = dot$_2$ $\tilde{\lambda}_1$ = 0. Let us consider, as an example, this equation on $\tilde{q}_{ab}$. We have, from (105), dot$_2$ $\tilde{q}_{ab} = 2\, \tilde{\lambda}_2\, \tilde{p}_{ab}$, and so, from (106), dot$_1$ dot$_2$ $\tilde{q}_{ab} = 2\, \tilde{\lambda}_2\, \tilde{D}_a \tilde{D}_b \tilde{\lambda}_1 + \tilde{\lambda}_2\, \tilde{\lambda}_1$ (terms), where "terms" represents the last three terms on the right in (106). Subtracting the result of interchanging $\tilde{\lambda}_1$ and $\tilde{\lambda}_2$, we have (dot$_1$ dot$_2$ - dot$_2$ dot$_1$) $\tilde{q}_{ab} = 2\, \tilde{\lambda}_2\, \tilde{D}_a \tilde{D}_b \tilde{\lambda}_1 - 2\, \tilde{\lambda}_1\, \tilde{D}_a \tilde{D}_b \tilde{\lambda}_2$. But the right side is just $\mathcal{L}_{\tilde{w}}\, \tilde{q}_{ab}$. Similarly, one checks this equation on (109)-(110), on (111), and on (112)-(113). This property is, of course, no coincidence: Rather, it reflects an elementary fact about Lie derivatives, namely $\mathcal{L}_\xi \mathcal{L}_\eta - \mathcal{L}_\eta \mathcal{L}_\xi = \mathcal{L}_{\mathcal{L}_\xi \eta}$. In Minkowski space-time, our property becomes the statement that the commutator of two infinitesimal boosts is an infinitesimal rotation. This property, then, suggests the construction of $\mathscr{S}$ and the embeddings of the U_S therein.

Let V be a four-dimensional vector space with metric g_{ab} of Lorentz signature. Denote by $\mathscr{S}$ the submanifold of V consisting of all unit spacelike vectors, so $\mathscr{S}$ is a hyperboloid in V. [See Fig. 5.] A typical point in $\mathscr{S}$ is represented by $\underline{n}^a$, a unit spacelike vector in V. The induced metric of $\mathscr{S}$ is that given, at point $\underline{n}^a$ of $\mathscr{S}$, by $\underline{g}_{ab} = g_{ab} - \underline{n}_a \underline{n}_b$. This metric has signature $(-, +, +)$, and has constant positive curvature. A three-dimensional manifold $\mathscr{S}$ with metric, so constructed, will be called an <u>asymptotic geometry</u> (at spatial infinity). By a <u>cross-section</u> of this asymptotic geometry, we mean the intersection of $\mathscr{S}$ with a spacelike 3-plane in V through the origin. Each such cross-section, then, is a metric two-sphere.

Now let $\tilde{T}$ be a spacelike submanifold of our space-time such that $\tilde{T}$, $\tilde{q}_{ab}$, $\tilde{p}_{ab}$ admits an asymptote, let S be the corresponding point at infinity, and let U_S be the sphere of unit tangent vectors at S. Select any cross-section of $\mathscr{S}$, and identify U_S isometrically with this cross-section (each being a metric two-sphere). In this way, we fix the embedding of this particular U_S in $\mathscr{S}$. We must now specify the embeddings of other U_S's. Stated "infinitesimally", the prescription is this. Let λ be a function on T - S which characterizes evolution to a nearby $\tilde{T}'$, $\tilde{q}'_{ab}$, $\tilde{p}'_{ab}$ with nearby asymptote, T', q'_{ab}. Then, as we have seen in the previous section, it is appropriate to demand that $\Omega^{1/2}\lambda$ vanish at S and be C^1 there. Ex-

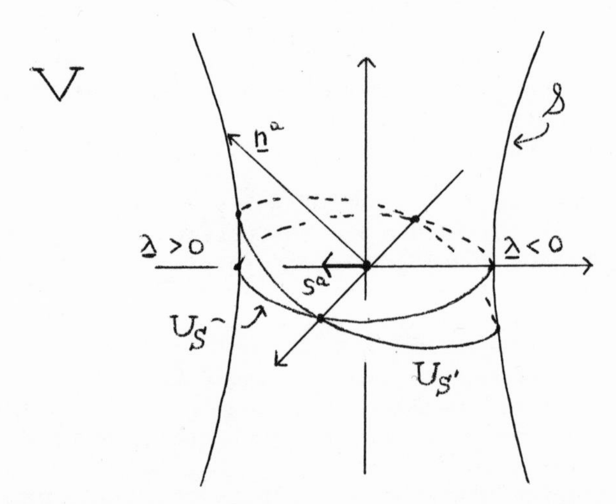

Fig. 5. The asymptotic geometry at spatial infinity. The two cross-sections, U_S and U_S', are related to each other by evolution using the function $\underline{\lambda} = s_a \underline{n}^a$ on U_S.

pressing this asymptotic behavior in the present terminology, it becomes the requirement that lim λ exist (as a direction-dependent scalar, i.e., as a scalar field on the manifold U_S) and be a "dipole function", i.e., be of the form $\underset{\sim}{\lambda} = s_a \underset{\sim}{n}^a$ for some s^a in V_S. We now consider a new, nearby, cross-section in $\mathcal{S}$, obtained by moving, at each point $\underset{\sim}{n}^a$ of U_S, a distance $\underset{\sim}{\lambda}(\underset{\sim}{n})$ along the normal to U_S at that point. [Fig. 5.] That the resulting submanifold of $\mathcal{S}$ is again a cross-section is guaranteed by the dipole behavior of $\underset{\sim}{\lambda}$. This cross-section we identify with $U_{S'}$. Thus, $\underset{\sim}{\lambda} = \lim \lambda$ plays the role of the evolution function from cross-section to cross-section in $\mathcal{S}$. One now iterates these infinitesimal evolutions. More precisely, let $\tilde{T}_t$ be a family of space-like submanifolds in $\tilde{M}$, each of which admits an asymptote, which is such that $\tilde{T}_0$ is our original $\tilde{T}$, and which is such that, for each t, the corresponding function λ (possibly depending on t) satisfies our asymptotic condition: $\lim \lambda = \underset{\sim}{\lambda}(\underset{\sim}{n}) = s_a \underset{\sim}{n}^a$, with s_a also possibly depending on t. Associate with this family, as above, a corresponding family of cross-sections, C_t, with C_0 our original cross-section. In this way, we associate, with each of the $\tilde{T}_t$, a cross-section. The "commutation of evolutions" in the physical space-time is now reflected by a similar formula in $\mathcal{S}$. Just as in $\tilde{M}$, one can associate operators, dot_1 and dot_2, in U_S with dipole functions, $\underset{\sim}{\lambda}_1$ and $\underset{\sim}{\lambda}_2$, on U_S; just as before, we shall have $(\text{dot}_1\, \text{dot}_2 - \text{dot}_2\, \text{dot}_1) = \mathcal{L}_w$, where w^a is the vector field on U_S given by $w^a = \underset{\sim}{g}^{ab} (\underset{\sim}{\lambda}_2\, \partial_b\, \underset{\sim}{\lambda}_1 - \underset{\sim}{\lambda}_1\, \partial_b\, \underset{\sim}{\lambda}_2)$. This observation guarantees that our choice of embeddings is consistent, e.g., in the sense that two families $\tilde{T}_t$ of submanifolds leading from $\tilde{T}$ to the same final submanifold $\tilde{T}'$ will produce families of cross-sections leading to the same final cross-section. Consider, as a special case, $\tilde{\lambda} = \text{const}$. Then $\lambda = \Omega^{1/2}$ (const), and so $\underset{\sim}{\lambda}(\underset{\sim}{n}) = \lim \lambda = 0$. Thus, in this example, our two nearby submanifolds, which are asymptotically "parallel", give rise to the same cross-section. It is only the "asymptotic tipping" of one submanifold relative to another - the "asymptotic boost" which takes one to another - which determines the relationship between the corresponding cross-sections.

In any case, it is by the prescription above that we introduce $\mathcal{S}$ and the embeddings of the U_S's in $\mathcal{S}$. It is unfortunate that the construction is not very precise. This, at least, is not a problem from the point of view of the formal mathematical structure: One can work with a single $\tilde{T}$, a single asymptote, a single U_S, and simply embed this U_S as a cross-section of $\mathcal{S}$ as above. The fields on that U_S will then give rise to fields on $\mathcal{S}$. It is, however, a problem in that it weakens the motivation. We shall remark, in Sect. IV, on some possibilities for carrying out the construction properly.

The three-dimensional manifold $\mathcal{S}$ is analogous to the three-dimensional manifold $\mathcal{I}$ in the null case. The universal field on $\mathcal{S}$ - analogous to g_{ab} and n^a on $\mathcal{I}$ - is $\underset{\sim}{g}_{ab}$, the induced metric on $\mathcal{S}$. [To push the analogy, one could also regard $\underset{\sim}{n}^a$, the "position vector" for $\mathcal{S}$, as a second universal field.] Just as one thinks of $\mathcal{I}$ as "a null surface off at null infinity", so one may regard $\mathcal{S}$ as "a timelike surface off at spatial infinity". In the null case the differential geometry of $\mathcal{I}$, $\Gamma^{ab}{}_{cd}$ was a rather complicated business. In the spatial case, by contrast, one has a three-dimensional manifold with invertible metric. Thus, one has a unique derivative operator, one can raise and lower indices, etc. - one has the standard differential geometry of a pseudo-Riemannian space.

What happened to the gauge-freedom of the null case? Let $\tilde{T}$, $\tilde{q}_{ab}$, $\tilde{p}_{ab}$ have asymptote T, q_{ab}, Ω, and let T, $\omega^2 q_{ab}$, $\omega\Omega$ be another asymptote, where ω is C^2 at S. Then $D_a D_b (\omega\Omega) = \omega D_a D_b \Omega + 2 D_{(a}\omega D_{b)}\Omega + \Omega D_a D_b \omega$. By the second condition for an asymptote, we must have $D_a D_b (\omega\Omega) = 2\omega^2 q_{ab}$ at S. Substituting, we obtain $\omega q_{ab} = \omega^2 q_{ab}$ at S, or $\omega = 1$ at S. Thus, there is no gauge-freedom in the spatial case. In the null case, the gauge freedom was that of conformal transformations on the asymptotic geometry. "If there were gauge-freedom" in the spatial case, it would also be that of conformal transformations. However, since the metric of $\mathcal{S}$ is invertible, one can uniquely scale that metric by requiring that $\mathcal{S}$ be the unit hyperboloid. This unique scaling was essentially introduced by the equation $D_a D_b \Omega = 2 q_{ab}$ of the second condition.

In contrast to the null case, the global structure of the asymptotic geometry in the spatial case is quite simple: $\mathcal{S}$ is the hyperboloid, with its induced metric, of unit spacelike vectors in a four-dimensional vector space with Lorentz metric. In particular, $\mathcal{S}$ has the topology of $S^2 \times R$.

Let $\mathcal{S}$, $\underline{g}_{ab}$ be an asymptotic geometry, embedded as a hyperboloid in V, g_{ab}. By a <u>symmetry</u> of $\mathcal{S}$, $\underline{g}_{ab}$, we mean a diffeomorphism from $\mathcal{S}$ to $\mathcal{S}$ which sends $\underline{g}_{ab}$ to itself. By an <u>infinitesimal symmetry</u>, we mean a Killing field on $\mathcal{S}$, $\underline{g}_{ab}$. The former form a group; the latter a Lie algebra. Let ψ be any metric-preserving isomorphism from the vector space V to itself. Then this ψ takes unit spacelike vectors in V to unit spacelike vectors, and so defines a diffeomorphism on $\mathcal{S}$. Since this ψ leaves invariant the metric g_{ab} of V, it also leaves invariant the metric $\underline{g}_{ab}$ of $\mathcal{S}$. Thus, we obtain a symmetry. Clearly, every symmetry is of this form. Thus, the group of symmetries on $\mathcal{S}$, $\underline{g}_{ab}$ is precisely the group of all metric-preserving isomorphisms on V, i.e., precisely the Lorentz group (acting on V). The infinitesimal symmetries are similarly described in terms of V. Let F_{ab} be an antisymmetric tensor over the vector space V. Then, for each unit spacelike $\underline{n}^a$ in V, the vector $F^a{}_b \underline{n}^b$ in V is orthogonal to $\underline{n}^a$, by antisymmetry, and so can be regarded as a tangent vector to $\mathcal{S}$ at the point $\underline{n}^a$ of $\mathcal{S}$. Thus, we obtain a vector field on the three-dimensional manifold $\mathcal{S}$. This field is a Killing field in $\mathcal{S}$, $\underline{g}_{ab}$ (indeed, its derivative, evaluated at a point of $\mathcal{S}$, is just F_{ab}, projected into $\mathcal{S}$ at that point, whence its symmetrized derivative vanishes), and, since the vector space of these is six-dimensional (the dimension of the vector space of antisymmetric tensors over V), this is the most general Killing field. Thus, the infinitesimal symmetries are identified with antisymmetric tensors over V ("infinitesimal Lorentz transformations"), and the Lie algebra of infinitesimal symmetries is the Lie algebra of the Lorentz group.

In the null case, the Lie algebra of infinitesimal symmetries contained an infinite-dimensional abelian ideal, the infinitesimal supertranslations. The quotient of the Lie algebra of infinitesimal symmetries by the ideal of infinitesimal supertranslations was isomorphic (in the Minkowskian case) with the Lie algebra of the Lorentz group. In the spatial case, by contrast, the infinitesimal supertranslations are missing entirely, with only the quotient retained as the Lie algebra of infinitesimal symmetries. There are at least two ways of understanding this phenomenon. Mathematically, the difference arises because the metric of $\mathcal{J}$ is not invertible, a feature which would normally be expected to result in an infinite-dimensional Lie algebra of infinitesimal symmetries. The metric on $\mathcal{S}$, however, is invertible: Necessarily, therefore, the Lie algebra of infinitesimal symmetries is finite-dimensional. Physically, the difference is that the ideal of infinitesimal translations in the Poincaré Lie algebra is "enlarged" to the infinitesimal supertranslations in the null case, and "reduced" to nothing in the spatial case. The latter, at least, is perhaps not unexpected. Dynamical effects inside the physical space-time would not be expected to propagate to spatial infinity. "Infinitesimal time-translations generate the dynamics". One might thus expect that infinitesimal time-translations in the physical space-time, and so, by taking linear combinations, all infinitesimal translations, would not have any effect at spatial infinity. In any case, the relationship between the Lie algebra of infinitesimal symmetries and the Poincaré Lie algebra is rather simpler in the spatial than in the null case. As a result, the interpretation of quantities constructed at spatial infinity will also be somewhat simpler.

4. The Physical Fields at Spatial Infinity

We now have the notion of an asymptotic geometry at spatial infinity - the three-dimensional manifold $\mathcal{S}$ with its universal field, the metric $\underline{g}_{ab}$. We further have a mechanism by which physical fields in the original space-time are to yield fields on this asymptotic geometry: Obtain for the fields an initial-value formulation; assign to each field a dimension and perform a conformal transformation; choose a spacelike submanifold $\tilde{T}$ with asymptote, and apply "lim" to obtain direction-dependent tensors at S; resolve these direction-dependent tensors into direction-dependent tensors tangent to the sphere U_S; embed U_S as a cross-section of $\mathcal{S}$, and express the resulting tensor fields on U_S as tensor fields on $\mathcal{S}$. The previous section consisted, for the most part, of application of this general program to the "physical field" $\tilde{g}_{ab}$, resulting in the metric g_{ab} of $\mathcal{S}$. We now wish to apply this program to the other fields - the Klein-Gordon, Maxwell, and gravitational.

Consider first the Klein-Gordon field. In this case we obtain, by Sect. III.1, two fields on $\tilde{T}$, and, assigning dimensions -1/2 and -1, fields φ and ψ on T - S. There are no constraint equations; the evolution equations are (115) and (116). We say that our Klein-Gordon field is asymptotically <u>regular</u> (with respect to this asymptote) if φ and ψ are $C_S{}^n$ for sufficiently large n. As in the null case, one must provide some justification for this definition. The evidence in its favor consists of the same items as those in favor of the definition of an asymptote in the spatial case: It admits "reasonable" examples (e.g., Klein-Gordon fields which "fall to zero asymptotically as 1/r"), and it is preserved by the evolution. Again, the evidence is somewhat weaker than that in the null case, for one has nothing analogous to Theorem 8.

We now take the limit at spatial infinity. Set $\underline{\varphi}$ = lim φ, and $\underline{\psi}$ = lim ψ. Each is a direction-dependent scalar at S, i.e., a scalar field on the two-sphere U_S. There are no constraints connecting these two fields on U_S. We now let $\tilde{T}_t$ be a family of such submanifolds, and let λ be the function, for each t, which describes the rate of change of the submanifold with t. Then the rates of change of φ and ψ are given by (115) and (116). To obtain the rates of change of $\underline{\varphi}$ and $\underline{\psi}$, we apply "lim" to these equations. As we saw in the previous section, lim $\Omega^{-1}\dot{\Omega}$ = 0. Thus, we obtain

$$\dot{\underline{\varphi}} = \underline{\lambda}\,\underline{\psi} \tag{126}$$

from (115). We next apply "lim" to (116). The first term on the right then vanishes as before. The second term yields $\partial^m(\underline{\lambda}\partial_m\underline{\varphi})$. For the third term, we note that $\Omega^{1/2}\,D_m\,n^m = \Omega^{1/2}\,D_m(1/2\,\Omega^{-1/2}\,D^m\Omega) = 1/2\,D_m\,D^m\Omega - n_m\,n^m$. Applying "lim", using the second condition in the definition of an asymptote, this yields 3 - 1 = 2. Thus, the third term on the right in (116) yields $-\underline{\lambda}\,\underline{\varphi}$. The fourth and fifth terms on the right in (116) yield zero. For example, for the fourth, lim $\Omega^{1/2}\,D_m\lambda = \partial_m\underline{\lambda}$, while $\underline{n}^m\,\partial_m\underline{\lambda} = 0$. The sixth term on the right also yields zero, for λ, ψ, and p are bounded near S, while lim $\Omega^{1/2} = 0$. Finally, for the last term in (116), we assume for the present that lim $\Omega^{-1}\,(\tilde{L}_{mn}\,\tilde{g}^{mn}) = 0$. This is essentially an assumption about the asymptotic behavior of the stress-energy. We shall return to this issue when we come to the gravitational case. Thus, "lim" of (116) yields

$$\dot{\underline{\psi}} = \partial^m(\underline{\lambda}\;\partial_m\underline{\varphi}) - \underline{\lambda}\,\underline{\varphi}. \tag{127}$$

Eqns. (126) and (127), then, give the rate of change of the asymptotic fields.

Regard U_S as embedded as a cross-section of our asymptotic geometry $\mathcal{S}$. The above yields two scalar fields, $\underline{\varphi}$ and $\underline{\psi}$, on this U_S. Consider now a nearby cross-section, related to U_S to first order by a scalar field $\underline{\lambda}$ on U_S of the form $\underline{\lambda} = s_a \underline{n}^a$ with s^a in V_S. Then (126) and (127) define $\underline{\varphi}$ and $\underline{\psi}$, to first order, on this nearby cross-section. More generally, integrating (126) and (127), we obtain a pair of scalar fields on every cross-section. In short, we have an algorithm which yields pairs of scalar fields on cross-sections. We wish to characterize this state of affairs in terms of fields on the manifold $\mathcal{S}$. To this end, denote by α that scalar field on $\mathcal{S}$ having the following two properties: i) α is a solution of

$$\underline{D}_m \underline{D}^m \alpha = \alpha \tag{128}$$

on $\mathcal{S}$, where $\underline{D}_a$ denotes the derivative operator on $\mathcal{S}$, $\underline{g}_{ab}$, and where indices are raised and lowered with this metric, and ii) on U_S, the value of α is $\underline{\varphi}$, and the value of the normal derivative of α is $\underline{\psi}$. Since the second property provides precisely the initial data appropriate to (128), these properties define α uniquely. We may now introduce a second algorithm for obtaining pairs of scalar fields on cross-sections: Given any cross-section, evaluate α on that cross-section, and evaluate its normal derivative. We now have two methods which yield, on each cross-section, a pair of scalar fields, one via (126) and (127), and the other via evaluation of α and its normal derivative. How do the two pairs of scalar fields compare with each other? We claim: They are identical. Indeed, Eqns. (126) and (127) are precisely the evolution equations, in $\mathcal{S}$, for (128). [We have essentially already done this calculation. See, e.g., (109) and (110).] Put another way, the values of α and its normal derivative on a cross-section nearby to U_S are determined from these on U_S by the evolution equations for Eqn. (128), i.e., by (126) and (127). But (126) and (127) also determine $\underline{\varphi}$ and $\underline{\psi}$ on this nearby cross-section. Since initially, on U_S, α is the value of $\underline{\varphi}$ and its normal derivative the value of $\underline{\psi}$, this relationship will be maintained to all cross-sections.

We conclude, then, that we do not have to deal with $\underline{\varphi}$, $\underline{\psi}$, or $\underline{\lambda}$, with cross-sections and evolution, or with (126) and (127). Rather, the complete asymptotic description at spatial infinity of our Klein-Gordon field is provided by a scalar field α on $\mathcal{S}$ satisfying (128). From this α, in particular, we can determine $\underline{\varphi}$ and $\underline{\psi}$ for any cross-section.

The procedure in the Maxwell case is similar, but slightly more complicated. Again, for asymptotic regularity, we require that E^a and B^a be $C_S{}^n$. Set $\underline{E}^a = \lim E^a$ and $\underline{B}^a = \lim B^a$. Each is a direction-dependent vector at $\bar{S}$. These direction-dependent vectors, however, are now not arbitrary, for the original E^a and B^a satisfied the constraint equations, (117). Multiplying each equation in (117) by

$\Omega^{1/2}$, and applying "lim", we obtain

$$\partial_a \underline{E}^a = 2\, \underline{E}^a \underline{n}_a, \qquad \partial_a \underline{B}^a = 2\, \underline{B}^a \underline{n}_a. \tag{129}$$

These are the constraints. Applying "lim" to the evolution equations (118) and (119), the first and fourth terms on the right vanish, and we obtain simply

$$\dot{\underline{E}}^a = \underline{\epsilon}^{amn} \partial_m (\underline{\lambda}\, \underline{B}_n) - \underline{\lambda}\, \underline{\epsilon}^{amn} \underline{n}_m \underline{B}_n, \tag{130}$$

$$\dot{\underline{B}}^a = -\underline{\epsilon}^{amn} \partial_m (\underline{\lambda}\, \underline{E}_n) + \underline{\lambda}\, \underline{\epsilon}^{amn} \underline{n}_m \underline{E}_n, \tag{131}$$

where we have set $\underline{\epsilon}^{amn} = \lim \epsilon^{amn}$, the direction (in-)dependent alternating tensor over V_S.

The next step is to resolve $\underline{E}^a$ and $\underline{B}^a$ into tensor fields tangent to U_S. Set $\mu = \underline{E}^a \underline{n}_a$, a direction-dependent scalar at S; we then have $\underline{E}^a = \mu\, \underline{n}^a + \kappa^a$, where the direction-dependent vector κ^a satisfies $\kappa^a \underline{n}_a = 0$. That is, we obtain from $\underline{E}^a$ a scalar field and a vector field on the two-dimensional manifold U_S. We now substitute into the first equation (129). For the left side, we have $\partial_a \underline{E}^a = \partial_a (\mu\, \underline{n}^a + \kappa^a) = 2\mu + \partial_a \kappa^a$. Since the right side of our equation is 2μ, this equation becomes simply: $\partial_a \kappa^a = 0$. Set $\underline{\epsilon}_{ab} = \underline{\epsilon}_{abm} \underline{n}^m$, a direction-dependent tensor tangent to U_S which represents the alternating tensor of this manifold. Then the equation $\partial_a \kappa^a = 0$ can be rewritten $(\delta_a{}^m - \underline{n}_a \underline{n}^m)(\delta_b{}^n - \underline{n}_b \underline{n}^n)$ $\partial_{[m} (\underline{\epsilon}_{n]p} \kappa^p) = 0$, i.e., can be rewritten as the statement that the curl of the vector field $\underline{\epsilon}_{np} \kappa^p$ on U_S vanishes. Since U_S is simply connected, this implies in turn that $\underline{\epsilon}_{am} \kappa^m$ on U_S is the gradient of some scalar field: $\underline{\epsilon}_{am} \kappa^m = \partial_a \nu$ for some ν. We conclude, then, that the most general solution of (129) is

$$\underline{E}^a = \mu\, \underline{n}^a - \underline{\epsilon}^{am} \partial_m \nu, \qquad \underline{B}^a = \sigma\, \underline{n}^a - \underline{\epsilon}^{am} \partial_m \tau, \tag{132}$$

where the second is obtained, similarly, from the second equation (129).

Thus, because of (129), each of $\underline{E}^a$ and $\underline{B}^a$ defines a pair of scalar fields on U_S. The fields μ and σ are uniquely determined by $\underline{E}^a$ and $\underline{B}^a$, respectively, while ν and τ are determined only up to addition of an arbitrary constant. Physically, μ $(= \lim E^a n_a)$ and σ represent the "asymptotically radial parts" of the electric and magnetic fields, respectively, i.e., the "asymptotic Coulomb parts". The fields ν and τ, on the other hand, are potentials for the "asymptotic transverse parts".

We turn next to the evolution equations. Substituting (132) int (130), we obtain

$$\dot{\mu}\, \underline{n}^a - \underline{\epsilon}^{am} \partial_m \dot{\nu} = \underline{\epsilon}^{amn} \partial_m (\underline{\lambda}\, \sigma\, \underline{n}_n - \underline{\lambda}\, \underline{\epsilon}_{np}\, \partial^p \tau) \tag{133}$$

$$- \underline{\lambda}\ \underline{\epsilon}^{amn}\ \underline{n}_m\ (\sigma\ \underline{n}_n - \underline{\epsilon}_{np}\ \partial^p \tau).$$

The equations for $\dot{\mu}$ and $\dot{\nu}$ are obtained, respectively, by projecting (133) parallel and orthogonal to $\underline{n}^a$. Contracting (133) with $\underline{n}_a$, the left side becomes $\dot{\mu}$. The second term on the right gives zero, while, for the first term on the right, the term $\underline{\lambda}\sigma\ \underline{n}_n$ also gives zero. Thus, the right side gives $- \underline{n}_a\ \underline{\epsilon}^{amn}\ \partial_m(\underline{\lambda}\ \underline{\epsilon}_{np}\ \partial^p \tau)$ $= \partial_m\ (\underline{\lambda}\ \partial^m \tau)$. Contracting (133) with $\underline{\epsilon}_{ba}$, the left side gives $\partial_b\ \dot{\nu}$. The first term on the right becomes, eliminating the two $\underline{\epsilon}$'s, $- \underline{n}^m\ \partial_b\ (\underline{\lambda}\sigma\ \underline{n}_m - \underline{\lambda}\ \underline{\epsilon}_{mp}\ \partial^p \tau) = - \partial_p\ (\underline{\lambda}\sigma) - \underline{\lambda}\ \underline{\epsilon}_{bp}\ \partial^p \tau$. For the second term on the right, again eliminating the two $\underline{\epsilon}$'s, we obtain $\underline{\lambda}\ \underline{\epsilon}_{bp}\ \partial^p \tau$. Thus, we have $\partial_b \dot{\nu} = - \partial_b\ (\underline{\lambda}\sigma)$. Since ν is anyway determined only up to a constant, we may remove the "∂_b's" from this equation. Thus, Eqn. (133) yields the first and fourth equations of:

$$\dot{\mu} = \partial_m\ (\underline{\lambda}\ \partial^m \tau), \qquad \dot{\tau} = \underline{\lambda}\ \mu, \tag{134}$$

$$\dot{\sigma} = - \partial_m(\underline{\lambda}\ \partial^m \nu), \qquad \dot{\nu} = - \underline{\lambda}\ \sigma. \tag{135}$$

The second and third are obtained, similarly, beginning with (131). Eqns. (134) and (135), then, give the evolution of μ, τ, σ, and ν from one cross-section to the next.

We now note that each pair of equations, (134) and (135), has essentially the same structure as (126) and (127). Thus, just as in the Klein-Gordon case, we may identify τ and μ, respectively, with the value on a cross-section and the normal derivative on that cross-section, of a scalar field α, and $-\nu$ and σ, respectively, with the value and normal derivative of a scalar field β; where these two scalar fields on $\mathcal{S}$ satisfy:

$$\underline{D}_m\ \underline{D}^m\ \alpha = 0, \qquad \underline{D}_m\ \underline{D}^m\ \beta = 0. \tag{136}$$

The freedom to add an arbitrary constant to τ (resp., ν) is reflected, in terms of the fields on $\mathcal{S}$, as the freedom to add an arbitrary constant to α (resp., β), noting that the addition of a constant is consistent with (136).

Thus, the asymptotic Maxwell field is described by a pair of solutions of the wave equation on $\mathcal{S}$, with each solution specified only up to an arbitrary additive constant. Physically, the solution α combines the "asymptotic Coulomb part" of the electric field with a potential for the "asymptotic transverse part" of the magnetic field, and similarly for β. In effect, it is these combinations which couple to each other through the asymptotic Maxwell equations. Since "transverse parts" are usually associated with radiation, and since no radiation escapes to spatial infinity, one might have thought it reasonable to impose, as an additional condition for asymptotic regularity, the vanishing of the transverse parts. We now see that this would not be possible. In particular, asymptotic electric and magnetic fields which, for a given cross-section, are "pure Coulomb" (i.e., have $\nu = \Upsilon = 0$) will in general fail to have this property for another cross-section, e.g., by (134) and (135). Thus, one thinks of the transverse parts, not as radiation fields, but rather as complements of the Coulomb parts.

Finally, we consider the asymptotic gravitational field. We say that the gravitational field is asymptotically regular if E^{ab} and B^{ab}, the components of the Weyl tensor, are $C_S{}^n$ for sufficiently large n. We note here a significant difference from the null case, in which asymptotic regularity did not have to be imposed separately for the gravitational field, but rather was a consequence of the definition of an asymptote. To see how this difference arises, consider Eqns. (102) and (103). Ignoring the sources in (102), these equations express $\tilde{E}^{ab}$ and $\tilde{B}^{ab}$ in terms of $\tilde{q}_{ab}$ and $\tilde{p}_{ab}$: Hence, they yield formulae for E^{ab} and B^{ab} in terms of q_{ab} and p_{ab}. But these last two fields are those to which the definition of an asymptote refers. Thus, one could express asymptotic regularity in terms of q_{ab} and p_{ab}, and then include this requirement as an additional condition in the definition of an asymptote. This is essentially what is done in the null case, for there asymptotic conditions on the geometry and on the fields combine naturally into a single condition: smoothness of the unphysical metric. In the spatial case, by contrast, we cannot require smoothness of q_{ab} and p_{ab}: Rather, we wish only to require that certain differential combinations of q_{ab} and p_{ab} - those giving E^{ab} and B^{ab} - admit direction-dependent limits. It is more natural in the spatial case, therefore, to separate the asymptotic conditions on the fields from those on the geometry.

We must deal with one more preliminary issue in the gravitational case, namely, that of the asymptotic behavior of the sources in Eqns. (120)-(123). As usual, this is essentially a physical question. We wish to impose some condition sufficiently weak to permit interesting space-times, yet sufficiently strong to permit the study of the asymptotic gravitational field. We may obtain some evidence on this question as follows. We now have two examples of physical fields - Klein-Gordon and Maxwell - which can act as a source for the gravitational field, and on which we have a notion of asymptotic regularity. We can therefore ask, with these examples, assuming asymptotic regularity of the physical fields, for the asymptotic behavior of the sources in (120)-(123). We claim: For these examples, $\Omega^{-2} L_{abc}$, with each index either contracted with t^a or projected orthogonal to t^a, vanishes at S. In the case of Maxwell fields, for example, $\tilde{L}_{ab}$, projected orthogonally to t^a and assigned dimension -2, has a direction-dependent limit at S, whence $\tilde{\nabla}_{[a} \tilde{L}_{b]c}$, projected orthogonally to t^a and assigned dimension - 5/2, has a direction-dependent limit at S. Hence, $\Omega^{-2} L_{abc}$, projected orthogonally to t^a, vanishes at S. Similarly for other components, and for the Klein-Gordon field. Based on this evidence, then, we shall demand that the sources in (120)-(123) satisfy: $\Omega^{-2} L_{abc}$, with each index either contracted with t^a or projected orthogonally to t^a, vanishes at S. Note that the consequence of this demand is that the source-terms in (120)-(123) drop out in the asymptotic limit. [Eqns. (120) and (121) must be multiplied by $\Omega^{1/2}$ before taking the limit.] This behavior is in marked contrast to the null case, in which sources appear in the limit (e.g., in (60)). The physical reason, of course, is that the matter fields can contribute a flux of energy-momentum at null infinity, and thus give rise to dynamical changes in the asymptotic gravitational field. No fields produce a flux of anything at spatial infinity.

We now take the asymptotic limit of the gravitational field. Set $\underline{E}^{ab} = \lim E^{ab}$, and $\underline{B}^{ab} = \lim B^{ab}$, each a direction-dependent tensor at S. Since E^{ab} and B^{ab} are trace-free, so are $\underline{E}^{ab}$ and $\underline{B}^{ab}$. Multiplying (120) and (121) by $\Omega^{1/2}$ and applying "lim", we obtain the constraint equations on these tensors:

$$\partial_b \underline{E}^{ab} = 3\, \underline{E}^{ab}\, \underline{n}_b, \qquad \partial_b \underline{B}^{ab} = 3\, \underline{B}^{ab}\, \underline{n}_b. \tag{137}$$

We next decompose into tensors tangent to U_S. Set $\underline{E}^{ab} = \mu\, \underline{n}^a\, \underline{n}^b + 2\, \mu^{(a}\, \underline{n}^{b)} + \mu^{ab}$, with each of μ, μ^a, and μ^{ab} tangent to U_S. Since $\underline{E}^{ab}$ is trace-free, we have $\mu + \mu^m{}_m = 0$. As in the Maxwell case, we wish to simplify our components of $\underline{E}^{ab}$ using the constraint equation. Substituting our expression for $\underline{E}^{ab}$ into the first equation (137), and projecting parallel and orthogonal to $\underline{n}^a$, we obtain, respectively,

$$\partial_a\, \mu^a = 0, \qquad (\delta_a{}^b - \underline{n}_a\, \underline{n}^b)\, \partial_m\, \mu^{am} = 0. \tag{138}$$

As before, the first equation in (138) is precisely the condition that $\underline{\epsilon}_{am}\, \mu^m$ is the gradient of some scalar field. We use the second equation in (138) to simplify μ^{ab} as follows. Let s_a be any vector in V_S, and set $v_a = s_a - \underline{n}_a\, (s_m\, \underline{n}^m)$. Then we have $\partial_{(a}\, v_{b)} = -\, (\underline{g}_{ab} - \underline{n}_a\, \underline{n}_b)\, s_m\, \underline{n}^m - \underline{n}_{(a}\, v_{b)}$. [That is, v_a is a conformal Killing field on the two-sphere U_S.] Contracting (138) with v_a, we obtain $\partial_b\, (\mu^{ab}\, v_a) + \mu\, (s_m\, \underline{n}^m) = 0$. Integrating this equation over U_S, the first term, as a divergence, gives zero, and so we conclude that the integral of $\mu\, (s_m\, \underline{n}^m)$ over U_S vanishes. But this is precisely the statement that, when μ on U_S is expanded in spherical harmonics, it has no dipole part. But this in turn implies the existence of a scalar field Υ on U_S satisfying $(\partial_m\, \partial^m + 2)\, \Upsilon = \mu$, for, expanding each side in multipoles, the ℓth multipole of the left side is $-\ell(\ell + 1) + 2$ times the ℓth multipole of Υ, while the ℓth multipole of the right side is the ℓth multipole of μ. Equating these two determines the ℓth multipole of Υ in terms of that of μ for every ℓ except $\ell = 1$ (for which $-\ell(\ell + 1) + 2$ vanishes). But the first multipole of μ vanishes. Thus, not only can we solve our equation for Υ in terms of μ, but that solution is unique up to addition to Υ of a dipole term, i.e., of a direction-dependent scalar of the form $s_a \underline{n}^a$. Now consider $\kappa^{ab} = \mu^{ab} - \partial^{(a}\partial^{b)}\Upsilon - \underline{n}^{(a}\partial^{b)}\Upsilon + (\underline{q}^{ab} - \underline{n}^a \underline{n}^b)(\partial_m \partial^m \Upsilon + \Upsilon)$. Taking the trace, using $\mu^m{}_m = -\mu$ and $(\partial_m \partial^m + 2)\Upsilon = \mu$, we obtain $\kappa^m{}_m = 0$. Contracting with $\underline{n}_b$, we also obtain zero. Finally, computing the divergence of κ^{ab}, using (138) for the first term, we obtain $(\delta_a{}^b - \underline{n}_a \underline{n}^b) \partial_m \kappa^{am} = 0$. Thus, κ^{ab} is a symmetric, trace-free, divergence-free tensor field on the two-sphere U_S. Hence, its double-dual, $\underline{\epsilon}_{am} \epsilon_{bn} \kappa^{mn}$, is symmetric, trace-free, and has vanishing curl. But this implies that the double dual is zero (See "uniqueness" in the proof of Theorem 5, Sect. II.5.), and hence that $\kappa^{ab} = 0$. Thus, from the formula defining κ^{ab}, we obtain an expression for μ^{ab} in terms of Υ. Putting all this together, we have: The most general $\underline{E}^{ab}$ satisfying the first equation (137) is given, in terms of two scalar fields on U_S, by the first equation in

$$\begin{aligned}\underline{E}^{ab} = {} & (2\partial_m \partial^m \Upsilon + 3\Upsilon)\,\underline{n}^a \underline{n}^b + \partial^{(a}\partial^{b)}\Upsilon + \underline{n}^{(a}\partial^{b)}\Upsilon \\ & - \underline{q}^{ab}(\partial_m \partial^m \Upsilon + \Upsilon) + 2\,\underline{n}^{(a}\underline{\epsilon}^{b)m}\partial_m \sigma,\end{aligned} \tag{139}$$

$$\begin{aligned}\underline{B}^{ab} = {} & (2\partial_m \partial^m \varphi + 3\varphi)\,\underline{n}^a \underline{n}^b + \partial^{(a}\partial^{b)}\varphi + \underline{n}^{(a}\partial^{b)}\varphi \\ & - \underline{q}^{ab}(\partial_m \partial^m \varphi + \varphi) + 2\,\underline{n}^{(a}\underline{\epsilon}^{b)m}\partial_m \psi.\end{aligned} \tag{140}$$

Furthermore, $\underline{E}^{ab}$ determines σ and Υ uniquely up to addition of a constant to σ and addition of a term of the form $s_a \underline{n}^a$ to Υ. Similarly for the magnetic part of the Weyl tensor. These four scalar fields on U_S, then, describe completely the asymptotic gravitational field for the single cross-section U_S.

We next determine rate of change with respect to cross-section. Taking "lim" of (122) and (123), we obtain, respectively,

$$\dot{\underline{E}}^{ab} = \underline{\epsilon}^{mn(a}\partial_m(\underline{\lambda}\,\underline{B}_n{}^{b)}) + \underline{\epsilon}^{mn(a}(\partial_m \underline{\lambda} - \underline{\lambda}\,\underline{n}_m)\,\underline{B}_n{}^{b)}, \tag{141}$$

$$\dot{\underline{B}}^{ab} = -\underline{\epsilon}^{mn(a}\partial_m(\underline{\lambda}\,\underline{E}_n{}^{b)}) - \underline{\epsilon}^{mn(a}(\partial_m \underline{\lambda} - \underline{\lambda}\,\underline{n}_m)\,\underline{E}_n{}^{b)}. \tag{142}$$

Thus, to determine the rates of change of σ, Υ, φ, and ψ, we substitute (139) and (140) into (141) and (142), and take components parallel and orthogonal to $\underline{n}^a$. The manipulations are entirely straightforward: Substitute, take the appropriate projection, put everything inside the derivative on the right, and then eliminate all pairs of $\underline{\epsilon}$'s. The result is:

$$\dot{\Upsilon} = -\underline{\lambda}\,\psi, \qquad \dot{\psi} = -\partial_m(\underline{\lambda}\,\partial^m \Upsilon) - 3\underline{\lambda}\,\Upsilon, \tag{143}$$

$$\dot{\varphi} = \underline{\lambda}\,\sigma, \qquad \dot{\sigma} = \partial_m(\underline{\lambda}\,\partial^m \varphi) + 3\underline{\lambda}\,\varphi. \tag{144}$$

We note that these equations are consistent with the ambiguities of σ, τ, φ, and Υ. Thus, addition to Υ of a constant on the right in the first equation (143) results in addition to $\dot{\tau}$ of a dipole term; addition to τ of a dipole term on the right in the second equation (143) results in addition to $\dot{\Upsilon}$ of a constant.

Again, we may represent these functions on cross-sections by scalar fields on $\mathcal{S}$. We identify $-\tau$ and Υ, respectively, with the value and normal derivative of a scalar field α, and φ and σ with the value and normal derivative of a scalar field β, where α and β satisfy:

$$\underline{D}_m \underline{D}^m \alpha = -3\alpha, \qquad \underline{D}_m \underline{D}^m \beta = -3\beta. \tag{145}$$

The ambiguities of σ, τ, φ, and Υ are represented in terms of α and β as follows. Let u_a be any vector in the four-dimensional vector space V, and consider the function f on $\mathcal{S}$ whose value at the point $\underline{n}^a$ of $\mathcal{S}$ is $u_a \underline{n}^a$. Then, by direct computation, one checks that this f in $\mathcal{S}$ satisfies $\underline{D}_m \underline{D}^m f = -3f$. Furthermore, evaluating f on any cross-section, we obtain a dipole function, of the form $s_a \underline{n}^a$, and evaluating its normal derivative we obtain a constant. Thus, the ambiguity in τ and Υ in (143) represents precisely the freedom to add to α such a scalar field f (thereby preserving the first equation (145)), and, similarly, the ambiguity in φ and σ represents the freedom to add to β such an f.

Thus, the asymptotic gravitational field at spatial infinity is represented by a solution of (145) on $\mathcal{S}$, with each of α and β indeterminant up to a solution of the form $u_a \underline{n}^a$. One thinks of $\underline{E}^{ab} \underline{n}_a \underline{n}_b$ as representing the "asymptotic Coulomb part" of the electric part of the Weyl tensor, of $\underline{E}^{ab} \underline{n}_b$, projected orthogonal to $\underline{n}^a$, as the "asymptotic transverse part", and of $\underline{E}^{ab}$, projected orthogonal to $\underline{n}^a$, as a sort of "asymptotic super-transverse part". The calculation preceeding Eqn. (139) shows that the super-transverse part can be expressed in terms of the Coulomb part. What remains, then, is two independent parts of $\underline{E}^{ab}$. The field α combines a potential for the asymptotic Coulomb part of $\underline{E}^{ab}$ with a potential for the asymptotic transverse part of $\underline{B}^{ab}$, and similarly for β. Again, the fact that these particular combinations arise is a reflection of the coupling implicit in the field equation, in this case, the Bianchi identity on the Weyl tensor. Roughly speaking, all parts couple to all parts in the physical space-time (e.g., by (112) and (113)). In the asymptotic limit at spatial infinity, however, a decoupling takes place, which permits the simple representation in terms of α and β.

The fact that the asymptotic Klein-Gordon field is expressed as a single scalar field on $\mathcal{S}$, while Maxwell and gravitational each require two, reflects the number of helicities for the particles in each case. The formula for the numerical coefficients on the right in (128), (136), and (145) is coefficient = $-(\text{spin})^2 + 1$: in retrospect, the formula one might have expected.

Finally, as an example of all this, we compute the asymptotic gravitational field for the Schwarzschild space-time. Choose $\tilde{T}$, $\tilde{q}_{ab}$, $\tilde{P}_{ab}$ and its asymptote as in Sect. III.2. Then, by the time-reflection symmetry, we have $\tilde{B}^{ab} = 0$, and hence $B^{ab} = 0$. Computing $\tilde{E}_{ab} = \tilde{C}_{ambn}\, \tilde{t}^m\, \tilde{t}^n$ from the Weyl tensor we obtain, to first nonvanishing order in r, $\tilde{E}_{ab} = m\, r^{-3}\, (\tilde{q}_{ab} - 3\, \tilde{D}_a r\ \tilde{D}_b r)$. Note that the "Euclidean components" of $\tilde{E}_{ab}$ vanish asymptotically as r^{-3}, as one would expect, since we have assigned to $\tilde{E}_{ab}$ dimension $-3/2$. [The conformal factor goes to zero as r^{-2}.] Thus, E^{ab} admits a direction-dependent limit, and in fact we have $\underline{E}^{ab} = \lim E^{ab} = m\,(\underline{q}^{ab} - 3\, \underline{n}^a\, \underline{n}^b)$. Next, we compute σ, τ, $\overline{\varphi}$, and ψ from (139) and (140). Since $\underline{B}^{ab} = 0$, we have $\varphi = \psi = 0$. Since $\underline{E}^{ab}\, \underline{n}_b$ is a multiple of $\underline{n}^a$, we have $\sigma = 0$. Finally, inspecting (139), we see that $\tau = -\,\overline{m}$. Note that our four functions on U_S respect the spherical symmetry. Regard U_S as a cross-section of $\mathcal{S}$, and denote by w^a the unit normal in V to the plane of U_S, so w^a is a unit timelike vector in the four-dimensional vector space V. Then α is to be that solution of the first equation (145) on $\mathcal{S}$ such that the value of α on U_S is the constant m, and the normal derivative in $\mathcal{S}$ is zero; β is to be that solution of the second equation such that the value and normal derivative of β on U_S are both zero. The solution for β is clearly $\beta = 0$. The solution for α is that given by the formula $\alpha = m\,(1 + 2\,(w_a\, \underline{n}^a)^2)\,(1 + (w_a\, \underline{n}^a)^2)^{-1/2}$, as one verifies by substitution into the first Eqn. (145). [To discover this formula, one notes, by the symmetry, that α must be a function of $w_a\, \underline{n}^a$ Choosing a general function, and substituting into (145), one obtains an ordinary differential equation for the function.] Thus, the scalar fields on $\mathcal{S}$ describing the asymptotic gravitational field of the Schwarzschild space-time are $\beta = 0$ and α the formula above.

5. Physical Quantities at Spatial Infinity

In the asymptotic limit at spatial infinity, a Klein-Gordon fiel is described by a solution of (128), a Maxwell field by a solution of (136) up to constants, and the gravitational field by a solution of (145) up to solutions of the form $u_a\, \underline{n}^a$ with u_a in V. In this section, we construct, and interpret physically, various quantities from these fields.

We begin with the Maxwell field. Fix any cross-section C, and consider, given α and β satisfying (136),

$$Q = \int_C \underline{D}^m \alpha\ \underset{\sim}{\epsilon}_{mab}\ dS^{ab}, \qquad Q^* = \int_C \underline{D}^m \beta\ \underset{\sim}{\epsilon}_{mab}\ dS^{ab}, \tag{146}$$

where $\underset{\sim}{\epsilon}_{mab}$ denotes the alternating tensor of the 3-dimensional manifold $\mathcal{S}$. We first note that the numbers Q and Q* so obtained depend only on the asymptotic Maxwell field, i.e., that these numbers do not change under addition of a constant to α or β. Furthermore, since in each case the integrand has vanishing curl, these numbers are independent of cross-section. These numbers, we claim, are to be interpreted as the total electric charge and total magnetic charge, respectively, of the system. The first integral, for example, is just the integral over C of the normal derivative of α, i.e., the integral over C of the μ given by (132), i.e., the integral over C of $\underline{E}^a\, \underline{n}_a$. But this is just the limit at S of the integral in the physical space-time which one normally associates with the total charge. Similarly for the total magnetic charge. Thus, as one would have expected, the total electric and magnetic charge can be recovered from the asymptotic Maxwell field at spatial infinity.

We consider next the gravitational field. Let α satisfy (145), and consider the expression $U^{ab} = \underline{D}^a \underline{D}^b \alpha + \alpha\, \underline{g}^{ab}$, a symmetric tensor field on the manifold $\mathcal{S}$. From (145), we have $U^m{}_m = \underline{D}^m \underline{D}_m \alpha + 3\alpha = 0$. Taking a divergence, we have $\underline{D}_m U^{am} = \underline{D}_m \underline{D}^a \underline{D}^m \alpha + \underline{D}^a \alpha = \underline{D}^a \underline{D}_m \underline{D}^m \alpha + 2\, \underline{D}^a \alpha + \underline{D}^a \alpha = 0$, where we used (145) in the last step, and the fact that, since $\mathcal{S}$ is a hyperboloid in a Minkowskian vector space, its Ricci tensor is $2\, \underline{g}_{ab}$ in the second step. Thus, U^{ab} is symmetric and divergence-free. Consider now the special case in which $\alpha = u_a\, \underline{n}^a$, with u_a a vector in V. Then $\underline{D}_a \alpha = u_a - \underline{n}_a\, (u_m\, \underline{n}^m)$, and so $\underline{D}_a \underline{D}_b \alpha = -\, \underline{g}_{ab}\, (u_m\, \underline{n}^m)$. That is, we have $U^{ab} = 0$ for α of this form. We conclude, then, that our U^{ab} is invariant under addition to α of a solution of (145) of the form $u_a\, \underline{n}^a$, i.e., that U^{ab} depends only on the asymptotic gravitational field itself. Next, let s_a be any vector in V, and consider the vector field $s_a - \underline{n}_a\, (s_m\, \underline{n}^m)$ on $\mathcal{S}$. By the calculation we just did, this vector field is a conformal Killing field. Hence, $U^{ab}\, (s_b - \underline{n}_b\, (s_m\, \underline{n}^m))$ has vanishing divergence. Consider, then, the right side of

$$P^a s_a = -1/8\pi \int_C (\underline{D}^m \underline{D}^n \alpha + \alpha\, \underline{g}^{mn})\, (s_m - \underline{n}_m\, s_p\, \underline{n}^p)\, \underset{\sim}{\epsilon}_{nab}\, dS^{ab}, \qquad (147)$$

where C is any cross-section. By what we have just shown, the number resulting from this integral depends only on the asymptotic gravitational field, and is independent of the choice of cross-section C. Furthermore, this number is obviously linear in the (arbitrary) vector s_a in V. Hence, there is some vector P^a in V such that (147) holds for every s_a in V. This P^a is called the <u>Arnowitt-Deser-Misner energy-momentum vector</u>.

This P^a is to be interpreted as the "total energy-momentum of the system, measured at spatial infinity". As in the null case, we must justify this interpretation. There are, as far as I am aware, just two pieces of evidence. First, one may verify that one obtains the "expected" P^a in familiar examples. As an example of these examples, we consider the Schwarzschild space-time. In this case, $\alpha = m\, (1 + 2\, (w_m\, \underline{n}^m)^2)\, (1 + (w_m\, \underline{n}^m)^2)^{-1/2}$, where w^a is the unit timelike vector orthogonal to the plane of the cross-section U_S

associated with the spacelike submanifold "t = const" in the Schwarzschild space-time. We compute P^a in this case. Since we may perform the integral (147) over any cross-section C, let us choose $C = U_S$. Then, by direct computation, we have, at points of C, $\underline{D}_a \underline{D}_b \alpha + \alpha \underline{g}_{ab} = 3\, m\, w_a w_b + m\, \underline{g}_{ab}$. Hence, $(\underline{D}^m \underline{D}^n \alpha + \alpha\, \underline{g}^{ab})(s_m - \underline{n}_m (s_p \underline{n}^p))\, w_n = -2m\, s_a w^a$. But, since w_n is the unit normal to the cross-section C, the above is the integrand in (147). Hence, the right side is $m\, s_a w^a$, and so we obtain $P^a = m\, w^a$. Thus, for the Schwarzschild space-time, the energy-momentum vector P^a is timelike, with norm $-m^2$, and is orthogonal to the plane of the cross-section U_S associated with the "t = const" spacelike submanifolds of the space-time.

Returning to the general case, this calculation suggests the following additional interpretation of P^a. Let $\tilde{T}$ be any spacelike submanifold, with asymptote, of our space-time. Then this $\tilde{T}$ leads to some cross-section, U_S, of $\mathcal{S}$. Denote by w^a the unit normal to the plane of U_S. One thinks of $\tilde{T}$ as defining an "asymptotic rest state" in the space-time, and of w^a, a unit timelike vector in V, as the associated "asymptotic four-velocity". This w^a can now be used to decompose P^a. The component parallel to w^a, $-P^a w_a$, is interpreted as the "total energy" of the system, measured with respect to this "asymptotic four-velocity", while the component perpendicular $P^a + w^a (P^m w_m)$, is interpreted as the "total momentum". This decomposition is of course identical to that of the energy-momentum of a particle with respect to an observer's four-velocity. Then, since P^a is just a fixed vector in V, we have that "energy and momentum transform like a four-vector under asymptotic boosts". In this language, our result for the Schwarzschild space-time is that, choosing $\tilde{T}$ orthogonal to the timelike Killing field in the space-time, we obtain energy m and momentum zero. This, of course, is the answer one would have expected.

Similarly, one computes P^a for other space-times, e.g., the Kerr and Weyl, and in each case obtains the result compatible with one's physical understanding of the space-time.

The other piece of evidence involves, more generally, space-times which are stationary. For such space-times, one "knows" already what the total energy should be, namely that as computed from the timelike Killing field. Choosing $\tilde{T}$ asymptotically orthogonal to the Killing field, and taking the limit of the energy integral as defined by the Killing field, one obtains precisely the same number as the total energy defined above for this $\tilde{T}$. Thus, one again obtains the "right answer".

Unfortunately, the evidence for our interpretation of P^a in the spatial case is somewhat weaker than the evidence for the interpretation of the corresponding quantity, the Bondi energy-momentum, in the null case. There, the energy-momentum depended on the choice of cross-section, and one could write down a formula, essentially (91), for the rate of change of this energy-momentum with respect to cross-section. The result was an expression containing, among other things, a quantity one could interpret as the "flux of energy-momentum of matter to null infinity". Nothing similar is available in the spatial case, for here P^a is independent of cross-section, as one might expect, for there is no flux of energy-momentum at spatial infinity. There is even a further sense in which the link between P^a and the physics is weaker in the spatial case than the corresponding link in the null case. In the latter, Killing fields in the physical space-time which behave asymptotically as translations give rise to infinitesimal translations on $\mathcal{I}$. The Bondi energy-momentum, for fixed cross-section, then emerges as a linear mapping from the vector space of infinitesimal translations to the reals. In this sense, then, the infinitesimal translations are the generators of the Bondi energy-momentum. In the spatial case, Killing fields on the physical space-time which behave asymptotically as translations essentially "vanish asymptotically at spatial infinity". [It is true that one could associate, with each such Killing field on the physical space-time, a conformal Killing field on $\mathcal{S}$, a field of the form $s^a - \underline{n}^a (s_m \underline{n}^m)$. Such an association, however, would yield little evidence for our interpretation of P^a. For example, this association does not preserve the Lie-algebra structures: Whereas asymptotic infinitesimal translations in the physical space-time will all commute with each other, conformal Killing fields on $\mathcal{S}$ do not.] In short, the Bondi energy-momentum is more closely linked to possible symmetries in the physical space-time than is P^a.

In any case, since one desires something one can call the total energy-momentum at spatial infinity, since the problems of evidence are attributable more to the structure one expects at spatial infinity than to P^a itself, and since one lacks any better candidates, it seems reasonable to so interpret P^a.

Why could one not carry out the integral (147) with β instead of α, obtaining a sort of "magnetic energy-momentum"? Of course one could, but it turns out that, with slightly stronger asymptotic conditions, the corresponding integral vanishes. Let us suppose that p_{ab} is $C_S{}^1$. Consider Eqn. (103). Applying our conformal transformation to this equation to eliminate the tildes, and then applying "lim", we obtain an expression for $\underline{B}_{ab}$ in terms of a potential obtained from lim p_{ab}. Expressing the integral (147) for β in terms of φ and σ, and using the existence of this potential, we find that the integral vanishes. This is analogous to the calculation in electrodynamics in which the existence of a vector potential denies the possibility of magnetic charge, the only difference being that the potential is here forced upon one from the geometry. The same thing happens, but in a different way, in the null case. Consider Eqn. (83). That this vector, inserted into (87), gives rise to a total energy-momentum rather than a "magnetic energy-momentum" is due to the presence of K^{am}, rather than $*K^{am}$, in the first term. But the second term in (83) is necessary in order that P^a have the required properties, and in particular this second term involves the news, N_{ab}, which satisfies (68). One is thus unable to write a corresponding formula with $*K^{am}$, because one lacks a corresponding "$*N_{ab}$" satisfying the equation analogous to (68).

Thus, one obtains at spatial infinity at least total charge and total energy-momentum - the same two quantities one obtains at null infinity. The structure of the equations in the null case rather suggested that these were the only physically interesting quantities which could be there obtained. Are there any other "quantities" in the spatial case? Although the answer is by no means clear, most likely there are not. Consider, as an example, the following class of possibilities. Let α be a scalar field on satisfying $\underline{D}_m \underline{D}^m \alpha = \kappa \alpha$ for some number κ, fix a positive integer n, and define

$$U^{ab} = n\,\alpha^{n-1}\,\underline{D}^a\,\underline{D}^b\alpha - (n\kappa + 2)\,\alpha^n\,\underline{g}^{ab} + n(n-1)\,\alpha^{n-2}\,\underline{D}^a\alpha\,\underline{D}^b\alpha - n(n-1)\,\alpha^{n-2}\,\underline{g}^{ab}\,\underline{D}_m\alpha\,\underline{D}^m\alpha \,. \tag{148}$$

Computing the divergence of U^{ab}, we obtain $\underline{D}_b U^{ab} = 0$. Hence, for ξ^b any Killing field on $\mathcal{S}$ - any vector field of the form $F^{ab}\,\underline{n}_b$ with F^{ab} an antisymmetric tensor over V - $U^{ab}\,\xi_b$ will have vanishing divergence. The integral of this vector field over any cross-section will therefore be independent of cross-section, and hence this integral will be of the form $X_{ab}\,F^{ab}$ for some tensor X_{ab} over V. What about these "quantities"? It turns out that they are all zero, for we have $U^{ab}\,\xi_b = \underline{D}_b\,(2n\,\alpha^{n-1}\,\underline{D}^{[a}\alpha\,\xi^{b]} - \alpha^n\,\underline{D}^{[a}\,\xi^{b]})$, whence the integral over a cross-section will always give zero. [Note in particular that, for n = 1 and $\kappa = -3$, this U^{ab} is the same as the field we introduced in Eqn. (147). Thus, the right side of (147) vanishes when the vector field on the right is replaced by a Killing field. Nothing was lost, then, by choosing $s_a - \underline{n}_a\,(s_p\,\underline{n}^p)$ for the vector field on the right.] Does there exist some theorem to the effect that charge and energy-momentum are the only "quantities", as tensors over V, which can be constructed from these asymptotic fields at spatial infinity? One might imagine proceeding as follows. Regard each of (128), (136), and (145) as defining a representation of the Lorentz group. [The underlying vector space, in each case, is the vector space of solutions of the equations; the action of the Lorentz group arises from its action as isometries on $\mathcal{S}$.] The last two representations, at least, are reducible. For example, the subspace of the vector space of solutions of (136) or (145) consisting of those with $\beta = 0$ is invariant. Furthermore, the subspaces consisting of solutions with zero charge, or with zero energy-momentum, are invariant. One might think of trying to prove that these are the only reductions, and then that this result denies the existence of further "quantities" at spatial infinity.

IV. CONCLUSION

This section consists of general remarks - principally on various difficulties, outstanding issues, areas for further work, etc.

Are the definitions of an asymptote in the two cases "reasonable", in that they strike a fair balance between weakness, so that interesting space-times are included, and strength, so that interesting asymptotic structure is obtained? In the null case, the answer on balance seems to be yes. It might be of interest to resolve Conjecture 1, or at least obtain some result along these lines, but indications are that the result will be positive. The spatial case seems to depend more heavily on examples and less heavily on external criteria. It is, however, not clear how one should proceed to find stronger evidence, for the difficulties seem, at least in part, to be implicit in the notion of a limit at spatial infinity rather in the details of a particular definition. It would perhaps be of interest to look for further evidence in the spatial case - if one could think of an idea as to what that evidence might be. It does appear, however, that neither notion is likely to turn out to be inappropriate in a substantive way.

In both the null and the spatial case, one begins with the physical space-time and ends with a certain abstract three-dimensional manifold on which there are various tensor fields. It seems almost certain that in each case the final end product is the "correct" one. In the null case, furthermore, the procedure which leads from the physical space-time to the abstract three-manifold with fields has a certain simplicity and elegance. In the spatial case, the corresponding procedure is anything but simple: The use of the initial-value formulation, of "lim", and of direction-dependent tensors are all rather awkward. There are a number of concrete manifestations of this. Why should the initial-value formulations have anything to do with asymptotic flatness at spatial infinity? Why do one's equations contain all those terms which make no contribution at spatial infinity? Why does one have to worry about differentiability class, something which almost certainly has no physical significance? Furthermore, there are a number of "facts" in the spatial case which seem difficult to prove. Consider, for example, the following: Let $\tilde{\xi}^a$ be a Killing field on the physical space-time, and resolve $\tilde{\xi}^a$ in terms of the normal $\tilde{t}^a$ to $\tilde{T}$ by $\xi^a = \tilde{\mu}\tilde{t}^a + \tilde{\nu}^a$, with $\tilde{\nu}^a$ orthogonal to $\tilde{t}^a$. Then, assigning each of $\tilde{\mu}$ and $\tilde{\nu}^a$ dimension - 1/2, each is C_S^1. [This is the essential step in proving the "obvious" result that Killing fields in the physical space-time give rise to infinitesimal symmetries on $\mathcal{S}$.] The analogous result in the null case is Theorem 3, and this is easy to prove, using the fact that a conformal Killing field on an open subset of a space-time has a smooth extension to its closure. The analogous method in the spatial case runs into all sorts of complications. One would begin by writing Killing's equation in terms of

$\tilde{\mu}$ and $\tilde{\nu}^a$ on $\tilde{T}$, and performing a conformal transformation. One would expect to prove the result from these differential equations. However, the details of the differentiability structures of q_{ab} and p_{ab} seem to play an important role, and anyway one is left with a rather complicated question involving smoothness of solutions of differential equations. Why are not such results easy? [The whole situation is perhaps analogous to that which would result if one tried to define "continuity" for a function of two real variables, but, having available only a notion of continuity for functions of one variable, insisted on expressing it in terms of this notion.]

At least two ideas have been suggested for improving the transition from the physical space-time to the final manifold $\mathcal{S}$ with its tensor fields. Sommers has suggested dispensing with the initial-value formulation entirely, instead simply "attaching" $\mathcal{S}$ to the physical space-time as $\mathcal{I}$ is attached in the null case. The various tensor fields on $\mathcal{S}$ would then arise directly from the fields on the physical space-time. One might expect that the difficulties with differentiability class would also disappear, for one could simply require smoothness in the space-time with its boundary attached. It is possible that the final formulation will nonetheless not be as simple as that for the null case, however, for the example of Minkowski space-time rather suggests that it will not be possible to impose, as one of one's conditions in the definition, that the conformally transformed metric be continuously extendible to the boundary. Ashtekar has suggested a formulation in which the "boundary at spatial infinity" begins as a single point. There would be no initial-value formulation, and the definition of an asymptote would require essentially the possibility of adding to the physical space-time this point, such that, among other things, the conformally transformed metric has a continuous extension to a tensor at this point. The manifold of unit spacelike vectors at this point would then give $\mathcal{S}$, and fields on $\mathcal{S}$ would be obtained using (presumably, direction-dependent) limits. This formulation has the advantage that one deals with an unphysical metric everywhere finite and invertible, but the disadvantage that, as examples suggest, one will not be able to require C^∞ differentiable structure. Either program looks as though it might be a considerable improvement over the present one. Is there some formulation which combines the best of both?

Another issue of some possible interest is that of whether or not we introduce "enough" asymptotic fields. There is apparently no reason in principle why one could not introduce additional physical fields on the two abstract manifolds. In the null case, one would simply take more derivatives of the fields in the unphysical space, arrange the indices properly, and apply ζ^*; in the spatial case, one would take more derivatives and apply "lim". The advantage of additional fields is of course that they permit the discussion of additional aspects of the asymptotic structure. There are, however, also some disadvantages. First, additional fields normally lead to a larger gauge group and more complicated gauge behavior, for one must deal with higher derivatives of the conformal factor. It may very well be that one soon reaches the point at which one's additional fields are "pure gauge", and so say nothing about the physical space-time. Furthermore, the interpretation of such fields normally becomes more and more complicated, and their physical significance of less and less interest. For example, such fields can refer to back-scattered radiation in the null case, and to the details of radiation emitted in the limit of the distant past in the spatial case, neither of which seems to be of as much interest as the information in the primary fields. An outstanding example of the reason why one might wish to investigate additional fields involves angular momentum. At both spatial and null infinity, we introduced a vector we called "total energy-momentum"; in neither case did we discuss angular momentum. The reason is that, as one sees by linearization, one does not expect to find any expression in terms of the present fields on $\mathcal{I}$ or $\mathcal{S}$ which one could call the "total angular momentum" of the system. Roughly speaking, the terms in the asymptotic Weyl tensor which describe the angular momentum of the system fall to zero too quickly to be registered by the present fields on $\mathcal{I}$ or $\mathcal{S}$. It is certainly possible that additional fields would lead to a reasonable candidate for such a quantity - but also possible that it will be completely obscured by radiation. It may be that a total angular momentum can be introduced, but only after imposing additional asymptotic conditions, which possibly effectively eliminate certain types of radiation. It seems remarkable that such a direct physical question - Does there or does there not exist a sensible notion of total angular momentum for an isolated system in general relativity? - apparently remains unanswered.

There is a large class of open questions concerning the relation between asymptotic structure at spatial and at null infinity. It is not known, for example, whether or not asymptotic flatness at one regime implies it at the other, say, imposing Einstein's equation with zero source. [It is easy to find counterexamples if one requires nothing of the stress-energy.] One would perhaps expect that the two notions should be equivalent in some sense. A positive result along these lines has not even been obtained, as far as I am aware, in the very much simpler special case in which the space-time is assumed to be stationary. A proof in this case should be relatively easy: It would perhaps be of interest to obtain it. Closely related to this question is that of the relationship between the Bondi energy-momentum and the Arnowitt-Deser-Misner energy-momentum. One would expect that a physical system just "has" a total energy-momentum, and so that these two quantities would agree when both are applicable. One has to formulate the question with a little care, because the energy-momentum at null infinity, but not spatial, depends on cross-section. The appropriate comparison is of the energy-momentum at spatial infinity with the energy-momentum at null infinity obtained in the limit (if it exists!) as the cross-section on the future boundary $\mathscr{I}$ moves into the past. There is a further complication as to what "equal" is to mean, since the vectors lie in different vector spaces. Is there some theorem to the effect that, possibly with stronger asymptotic conditions, the appropriate limit of the Bondi energy-momentum equals the Arnowitt-Deser-Misner energy-momentum? More generally, is there some treatment of asymptotic structure which combines the null and spatial regimes into a single framework?

A more difficult, and perhaps more vague, issue concerns the relationship between the asymptotic fields and the details of the system whose asymptotic structure is being discussed. In Newtonian gravitation, for example, the total mass of a system, measured asymptotically, is equal to the integral over the system of its mass density; in electrodynamics in flat space-time, the radiation field can be expressed in terms of the sources. Are any analogous formulae available in general relativity? How is the gravitational radiation field at null infinity related to the sources? Is there a reasonably simple expression for the energy-momentum, determined asymptotically, in terms of the stress-energy distribution? These are of course old questions - and apparently very difficult ones. Einstein's equation is a complicated, highly nonlinear partial differential equation. Perhaps we are in some sense lucky that it yields to any notion of "asymptotically flat", and are asking too much to now expect that the simple asymptotic quantities be simply related to the internal structure of the system itself. There is, however, at least one more concrete manifestation of this issue - a question which should be resolvable one way or the other. The formula in Newtonian gravitation relating the total mass, measured asymptotically in terms of the gravitational field, to the mass density implies: If the mass density is non-negative, so is the total mass. The analogous statement in general relativity would read: In the absence of singularities in an appropriate sense (to exclude those having "negative mass"), if the stress-energy of the space-time satisfies an energy condition, then the total energy-momentum, determined asymptotically, is future-directed timelike. This is really two statements, one for null infinity and one for spatial. Both are open. If, for example, one could prove the statement for null infinity, and also that the appropriate limit of the energy-momentum at null infinity is that at spatial infinity, then one would also have a proof for spatial infinity. The existence of such an open - and seemingly so elementary - question perhaps at least suggests that more could be done regarding the relation of asymptotic structure to internal structure.

We have discussed asymptotic structure in two regimes: at null infinity and at spatial infinity. Presumably, the former was introduced in order to be able to discuss radiation - in particular, gravitational radiation - the latter, in response to our familiarity with non-relativistic systems. There also exists a treatment of asymptotic flatness at "timelike infinity" [13]. However, none of these choices seems to be demanded, particularly, by an attempt to capture the notion of an "isolated system". Might there be any other asymptotic regimes of physical interest?

I wish to thank Abhay Ashtekar, John Friedman, Bernd Schmidt, Paul Sommers, Bob Wald, and Basilis Xanthopolous for helpful discussions.

APPENDIX: CONFORMAL TRANSFORMATIONS

We summarize a few definitions, facts, and formulae connected with conformal transformations.

Let $\tilde{M}$ be an n-dimensional manifold, and let there be given on $\tilde{M}$ an invertible metric $\tilde{g}_{ab}$ of some signature. By a <u>conformal transformation</u> on $\tilde{M}$, $\tilde{g}_{ab}$, we mean the introduction of a new metric, $g_{ab} = \Omega^2 \tilde{g}_{ab}$, on $\tilde{M}$, where Ω, the <u>conformal factor</u>, is some (normally smooth) positive scalar field on $\tilde{M}$. Since now either $\tilde{g}_{ab}$ or g_{ab} could be used to raise and lower indices of tensor fields on $\tilde{M}$, some convention is needed to avoid confusion as to which metric is being so used. We adopt the following convention. Let there be given a tensor field $\tilde{\alpha}^{a..c}{}_{b..d}$ on $\tilde{M}$, with u contravariant indices and d covariant indices. Let there further be given a real number s, called the <u>dimension</u> of $\tilde{\alpha}$. We then set

$$\alpha^{a..c}{}_{b..d} = \Omega^{s-u+d}\, \tilde{\alpha}^{a..c}{}_{b..d}. \tag{A1}$$

Thus, the number s, together with the index structure of $\tilde{\alpha}$, provide the instructions for the introduction of a new tensor field on $\tilde{M}$. We further adopt the convention that the indices of tensor fields with a tilde are to be raised and lowered with $\tilde{g}_{ab}$; those without, with g_{ab}. The presence of "u-d" in the exponent in (A1) now ensures that no ambiguities will result, provided that all presentations of $\tilde{\alpha}$, e.g., $\tilde{\alpha}_{a..cb..d}$, $\tilde{\alpha}^{a..cb..d}$, etc., are assigned the same dimension as $\tilde{\alpha}^{a..c}{}_{b..d}$. For example, let $\tilde{V}^a$ have dimension -3, so $V^a = \Omega^{-4}\tilde{V}^a$. Then, assigning $\tilde{V}_a = \tilde{g}_{ab}\tilde{V}^b$ also dimension -3, we shall have $V_a = \Omega^{-2}\tilde{V}_a$. But this is consistent with our convention, for we now have $V_a = g_{ab}V^b$. These dimensions in fact correspond to physical dimensions (in gravitational units; everything in seconds). We note further that dimension is unchanged under contraction, and is additive under outer product. The metric $\tilde{g}_{ab}$ itself, under (A1), has dimension zero, as does the alternating tensor. [Thus, "raising and lowering of indices does not change dimension" is a special case.] The conclusion from all this is that one can check the "dimensional correctness" of an equation by combining the dimensions of each factor on each side: One need not be concerned with contractions or index-locations.

Each of $\tilde{g}_{ab}$ and g_{ab} gives rise to a derivative operator, $\tilde{\nabla}_a$ and ∇_a, respectively. To find the relationship between these, we first note that, given any two derivative operators, we have, for some tensor field $C^m{}_{ab} = C^m{}_{(ab)}$ on M,

$$\tilde{\nabla}_a \alpha^{b..c}{}_{d..e} = \nabla_a \alpha^{b..c}{}_{d..e} + C^b{}_{am} \alpha^{m..c}{}_{d..e} + \dots + C^c{}_{am} \alpha^{b..m}{}_{d..e} - C^m{}_{ad} \alpha^{b..c}{}_{m..e} - \dots - C^m{}_{ae} \alpha^{b..c}{}_{d..m} \quad \text{(A2)}$$

for every tensor field $\alpha^{b..c}{}_{d..e}$ on $\tilde{M}$. [Note: One "C-term" on the right for each index of α.] In particular, it suffices, in order to specify the relation between the actions of the derivative operators on an arbitrary tensor field, to specify that relation only for an arbitrary vector field. To find $C^m{}_{ab}$ in the present case, we use the defining equation for the derivative operator in terms of the metric. We have $0 = \tilde{\nabla}_a \tilde{g}_{bc} = \tilde{\nabla}_a (\Omega^{-2} g_{bc}) = \nabla_a (\Omega^{-2} g_{bc}) - 2\, C^m{}_{a(b} (\Omega^{-2} g_{c)m}) = -2\,\Omega^{-3} (\nabla_a \Omega)\, g_{bc} - 2\,\Omega^{-2} C^m{}_{a(b}\, g_{c)m}$, where we used (A2) in the third step. Solving for $C^m{}_{ab}$, we have

$$C^m{}_{ab} = -\Omega^{-1} (2\, \delta^m{}_{(a} \nabla_{b)} \Omega - g_{ab}\, g^{mn} \nabla_n \Omega). \quad \text{(A3)}$$

Thus, Eqn. (A1) provides the instructions for removing tildes from the basic tensor fields, and (A2) and (A3) the instructions for removing tildes from the derivative operator. Using these, we may convert any expression, and hence any equation, involving tilded variables to one involving variables without tildes. For example, let $\tilde{v}^a$ have dimension -3. Then

$$\begin{aligned} \tilde{\nabla}_a \tilde{v}^b &= \nabla_a (\Omega^4 v^b) + C^b{}_{am} (\Omega^4 v^m) \\ &= \Omega^4 \nabla_a v^b + 3\,\Omega^3 v^b \nabla_a \Omega - \Omega^3 \delta^b{}_a v^m \nabla_m \Omega \\ &\quad + \Omega^3 v_a \nabla^b \Omega, \end{aligned} \quad \text{(A4)}$$

where we have used (A2) in the first step and (A3) in the second.

Since the curvature tensor is obtained from the derivative operator, and since we know the relationship between $\tilde{\nabla}_a$ and ∇_a, we can find the relationship between the curvature tensors, $\tilde{R}_{abcd}$ and R_{abcd}, respectively. For any field k_c, we have

$$\begin{aligned} 1/2\, \tilde{R}_{abc}{}^d k_d &= \tilde{\nabla}_{[a} \tilde{\nabla}_{b]} k_c = \tilde{\nabla}_{[a} (\nabla_{b]} k_c + C^m{}_{b]c} k_m) \\ &= \nabla_{[a} (\nabla_{b]} k_c + C^m{}_{b]c} k_m) + C^n{}_{[ab]} (\nabla_n k_c + C^m{}_{nc} k_m) \\ &\quad + C^n{}_{c[a} (\nabla_{b]} k_n + C^m{}_{b]n} k_m) \end{aligned} \quad \text{(A5)}$$

$$= 1/2\ R_{abc}{}^{d}\ k_d + \nabla_{[a}\ C^{m}{}_{b]c}\ k_m + C^{n}{}_{c[a}\ C^{m}{}_{b]n}\ k_m,$$

where we used (A2) in the second and third step. Since k_d is arbitrary, we have

$$\tilde{R}_{abc}{}^{d} = R_{abc}{}^{d} + 2\ \nabla_{[a}\ C^{d}{}_{b]c} + 2\ C^{n}{}_{c[a}\ C^{d}{}_{b]n}. \tag{A6}$$

Substituting (A3), we obtain the formula relating the curvature tensors. It follows in particular that the Weyl tensor has dimension -2 (the "correct" dimension for curvature), and that the Ricci tensors and scalar curvatures are related by:

$$\begin{aligned}\tilde{R}_{ab} = {} & R_{ab} + (n-2)\ \Omega^{-1}\ \nabla_a\ \nabla_b\Omega + \Omega^{-1}\ g_{ab}\ \nabla^{m}\ \nabla_m\Omega \\ & - (n-1)\ \Omega^{-2}\ g_{ab}\ (\nabla^{m}\Omega)\ (\nabla_m\Omega),\end{aligned} \tag{A7}$$

$$\tilde{R} = \Omega^{2}\ R + 2(n-1)\Omega\ \nabla^{m}\ \nabla_m\Omega - n(n-1)\ (\nabla^{m}\Omega)\ (\nabla_m\Omega). \tag{A8}$$

REFERENCES

1. R. Arnowitt, S. Deser, C. W. Misner, Phys. Rev. 118, 1100 (1960); 121, 1556 (1961); 122, 997 (1961).
2. R. Geroch, J. Math. Phys. 13, 956 (1972).
3. H. Bondi, M. G. J. Van Der Berg, A. W. K. Metzner, Proc. Roy. Soc. A269, 21 (1962).
4. R. Sachs, Proc. Roy. Soc. A270, 193 (1962).
5. R. Penrose, Proc. Roy. Soc. A284, 159 (1965).
6. L. Tamburino, J. Winicour, Phys. Rev. 150, 1039 (1966).
7. R. Sachs, Phys. Rev. 128, 2851 (1962).
8. J. Winicour, J. Math. Phys. 9, 861 (1968).
9. J. Winicour, Phys. Rev. D3, 840 (1971).
10. D. Brill, S. Deser, Ann Phys. 50, 548 (1968).
11. B. Schmidt, M. Walker, P. Sommers, J. Gen. Rel. & Grav. 6, 489 (1975).
12. E. T. Newman, R. Penrose, Proc. Roy. Soc. A305, 175 (1968).
13. D. Eardley, R. Sachs, J. Math. Phys. 14, 209 (1973).

THE PRODUCTION OF ELEMENTARY PARTICLES BY STRONG GRAVITATIONAL FIELDS*

Leonard Parker

Department of Physics

University of Wisconsin-Milwaukee

CONTENTS

1. INTRODUCTION

The creation of particles by gravitational fields is a natural consequence of quantum field theory in curved space-time. It occurs in particle-antiparticle pairs in the models considered, and does not violate the local conservation laws. This process has important consequences in cosmological and black hole metrics.

In these lectures, the material on quantized fields and particle production in curved space-time will be covered in sufficient detail to serve as an introduction for the uninitiated, as well as to bring one to the level of current research in the field. We will deal with quantized particle fields in the presence of gravity characterized by a classical metric. In this way one can go quite far (to the dimensions characterized by the gravitational constant G, ħ, and c, known as the Planck dimensions) without having to face the uncertainties and difficulties associated with the quantization of the gravitational field itself. Before turning to quantum field theory in curved space-time, let me discuss the significance of this particle creation in cosmological and black hole space-times, and give a heuristic overview of some of the main results.

Quantum field theory predicts that significant pair creation caused by the expansion of the universe[1-9] will occur in the very early stages of the expansion (except in the particular model in which the equations governing the particle fields are all conformally invariant and the expansion is perfectly isotropic with no event horizons). If this particle creation at very early times is characterized by the fundamental constants G, $\hbar$, and c, then it would have two important consequences. First, the reaction back of the created particles on the metric would almost instantaneously bring about isotropic expansion from initially anisotropic expansion rates.[7,11] Second, the initial production of pairs, and the subsequent loss of information about the correlation of the members of each pair due to scattering, decays, and large spatial separations, would produce entropy of the order of magnitude demanded by the observed 3°K black-body radiation.[7,12,13] The entropy is of significance here because it is approximately conserved in a comoving volume element as the universe expands (even before the decoupling of matter and radiation.)[14]

Quantum field theory also predicts that black holes of mass less than about 10^{15} gm will create particles (including radiation) at a significant rate, for non-rotating[15-17] as well as rotating[18-24] black holes. This permits one to set an upper limit (of about 10^{-37} gm/cm^3) on the present average density of such black holes.[25,26] Their mass is evidently too small to result from stellar evolution. However, they might be formed in the very early universe.[27,28] The rate of formation of primordial black holes from density fluctuations in the very early universe is sensitive to the equation of state of the hot "gas" of hadrons present at such early times ($t \sim 10^{-4}$ sec). The above upper limit on the density seems to favor a hard equation of state like $p = \rho/3$ over a soft one like $p = \rho_* \ln(\rho/\rho_*)$ [where p = pressure, ρ = density, ρ_* = constant $\sim 10^{14}$ gm/cm^3].[26] A hard equation of state tends to prevent the formation of density fluctuations large enough to produce such black holes.

Heuristic Discussion, Cosmology

It is useful for the purpose of orientation to outline the main results by means of heuristic and dimensional arguments leading to order of magnitude estimates. The real justification is given in the calculations in the later sections. In the present units, $\hbar = c = 1$.

In many cosmological metrics, the frequency or energy ω of the quantized field is analogous to the frequency of a quantized simple pendulum, and the volume V of a co-moving volume element is analagous to the length ℓ of the pendulum's suspending string. The number of quanta or particles of energy ω in the field is analogous to the quantum number of the energy level of the pendulum (simple harmonic oscillator). If the pendulum is originally in its ground state and the length ℓ is changed at a rate $\dot{\ell}$, then the probability of excitation approaches unity when the average frequency ω of the pendulum satisfies $\omega \lesssim |\dot{\ell}/\ell|$. Similarly, in the expanding (or contracting) universe, one expects (except in special cases) that the probability of pair production approaches unity for all energies such that

$$\omega \lesssim |\dot{V}/V| \quad . \tag{1.1}$$

If the classical metric has a cosmological singularity at $t = 0$, then typically $|\dot{V}/V| \sim t^{-1}$. Thus, it is the particle creation at the earliest times which is most significant. If no particles are present at some early time t_o, then the particle production occurring near time t_o will dominate that at later times, and pairs will be created for all energies $\omega \lesssim t_o^{-1}$. If t_o^{-1} is large with respect to the elementary particle rest masses and interaction energies, then from dimensional considerations it is clear that the energy density ρ of the particles created at about time t_o will be

$$\rho \sim t_o^{-4} \quad . \tag{1.2}$$

This density will of course decrease as the universe expands, but will nevertheless dominate the density of particles created at later times, since that latter density decreases at least as fast as t^{-4}. The time t_o is usually assumed on the basis of dimensional considerations to be of the order of the time known as the Planck time t_p, characterized by the fundamental constants G, ħ, and c,

$$t_o \sim t_p = (G\,\hbar/c^5)^{\frac{1}{2}} = 5.4\times 10^{-44}\ \text{sec.} \tag{1.3}$$

As the cosmological singularity is approached, quantum effects become increasingly important. For example, there exist models in which, for certain choices of the quantum state of the particle field, a contracting universe will avoid the singularity and pass over into an expansion phase.[29] The avoidance was caused in that case by the coupling of the expectation value of the quantum field's energy-momentum tensor to the classical metric through Einstein's equation. The maximum density of created particles was given by Eq. (1.2) with t_o of the order of the Compton time. However, in the particular case considered, not all the matter present came from the particle creation mechanism, and the avoidance of the singularity occured at about the Compton time associated with the quantized field. It is therefore not inconceivable that more general effects, not particularly dependent on the quantum state of the system, may occur at still earlier times which may serve to effectively suppress the particle creation. Since the fundamental constants entering into the relevant equations are evidently G, ħ, and c, one expects the characteristic time at which such a suppression of particle production would occur is probably of the order of t_p, so that t_o would be given by Eq. (1.3).

The production of particles by strong gravitational fields is analogous to pair production by strong electric fields,[30-32] and a similar heuristic picture can be used to a certain extent. Because the gravitational field couples to all kinds of particles, one can think of short-lived virtual pairs being created in the vacuum, just as in quantum electrodynamics, but without the need for electric charge. For example, one can also think of virtual pairs of massless neutral particles (photons, gravitons, neutrinos) existing in the vacuum. The maximum lifetime of a virtual pair is limited by the uncertainty principal to a time of order ω^{-1}, where ω is the energy of each member of the pair. (We ignore numerical factors in the following rough estimates, and use units with $\hbar = c = 1$). Let $\vec{F}$ be the differential gravitational force tending to separate the members of a pair (it is related to the Riemann tensor through the equation of geodesic deviation). If $\vec{F}$ is strong enough to give the members of a properly oriented virtual pair an energy of order ω in a time of order ω^{-1}, the pair can become real, and significant pair creation from vacuum will occur. Since the particles can travel no further than a distance ω^{-1} in a time ω^{-1}, the work done on the members of the pair must satisfy

$$\int_0^{\omega^{-1}} \vec{F}\cdot d\vec{\ell} \gtrsim \omega \tag{1.4}$$

for significant pair production to occur. This heuristic picture can be applied to both the cosmological and black hole situations to get a rough idea of the kind of results that the machinery of quantum field theory predicts.

Let us first briefly consider the cosmological case from this viewpoint.[7] A simple example of an anisotropically expanding universe is the Kasner metric

$$ds^2 = dt^2 - t^{2p_1} d\xi^2 - t^{2p_2} d\eta^2 - t^{2p_3} d\zeta^2 , \tag{1.5}$$

which is a vacuum solution of the Einstein equations if the real numbers p_1, p_2, p_3 satisfy the relations

$$p_1 + p_2 + p_3 = p_1^2 + p_2^2 + p_3^2 = 1 . \tag{1.6}$$

Thus, the universe expands along two axes and contracts along the third. For example, $p_3 = -\frac{1}{3}$, $p_1 = p_2 = \frac{2}{3}$ satisfy the above relations. The case $p_3 = 1$, $p_1 = p_2 = 0$ is the degenerate Kasner universe, which is equivalent to a region of Minkowski space in a curvilinear coordinate system. A free particle on a geodesic at fixed spatial coordinates (ξ,η,ζ), has a proper distance from the origin given by $(x,y,z) = (t^{p_1}\xi,\ t^{p_2}\eta,\ t^{p_3}\zeta)$, and experiences an effective gravitational acceleration which has a component directed away from the origin only along the axis (say the z axis) with respect to which a contraction is occuring (i.e. the proper distance from the origin is decreasing). A virtual pair produced at the origin and oriented along the z axis will therefore tend to be pulled apart by an effective force of order $\omega p_3(p_3-1)t^{-2}z$ (which vanishes in the degenerate Kasner). Hence, at time t significant pair production by the gravitational field will occur according to Eq. (1.4) for all energies such that

$$\omega \lesssim t^{-1} , \tag{1.7}$$

which agrees with the earlier discussion based on the analogy to the harmonic oscillator with a changing frequency. At sufficiently early times that the rest mass m can be neglected with respect to t^{-1}, it follows from dimensional argument that the energy density of the newly created particles is

$$\rho \sim t^{-4} . \tag{1.8}$$

If no particles are present at some early time t_o, then the main particle production will occur near time t_o, and the maximum energy density of created particles will be given by Eq. (1.2). The particles created at later times will have an energy density which is small compared to the particles created near t_o, as discussed following Eq. (1.2) [for an approximately Kasner universe $V \propto t$, so that the density of the relativistic particles created at t_o will fall as $t^{-4/3}$, while that of the newly created particles is proportional to t^{-4}, if one takes for example approximately isotropic principal pressures].

Zeldovich[7] pointed out that if t_o were of the order of t_p [Eq. (1.3)], then the created particles would be expected to very rapidly isotropize the expansion rates, thus dynamically producing isotropic expansion from anisotropic initial conditions. This was investigated numerically by Lukash and Starobinsky,[11] who integrated (with suitable approximations) the Einstein equations for the anisotropically expanding metric

$$ds^2 = dt^2 - a_1{}^2(t)d\xi^2 - a_2{}^2(t)d\eta^2 - a_3{}^2(t)d\zeta^2 , \qquad (1.9)$$

which take the form

$$\sum_{i=1}^{3} \ddot{a}_i/a_i = -8\pi G\rho , \qquad (1.10)$$

$$\frac{\ddot{a}_1}{a_1} + \frac{\dot{a}_1\dot{a}_2}{a_1a_2} + \frac{\dot{a}_1\dot{a}_3}{a_1a_3} = 8\pi GP_1 \quad \text{et cyc.} \qquad (1.11)$$

(Here dot denotes d/dt and P_i are the principal pressures.) They took $a_1 = a_2 \neq a_3$ for simplicity and assumed initial conditions on the metric near t_o corresponding to a Kasner universe with $p_1 = p_2 = 2/3$ and $p_3 = -1/3$. (The coordinates were scaled so that at t_o one had $a_1 = a_2 = a_3$). They treated the matter created near t_o as a classical relativistic gas of density $\rho \sim t_o{}^{-4}$ (put in as an initial condition) ignoring the reaction back of the particle creation process itself on the metric. For that to be consistent it was necessary to take t_o large with respect to t_p. For $t \geq t_o$, the system was treated entirely classically with no further particle production. Integration of the Einstein equations with the above initial conditions imposed at t_o gives the following results (see Fig. 1).

For the interval $t_o \leq t \lesssim t_o\,(t_o/t_p)^{3/2}$ the expansion continues approximately as a Kasner solution with $a_1 = a_2 \propto t^{2/3}$ and $a_3 \propto t^{-1/3}$. At $t \approx t_o\,(t_o/t_p)^{3/2}$ the presence of the matter reacts back on the expansion in such a way that a_3 begins to

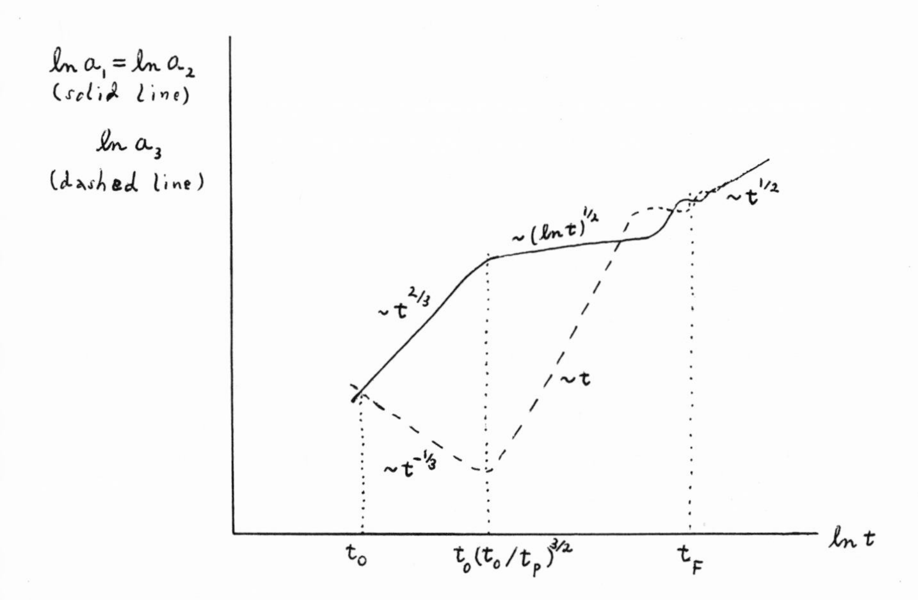

Fig. 1. Damping of anisotropy (V.N. Lukash and A.A. Starobinsky, Ref. 11)

expand rapidly while the rate of expansion of $a_1 = a_2$ is slowed. After a_3 becomes larger than a_1 the situation is reversed, so that a_3 remained almost constant while a_1 increases. The expansion parameters a_1 $(=a_2)$ and a_3 undergo two or three such quasi-oscillations of diminishing amplitude until the time (in the approximation of non-interacting particles)

$$t_F \approx t_o\{(t_o/t_p) \ln (t_o/t_p)\}^3 , \tag{1.12}$$

at which point $|\ln(a_3/a_1)| < 1$, and one has an effectively isotropic expansion with

$$a_1 = a_2 \approx a_3 \propto t^{1/2} \tag{1.13}$$

corresponding to the Friedmann universe dominated by relativistic particles and radiation. (The isotropization occurs even more rapidly if there are interactions strong enough to thermalize the particle distribution almost immediately.) By extrapolating these results back to t_o not much larger than t_p one infers that the isotropization should occur almost instantaneously in that case. Observational limits on the temperature anisotropy of the 3^oK black-body radiation together with theoretical studies of the evolution of that radiation in homogeneous anisotropically expanding models evidently require that $t_F \lesssim 10^7\ t_p \sim 10^{-36}$ sec.[11,33]

For t_o near t_p the reaction back of the created particles is almost instantaneous, so that one must consider the coupling between the metric and the particle field even during the process of particle creation. One way of doing this, which stops short of quantizing the gravitational field itself, is to use the expectation value $\langle T_{\mu\nu}\rangle$ of the energy-momentum tensor of the quantized field as the source of the gravitational field in Einstein's equation,[34]

$$G_{\mu\nu} = -8\pi G\langle T_{\mu\nu}\rangle \ . \tag{1.14}$$

Although there are cases in which it seems unreasonable to generate the gravitational field from an average of the matter distribution, the use of Eq. (1.14) appears justified for homogeneous cosmological models. In order to use Eq. (1.14) one must define a suitable expression for $\langle T_{\mu\nu}\rangle$ in such a metric through regularization or renormalization.[8,23,35-39] One would hope to show that rapid isotropization occurs for essentially all reasonable choices of the initial state vector.

For the above type of models in which the particle production is suppressed for $t \lesssim t_p$, it is of interest to make a definite calculation of the entropy produced and of the spectrum of created particles.[12,13] Although the original spectrum will not in general survive the thermalizing mechanisms which occur before the decoupling of matter and radiation, the entropy created in any co-moving volume will be approximately conserved.[14,40,41] For a class of models in which the space-time is assumed to approach empty Minkowski space in the distant past and joins smoothly to either an anisotropically or isotropically expanding universe, one finds that the original spectrum of created particles is that of black-body radiation to good approximation, even in the absence of interactions between the particles (in the anisotropically expanding universe with no particle interactions the radiation is Doppler shifted anisotropically). Entropy is generated because the particles are created in pairs, and the correlations between the members of a pair is effectively lost in any realistic model because of interactions, decays, scattering, or large spatial separation. Most of the particle production occurs during the initial stage of the expansion. If that initial stage is characterized by G, $\hbar$, and c, then one generates from the originally empty flat space-time an expanding universe dominated by relativistic particles and radiation, with a temperature of order $\hbar\, t_p^{-1}\, k_B^{-1} \sim 10^{32}\,^{o}K$ at $t \sim t_p$ (where k_B = Boltzmann's constant). Assuming that the Einstein equations are valid for $t \gtrsim t_p$ the expansion will rapidly become a Friedmann expansion dominated by relativistic particles and radiation, even if the expansion rates were originally anisotropic. The entropy produced is consistent with the requirements of the Einstein equations, and of the observed $3^{o}K$ radiation. Thus, the particle creation process occuring near t_p seems to offer a natural explanation for the origin of the entropy now observed in the $3^{o}K$ black-body radiation. Any other process which leads to a universe dominated by relativistic particles and radiation at early times, and which is consistent with Einstein's equations, would also agree with observations. However, the particle production process seems to offer the simplest explanation of the origin of the required entropy, and at the same time offers a dynamical explanation of the isotropy of the expansion.

The fact that the particle production process yields equal quantities of matter and antimatter is not necessarily inconsistent with the usual picture of the hot early universe, since in thermal equilibrium it would contain almost equal quantities of matter and antimatter. It would only require a small fractional excess of baryons over antibaryons to lead to a present universe filled with matter (as opposed to antimatter), even in the absence of matter-antimatter separation mechanisms (for discussion of such mechanisms and references see Ref. 41). The necessary baryons could be introduced in the initial state, or might result from interactions which violate both CP and baryon number conservation in the early universe.

Heuristic Discussion, Black Holes

Let me now apply the previous heuristic method to estimate the energy flux one would expect to be created by the gravitational field of a spherical black hole. I will ignore all numerical factors, and for simplicity also the red shift of the emitted radiation. For a more sophisticated approach including red shift see Eqs. (5.15)-(5.17) in Section 5. The real justification for the results lies in a calculation involving quantum field theory, first carried out by Hawking[15,16] (see Section 4).

Consider a Schwarzschild black hole of mass M.

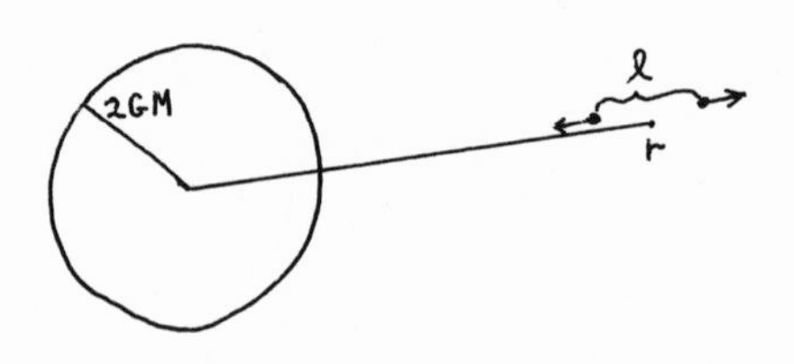

Pair creation by non-rotating black hole

For a radially oriented virtual pair (of photons, for example), the graviational tidal force is such as to separate the members of the pair. If ℓ is the separation of a pair at average radial distance r, then the tidal force tending to separate the members of a pair is of order

$$F \sim \omega \frac{GM}{r^3} \ell \tag{1.15}$$

where ω is the energy of each particle, as one can see either by using the formula for geodesic deviation, or the Newtonian picture. Recall that the condition for particles of energy ω to be produced at a significant rate is that the energy ω be imparted to the pair in a distance of ω^{-1} or less [Eq. (1.4)]:

$$\int_0^{\omega^{-1}} F d\ell \gtrsim \omega \ .$$

This condition yields with the above expression for F,

$$GMr^{-3}\omega^{-1} \gtrsim \omega \ . \tag{1.16}$$

For the outgoing photon to reach infinity one must have $r \gtrsim 2GM$. Hence, there will be a significant creation rate for all ω such that

$$\omega \lesssim (GM)^{-1} \ . \tag{1.17}$$

A derivation based on the uncertainty principle, and including the gravitational red shift, is given in Section 5, Eqs. (5.15)-(5.17). As will be discussed in Section 5, the incoming particle actually decreases the mass of the black hole, i.e. in the curved space-time the incoming particle carries negative energy relative to infinity.

To estimate the luminosity L or power radiated by the black hole, roughly approximate the spectrum by taking $dL/d\omega$, the luminosity per unit frequency, to be independent of ω up to the frequency $\omega \sim (GM)^{-1}$ given by Eq. (1.17), after which it falls rapidly to zero. (The actual spectrum turns out to be thermal in nature, but the present approximation is good enough for our rough estimate.) For $\omega < (GM)^{-1}$ the constant value of $dL/d\omega$ must be formed from GM, c = 1, and $\hbar = 1$. The only possibility is

$$\frac{dL}{d\omega} \sim \begin{cases} (GM)^{-1} & \text{for} \quad \omega \lesssim (GM)^{-1} \\ 0 & \text{for} \quad \omega > (GM)^{-1} \end{cases} . \tag{1.18}$$

Then

$$L \sim \int_0^{(GM)^{-1}} (GM)^{-1} d\omega \sim (GM)^{-2} , \tag{1.19}$$

or in cgs units

$$L \sim \hbar c^6 (GM)^{-2} . \tag{1.20}$$

We have neglected numerical factors and the red shift of the emitted radiation, which would tend to lower the above value of L by several orders of magnitude. The quantum field theory calculation gives

$$L \approx 10^{-4} \hbar c^6 (GM)^{-2} \approx L_\odot \left(\frac{2\times 10^6 \text{gm}}{M}\right)^2 , \tag{1.21}$$

where $L_\odot = 3.9 \times 10^{33}$ erg sec^{-1}.

The actual spectrum is found to be thermal in nature (it would be that of a perfect black-body except for backscattering from the curved geometry). To estimate the temperature T, treat the black hole as a black-body radiator of radius 2GM. Then its luminosity (with $\hbar = c = 1$) is

$$L \sim k_B^{\,4}\, T^4 (GM)^2 . \tag{1.22}$$

Comparison with Eq. (1.20) gives

$$T \sim (k_B\, GM)^{-1} = \hbar c^3\, (k_B GM)^{-1} . \tag{1.23}$$

The temperature increases as the mass decreases because the tidal force producing the pairs increases. The exact calculation based on quantum theory yields (see Section 4)

$$T = \frac{\hbar c^3}{8\pi k_B GM} = (1.2 \times 10^{26}\ {}^{o}K)\left(\frac{1\ gm}{M}\right) . \tag{1.24}$$

A solar mass black hole has a temperature of only about 10^{-7} ^{o}K. For $M \sim 10^{14}$ gm one has $T \sim 10^{12}$ ^{o}K. Clearly, the particle creation process is significant only for the less massive black holes which may have formed in the early universe.

As the black hole radiates, its mass must decrease by conservation of energy. Therefore, its temperature increases, so that the final stage of the process is explosive. The rate of mass loss must equal the luminosity, or

$$\frac{d}{dt}(M_c^2) \approx -10^{-4} \frac{\hbar c^6}{G^2 M^2} \tag{1.25}$$

from which it follows that the lifetime of a black hole of mass M will be

$$\tau \approx 10^4 \frac{G^2 M^3}{\hbar c^4} = 10^{-26}\ \sec \left(\frac{M}{1\ gm}\right)^3 . \tag{1.26}$$

Thus, a black hole of mass less than about 2×10^{14} gm has a lifetime of less than 10^{17} sec so that such a black hole present in the early universe would have radiated away its mass by now. A black hole with $Mc^2 \approx 10^{29}$ erg will radiate away its energy in a time of about 0.1 sec, or less, depending on the number of types of different elementary particles it can radiate.

The particle creation process is relevant to a generalization of the second law of thermodynamics which includes black holes as well as other forms of matter. Before the discovery of radiation by black holes Bekenstein[42-44] showed, on the basis of Hawking's area theorem[45] and by considering specific examples, that a generalized entropy which does not decrease could be defined by adding to the entropy of the external matter a "black hole entropy" proportional to the area of the event horizon of the black hole, and that the temperature of a black hole was a multiple of the surface gravity. The radiation process determines the proportionality constant for that entropy. When the radiation process occurs quasi-statically one can write for the entropy change of the black hole

$$dS = \frac{d(Mc^2)}{T} . \tag{1.27}$$

Since the area of the event horizon is $A = 4\pi(\frac{2GM}{c^2})^2$ and $T = \frac{\hbar c^3}{8\pi k_B GM}$, this is

$$dS = \frac{1}{4} k_B \left(\frac{c^3}{G\hbar}\right) dA . \tag{1.28}$$

Taking $S \to 0$ when $A \to 0$, one has for the black hole entropy

$$S = \frac{1}{4} k_B \left(\frac{c^3}{G\hbar}\right) A . \tag{1.29}$$

(Of course, if a naked singularity remains after the black hole evaporates, one may have to attribute an additional entropy to it.) The entropy in Eq. (1.29) is typically quite large.

Remarks

Particle production near the cosmological singularity (and at a lower rate in the present universe) is the result of a mixing of positive and negative frequency components of the particle field caused by the time-dependence or curvature associated with the metric. In the case of a non-rotating black hole, the mixing of positive and negative frequency components, and hence the particle creation, results from the presence of an event horizon. (For a collapsing body one will of course also have a mixing due to the time-dependence of the metric.) Recently, Unruh[46] has suggested, on the basis of earlier work by Fulling[47,48] and through the analysis of particle detectors, that accelerated particle detectors even in empty flat space-time have some probability of excitation, corresponding to detection of particles. Thus, the detection of particles would depend on the world line of the observing apparatus. Gibbons and Hawking[49] have generalized those results to a class of cosmological metrics corresponding to certain solutions of Einstein's equation with a cosmological constant. In those universes there are geodesics which have an associated event horizon. Particle detectors on such geodesics evidently would observe a flux of particles due to the event horizon, analogous to that from a black hole. (For related results see Ref. 50.) The temperature of that radiation would be below the limits of observation in a universe of realistic dimensions.

The first to realize that a mixing of positive and negative frequency waves will occur in an expanding universe was Schrödinger.[9,10] He considered the propagation of classical waves, and showed that a wave traveling in a given direction will produce a weaker wave traveling in the opposite direction, as though reflection from the expanding space were occuring. However, in contrast to what occurs in true reflection, the amplitude of the forward moving wave is increased rather than diminished by the process. He surmized (without second quantization) that this phenomenon corresponded to the creation of a pair, with one member moving in the forward direction of the wave and the second member in the backward direction. Imamura[51] considered a universe whose radius jumps instantaneously from one value to another, and showed by means of quantum field theory that an infinite number of particles would be created by the discontinuity. Later, on the basis of quantum field theory, the creation of particles by the expanding universe was independently rediscovered and investigated for isotropic expansions by Parker,[1-4] and by Sexl and Urbantke.[5,6]

The method used in Ref. 1-4 was applicable to strong gravitational fields. It was extended to anisotropically expanding universes by Zeldovich[7] and Zeldovich and Starobinsky.[8]

The quantum field theoretical framework is developed in Section 2. The cosmological case is discussed further in Section 3, and the black hole case in Section 4. In section 5 a number of topics are considered, such as particle production by singularities, and conservation of energy-momentum and pair creation. Part of the material in Sections 3 and 4 was also presented in Ref. 12. The reader should keep in mind that the material in Section 2 forms the basis for the calculations in Sections 3 and 4. However, Sections 3 and 4 are independent of one another, and the second part of Section 5 (on conservation laws and pair creation) is to a large extent self-contained. Although the present lectures are largely a review, I have included some unpublished ideas and calculations which are to my knowledge new.

2. QUANTIZED FIELDS IN CURVED SPACE-TIME

In curved space-time, the simplest generalization of the Klein-Gordon equation is obtained by replacing derivatives ∂_μ by covariant derivatives ∇_μ. This procedure gives the so-called minimal coupling to the gravitational field, but there is nothing that prevents one from adding a term proportional to the scalar curvature, i.e. $\nabla_\mu \nabla^\mu \phi + (m^2 + \gamma R)\phi = 0$ or

$$(-g)^{-\frac{1}{2}}\, \partial_\mu[(-g)^{\frac{1}{2}} g^{\mu\nu} \partial_\nu \phi] + (m^2 + \gamma R)\phi = 0 \;, \qquad (2.1)$$

where R is the scalar curvature, and γ is a real constant. (We use the following conventions: metric signature -2; $(\nabla_\mu \nabla_\nu - \nabla_\nu \nabla_\mu)\xi_\lambda = \xi_\sigma R^\sigma{}_{\lambda\mu\nu}$; $R_{\mu\nu} = R^\lambda{}_{\mu\lambda\nu}$; $R = R^\lambda{}_\lambda$.) The field ϕ describes particles of spin 0. We will canonically quantize that field in order to apply the theory to particle production in cosmology and gravitational collapse. It would introduce unnecessary complications to discuss higher spin fields here.[52-56] We will not introduce non-gravitational interactions (for an interesting new aspect of such interactions in curved space-time see Adler, Lieberman, Ng, and Tsao, Ref. 152, and Adler, Ref. 153).

If g_1 and g_2 are two solutions of Eq. (2.1), then one can define a conserved scalar product

$$(g_1, g_2) = -i\int_S (g_1 \partial^\mu g_2{}^* - g_2{}^* \partial^\mu g_1)(-g)^{\frac{1}{2}}\, dS_\mu \,, \tag{2.2}$$

where dS_μ is the future directed surface element of the 3-dimensional hypersurface S which is taken to be a complete Cauchy hypersurface for Eq. (2.1). If the coordinates are chosen so that S is a constant t hypersurface, then $dS_o = dx^1 dx^2 dx^3$ and $dS_i = 0$ $(i = 1, 2, 3)$. The scalar product (2.2) is conserved under smooth displacements and deformations of S. This follows from the self-adjointness of the operator in Eq. (2.1) by means of the 4-dimensional divergence theorem and integration by parts. This scalar product satisfies $(g_1,g_2)^* = (g_2,g_1)$, but is not positive definite.

The Lagrangian density of the neutral (hermitian) scalar field is

$$\mathcal{L} = \frac{1}{2}(-g)^{\frac{1}{2}} [g^{\mu\nu}\, \partial_\mu\phi\, \partial_\nu\phi - (m^2 + \gamma R)\phi] \,. \tag{2.3}$$

The canonical momentum conjugate to ϕ is

$$\pi = \partial\mathcal{L}/\partial(\partial_o\phi) = (-g)^{\frac{1}{2}}\, g^{o\mu}\, \partial_\mu\phi \,. \tag{2.4}$$

(The charged scalar field can be written as a superposition of hermitian fields, so that one can easily generalize these results to the charged field.) To quantize the field one now regards ϕ and π as operators (operator valued distributions) and imposes the canonical commutation relations

$$[\phi(\vec{x},t),\ \phi(\vec{x}',t)] = 0\ , \quad [\pi(\vec{x},t),\ \pi(\vec{x}',t)] = 0$$

and

$$[\phi(\vec{x},t),\ \pi(\vec{x}',t)] = i\delta(\vec{x}-\vec{x}')\ , \tag{2.5}$$

in analogy to the position and momentum of a particle in quantum mechanics. The Heisenberg equations of motion obtained from the Hamiltonian

$$H(t) = \int d^3x\ (\pi\partial_o\phi - \mathcal{L}) \tag{2.6}$$

are equivalent to the original equation of motion (2.1). For any function F which can be written as a power series in terms of ϕ, $\partial_j\phi$ $(j = 1,2,3)$, and π, one has[57]

$$[F,H]_t + i\ \partial F/\partial t = i\ dF/dt\ , \tag{2.7}$$

where $\partial F/\partial t$ is with respect to explicit time dependence, if any, not contained in the dependence of F on ϕ, $\partial_j\phi$, and π. (For $F = \pi$, one gets the original field equation.) The canonical commutation relations (2.5) are propagated consistently by the equation of motion. Thus, using Jacobi's identity[58]

$$-i\ \frac{d}{dt}\ [\phi,\pi] = [[H,\phi],\pi] + [\phi,[H,\pi]] = [H,[\phi,\pi]] = 0\ , \tag{2.8}$$

since $[\phi,\pi]$ is not an operator (here ϕ and π are evaluated at two spatial points at time t, and H refers to H(t)). Because of the freedom in choosing a coordinate system, it follows that if the canonical commutation relations (2.5) hold on one spacelike Cauchy hypersurface then they hold on any such hypersurface. Therefore, they hold in any coordinate system with spacelike constant time hypersurfaces, so that the field algebra and equation of motion form a generally covariant system.[59,60]

The field operators act on state vectors $|\ \rangle$ which describe the possible states of the system. The construction of a state vector space (Fock space) in curved space-time is not as straightforward or unambiguous as the construction of the field algebra. First let us consider the construction in flat space-time.

Minkowski Space

In flat space-time one now defines a set of positive frequency solutions f_j of Eq. (2.1) as proportional to the plane waves $\exp[i(\vec{k}\cdot\vec{x}-\omega t)]$ of definite momentum $\vec{k}$ and energy $\omega = (k^2 + m^2)^{\frac{1}{2}}$, or as superpositions of such plane waves. These solutions are normalized so that

$$(f_j, f_\ell) = \delta_{j\ell} . \tag{2.9}$$

It follows that

$$(f_j^*, f_\ell^*) = -\delta_{j\ell} , \quad (f_j, f_\ell^*) = 0 . \tag{2.10}$$

The f_j are chosen so that the f_j and f_j^* are a complete set for expanding any solution of (2.1). Thus, one can write the hermitian field ϕ as

$$\phi = \sum_j (a_j f_j + a_j^\dagger f_j^*) \tag{2.11}$$

where the a_j and $a_j^\dagger$ are operators with no space or time dependence. The canonical relations (2.5) imply that the

$$[a_j, a_\ell^\dagger] = \delta_{j\ell} , \quad [a_j, a_\ell] = 0 . \tag{2.12}$$

(To derive Eqs. (2.12) use $a_j = (\phi, f_j) = -i\int_{S_t} (\phi\sqrt{-g}\, f_j^* - f_j^*\pi)\, d^3x$, where S_t is a constant t hypersurface. The same derivation holds in curved space-time.) The operator a_j is called an annihilation operator for particles described by the single-particle wave function f_j because of the operator's action on the basis states defined below. Similarly, the hermitian adjoint operator $a_j^\dagger$ is called a creation operator for such particles. The a_j and $a_j^\dagger$ resemble ladder operators for an harmonic oscillator, and the particles are analogous to quanta of the oscillator, the ground state being the state with no quanta or the vacuum state. One defines the vacuum state by

$$a_j|0\rangle = 0 \qquad \text{for all } j \ , \tag{2.13}$$

and proceeds to build the space of state vectors as for the harmonic oscillator. The state containing n_j particles, each described by wave function f_j, n_ℓ described by f_ℓ, etc. is

$$|n_j, n_\ell, \ldots\rangle = (n_j!n_\ell!\ldots)^{-\frac{1}{2}}(a_j^\dagger)^{n_j}(a_\ell^\dagger)^{n_\ell}\ldots|0\rangle \ . \tag{2.14}$$

The set of states described by (2.14) which contain a finite total number of particles form the basis of a vector space called the Fock space of the system. It is convenient to choose the f_j so that the set of indices represented by j are discrete, or if the plane waves $\exp[i(\vec{k}\cdot\vec{x}-\omega t)]$ are used, to impose periodic boundary conditions in a cube of side L, so that $k_j = 2\pi n_j L^{-1}$ (with n_j an integer) and ω ranges over a discrete spectrum. This imposition of periodic boundary conditions is regarded as a convenient mathematical device without physical significance, as the quantity L is to approach ∞ after quantities of physical interest have been calculated. The operator

$$N_j = a_j^\dagger a_j \tag{2.15}$$

corresponds to the number of particles of type f_j. If one is dealing with plane waves with periodic boundary conditions, then

$$N_{\vec{k}} = a_{\vec{k}}^\dagger\, a_{\vec{k}} \tag{2.16}$$

corresponds to the number of particles of momentum $\vec{k}$ in the volume L^3, and the Hamiltonian of Eq. (2.6) takes the form

$$H = \frac{1}{2}\sum_{\vec{k}} \omega_k \left\{a_{\vec{k}}^\dagger\, a_{\vec{k}} + a_{\vec{k}}\, a_{\vec{k}}^\dagger\right\} . \tag{2.17}$$

This operator has infinite expectation values even in the vacuum state. It is replaced by an operator that has the same algebraic properties by subtracting off the vacuum expectation value of (2.17), i.e. $\frac{1}{2}\sum_{\vec{k}} \omega_k$, mode by mode. This "regularized" Hamiltonian is "normal ordered" in the sense that annihilation operators appear to the right of creation operators:

$$H_{reg} = \sum_{\vec{k}} \omega_k \, a^\dagger_{\vec{k}} \, a_{\vec{k}} \,, \qquad (2.18)$$

so that the vacuum expectation value vanishes.

Curved Space-Time

In curved space-time, one encounters difficulties in defining physically relevant positive frequency solutions. In addition, one can no longer regularize the Hamiltonian and the energy-momentum tensor by simply normal ordering. Even in a static or stationary space-time in which one can attempt to define positive frequency solutions in the same way as in Minkowski space (i.e. through $\exp(-i\omega t)$ time-dependence), one cannot be confident that normal ordering of H is the correct way to define the regularized Hamiltonian because there may exist a non-vanishing vacuum energy resulting from the curvature or the topological properties of the space-time, in analogy with the vacuum energy between conducting plates associated with the Casimir effect,[98-102] as discussed in Sections 3 and 5.

One cannot define the space of positive frequency solutions simply by demanding that Eq. (2.9) be satisfied in the curved space-time by the basis wave functions. For example, if we let

$$f'_j = \alpha_j f_j + \beta_j f^*_j \qquad (2.19)$$

with

$$|\alpha_j|^2 - |\beta_j|^2 = 1 \,, \qquad (2.20)$$

then one will have

$$(f'_j, f'_\ell) = \delta_{j\ell} \tag{2.21}$$

although f'_j is not pure positive frequency with respect to the f_j basis. That the f_j and f'_j cannot describe the same particles with respect to a given observer can be seen as follows. Expand the field in terms of the f'_j and the f_j, namely

$$\phi = \sum_j (a'_j f'_j + a'^\dagger_j f'^*_j) , \tag{2.22}$$

and

$$\phi = \sum_j (a_j f_j + a^\dagger_j f^*_j) . \tag{2.23}$$

Then

$$a'_j = (\phi, f'_j) = \sum_\ell ([a_\ell f_\ell + a^\dagger_\ell f^*_\ell], [\alpha_j f_j + \beta_j f^*_j]) , \tag{2.24}$$

or

$$a'_j = \alpha^*_j a_j - \beta^*_j a^\dagger_j . \tag{2.25}$$

Similarly

$$a'^\dagger_j = \alpha_j a^\dagger_j - \beta_j a_j . \tag{2.26}$$

Such a transformation of creation and annihilation operators is known as a Bogoliubov transformation.[61] By virtue of Eq. (2.20) it follows that both the primed and unprimed operators satisfy commutation relations of the form (2.12). In the usual way, one can construct the Fock space $\mathcal{H}'$ on $|0'\rangle$ (where $a'_j|0'\rangle = 0$ for all j), and the Fock space $\mathcal{H}$ on $|0\rangle$ (where $a_j|0\rangle = 0$ for all j). The average number of primed particles present in the unprimed vacuum state is

$$N' = \sum_j \langle 0|a'^\dagger_j a'_j|0\rangle = \sum_j |\beta_j|^2 . \tag{2.27}$$

(If N' diverges, then the primed and unprimed vacuum states are not even related by a unitary transformation, i.e. one has unitarily inequivalent representations of the canonical commutation relations.) The number of unprimed particles in the unprimed vacuum is zero. Thus, the primed and unprimed particles cannot both be measured at a given time by the same apparatus. In Section 3, we show that the physically relevant creation and annihilation operators at different times in an expanding universe are related by a Bogoliubov transformation.[1-4]

However, the problems associated with the definition of physically relevant positive frequency solutions already arise in flat space-time, as shown by Fulling.[47,48] A Lorentz transformation does not mix positive and negative frequencies. However, there is another coordinate transformation from Minkowski coordinates to a curvilinear coordinate system in flat space-time, known as Rindler coordinates,[62] in which the metric is static. The time in these coordinates is proportional to the proper time measured by a set of accelerating clocks. Let f_j' be the positive frequency solutions defined with respect to the Rindler time coordinate. These solutions may be found by solving the spin-0 field equation (2.1) in Rindler coordinates through separation of variables. Let f_j be the usual positive frequency solutions based on the Minkowski time coordinate. Fulling showed that the f_j' are related to the f_j by a transformation which mixes positive and negative frequencies,

$$f_j' = \sum_\ell (\alpha_{j\ell}\, f_\ell + \beta_{j\ell}\, f_\ell^*) \tag{2.28}$$

with

$$\sum_\ell (\alpha_{j\ell}\, \alpha_{k\ell}^* - \beta_{j\ell}\, \beta_{k\ell}^*) = \delta_{jk} \tag{2.29}$$

and

$$\sum_\ell (\alpha_{j\ell}\, \beta_{k\ell} - \beta_{j\ell}\, \alpha_{k\ell}) = 0 \ . \tag{2.30}$$

The α's and β's may be calculated using

$$\alpha_{j\ell} = (f_j', f_\ell) \ , \qquad \beta_{j\ell} = (f_j', f_\ell^*) \ . \tag{2.31}$$

Equations (2.29) and (2.30) preserve the orthonormality relations

$$(f'_j, f'_k) = \delta_{jk} , \qquad (f'_j, f'^*_k) = 0 . \tag{2.32}$$

One finds in analogy with Eqs. (2.25) and (2.26) that

$$a_\ell = \sum_j (\alpha_{j\ell}\, a'_j + \beta^*_{j\ell}\, a'^\dagger_j) \tag{2.33}$$

and

$$a'_j = \sum_\ell (\alpha^*_{j\ell}\, a_\ell - \beta^*_{j\ell}\, a^\dagger_\ell) . \tag{2.34}$$

(For example, Eq. (2.34) is obtained by noting that Eq. (2.22) implies that $a'_j = (\phi, f'_j)$. Replacing ϕ by Eq. (2.23) and f'_j by Eq. (2.28) yields Eq. (2.34).) Thus, the primed and unprimed creation and annihilation operators are related by a somewhat more general Boboliubov transformation than in Eqs. (2.25) and (2.26). In the Minkowski vacuum defined by

$$a_\ell |0\rangle = 0 \qquad \text{for all } \ell , \tag{2.35}$$

one has the average number of primed particles in mode j given by

$$N'_j = \langle 0 | a'^\dagger_j\, a'_j | 0 \rangle = \sum_\ell |\beta_{j\ell}|^2 . \tag{2.35}$$

More precisely, Rindler coordinates v,z are related to Minkowski coordinates t,x (2-dimensional case) through ($z > 0$, $-\infty < v < \infty$)

$$t + x = z e^{v} , \qquad t - x = -z e^{-v} , \tag{2.37}$$

so that

$$ds^2 = z^2 dv^2 - dz^2 . \tag{2.38}$$

The metric coefficients are independent of the time coordinate v. The world line z = const. is that of an accelerating clock with instantaneous velocity at time v given by $dx/dt = \tanh v$ (see Fig. 3).

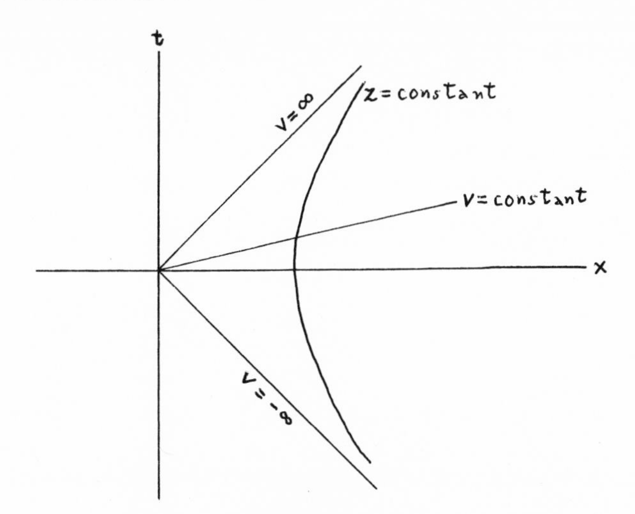

Fig. 2. Path of accelerated clock (z=const) measuring Rindler time

The world lines of the accelerated clocks in Rindler coordinates have common future and past event horizons. They cannot influence events below the past even horizon ($v = -\infty$ in Fig. 2 and its linear extension), and cannot be influenced by events above the future event horizon ($v = \infty$ in Fig. 2 and its linear extension).

Fulling solved the field equation (2.1) in Rindler coordinates (v,z) for the stationary wave functions of time-dependence $\exp(jv)$. He considered the case with mass $m \neq 0$, and found that

$$\alpha_{jk} = [2\pi\omega_k(1 - e^{-2\pi j})]^{-\frac{1}{2}} [(\omega_k + k)/m]^{ij}$$

$$\beta_{jk} = [2\pi\omega_k(e^{2\pi j} - 1)]^{-\frac{1}{2}} [(\omega_k + k)/m]^{ij} ,$$

where $\omega_k = (k^2 + m^2)^{\frac{1}{2}}$ is the energy in Minkowski coordinates, and the variable j runs continuously from 0 to ∞ ($i = \sqrt{-1}$ above). Hence

$$|\alpha_{jk}|^2 = |\beta_{jk}|^2 \exp(2\pi j) . \qquad (2.39)$$

From Eq. (2.29) one has ($\sum_k \to \int_0^\infty dk$ in this 2-dimensional example),

$$1 = \int_0^\infty dk \, (|\alpha_{jk}|^2 - |\beta_{jk}|^2) = (e^{2\pi j} - 1)\int_0^\infty dk \, |\beta_{jk}|^2 ,$$

or

$$N'_j = \int_0^\infty dk \, |\beta_{jk}|^2 = (e^{2\pi j} - 1)^{-1} . \tag{2.40}$$

This is a black-body spectrum (in 1 spatial dimension) as noted by Unruh.[46] An accelerating observer on the world line z = const. has proper time $\tau = zv$ (see Eq. (2.38)), so that the energy conjugate to τ is $E_j = j/z$. Hence, from Eq. (2.40) the temperature of the primed particles is

$$T' = (2\pi \, k_B \, z)^{-1} \, \hbar c . \tag{2.41}$$

The maximum acceleration d^2x/dt^2 of the observer on the world line z = const. is $a = c^2 z^{-1}$. For $a = \gamma g$ (γ a number, g = 980 cm/sec^2), I find $T' = \gamma(4 \times 10^{-20})^\circ K$. It would require a $\sim 10^{20}$ g to produce a temperature for the primed particles of $T' \sim 1^\circ K$.

What is the meaning, if any, of these primed particles? Recently, Unruh[46] has made a fundamental suggestion which may resolve that question. The suggestion is that a localized particle detector (e.g., an atom) will detect the particles corresponding to positive frequency solutions defined with respect to the proper time along the world line of the detector. Unruh supports this idea with two specific models of particle detectors. One may now find the meaning of the a'_j operators. In accordance with standard field theory, the a'_j annihilate particles corresponding to the single particle wave function f'_j. But the f'_j are positive frequency with respect to the Rindler time coordinate (corresponding to the proper time of the set of accelerating observers). Therefore, the a'_j must correspond to the particles which would be detected by measuring devices moving along the above set of accelerating world lines z = const.

If a particle described by wave function f'_j is absorbed by one of those detectors from the original Minkowski vacuum $|0\rangle$, then the new state of the field after detection is proportional to

$$a'_j|0\rangle = - \sum_{\ell} \beta^*_{j\ell} \, a^{\dagger}_{\ell}|0\rangle \ , \tag{2.37}$$

which contains one unprimed particle described by a wave function proportional to $\sum_{\ell} \beta^*_{j\ell} f_{\ell}$. That means, as Unruh pointed out, that when one of the accelerating detectors jumps into an excited state (detection of a primed particle), the inertial detectors will have a finite probability of detecting a newly created particle. From the point of view of the inertial observer, the energy of excitation of the detector as well as that of the newly created particle must come from somewhere. Possibly it comes from the kinetic energy of the detector or from the system producing the acceleration. Such questions of energy balance and consistency have not yet been sufficiently analyzed. If DeWitt's suggestion is adopted that the Rindler coordinates really describe the physics of a mirror moving along the common event horizon of the system of accelerating observers, then the mirror would also be available to lose energy. Davies and Fulling considered this example in two dimensions using the energy-momentum tensor obtained through point-splitting regularization.[63]

Unruh[46] also investigated the question of what particle detectors supported in a Schwarzschild gravitational field (in two dimensions) would measure. He found that if one specifies the state vector by the condition that no particles are detected coming out of the past event horizon of the black hole by a detector using the affine parameter on the past horizon as its "time" coordinate for defining positive and negative frequencies, then a particle detector at a constant radius far from the black hole will observe the thermal radiation of Hawking. (The massless detectors on the incoming null geodesics of the past horizon can possibly be replaced by detectors on timelike incoming geodesics arbitrarily close to the past horizon.) The detector in this case is detecting a particle coming from a pair creation process near the event horizon, while the other particle enters the black hole where it has negative energy relative to infinity. Thus, the mass of the black hole is reduced by the energy absorbed by the detector.

Very recently Gibbons and Hawking[49] have considered cosmological spacetimes, such as de Sitter space, in which observers on geodesic world lines have event horizons. They used the path integral technique applied by Hartle and Hawking[64] to the Schwarzschild metric. They find, using the proper time along each geodesic to define particles, that each such observer will detect thermal radiation coming from his event horizon. This radiation is highly observer-dependent since the event horizons do not have an absolute significance as in the Schwarzschild case. The approach of Ref. 49 has been extended to other examples by Lapedes.[65] A general viewpoint leading to similar results was given by Israel.[66] An approach yielding no particle creation in de Sitter space is given by Candelas and Raine.[130] Alternative vacuum states in de Sitter and other space-times are discussed by Fulling.[67]

In a general non-stationary space-time, a localized particle detector in a local inertial frame will clearly respond to particles corresponding to positive frequency solutions with respect to the local inertial time coordinate, or equivalently, the proper time of the detector. But because the local inertial frame is limited to linear space-time dimensions of order d, where $d \approx \text{Max of } |R_{\mu\nu\lambda\sigma}|^{-\frac{1}{2}}$, the positive frequency solutions are not uniquely determined for energies $\omega \lesssim d^{-1}$. This corresponds to an energy density uncertainty of order $\int_0^{d^{-1}} \omega(\omega^2 d\omega) \sim d^{-4}$. This energy density uncertainty leads to various particle creation and vacuum polarization effects. When event horizons are present, as in the Schwarzschild and de Sitter space-times, it evidently is observed as thermal radiation. In non-stationary space-times without event horizons, it also manifests itself through particle creation processes. In Section 3, we consider such a non-stationary space-time. In particular, we consider the quantized scalar field described by Eqs. (2.1)-(2.8) in a cosmological metric, and explain how particles are created as a result of the time-dependence of the metric.

3. PARTICLE CREATION BY THE EXPANDING UNIVERSE

General Considerations

Let us consider a quantized neutral scalar (or pseudoscalar) field satisfying Eq. (2.1) in the isotropic expanding universe with Euclidean 3-space,

$$ds^2 = dt^2 - a^2(t) \sum \delta_{ij}\, dx^i\, dx^j , \qquad (3.1)$$

where a(t) will be considered as a prescribed function of the time coordinate t, which corresponds to the proper time of the set of geodesics x^i = constant. The scalar field in the Robertson-Walker universeswith positive and negative spatial curvature is discussed in Ref. 35. In a self-consistent calculation, a(t) would be a solution, for example, of the coupled Einstein-spin-0 field equations.

Definitions of positive frequency solutions (i.e. of particles) during the actual expansion or in general curved space-times have been attempted by Parker,[1,2] Fulling,[47] Parker and Fulling,[35] Unruh,[46] Ashtekar and Magnon,[60] and Woodhouse.[68] Instantaneous diagonalization of the Hamiltonian to define particles has been used by Imamura,[51] Grib and Mamayev,[69,70] Berger,[71-74] Castagnino, Verbeure, and Weder,[75,76] and Frolov, Mamayev, and Mostepanenko.[54] The use of quantum numbers which emerge naturally from an analysis of the particle field equation near the cosmological singularity to specify the initial state of the field (as opposed to the use of particle number for that purpose) has been pioneered by Misner[77] and Berger.[74] Chitre and Hartle[79] have applied the Feynman path integral technique to models with a cosmological singularity, in analogy to the approach used by Hartle and Hawking[64] for the Schwarzschild metric. Nariai and Tanabe[154,155] have also considered propagators in expanding universes.

For a slowly expanding universe, Parker (Ref. 1, Chapter V, and Ref. 2) analyzed the question of defining positive frequency solutions (i.e. creation and annihilation operators for physical particles) on the basis of the requirements of measurement theory (for detectors moving on the geodesics x^i = const.) and found that in the spin-0 case an extended WKB approximation was appropriate for defining positive frequency. He also showed that this definition leads to a finite and small particle production rate in the present expanding universe, and carried out the analysis for the spin-$\frac{1}{2}$ Dirac field as well. Fulling and Parker[35] showed that the method of defining positive frequency through an extended WKB approximation also leads to a regularization of the energy-momentum tensor, which can be extended to the rapid phases of the expansion, and is related to a regularization technique developed by Zeldovich and Starobinsky.[8] Recently, Woodhouse[68] has used a closely related approach based on the WKB approximation for defining positive frequency solutions in a region of small curvature in a generally covariant manner. That method evidently reduces to the lowest order WKB definition of Refs. 1 and 35 in a slowly expanding universe.

In order to obtain results which are entirely independent of the definition of particles during the expansion of the universe, we will consider a statically bounded expansion

$$a \to \begin{cases} a_1 & \text{as } t \to -\infty \\ a_2 & \text{as } t \to +\infty \end{cases} \qquad (a_1, a_2 > 0) \ . \tag{3.2}$$

During the periods when a(t) is constant, the theory reduces to that of special relativity and the interpretation is unambiguous (in inertial frames). One could equally well replace the periods when a(t) is constant by periods when the WKB definition of positive frequency is valid. The geodesic world lines in statically bounded expansions do not possess event horizons.

Conformal Coupling

Let me first discuss the case when $\gamma = 1/6$ and $m = 0$ in Eq. (2.1). In this case no statically bounded expansion will create particles, and one can unambiguously define positive frequency solutions and annihilation operators during the expansion of the universe[1,3] for observers on the geodesics x^i = const. The proof is based on the conformal invariance of the field equation. Under the conformal transformation

$$g_{\alpha\beta}(x) \to \tilde{g}_{\alpha\beta}(x) = \Omega^{-2}(x)\, g_{\alpha\beta}(x) \tag{3.3}$$

one has[80,57]

$$\tilde{\nabla}^{\mu}\tilde{\nabla}_{\mu}\tilde{\phi} + \frac{1}{6}\tilde{R}\tilde{\phi} = \Omega^{3}(\nabla^{\mu}\nabla_{\mu}\phi + \frac{1}{6}R\phi) \tag{3.4}$$

where

$$\tilde{\phi} = \Omega\phi \tag{3.5}$$

Hence, if the conformal field equation (2.1), which is

$$\nabla^{\mu}\nabla_{\mu}\phi + \frac{1}{6}R\phi = 0 \tag{3.6}$$

holds in the original metric, it also holds for the conformally transformed quantities. By choosing $\Omega = a(t)$ and using time coordinate $x^0 = \int^t a^{-1}(t')dt'$ one maps the problem for the metric of Eq. (3.1) into that of a field $\tilde{\phi}$ in Minkowski space with the initial and final flat regions mapped into $x^0 \to -\infty$ and $x^0 \to +\infty$ respectively. In the Minkowski space formulation it is trivial to see that if the state vector is specified by the condition that no particles are present as $x^0 \to -\infty$, then no particles will be present as $x^0 \to +\infty$. Thus, there is no production of conformally invariant particles for any statically bounded expansion, and the positive frequency solution in the original space-time is obtained from that in Minkowski space through Eq. (3.5). Thus, they are proportional to

$$a^{-1}(t)\; e^{-ik\int^t a^{-1}(t')dt'}\; e^{i\vec{k}\cdot\vec{x}} \; .$$

The absence of conformally invariant particle creation in any statically bounded open or closed Robertson-Walker expansion holds if there is none in the static open or closed Robertson-Walker expansion, since one can use the same $\Omega = a(t)$ to transform to the corresponding static space-time. (One could alternatively transform those universes to flat space-time, but the Ω involved is globally too complicated for one to be confident of conclusions involving non-local processes such as particle creation.) The above results also hold for neutrinos and photons obeying conformally invariant field equations (which are the simplest covariant generalizations of the corresponding equations in flat space-time). An alternative derivation based on tracelessness of the energy-momentum tensor is given below [Eqs. (3.98)-(3.110)].

The above results concerning the absence of particle creation for conformally invariant fields in Robertson-Walker universes are restricted to arbitrary statically bounded expansions. The geodesics in any such universe do not possess event horizons. Therefore, there is no contradiction between our results and the recent results of Gibbon's and Hawking according to which observers on geodesics possessing event horizons will observe a flux of particles.

It should be noted that certain methods of regularizing the energy-momentum tensor seem to give rise to a non-zero trace of $T_{\mu\nu}$ in a conformally invariant theory where the formal trace vanishes,[81-84] while other methods preserve the tracelessness of the original theory, at least in the absence of non-gravitational interactions.[164,8,35-37,99] The appearance of a non-zero $T_\mu{}^\mu$ would not affect the above conclusions concerning the absence of conformally invariant particle creation resulting from a statically bounded expansion (unless the original conformally invariant field equations were altered).

For gravitons treated as perturbations on a Robertson-Walker background, the linearized Einstein equations are not conformally invariant, in contrast to the field equations for neutrinos and photons. (In fact, the conformally invariant spin-2 field equation in the Robertson-Walker background evidently does not satisfy certain consistency conditions of Buchdahl. The conclusions in Refs. 1-3 concerning the absence of graviton production in isotropic universes were based on the conformally invariant spin-2 field equation.[80]) The linearized Einstein equations do yield graviton production in isotropic expansions, as pointed out by Grishchuk.[85,86] (Graviton production in the inhomogeneous Gowdy[87] universe has been analyzed by Berger[71-73] using the Hamiltonian formalism.)

With the present background metric of Eq. (3.1), if one imposes the conditions that the metric perturbations $h_\mu{}^\nu$ satisfy $h_\mu{}^\nu{}_{;\nu} = 0$, $h_\nu{}^\nu = 0$, and $h_{\mu o} = 0$, then the non-vanishing components of $h_\mu{}^\nu$ satisfy (Ref. 85, 86, or Ref. 14, Eq. (15.10.39) with $\eta = 0$, index on h_{ij} raised, and τ times introduced),

$$\frac{d^2}{d\tau^2} h_i{}^k - a^4 \sum_{\ell=1}^{3} \frac{d^2}{dx^{\ell 2}} h_i{}^k = 0 , \qquad (3.7)$$

where $d\tau = a^{-3}dt = a^{-2}d\eta$. This equation is identical with Eq. (2.1) expressed in terms of τ, and with $m = 0$ and $\gamma = 0$. Thus, each component of $h_i{}^k$ satisfies the minimally coupled scalar wave equation. Since Einstein's equations imply that gravitons considered as perturbations on the Robertson-Walker background do not obey conformally invariant field equations, there seems little reason to require that spin-0 particles such as pions should obey the form of Eq. (2.1) which is conformally invariant when m is neglected (i.e. with $\gamma = 1/6$).

Minimal Coupling

Therefore, let me now turn to Eq. (2.1) with $\gamma = 0$, the so-called minimal coupling to the gravitational field,

$$g^{\mu\nu}\nabla_\mu\nabla_\nu\phi + m^2\phi = 0 \ . \tag{3.8}$$

(Unlike the spin-0 case, the minimally coupled neutrino and photon fields do obey conformally invariant field equations.) Because of the translational invariance of the metric (3.1) in 3-space, one has a conserved quantity analogous to linear momentum, namely

$$K_j = \frac{1}{2}\int d^3x\ (\pi\partial_j\phi + \partial_j\phi\pi) \ . \tag{3.9}$$

One can show that K_j is the generator of coordinate translations,

$$[K_j, F] = idF/dx^j \ . \tag{3.10}$$

Since a coordinate translation by dx^j corresponds to a translation by the physical length $a(t)dx^j$, the operator corresponding to such a physical translation is

$$K_j/a(t) \ . \tag{3.11}$$

It is that operator which should be identified with the physical momentum. The physical momentum is proportional to $a^{-1}(t)$ just as in classical general relativity.

For the present metric the scalar field equation (2.1) with $\gamma = 0$ is

$$a^{-3}\frac{\partial}{\partial t}(a^3\frac{\partial}{\partial t}\phi) - \sum a^{-2}\frac{\partial^2}{\partial x^{j\,2}}\phi + m^2\phi = 0 \ . \tag{3.12}$$

The space and time variables are conveniently separated by writing the field in the form

$$\phi = (2\pi)^{-3/2} \int d^3k \; [A_{\vec{k}} \, e^{i\vec{k}\cdot\vec{x}} \, \psi_k(\tau) + \text{H.C.}] \,, \tag{3.13}$$

where $k = |\vec{k}|$, and H.C. denotes the hermitian conjugate. The function ψ_k satisfies

$$\frac{d^2\psi_k}{d\tau^2} + a^6 \, \omega_k^{\,2} \, \psi_k = 0 \,, \tag{3.14}$$

with

$$\omega_k = \left(\frac{k^2}{a^2} + m^2\right)^{\frac{1}{2}} , \tag{3.15}$$

and

$$\tau = \int^t a^{-3}(t')dt' \,. \tag{3.16}$$

The $A_{\vec{k}}$ are constant operators which must satisfy the usual commutation rules for annihilation operators by virtue of the canonical commutators of Eq. (2.5). Thus,

$$[A_{\vec{k}}, A_{\vec{k}'}] = 0 \,, \quad [A_{\vec{k}}, A_{\vec{k}'}^{\dagger}] = \delta(\vec{k}-\vec{k}') \,. \tag{3.17}$$

The Wronskian condition that must be imposed for the consistency of Eqs. (3.17) and (2.5), or so that $(f_{\vec{k}}, f_{\vec{k}'}) = \delta(\vec{k}-\vec{k}')$ with $f_{\vec{k}} = (2\pi)^{-3/2} \, e^{i\vec{k}\cdot\vec{x}} \, \psi_k(\tau)$, is

$$\psi_k \frac{d\psi_k^{\,*}}{d\tau} - \psi_k^{\,*} \frac{d\psi_k}{d\tau} = i \,. \tag{3.18}$$

For a statically bounded expansion, Eq. (3.2), one has

$$a \rightarrow \begin{cases} a_1 & \text{as } \tau \rightarrow -\infty \\ a_2 & \text{as } \tau \rightarrow +\infty \end{cases} \,. \tag{3.19}$$

During the periods when a is constant, the theory reduces to that of special relativity and the interpretation is unambiguous. (One can replace the periods when a is constant by periods when the WKB approximation is valid, and use the WKB definition of positive and negative frequencies.) We suppose that for large negative τ

$$(\psi_k)_{\tau\to-\infty} = (2a_1{}^3\omega_k)^{-\frac{1}{2}} \exp(-ia_1{}^3\omega_k\tau) \ , \tag{3.20}$$

where ω_k is evaluated for large negative τ (when $a = a_1$). Since τ is given by $a_1^{-3}t$ to within a constant when $a = a_1$, the field expansion (3.13) yields as a consequence of (3.20),

$$(\phi)_{\tau\to-\infty} = (2\pi)^{-3/2}\int d^3p_I(2\omega_k)^{-\frac{1}{2}}[A_{\vec{p}_I}\, e^{i\vec{p}_I\cdot\vec{x}_I}\, e^{-i\omega_k t} + \text{H.C.}] \ , \tag{3.21}$$

where

$$\vec{p}_I = \vec{k}/a_1, \qquad \vec{x}_I = a_1\vec{x} \tag{3.22}$$

and

$$A_{\vec{p}_I} = a_1{}^{3/2} A_{\vec{k}} \ . \tag{3.23}$$

The rescaled coordinate $\vec{x}_I$ is the usual Minkowski coordinate in the initial space-time, p_I is the physical momentum, and by virtue of the commutation relations of the $A_{\vec{k}}$, the operators $A_{\vec{p}_I}$ obey the boson commutation relations with the correct δ-function $\delta(\vec{p}_I-\vec{p}_I') = a_1{}^3\delta(\vec{k}-\vec{k}')$. It is clear from the field expansion (3.21) that, with the initial condition of Eq. (3.20), the $A_{\vec{p}_I}$ or $A_{\vec{k}}$ correspond to physical particles in the initial Minkowski space.

The function $\psi_k(\tau)$ evolves in accordance with Eq. (3.14). For large τ the space-time is again flat ($a = a_2$), and one finds that in general ψ_k is a superposition of both positive and negative frequency components,

$$(\psi_k)_{\tau\to\infty} = (2a_2{}^3\omega_k)^{-\frac{1}{2}}[\alpha_k \exp(-ia_2{}^3\omega_k\tau) + \beta_k \exp(ia_2{}^3\omega_k\tau)], \tag{3.24}$$

where α_k and β_k are complex constants, and ω_k is evaluated at large τ when $a = a_2$. Using this expression for ψ_k in the field expansion (3.13) for large τ, and regrouping according to positive and negative frequencies, one obtains a form for ϕ like Eq. (3.21), but with a_1 replaced by a_2, and the annihilation operator for physical particles at late times given by

$$a_{\vec{k}} = \alpha_k A_{\vec{k}} + \beta_k^* A_{-\vec{k}}^\dagger . \tag{3.25}$$

The $a_{\vec{k}}$ and $a_{\vec{k}}^\dagger$ obey the correct commutation relations

$$[a_{\vec{k}}, a_{\vec{k}'}] = 0 , \qquad [a_{\vec{k}}, a_{\vec{k}'}^\dagger] = \delta(\vec{k}-\vec{k}') \tag{3.26}$$

by virtue of the Wronskian condition (3.18) which demands that

$$|\alpha_k|^2 - |\beta_k|^2 = 1 . \tag{3.27}$$

It is noteworthy that the fermion anti-commutation relations would not be propagated consistently from the initial to the final Minkowski space. For the Dirac equation one finds that the reverse situation holds, in that only the fermion anti-commutation relations are propagated consistently. These results can be used as the basis of an independent derivation of the relation between spin and statistics.[3,4]

Let us now suppose that no particles are present in the initial Minkowski space,

$$A_{\vec{k}}|0\rangle = 0 \qquad \text{for all } \vec{k} , \tag{3.28}$$

where $|0\rangle$ is the state of the system in the Heisenberg picture. Are there particles present in the final Minkowski space? For convenience I will impose periodic boundary conditions on $\phi(\vec{x},t)$ in a spatial cube of coordinate length L, so that $\vec{k}$ takes on discrete values. The limit $L \to \infty$ is to be taken after the physically significant quantities have been calculated. This is a purely mathematical device with no physical influence on the results.

The expectation value of the number of particles in mode $\vec{k}$ at late times is

$$\langle N_{\vec{k}} \rangle_{\tau\to\infty} = \langle 0 | a_{\vec{k}}^{\dagger} a_{\vec{k}} | 0 \rangle = |\beta_k|^2 , \tag{3.29}$$

where I have used Eqs. (3.25) and (3.28). One can show that β_k vanishes in the limit of an infinitely slow expansion (with all derivatives of $a(\tau)$ continuous and small), so that the particle number is an adiabatic invariant. However, in the general case β_k does not vanish, so that there are particles present in the final Minkowski space. These particles must have been created during the expansion of the universe from a_1 to a_2. Therefore, in any correct treatment, particles must be created during the expansion of the universe (except in special cases).

In fact, during a rapid expansion in which particle creation is occuring the concept of particle number loses its operational basis. If Δt is the interval of time over which a measurement of the particle number is made, then one cannot make Δt too small or a significant number of particles will be created by the measurement process because of the time-energy uncertainty relation. On the other hand, one cannot make Δt too large or a significant number of particles will be created by the expansion of the universe during the measurement process. Therefore, when the particle creation process is rapid, the irreducible imprecision in the measurement of particle number will become very great. This situation is reflected in the theory by the fact that the definition of positive and negative frequency parts of the field becomes ambiguous and imprecise.

However, because the particle number is an adiabatic invariant one has no difficulty in generalizing the Minkowski space definition of particles to a slow expansion. The initial and final periods in which $a(\tau)$ is constant could just as well be replaced by periods in which the expansion is slow, in the sense that the WKB solutions of Eq. (3.14) are a valid approximation to the function ψ_k. The positive and negative frequency WKB solutions replace the special relativistic positive and negative frequency solutions when $a(\tau)$ varies sufficiently slowly, since those WKB solutions would pass over into the usual positive and negative frequency solutions if $a(\tau)$ were to gradually approach a constant value (and particle number is adiabatically constant).

One possible method of giving the field theory a particle interpretation during the expansion of the universe is to use the creation and annihilation operators which instantaneously diagonalize the Hamiltonian H(t) to define particles at time t, as noted previously (in the second paragraph of Section 3). For the minimally coupled scalar field it was shown in Ref. 1 (Appendix AI), and Refs. 75 and 76 that specification of particle number by the above method of Hamiltonian diagonalization leads to an infinite rate of particle creation when summed over modes. For the conformally coupled scalar field with zero mass, on the other hand, Hamiltonian diagonalization leads to the same creation and annihilation operators as were obtained by our previous argument based on conformal transformations (since one can conformally transform the Hamiltonian from the flat or static space-time), and hence also gives zero particle production in isotropically expanding universes. For the conformally coupled scalar field with non-zero mass, Hamiltonian diagonalization was used in Ref. 54 to define the initial vacuum state near the cosmological singularity in Robertson-Walker universes, and finite results for the particle density observed at late times were obtained. They also obtained specific finite results for the Dirac field with non-zero mass (the spin-$\frac{1}{2}$ field is also conformally invariant for zero mass). In Ref. 4, it was shown that the density of created particles at late times was finite for the Dirac field even if one starts at the cosmological singularity, but the actual value of the particle density was not found.

In general, the interpretation of the creation operators

which instantaneously diagonalize the Hamiltonian as corresponding to physical (observable) particles during the expansion of the universe is open to criticism, although the use of those operators as one possible and fairly natural means of specifying initial conditions cannot be criticized. For one thing, such an interpretation would evidently imply an infinite rate (summed over modes) of production of minimally coupled spin-0 particles, and evidently also of gravitons considered as perturbations of the Robertson-Walker metric. Furthermore, if a(t) were to *suddenly* become constant at its instantaneous value, then the particle number in mode $\vec{k}$ in the final Minkowski space would be equal to the particle number obtained using the creation and annihilation operators which diagonalize H(t) just before a(t) became constant. But one expects (except in special cases) particles to be created by the sudden stopping of the expansion a(t), since $\ddot{a}$ becomes infinite at that time and the differential gravitational forces become infinite. Therefore one would not expect the number of physical particles just before the expansion is suddenly stopped to equal the number in the final Minkowski space, so that the creation and annihilation operators which diagonalize H(t) evidently do not correspond to physical particles present at time t (for further reasons against such an identification see Parker, Ref. 1, Appendix AI, and Raine and Winlove, Ref. 88).

Our results will be independent of any definition of particles during the stage of significant particle production (if indeed any such definition is physically meaningful except in special cases). We will suppose that a(t) varies smoothly, and will compare particle number in initial and final regions in which the expansion is adiabatic or static.

Probability Distribution and Average Number of Created Particles

The probability distribution of the created particles can be obtained as follows.[1,3] Let $|0)$ be the vacuum in the final

Minkowski space. It is defined by

$$a_{\vec{k}}|0) = 0 \qquad \text{for all } \vec{k} . \tag{3.30}$$

We must calculate the matrix elements of the initial vacuum $|0\rangle$ with the basis vectors of the Fock space constructed from the final vacuum $|0)$. (We are still working with the discrete representation.)

First consider the scalar product of $|0\rangle$ with the state

$$|n,n) = (n!)^{-1}(a^{\dagger}_{-\vec{k}})^{n}(a^{\dagger}_{\vec{k}})^{n}|0) \tag{3.31}$$

containing n particles in mode $-\vec{k}$ and n in mode $\vec{k}$ in the final Minkowski space. It follows from Eqs. (3.25) and (3.28) that

$$a_{\vec{k}}|0\rangle = (\beta^{*}_{k}/\alpha^{*}_{k})a^{\dagger}_{-\vec{k}}|0\rangle . \tag{3.32}$$

Hence,

$$(n,n|0\rangle = (n!)^{-1}(\beta^{*}_{k}/\alpha^{*}_{k})^{n}(0|(a_{-\vec{k}})^{n}(a^{\dagger}_{-\vec{k}})^{n}|0\rangle$$

$$= (n!)^{-\frac{1}{2}}(\beta^{*}_{k}/\alpha^{*}_{k})^{n}(0,n|(a^{\dagger}_{-\vec{k}})^{n}|0\rangle , \tag{2.33}$$

where $(0,n|$ is the state containing n particles in mode $-\vec{k}$ in the final Minkowski space. Repeated use of the relation

$$(0,n|a^{\dagger}_{-\vec{k}} = n^{\frac{1}{2}}(0,n-1| \tag{3.34}$$

then yields

$$(n,n|0\rangle = (\beta^{*}_{k}/\alpha^{*}_{k})^{n}(0|0\rangle . \tag{3.35}$$

Matrix elements of the form $(m,n|0\rangle$ vanish when $m \neq n$ because if the above steps are carried out one is left at the end with $(0|a^{\dagger}_{-\vec{k}}|0\rangle$ if $m > n$, or $(0|a^{\dagger}_{\vec{k}}|0\rangle$ if $m < n$, and those matrix elements vanish in accordance with Eq. (3.30). Therefore, Eq. (3.35) gives the most general non-vanishing matrix element of a final state containing only particles in modes $\vec{k}$ or $-\vec{k}$ with the initial vacuum. Using these results one can show that the most general non-vanishing matrix element of a final basis state with the initial vacuum is of the form

$$(\{n_{\vec{k}}\}|0\rangle = \prod_{\vec{k}} (\beta_k^*/\alpha_k^*)^{n_{\vec{k}}} (0|0\rangle \tag{3.36}$$

where $|\{n_{\vec{k}}\})$ is a final state containing $n_{\vec{k}}$ pairs, with one particle in mode $\vec{k}$ and the other in mode $-\vec{k}$, for a set of different modes $\vec{k}$ (i.e. $\{n_{\vec{k}}\} = \{n_{\vec{k}_1}, n_{\vec{k}_2}, \ldots\}$). To find $|(0|0\rangle|$, use Eq. (3.36) to write

$$\begin{aligned} 1 = \langle 0|0\rangle &= |(0|0\rangle|^2 \sum_{\{n_{\vec{k}}\}} \prod_{\vec{k}} |\beta_k/\alpha_k|^{2n_{\vec{k}}} \\ &= |(0|0\rangle|^2 \prod_{\vec{k}} \sum_{n=0}^{\infty} |\beta_k/\alpha_k|^{2n} \\ &= |(0|0\rangle|^2 \prod_{\vec{k}} |\alpha_k|^2 , \end{aligned}$$

where the first sum was over all sets $\{n_{\vec{k}}\}$, and I have used $|\alpha_k|^2 - |\beta_k|^2 = 1$. Thus,

$$|(0|0\rangle|^2 = \prod_{\vec{k}} |\alpha_k|^{-2} = \exp[\sum_{\vec{k}} \ln (|\alpha_k|^{-2})] , \tag{3.37}$$

and

$$|(\{n_{\vec{k}}\}|0\rangle|^2 = \prod_{\vec{k}} (|\beta_k/\alpha_k|^{2n_{\vec{k}}} |\alpha_k|^{-2}) . \tag{3.38}$$

It follows that the production of pairs in two different modes are independent events, and that the probability $P_n(\vec{k})$ of observing n particles in mode $\vec{k}$ at large τ is

$$P_n(\vec{k}) = |\beta_k/\alpha_k|^{2n} |\alpha_k|^{-2} . \tag{3.39}$$

This result can be used to find the average number of particles $\langle N_{\vec{k}} \rangle$ present in mode $\vec{k}$ in the volume $[La_2]^3$:

$$\langle N_{\vec{k}} \rangle = \sum_{n=0}^{\infty} n \, P_n(\vec{k}) = |\beta_k|^2 , \qquad (3.40)$$

which agrees with Eq. (3.29). The average particle density, summing over all modes and going to the continuum limit, is

$$\langle N \rangle = \lim_{L\to\infty} [La_2]^{-3} \sum_{\vec{k}} \langle N_{\vec{k}} \rangle = [2\pi^2 \, a_2{}^3]^{-1} \int_0^\infty dk \; k^2 |\beta_k|^2 , \qquad (3.41)$$

where I have used the standard result that $L^{-3} \sum_{\vec{k}} \to (2\pi)^{-3} \int d^3k$ in the continuum limit.

Suppose the initial Minkowski space is not empty, but is described by a statistical mixture of pure states, each of which contains definite numbers of particles in each mode. Then the statistical density matrix ρ is diagonal in the representation whose basis consists of the eingenstates of the operators $A^{\dagger}_{\vec{k}} A_{\vec{k}}$, and the operator ρ must contain an equal number of operators $A_{\vec{k}}$ and $A^{\dagger}_{\vec{k}}$. For example, if ρ represented an equilibrium distribution in the initial Minkowski space one might expect ρ to be a function of the initial free particle Hamiltonian, which is diagonal in the above representation. In such cases, one finds for the average number of particles in mode $\vec{k}$ and volume $[La_2]^3$ present at late times[1,3]

$$\langle N_{\vec{k}} \rangle_{T\to\infty} = \mathrm{Tr}\, \rho \, a^{\dagger}_{\vec{k}} a_{\vec{k}} = \langle N_{\vec{k}} \rangle_{T\to-\infty} + |\beta_k|^2 [1 + 2\langle N_{\vec{k}} \rangle_{T\to-\infty}], \qquad (3.42)$$

where $\langle N_{\vec{k}} \rangle_{T\to-\infty} = \mathrm{Tr}\, \rho \, A^{\dagger}_{\vec{k}} A_{\vec{k}}$ is the number present initially in mode $\vec{k}$ and volume $[La_1]^3$. One can go the continuum limit as in Eq. (3.41) to obtain the final particle density. Comparison of Eq. (3.42) with Eq. (3.40) shows that the initial presence of bosons tends to stimulate the creation of bosons. The reverse is true for fermions.

Black-Body Distribution Resulting From Smooth Initial Conditions

Later I will be considering a model of the very early universe in which the significant particle production occurs near the Planck time $t_p = G^{\frac{1}{2}}$, so that the average energy of the initially created particles will be of the order of t_p^{-1}, which is very large with respect to any of the elementary particle rest masses. Therefore, I will restrict myself to the case when the mass m of the particle is negligible or vanishes. Then Eq. (3.14) can be written as

$$\frac{d^2\psi_k}{d\tau^2} + k^2 a^4 \psi_k = 0 . \tag{3.43}$$

In order to obtain a more explicit expression for the average particle number $\langle N_{\vec{k}} \rangle$ and the probability distribution $P_n(\vec{k})$ for a smooth expansion from a_1 to a_2, let me first consider a function for $a^4(\tau)$ appearing in Eq. (2.43) which is general enough that it includes a number of adjustable parameters, so that its shape can be altered considerably, but for which the exact solution of Eq. (3.43) is known. Namely,

$$a^4(\tau) = a_1^4 + e^{\zeta}[(a_2^4 - a_1^4)(e^{\zeta} + 1) + b](e^{\zeta} + 1)^{-2} , \tag{3.44}$$

where

$$\zeta = \tau s^{-1} . \tag{3.45}$$

Here b, a_1, a_2, and s are adjustable constants, all of which will be taken as positive, with $a_2 > a_1$. This function and all its derivatives are continuous, and it approaches a_1 as $\tau \to -\infty$ and a_2 as $\tau \to +\infty$. Its rate of change and shape can be varied considerably by changing the various constants. Our main result will be independent of the values of a_2 and b. The above function was first used by Eckart and Epstein in connection with above-barrier quantum mechanical scattering and the reflection of radio waves by a medium with a smoothly changing index of refraction.

The exact solution of Eq. (3.43) is given in the present case by Epstein[89] and Eckart.[90] One readily finds that Eq. (21) of Ref. 89 implies

$$\left|\frac{\beta_k}{\alpha_k}\right|^2 = \frac{\sin^2 \pi d + \sinh^2[\pi sk(a_1{}^2 - a_2{}^2)]}{\sin^2 \pi d + \sinh^2[\pi sk(a_1{}^2 + a_2{}^2)]} , \tag{3.46}$$

where d is a real number involving the constant b. As will become clear below, for the type of cosmological model under consideration with a_2 large with respect to a_1 one has $\pi sk(a_2{}^2 - a_1{}^2) >> 1$. Then to very good approximation Eq. (3.46) yields

$$|\beta_k/\alpha_k|^2 \cong \exp(-4\pi ska_1{}^2) . \tag{3.47}$$

This result is independent of the adjustable parameters b and a_2, so that one expects it to hold for a wide class of functions $a^4(\tau)$.

Most of the change in $|\beta_k|^2$, and thus the period of significant particle production, occurs when the dimensionless quantity

$$Q = \frac{d}{d\tau} (k^2a^4)/(k^2a^4)^{3/2} \tag{3.48}$$

is of order unity or larger. In the cosmological context, for which a_2 is large with respect to a_1, the quantity Q is large only during an early period when $e^{\tau s^{-1}}$ is small. During that period a^4 in Eq. (3.44) effectively has the form

$$a^4(\tau) = a_1{}^4 + a_0{}^4 e^{\tau s^{-1}} , \tag{3.49}$$

where a_0 is an adjustable parameter. Since for this form of $a^4(\tau)$ the quantity Q vanishes as $\tau \to \infty$ (as well as $-\infty$), one can calculate $|\beta_k/\alpha_k|^2$ in this case by making use of the WKB positive and negative frequency solutions at late τ. The form of $a^4(\tau)$ in Eq. (3.49) was suggested to me as an alternative to the Eckart-Epstein form by B.K. Berger, who considered the function in Eq. (3.49) in connection with particle creation in the degenerate Kasner universe, which is an anisotropically expanding universe (it is defined following Eq. (1.6)) equivalent to a part of Minkowski space in a non-Minkowskian coordinate system.[74] In that case Berger obtained a direction-dependent expression for the average particle number which resembled a black-body distribution along one direction. In the case of the degenerate Kasner universe we would not use the same definition of particles as was used by Berger, and would expect to obtain no particle production. In particular, the results of Fulling, Parker and Hu, Ref. 37, part IV, show that if the state is chosen to be the usual Minkowski vacuum, then there will be no particle flux for observers moving along the preferred geodesics in the degenerate Kasner universe (i.e. for observers whose proper time is the Kasner time coordinate). The difference between this latter result and that for Rindler coordinates can be understood because from the viewpoint of the inertial Minkowski coordinates, the world lines of clocks measuring Rindler time are accelerated, while those of clocks measuring Kasner time are not accelerated.

With $a^4(\tau)$ given by Eq. (3.49), the solution ψ_k of Eq. (3.43) which reduces to a positive frequency solution in the initial Minkowski space is

$$\psi_k = (2ka_1{}^2)^{-\frac{1}{2}} (k'a_0{}^2)^{ik'a_1{}^2}\, 2^{-ik'a_1{}^2}$$

$$\times\, \Gamma(1-ik'a_1{}^2)\, J_{-ik'a_1{}^2}(k'a_0{}^2\, e^{\tau'}) \tag{3.50}$$

where

$$k' = 2ks\ , \qquad \tau' = \tau s^{-1}/2\ , \tag{3.51}$$

and J is a Bessel function. For large τ, ψ_k must be a superposition of the positive and negative frequency WKB solutions of Eq. (3.43), namely

$$(\psi_k)_{\tau\to\infty} = (2ka_0{}^2 e^{\tau'})^{-\frac{1}{2}} \left\{\alpha_k \exp(-ik'a_0{}^2 e^{\tau'}) + \beta_k \exp(ik'a_0{}^2 e^{\tau'})\right\}, \tag{3.52}$$

where α_k and β_k must satisfy Eq. (3.27) because of the Wronskian condition of Eq. (3.18). Making use of the well-known asymptotic form of the Bessel function,

$$(J_\nu(z))_{z\to\infty} = (\pi z/2)^{-\frac{1}{2}} \cos(z - \tfrac{1}{2}\pi\nu - \tfrac{1}{4}\pi) \ , \tag{3.53}$$

I find that

$$\alpha_k = (2k'a_1^{\ 2})^{-\frac{1}{2}} (k'a_0^{\ 2})^{ik'a_1^{\ 2}} \, 2^{-ik'a_1^{\ 2}}$$

$$\times\, \Gamma(1-ik'a_1^{\ 2})\ \pi^{-\frac{1}{2}}\ e^{i\frac{\pi}{4}}\ e^{\frac{\pi}{2}k'a_1^{\ 2}} \ , \tag{3.54}$$

and

$$\beta_k = -i\alpha_k\ e^{-\pi k'a_1^{\ 2}} \ . \tag{3.55}$$

It follows that

$$|\beta_k/\alpha_k|^2 = \exp(-4\pi s k a_1^{\ 2}) \ , \tag{3.56}$$

which is the same result as in Eq. (3.47).

The exponential form of $|\beta_k/\alpha_k|^2$ in Eq. (3.56) or (3.47) clearly holds for a rather general class of functions $a^4(\tau)$ which smoothly approach a constant $a_1^{\ 4}$ as $\tau \to -\infty$. For the exponential result for $|\beta_k/\alpha_k|^2$ to hold it is necessary that $a(\tau)$ and all its derivatives be continuous, except in special cases when discontinuities occur when the phase of ψ_k has certain values. This follows from a theorem of Kulsrud.[91] It is not difficult to show that when Kulsrud's theorem is applied to Eq. (3.43) it implies that $|\beta_k/\alpha_k|^2$ is zero to as many orders in the rate of change of $a(\tau)$ as $a(\tau)$ has continuous derivatives. The quantity $|\beta_k/\alpha_k|^2$ is zero to higher order only in exceptional cases. Since the rate of change of $a(\tau)$ is proportional to s^{-1} in the previous examples, the exponential expressions in Eqs. (3.56) and (3.47) are zero to all orders in the rate of change of $a(\tau)$. To obtain such an exponential result for $|\beta_k/\alpha_k|^2$ it is clearly necessary because of Kulsrud's theorem that $a(\tau)$ and all its derivatives be continuous (except possibly for the exceptional cases noted above, which require special phase information and are not relevant to the types of models under consideration). The least restrictive conditions which are sufficient for the exponential form

$$|\beta_k/\alpha_k|^2 \cong \exp(-k \times \text{constant}) \tag{3.57}$$

to hold (with the constant proportional to the inverse of the rate of change of $a(\tau)$) are not known at this time. I conjecture that one set of sufficient conditions may be that Eq. (3.57) will hold if $a(\tau)$ and all of its derivatives are continuous, and a_2 is sufficiently large with respect to a_1, where $a(\tau)$ equals a_1 for large negative τ and equals a_2 for large positive τ (also $a(\tau)$ should remain finite and non-zero at all times). A reason in favor of the above conjecture is that phase-integral approximations seem to give $|\beta_k/\alpha_k|^2$ in the form of Eq. (3.57) for functions a of the above type. These phase-integral approximations are applicable to a wide class of functions.

Since the exponential form of $|\beta_k/\alpha_k|^2$ in Eqs. (3.56) and (3.47) seems to hold for a wide class of functions $a^4(\tau)$ which join sufficiently smoothly to a_1^4 as $\tau \to -\infty$, it is natural to ask if that form of $|\beta_k/\alpha_k|^2$ corresponds to a similarly general type of probability distribution for the particles present at late times. Substituting Eq. (3.56) into the probability distribution of Eq. (3.39) and the average number of Eq. (3.40), and making use of $|\alpha_k|^2 - |\beta_k|^2 = 1$, one finds that[13]

$$P_n(\vec{k}) = \exp(-n\mu k)[1 - \exp(-\mu k)] , \tag{3.58}$$

and

$$\langle N_{\vec{k}} \rangle = [\exp(\mu k) - 1]^{-1} , \tag{3.59}$$

where

$$\mu = 4\pi s a_1^2 . \tag{3.60}$$

The average particle density of Eq. (3.41) is

$$\langle N \rangle = [2\pi^2 a_2^3]^{-1} \int_0^\infty dk\, k^2 [\exp(\mu k) - 1]^{-1} . \tag{3.61}$$

The physical momentum at late times is $k/a_2 = P = 2\pi/\lambda$, where λ is the physical wavelength ($h = 2\pi$ in these units). Therefore in cgs units

$$\langle N \rangle = 4\pi \int_0^\infty d\lambda\, \lambda^{-4} [\exp(\mu a_2 hc/\lambda) - 1] . \tag{3.62}$$

The above probability distribution corresponds to that of black body radiation of temperature

$$T = (k_B \mu a_2)^{-1} = (4\pi k_B s a_1^2 a_2)^{-1} , \tag{3.63}$$

where k_B is Boltzmann's constant. The number density per unit wavelength which follows from Eq. (3.62) is half that of electromagnetic black-body radiation because the spin-0 particles have only one spin state. When the period in which $a(\tau)$ equals a_2 is replaced by a period in which the expansion is adiabatic, in the sense that the particle production has effectively ceased (i.e. Q of Eq. (3.48) is small), then the temperature at a time t, after the particle production has ceased, is given by

$$T(t) = (4\pi k_B s a_1^2 a(t))^{-1} , \tag{3.64}$$

since the momentum at time t is $k/a(t)$.

It would still be possible to distinguish between the above distribution and a thermal distribution, at least in principle, if interference effects could be produced in the final state, which would reveal that the state was a coherent superposition rather than a mixture. For example, one might try to show that the particle and antiparticle of a pair were correlated. However, that does not seem possible because the original particles (and antiparticles) will undergo interactions, decays, and annihilations in the early universe, which will effectively destroy the coherence, although presumably maintaining the black-body distribution in the resulting radiation.

Since the particle creation occurs mainly at very early times, and the above results are independent of the behavior of a(t) once the particle creation has effectively ceased, they will hold for a model universe in which a form of a(t) which leads to a black-body distribution of temperature given by Eq. (3.64) is joined smoothly to a classical Friedmann expansion. We will consider such a model below. If the initial stage of the expansion is less smooth than we have assumed, it would affect mainly the low frequency part of the spectrum, while the high frequency part would continue to resemble black-body radiation of the above temperature. The entropy of the radiation in a co-moving volume element (which is the conserved quantity) would be approximately the same as in the black-body case. Furthermore, interactions would be expected to thermalize the radiation, so that the results would be approximately the same as for the initial black-body distribution.

Relation to Einstein Equations and to 3°K Background Radiation

The previous results concerning particle production are of intrinsic interest. In order to see whether they may also have applications to cosmology, I will consider a simple cosmological model. The form of the metric is that of Eq. (3.1), a spatially flat Robertson-Walker universe. I assume that as the curvature invariants grow large enough to be characterized by $\hbar$, c and G (i.e. as the cosmological singularity is approached for large negative τ), the effective expansion parameter should be taken as smoothly approaching a constant value a_1. The behavior of $a(\tau)$ is assumed to be sufficiently smooth that the created particles have a black-body spectrum described by Eqs. (3.58), (3.59), and (3.64). (If the behavior is not so smooth, mainly the low frequency end of the spectrum is affected, but the main conclusions are the same.) As in the previous examples, the quantity s appearing in the expression for the temperature, Eq. (3.64), is roughly equal to the interval of τ-time during which significant particle production occurs. The justification for assuming that $a(\tau)$ smoothly approaches a constant value a_1 in terms of τ-time lies at a deeper largely unexplored level. Perhaps it can be understood on the basis of a "bouncing" cosmological model in which the singularity is avoided, or in terms of elementary particle theories in which it has been suggested that the gravitational interaction effectively disappears at the Planck dimensions.

It is also being assumed that the free field equations can be used in analyzing the particle creation process. This is certainly the simplest model one can consider, and may possibly be justified on the basis of the asymptotically free field theories in which the elementary constituents are effectively free with respect to sufficiently high energy interactions. In this model, I will suppose that there are a definite limited number of elementary fundamental fields. It would be interesting to see how these results would be modified by a composite particle model, and by various kinds of interactions.

Since the significant particle creation occurs when $a \approx a_1$, and s is roughly the τ-interval during which significant particle creation occurs, we have from Eq. (3.16)

$$s \approx a_1^{-3} \Delta t \tag{3.65}$$

where Δt is the corresponding period in t-time. On dimensional grounds, we assume that Δt is of the order of $t_p = (\hbar G/c^5)^{\frac{1}{2}} = 5.4 \times 10^{-44}$ sec. Then

$$s \approx a_1^{-3}\, t_p \quad . \tag{3.66}$$

It should be emphasized that the present considerations are largely of a dimensional nature, and are necessarily very rough. Related dimensional arguments appeared in Ref. 7 and Ref. 92, and are mentioned in Ref. 2. A new interesting feature of the present model is that one has an initial black-body spectrum and associated definite temperature and entropy density, even in the absence of classical thermalizing mechanisms, and prior to the possible action of such mechanisms.

If the initial state is taken to be the vacuum $|0\rangle$, then equal numbers of particles and antiparticles will be created. In order to explain the observed local excess of baryons over antibaryons which are left at the present time after annihilations, decays, and interactions have occured in the early universe, one must appeal to a matter-antimatter separation process or an interaction which does not conserve baryon number and also violates CP invariance.

Although matter-antimatter separation mechanisms have been considered (see the references in E.R. Harrison, Ref. 41), the possibility that matter and antimatter are created symmetrically but that there are interactions at very high energies which violate both CP invariance and baryon number conservation has not to my knowledge been investigated. (This is not the same as Hawking's suggestion that black holes creating particles may emit more matter than antimatter through interactions involving CP violation.) Alternatively, one could assume that the excess baryons were already present in the initial state. In that case Eq. (3.42) would be used to calculate the spectrum of the created particles and antiparticles. Since the actual number density of excess baryons is very small compared to the number density of the created particles (which in the present model eventually lead to the black-body radiation), the spectrum of the created particles will be almost unaffected by the presence of the initial baryons. For simplicity, I will take the initial state to be the vacuum $|0\rangle$ in our present considerations, and will avoid the question of the origin of the observed ratio of photons to baryons (i.e. the entropy per baryon).

From Eq. (3.64) and Eq. (3.66) it follows that when $a \approx a_1$ the initial temperature is of the order of $T \approx (4\pi k_B t_p)^{-1} \sim 10^{31}$ °K, the Planck temperature. Therefore, the created particles are highly relativistic and the expansion soon must become a radiation dominated Friedmann solution of Einstein's equations, namely

$$a(t) = C\, t^{\frac{1}{2}} \tag{3.67}$$

where C is an arbitrary constant (it drops out of the curvature invariants, which depend only on t). Since the curvature invariants become characterized by G, ħ and c when $t \approx t_p$, we take the minimum expansion parameter to be

$$a_1 \approx C\, t_p^{\frac{1}{2}} . \tag{3.68}$$

Using Eqs. (3.64), (3.66), (3.67) and (3.68), one finds that in the early phases of the expansion, the temperature is approximately

$$T(t) \approx (4\pi k_B)^{-1}\, t_p^{-1}\, (a_1/a(t)) \tag{3.69}$$

$$T(t) \approx (4\pi k_B)^{-1}\, t_p^{-1}\, (t_p/t)^{\frac{1}{2}} . \tag{3.70}$$

Is this result consistent in order of magnitude with the Einstein equations? We are already using those equations to give $a \propto t^{\frac{1}{2}}$ after the period of significant particle creation. However, it must be checked that our assumption, that most of the particle creation occurs near $t \approx t_p$ for a period of the order of t_p, leads to a density consistent with the Einstein equation

$$\left(\frac{d^2a}{dt^2}\right) a^{-1} = -\frac{8\pi G}{3}\, \rho , \tag{3.71}$$

where we have used $p = \rho/3$. Using Eq. (3.67) and putting $\rho = (\eta_a/2)\,\sigma T^4$, where η_a is the number of different kinds of particles produced initially (counting each state with different quantum numbers, like spin and charge, separately), and $\sigma = \pi^2 k_B{}^5/15$ is the Stefan-Boltzmann constant, one finds from Eq. (3.71) that

$$T(t) = 2(45\pi)^{\frac{1}{4}}\, \eta_a{}^{-\frac{1}{4}} (4\pi k_B)^{-1}\, t_p{}^{-1} (t_p/t)^{\frac{1}{2}} , \tag{3.72}$$

which is the same as the temperature obtained by means of the particle creation process, Eq. (3.70), except for a factor of approximately $7\,\eta_a{}^{-\frac{1}{4}}$. (The quantity t_p arises here because $G = t_p{}^2$.) This agreement is satisfactory, in view of the fact that Eq. (3.70) was obtained from a rough order-of-magnitude estimate. The dependence on $\eta_a{}^{-\frac{1}{4}}$ might possibly arise in the expressions for a_1 and s if the early behavior of $a(\tau)$ is dependent on quantum effects involving the number of fundamental fields. These results at early times are the same in the spatially curved Robertson-Walker models. Before a more precise expression for the temperature can be obtained from the particle creation process much more must be understood about mechanisms which govern the expansion when the curvature invariants become very large. Presumably, whatever equations govern that earliest phase of the expansion, they must merge with the Einstein equations when t is large with respect to t_p, so that the expression for T(t) obtained from the particle creation process in a self-consistent theory must agree with Eq. (3.72) at such times. Therefore, our assumptions that s and a_1 are characterized by t_p should arise naturally in such a self-consistent treatment.

The temperature in Eq. (3.72) or (3.70) can be connected with the present temperature by using the conservation of entropy in a co-moving volume (Ref. 14, p. 589). At a time t during the radiation dominated phase of the expansion, the entropy in a volume $a^3(t)$ is

$$S = \eta(t)(2/3)\,\sigma T^3(t)a^3(t) \;, \tag{3.73}$$

where $\eta(t)$ is the number of constituents at time t. As the universe expands, the original particles decay, interact, and annihilate, leaving η_b components of the present black body radiation (counting each spin and charge state separately, and each fermion as 7/8 of a boson in forming the total entropy below). At the present time t_b the entropy in the expanded volume $a^3(t_b) = a_b{}^3$ is

$$S = \eta_b(2/3)\,\sigma T_b{}^3\, a_b{}^3 \;, \tag{3.74}$$

where T_b is the present temperature (we neglect differences in temperature of the various components which decouple at different times). Conservation of entropy in the co-moving volume then yields

$$T_b \approx (\eta(t)/\eta_b)^{1/3}\; T(t)\;\; (a(t)/a_b) \;. \tag{3.75}$$

For simplicity let me assume that the various factors involving η, the number of constituents at various times, are of order unity, since powers like $\frac{1}{3}$ or $\frac{1}{4}$ are involved in those factors. Then dropping such factors involving the η's, Eq. (3.72) and (3.75) yield (in cgs units)

$$T_b \approx 7(4\pi k_B)^{-1}\; \hbar t_p{}^{-1}\;\; (a_1/a_b) \;, \tag{3.76}$$

where I have used $a(t) \approx Ct^{\frac{1}{2}}$ and $a_1 \approx Ct_p{}^{\frac{1}{2}}$. This can be interpreted as stating that the present temperature is of the order of the red shifted Planck temperature. Let us estimate this numerically for the spatially flat model. Denoting the earliest time at which the universe becomes matter dominated by t_e (so that $a \propto t^{2/3}$ for $t > t_e$), one has approximately $a_1/a_b \approx (t_p/t_e)^{\frac{1}{2}}(t_e/t_b)^{2/3} = (3/2)^{2/3}(t_p/t_H)^{\frac{1}{2}}(t_e/t_H)^{1/6}$, where $t_H = a(da/dt)_b{}^{-1} = 3t_b/2$ is the Hubble time. With $t_H = 5.8 \times 10^{17}$ sec. one finds[13]

$$T_b \approx (t_e/t_H)^{1/6}\; (30^{\circ}K) \;, \tag{3.77}$$

which is within a factor of 10 of 3°K for all possible value of t_e greater than 6×10^5 sec ($\approx$8 days), and gives 3°K for $t_e \approx 6 \times 10^{11}$ sec ($\approx 2 \times 10^4$ yrs). To determine t_e in the present model one would have to predict the baryon number density (or entropy per baryon), a point which I discussed earlier.

Any theory which is consistent with Einstein's equations and which leads to black-body radiation prior to t_e will leads to Eqs. (3.76) and (3.77), since Eq. (3.72) (with η_a replaced by η at t_e) can be obtained from the Einstein equation (3.71) at a time t shortly before t_e. However, the particle creation process acting at $t \approx t_p$ leads to the interpretation of T_b in Eq. (3.76) as the red-shifted temperature of the particles created near t_p with a temperature of the order of the Planck temperature. In that sense, a model of the present kind (or possibly starting with an anisotropic expansion as below) seems to point toward a plausible explanation of the origin of the density and entropy of the black-body radiation. Even if the initial form of the expansion or the particle field equations were not of the type which yielded a black-body spectrum, one would still expect a created energy density and entropy of the correct order of magnitude, and the black-body spectrum could probably arise through interactions. However, the present model is perhaps the simplest of the models involving particle creation near the Planck time. As in the scalar case, one expects the production of an approximately black-body spectrum of gravitons with the same initial temperature as for the minimally coupled scalar particles. Thus, if one is justified in using the linearized equation, the presence of an isotropic approximately black-body distribution of gravitons would be expected. Their present temperature might be of the order of 1°K, as estimated by Weinberg (Ref. 14, p. 592) for classical gravitational waves propagating in an isotropic background metric (see also the discussion in Ref. 85).

Anisotropic Expansions With Smooth Initial Boundary Conditions

Particle production in anisotropically expanding universes was first considered by Zeldovich,[7] who suggested that the gravitational effect of the initially created particles would rapidly (in a time of order t_p) isotropize the expansion rates. This work was continued by Zeldovich and Starobinsky,[8] Lukash and Starobinsky,[11] and in related work by Berger,[74] Hu[93,94] and by Fulling, Hu, and Parker.[37,95] I will show here that with smooth initial conditions like those used in the isotropic case, the particles are created with a direction-dependent distribution resembling Doppler-shifted black-body radiation in an anisotropically expanding thermal cavity with perfectly smooth reflecting walls and no dust.[13] Just as the presence of dust or other interactions in the expanding cavity will bring about a true black-body spectrum, one would expect interactions to do the same in the anisotropically expanding universe, while the expansion itself was isotropized through the gravitational interaction of the created matter.

The metric under consideration is given by

$$ds^2 = dt^2 - \sum_j a_j^{\,2}(t)\, dx^{j^2} \quad . \qquad (3.78)$$

As in the isotropic case one can write the solution of the field equation

$$\nabla^\mu \nabla_\mu \phi = 0 \qquad (3.79)$$

in the form

$$\phi = (2\pi)^{-3/2} \int d^3k \, [A_{\vec{k}} \, e^{i\vec{k}\cdot\vec{x}} \, \psi_{\vec{k}}(\tau) + \text{H.C.}] \ , \qquad (3.80)$$

where $A_{\vec{k}}$ and $A_{\vec{k}}^\dagger$ obey the usual commutation relations, and $\psi_{\vec{k}}$ satisfies

$$\frac{d^2\psi_{\vec{k}}}{d\tau^2} + V^2\,\omega^2\,\psi_{\vec{k}} = 0\ , \tag{3.81}$$

with

$$V = \sqrt{-g} = a_1 a_2 a_3\ , \tag{3.82}$$

$$\tau = \int^t V^{-1}(t')dt' \tag{3.83}$$

and

$$\omega^2 = \sum k_j^{\,2}/a_j^{\,2}\ . \tag{3.84}$$

The Wronskian condition is

$$\psi_{\vec{k}}\, d\psi^*_{\vec{k}}/d\tau - \psi^*_{\vec{k}}\, d\psi_{\vec{k}}/d\tau = i\ . \tag{3.85}$$

We assume that because of physical effects occuring near an actual cosmological singularity, the a_j can be represented as smoothly attaining (or approaching) constant values a_{jI} $(j = 1,2,3)$ as $\tau \to -\infty$. The initial spacetime is Minkowskian, as can be seen by a simple rescaling of coordinates. One finds that the $A_{\vec{k}}$ are annihilation operators for particles in that initial Minkowski space if the following boundary condition on $\psi_{\vec{k}}$ is imposed,

$$(\psi_{\vec{k}})_{\tau\to-\infty} = (2\omega_I V_I)^{-\frac{1}{2}} \exp(-i\omega_I V_I \tau)\ , \tag{3.86}$$

where ω_I and V_I are the initial values of ω and V. As in the isotropic case, we take the state vector to be $|0\rangle$, characterized by the absence of particles in the initial spacetime,

$$A_{\vec{k}}|0\rangle = 0 \qquad \text{for all } \vec{k}\ . \tag{3.87}$$

For convenience, we let the a_j approach constant values a_{jF} as $\tau \to \infty$ (a condition which can be replaced by a continued expansion in which the particle production is negligible, without affecting the value of $|\beta_k/\alpha_k|^2$). Then at late times one again has a flat spacetime with an unambiguous particle interpretation. At late times

$$(\psi_{\vec{k}})_{\tau\to\infty} = (2\omega_F V_F)^{-\frac{1}{2}} \left\{ \alpha_{\vec{k}}\, e^{-i\omega_F V_F \tau} + \beta_{\vec{k}}\, e^{i\omega_F V_F \tau} \right\} , \tag{3.88}$$

with

$$|\alpha_{\vec{k}}|^2 - |\beta_{\vec{k}}|^2 = 1 . \tag{3.89}$$

The annihilation operators at late times are

$$a_{\vec{k}} = \alpha_{\vec{k}} A_{\vec{k}} + \beta_{\vec{k}}^{*} A_{-\vec{k}}^{\dagger} . \tag{3.90}$$

The quantity $V^2\omega^2$ appearing in Eq. (3.81) smoothly approaches the constant value $V_I{}^2\omega_I{}^2$ for large negative τ. If one supposes that its behavior is sufficiently smooth, then one will have the same result as was obtained from Eq. (3.43) in the isotropic case, namely

$$|\beta_{\vec{k}}/\alpha_{\vec{k}}|^2 \cong \exp(-4\pi s\ V_I\omega_I) , \tag{3.91}$$

where s is the characteristic τ-time interval during which significant particle creation occurs. If most of the particle creation occurs in a t-time interval of order t_p, during which time $V \approx V_I$ then Eq. (3.83) gives

$$s \approx V_I^{-1}\ t_p . \tag{3.92}$$

By the same method as in the isotropic case one finds that the probability of observing n particles in mode $\vec{k}$ at late times is

$$P_n(\vec{k}) = |\beta_{\vec{k}}/\alpha_{\vec{k}}|^{2n} |\alpha_{\vec{k}}|^{-2} = \exp(-n4\pi t_p\omega_I)[1 - \exp(-4\pi t_p\omega_I)] . \tag{3.93}$$

The average number of particles per unit volume in the state $|0\rangle$ at late times, when $V = V_F$, is

$$\langle N \rangle = (8\pi^3 V_F)^{-1} \int d^3k \; |\beta_{\vec{k}}|^2$$

$$= (8\pi^3 V_F)^{-1} \int d^3k \; [\exp(4\pi t_p \omega_I) - 1]^{-1} \; . \qquad (3.94)$$

The momentum components at late times are given by

$$P_{jF} = k_j / a_{jF} \; . \qquad (3.95)$$

Writing

$$\omega_I = \Big(\sum_{j=1}^{3} P_{jF}^{\;2} \, a_{jF}^{\;2} / a_{jI}^{\;2} \Big)^{\frac{1}{2}} \; , \qquad (3.96)$$

it is apparent that the above distribution is the same as would be obtained if one started with black-body radiation at initial temperature $T_I = (4\pi k_B t_p)^{-1}$ in a box of dimensions a_{jI} with perfectly reflecting smooth walls containing no dust, and expanded the cavity slowly and adiabatically until it had linear dimensions a_{jF}. The factors $a_{jF}^{\;2}/a_{jI}^{\;2}$ result from the Doppler shifting of the radiation traveling along the three axes of the box. As for the expanding cavity with dust or rough relfecting walls, one would expect sufficiently strong interactions (with interaction time short with respect to the characteristic expansion time) to maintain a true black body distribution, in accordance with the requirements of the second law of thermodynamics. As in the case of the anisotropically expanding box with interactions, the factor ω_I in the distributions of Eqs. (3.93) and (3.94) would be replaced by $\omega_F (V_F/V_I)^{1/3}$ in the actual black-body distribution, so that the temperature at late times would be given by

$$T_F = T_I (V_I/V_F)^{1/3} = (4\pi k_B t_p)^{-1} (V_I/V_F)^{1/3} \; . \qquad (3.97)$$

This result is analogous to Eq. (3.69) in the isotropic case. At the same time the expansion rates would be isotropized by the gravitational effect of the created particles. One would expect V_I to be of the order of $(C\, t_p^{\frac{1}{2}})^3$, where $a = C\, t^{\frac{1}{2}}$ for the Friedmann universe onto which the initial anisotropic expansion joins. The results will then be the same as in the isotropic case already discussed.

Tracelessness of $T_{\mu\nu}$, Particle Production, and Vacuum Polarization in Robertson-Walker Universes

Here I will show that particles having a covariantly conserved, traceless, energy-momentum tensor are not created in a Robertson-Walker expansion of the universe, at least in the absence of event horizons.[96] This is an alternative to the proof based on conformal invariance which was given previously. I will also apply these ideas to vacuum polarization.

Consider a quantized field which has an energy-momentum tensor $T_{\mu\nu}$ satisfying

$$T^{\mu}{}_{\mu} = 0 . \qquad (3.98)$$

This equation is formally satisfied by the neutrino and photon fields, and by the massless scalar field which satisfies the conformally invariant wave equation

$$(\nabla^{\mu}\nabla_{\mu} + \frac{1}{6} R)\phi = 0 , \qquad (3.99)$$

but not by the minimally coupled scalar field. In a Robertson-Walker universe (with positive, zero, or negative spatial curvature) the expectation values of $T_{\mu}{}^{\nu}$ must satisfy the homogeneity and isotropy conditions, which imply that the only non-vanishing expectation values are

$$\langle T_{o}{}^{o}\rangle = \rho , \qquad \langle T_{i}{}^{j}\rangle = -P\delta_{i}{}^{j} , \qquad (3.100)$$

where ρ and P are the energy-density and pressure. It has already been pointed out (several paragraphs after Eq. (3.7)) that certain methods of regularization may give rise to a trace anomally, even in the absence of non-gravitational interactions (Refs. 81-84; for such an anomaly in the presence of other interactions see Ref. 97). Such an anomally might possibly be related to various vacuum effects, as occur in de Sitter space and in the closed Einstein universe.[49,50,99-102] For the present discussion, it is assumed that no trace anomaly appears in the absence of non-gravitational interactions, which is consistent with other methods of regularization.[164,8,35-37,99]

The conservation law

$$\nabla_\nu \langle T_\mu{}^\nu \rangle = 0 \tag{3.101}$$

in the Robertson-Walker metrics

$$ds^2 = dt^2 - a^2(t) \sum h_{ij}\, dx^i\, dx^j \tag{3.102}$$

(h_{ij} metric of space of constant curvature) takes the well known form

$$\frac{d}{dt}(\rho a^3) + P \frac{d}{dt}(a^3) = 0 \,. \tag{3.103}$$

With the tracelessness condition $\langle T_\mu{}^\mu \rangle = 0$ or $P = \rho/3$, this equation implies (as is well known) that

$$\rho(t) = \text{constant} \times a^{-4}(t) \,. \tag{3.104}$$

This last equation implies that regardless of the form of a(t), and irrespective of any additional "energy conditions", if ρ vanishes at one time then it vanishes at all times. One must still consider the possibility of vacuum polarization according to which ρ does not vanish even in vacuum.

For the spatially flat Robertson-Walker universe one can suppos that during some interval a(t) is constant and one has no particles present in Minkowski space, so that $\rho = 0$ during that interval. It then follows that ρ remains zero in the vacuum state regardless of the subsequent (or previous) behavior of a(t). Hence, the above argument precludes both particle creation and vacuum polarization in the spatially flat Robertson-Walker universe if $\langle T_\mu{}^\mu \rangle = 0$.

In the case of negative spatial curvature, one can suppose that during some interval of time a(t) is proportional to t. In that case it is well known that the spacetime is equivalent to a part of Minkowski space in a non-Minkowskian coordinate system. Therefore, if one specifies that the state is the vacuum during such an interval, then $\rho = 0$ as for Minkowski space. Eq. (3.104) then implies that $\rho = 0$ in the vacuum state in the Robertson-Walker universe with negative spatial curvature, regardless of the form of a(t). Hence, there is no particle creation and no vacuum polarization in the open Robertson-Walker universe if $\langle T_\mu{}^\mu \rangle = 0$.

The closed Robertson-Walker universe is fundamentally different because there is no form of a(t) which makes the spacetime flat. In the case of the conformally invariant scalar field in the closed Einstein static universe (a = constant) there is evidently a non-vanishing vacuum polarization given by[99]

$$\rho_o = (480\ \pi^2 a^4)^{-1} . \tag{3.105}$$

It follows from Eq. (3.104) that the energy density of the conformal scalar field in that state has the form of Eq. (3.105) even when a is time dependent provided $T_\mu{}^\mu$ vanishes. On the basis of conformal invariance one can define positive and negative frequency solutions of Eq. (3.99) in the closed Robertson-Walker universe for any statically bounded form of a(t) and show that in the absence of event horizons there is no particle creation. Thus, Eq. (3.105) represents a vacuum energy even when a is time-dependent. Analogous conclusions hold for the neutrino and photon fields.

According to Ref. 49, a detector on a geodesic will observe a steady flux of particles if the geodesic has an event horizon. In the de Sitter universe, this will be a steady isotropic flux of black-body radiation. It is observer-dependent radiation, which is isotropic relative to two geodesic observers with non-zero relative velocity. Thus, it cannot be described by a single energy-momentum tensor. Furthermore, the ρ and $P(=\rho/3)$ for that steady radiation are constant, and hence do not satisfy the conservation law of Eq. (3.103) in the coordinate system of Eq. (3.1). Evidently, the radiation of Gibbons and Hawking does not directly correspond to the conserved $\langle T_{\mu\nu} \rangle$ we have been considering. Rather, it seems to be a consequence of the interaction of each detector with vacuum fluctuations.

This is perhaps supported by the following observation, which has not been noticed before. As is well known, a part of de Sitter space-time can be covered by coordinated $t, \vec{x}$ for which the metric takes the form of Eq. (3.1) with

$$a(t) = \exp[(\Lambda/3)^{\frac{1}{2}} t] , \tag{3.106}$$

where Λ is the cosmological constant. A geodesic observer at fixed spatial coordinates $\vec{x}$ has an event horizon of proper radius at time t_1 given by

$$d(t_1) = a(t_1) \int_{t_1}^{\infty} dt' a^{-1}(t') = (\Lambda/3)^{-\frac{1}{2}} , \tag{3.107}$$

which is independent of t_1 for the de Sitter universe. Gibbons and Hawking[49] find that such an observer will detect black-body radiation of temperature

$$T = (2\pi k_B)^{-1} (\Lambda/3)^{\frac{1}{2}} = (2\pi k_B d)^{-1} , \tag{3.108}$$

where k_B is Boltzmann's constant (and $\hbar = c = 1$). This radiation has a density (for spin-0 massless particles) of

$$\rho = \frac{1}{2} \sigma T^4 = \frac{1}{2} (\pi^2 k_B^4/15)(2\pi k_B d)^{-4} , \tag{3.109}$$

where σ is the Stefan-Boltzmann constant, and the factor of $\frac{1}{2}$ appears for spin-0 particles. This is

$$\rho = (480 \pi^2 d^4)^{-1} , \tag{3.110}$$

which is the same as the vacuum energy density (3.105) for a static Einstein universe of radius equal to the radius d of the event horizon. This may be merely a coincidence, or may reflect a deeper connection. [A derivation of Eq. (3.105) for the Einstein universe is given below (Eq. (3.132)), and a heuristic argument suggesting why Eq. (3.110) and Eq. (3.105) should perhaps be related is given in the latter part of Section 5.]

Finally, let me comment on the relation between particle creation and the dominant energy condition. Hawking[103] showed that in general if $T_{\mu\nu}$ satisfies a condition, called the dominant energy condition (DEC), then there can be no particle production. In the present Robertson-Walker context that condition is $|P| < |\rho|$ (and $\rho > 0$, which is not needed in this case). If initially $\rho = 0$ at $t = t_1$, then the DEC requires that $P = 0$ at t_1. But then Eq. (3.103) implies that $\rho(t_1 + dt) = 0$. The DEC then implies that $P(t_1 + dt) = 0$, and the process can be repeated ad infinitum implying $\rho = 0$, $P = 0$ for all time, and hence no particle production. Zeldovich and Pitaevskii[104] showed explicitly that the dominant energy condition is indeed violated by the particle creation process, even though $T_{\mu\nu}$ continues to be covariantly conserved.

Regularization and Renormalization of the Energy-Momentum Tensor

In the previous work the reaction of the created particles back on the metric has not been explicitly considered. One approach to finding the reaction back, which is essentially classical, has been used by Lukash and Starobinsky[11] to show that an initially anisotropic expansion is rapidly isotropized by the gravitational effect of the particles created near the Planck time. In that approach, the initial particle creation process is treated ignoring the reaction back on the metric. Then the reaction back is obtained by regarding the created particles as a classical relativistic gas and solving the classical Einstein equations with that source. A second approach which also stops short of quantizing the gravitational field is to couple the quantized matter fields described by energy-momentum tensor $T_{\mu\nu}$ to the classical Einstein tensor $G_{\mu\nu}$ $(= R_{\mu\nu} - \frac{1}{2} g_{\mu\nu}R)$ through the simplest semi-classical generalization of Einstein's equation

$$G_{\mu\nu} = -8\pi G\langle T_{\mu\nu}\rangle \ . \tag{3.111}$$

[The conventions are $(\nabla_\mu\nabla_\nu - \nabla_\nu\nabla_\mu)\xi_\lambda = \xi_\sigma R^\sigma{}_{\lambda\mu\nu}$, and $R_{\mu\nu} = R^\lambda{}_{\mu\lambda\nu}$, where ξ_λ is an arbitrary vector field. The metric signature is -2.] This equation was used in connection with spherically symmetric static gravitational fields by Møller,[34] Bonazzola and Pacini,[104] and Ruffini and Bonazzola.[105] In the case of an isotropic universe, Parker and Fulling[29] have used Eq. (3.111) to construct a model in which a closed collapsing Friedmann universe avoids the singularity and passes over into an expansion phase. Evidently Eq. (3.111) gives a reasonable description of the gravitational field only in those cases in which the actual distribution of matter resembles the average distribution. In particular, such a description seems reasonable for homogeneous cosmological models.

The formal expression for $T_{\mu\nu}$ is

$$T_{\mu\nu} = 2(-g)^{-\frac{1}{2}} \frac{\delta S}{\delta g^{\mu\nu}} , \tag{3.112}$$

where $S = \int \mathcal{L}\, d^4x$ is the action, and $\mathcal{L}$ is the Lagrangian density of the quantized field. For example, using Eq. (2.3) for the minimally coupled scalar field $(\gamma = 0)$ one has

$$T_{\mu\nu} = \frac{1}{2}(\partial_\mu\phi\partial_\nu\phi + \partial_\nu\phi\partial_\mu\phi) - \frac{1}{2} g_{\mu\nu}(g^{\alpha\beta}\partial_\alpha\phi\partial_\beta\phi - m^2\phi^2) \ . \tag{3.113}$$

The first problem one encounters in attempting to use Eq. (3.111) is that the expectation values of $T_{\mu\nu}$ in general are infinite. This problem is more severe than in special relativity, where normal ordering or a single vacuum energy subtraction is sufficient. For example, it is clear that in a time-dependent metric in which positive and negative frequency solutions become mixed, i.e. particle creation occurs, there is in general no preferred time at which one should perform the normal ordering of the creation and annihilation operators in $T_{\mu\nu}$. Therefore more sophisticated methods are required. These techniques of obtaining suitable finite expressions from the formal expressions for $\langle T_{\mu\nu} \rangle$ are known as regularization and renormalization. In regularization, the divergent quantities are replaced by well-defined expressions in a manner consistent with the physical basis of the theory (for example, so as to preserve the covariant conservation of $\langle T_{\mu\nu} \rangle$). In renormalization, the infinities are absorbed into the physical constants such as the gravitational constant, or are cancelled by a finite number of suitable counterterms in the Lagrangian. The difficulties encountered in these procedures are not as great as in the case when the gravitational field is also quantized (a case that will not be discussed here). A number of different approaches have yielded finite expressions for $\langle T_{\mu\nu} \rangle$. The various methods seem to give the same or nearly the same results when applied to the same problem, but the subject is still in a state of development so that a definitive discussion is not yet possible.

I will only give a brief discussion of several methods which have been developed in the context of the type of cosmological metrics considered earlier. In particular, I will discuss the work of Zeldovich and Starobinsky,[8] Parker and Fulling,[35] Fulling and Parker,[36] Fulling, Parker and Hu,[37] and Ford.[99,100] In his excellent review article, DeWitt[23] shows how the background field method developed by DeWitt[106] can be used to understand the above techniques from a more general coordinate-space point of view. The Schwinger-DeWitt approach has been further developed by Christensen.[38,39] Other techniques for dealing with divergent parts of $T_{\mu\nu}$ have been used by Utiyama and DeWitt,[107] Deser and van Nieuwenhuizen,[108,109] t'Hooft and Veltman,[110,111] Halpern,[112] Candelas and Raine,[130] Dowker and Critchley,[101,102] Streeruwitz,[113] Davies and Fulling,[83] Davies, Fulling, and Unruh,[82] Adler,

Lieberman, and Ng,[164] and others.

Suppose now that the spacetime is described by a cosmological metric, such as Eq. (3.78) or one of the Robertson-Walker metrics, in which the time-dependence of the quantized field can be separated out, and the field can be written in a mode decomposition like Eq. (3.80) or (3.13). Then one can write the vacuum expectation value of the energy-momentum as a sum or integral over modes,

$$\langle T_{\mu\nu}\rangle = \int d^3k\, T_{\mu\nu}(\vec{k},m) \ , \tag{3.114}$$

where m is the mass of the field. Zeldovich and Starobinsky[8] define

$$T^{(n)}_{\mu\nu}(\vec{k}) = \frac{1}{n} T_{\mu\nu}(n\vec{k},nm) \ . \tag{3.115}$$

In their so-called "n-wave regularization" method, the regularized tensor $\langle T_{\mu\nu}\rangle_{reg}$ is given by

$$\langle T_{\mu\nu}\rangle_{reg} = \int d^3k \lim_{n\to\infty} \left[T_{\mu\nu}(\vec{k},m) - T^{(n)}_{\mu\nu}(\vec{k}) - \frac{\partial T^{(n)}_{\mu\nu}(\vec{k})}{\partial(n^{-2})} - \frac{1}{2}\frac{\partial^2 T^{(n)}_{\mu\nu}(\vec{k})}{\partial(n^{-2})^2} \right] . \tag{3.116}$$

The subtracted terms serve to remove the parts of $T_{\mu\nu}(\vec{k},m)$ which would diverge when integrated over $\vec{k}$, while maintaining the covariant conservation of $T_{\mu\nu}$. They applied this expression to the metric of Eq. (3.80) of a spatially flat anisotropically expanding universe.

In the so-called "adiabatic regularization" method[35,36,37] a physical basis for the above result was found, and the method was extended also to metrics with curved 3-space. In addition, the form of the subtracted terms was shown to be related to geometric tensors like the Einstein tensor $G_{\mu\nu}$, so that contact with covariant renormalization techniques could be made.

The basic idea of the adiabatic regularization method in the case of a flat spatial metric is to introduce a slowness parameter T, such that when T is large the expansion of the universe is slow. The terms $T_{\mu\nu}(\vec{k},m)$ in the mode sum (3.114) are expanded in powers of T^{-1}, and the minimum number of terms in that series are subtracted from the original expression for $T_{\mu\nu}(\vec{k},m)$ to insure that the adiabatic limit ($T \to \infty$) of the regularized energy-momentum $\langle T_{\mu\nu}\rangle_{reg}$ is well-defined and is equal to the known Minkoski space result. These subtractions preserve the relations $\nabla_\nu \langle T_\mu{}^\nu\rangle_{reg} = 0$ for all values of T. It is assumed that the same subtractions hold when T is set equal to unity so that the original problem is recovered. The resulting expression for $\langle T_\mu{}^\nu\rangle_{reg}$ can be shown to agree with that of Zeldovich and Starobinsky with their n replaced by the slowness parameter T. It is important to note that the final expression is well-defined and does not involve an infinite series. The expansion in powers of T^{-1} was only used to find the form of the several terms which are subtracted from $T_{\mu\nu}(\vec{k},m)$. For a statically bounded expansion the regularized energy-momentum tensor obtained in this way is normal-ordered in terms of the appropriate creation and annihilation operators in both the initial and final Minkowski spaces, but is not normal-ordered at all during the intermediate expansion. That is because the subtractions reduce to the appropriate form in each of the Minkowski regions. In the case when the 3-space is curved, the "slowness parameter" is also associated with the spatial curvature, so that in the adiabatic limit, $T \to \infty$, one approaches flat spacetime.

To be specific consider the minimally coupled scalar field in the spatially flat Robertson-Walker metric of Eq. (3.1). The Hilbert space of states is taken as the Fock space constructed from the $A_{\vec{k}}$ and $A_{\vec{k}}^\dagger$ appearing in the field expansion of Eq. (3.13). (The A_k will not correspond to physical particles unless the expansion parameter is very slowly changing.) The expectation values of $T_o{}^o$ consist of a finite part depending on the state, and a divergent part ρ_o which is the same for all states, namely[35]

$$\rho_o = (4\pi^2)^{-1} \int dk\ k^2 \ (|\partial_o \psi_k|^2 + \omega_k{}^2 |\psi_k|^2) \ , \tag{3.117}$$

where ψ_k satisfies Eq. (3.14). There is a similar divergent expression in the pressure. To apply adiabatic regularization replace $a(\tau)$ in Eqs. (3.14) and (3.15) by $a(\tau/T)$. One then has a one-parameter family of solutions $\psi_k(\tau,T)$ of Eq. (3.14). Impose the boundary condition that $\psi_k(\tau,T)$ reduce to a positive frequency solution to order T^{-4} in the limit of large T, or equivalently, that it be positive frequency to order T^{-4} during any interval in which the expansion parameter becomes constant. That condition can be imposed, and in fact will automatically be satisfied if ψ_k is taken to be positive frequency during a single static or quasi-static interval, because Kulsrud's theorem[91] (see discussion following Eq. (3.56)) insures that $|\beta_k|$ vanishes to order T^{-4} during any static interval (I am assuming that $a(\tau)$ is continuous and has continuous derivatives up to at least the fourth order). To find the expansion of the integrand of ρ_o in powers of T^{-1}, it is convenient to use the positive frequency extended WKB approximation to fourth order in T^{-1}, which is known to satisfy the boundary condition on ψ_k. One finds to fourth order that[35]

$$|\partial_o \psi_k|^2 + \omega_k{}^2 |\psi_k|^2 \cong (a^3\omega)^{-1} [\omega^2 + \frac{1}{8} Z^2 + \frac{1}{8} Z\dot{\varepsilon}_2$$
$$- \frac{1}{16} Z^2 \varepsilon_2 + \frac{1}{8} \omega^2 \varepsilon_2{}^2] \tag{3.118}$$

where the dot denotes differentiation with respect to t,

$$Z = 3 \frac{\dot{a}}{a} + \frac{\dot{\omega}}{\omega} \tag{3.119}$$

and

$$\varepsilon_2 = -(a^3\omega)^{-3/2} \ \partial_\tau \left\{ (a^3\omega)^{-1} \partial_\tau [(a^3\omega)^{\frac{1}{2}}] \right\} \ . \tag{3.120}$$

(The slowness parameter T has been set equal to unity.) If one denotes the terms on the right of Eq. (3.118) by $\rho_{vac}(k)$, then the regularized energy density has the form

$$\langle T_o{}^o\rangle_{reg} = (4\pi^2)^{-1} \int dk\, k^2 \Big\{ |\partial_o \psi_k|^2 + \omega_k{}^2 |\psi_k|^2 - \rho_{vac}(k) \Big\} + \int d^3k\, \rho_1(\vec{k}) \ . \qquad (3.121)$$

where $\int d^3k\, \rho_1(\vec{k})$ is a finite term dependent on the state vector. One obtains a similar expression for the pressure. In a statically bounded expansion, if one specifies that $\psi_k(\tau)$ is a positive frequency solution in the initial Minkowski space (see Eq. (3.20)) and that no particles are present initially, i.e. $A_{\vec{k}}|0\rangle = 0$ for all $\vec{k}$, then $\rho_1(\vec{k})$ is zero in Eq. (3.121), and one finds and in the final Minkowski space the only contribution to $\langle T_o{}^o\rangle_{reg}$ comes from the particles created during the expansion of the universe. During the expansion of the universe (whether or not it has static intervals) Eq. (3.121) and the corresponding regularized pressure contain contributions both from particle creation and vacuum polarization, and serve as the source of the gravitational field through Eq. (3.111) in the semi-classical theory.

In order to strengthen the regularization procedure by considerations related to covariance, Fulling and Parker[36] compared the terms on the right of Eq. (3.118), i.e. $\rho_{vac}(k)$, with tensors formed from the metric. They found that when the physical momentum $\vec{k}/a$ was used as the variable of integration in (3.121), then those subtracted terms took the form

$$(4\pi^2)^{-1} \int dk\, k^2\, \rho_{vac}(k) = \lambda_1 g_o{}^o + \lambda_2 G_o{}^o + \lambda_3 H_o{}^o + f(m^2), \qquad (3.122)$$

where λ_1, λ_2, and λ_3 are independent of $a(\tau)$, but diverge in the momentum as ∞^4, ∞^2, and $\ln\infty$, respectively, and $f(m^2)$ is a finite term which vanishes when m vanishes, but cannot be expressed as a tensor formed from $a(\tau)$ and its derivatives. Here $G_o{}^o$ refers to the Einstein tensor, and

$$H_0^{\ 0} = 2R(R_0^{\ 0} - \frac{1}{4} R) + 2(\nabla_0 \nabla^0 R - g^{\alpha\beta} \nabla_\alpha \nabla_\beta R) \tag{3.123}$$

is the 0-0 component of the tensor $H_\mu^{\ \nu}$ which satisfies $\nabla_\nu H_\mu^{\ \nu} = 0$ and is formed from terms which are no more than quadratic in the Riemann tensor (in the Robertson-Walker metrics only one such tensor appears).

These are the same kind of divergent terms found in Utiyama and DeWitt.[107] Following their procedure, one can remove those terms from the formal expression for $\langle T_0^{\ 0} \rangle$ by renormalization instead of simply subtracting them as in adiabatic regularization. One writes the identity

$$\begin{aligned} \langle T_0^{\ 0} \rangle &= \langle T_0^{\ 0} \rangle_{reg} + (4\pi^2)^{-1} \int dk\, k^2\, \rho_{vac}(k) \\ &= \langle T_0^{\ 0} \rangle_{reg} + \lambda_1 g_0^{\ 0} + \lambda_2 G_0^{\ 0} + \lambda_3 H_0^{\ 0} + f(m^2) \ . \end{aligned} \tag{3.124}$$

Next introduce counterterms into the Einstein equation, so that it becomes

$$G_\mu^{\ \nu} + \Lambda_0 g_\mu^{\ \nu} + \sigma_0 H_\mu^{\ \nu} = -8\pi G_0 \langle T_\mu^{\ \nu} \rangle \ . \tag{3.125}$$

Substituting Eq. (3.124) into the 0-0 component of Eq. (3.125) one can now absorb the terms involving λ_1, λ_2, and λ_3 into a renormalization of the coupling constants Λ_0, σ_0, and G_0, with the result

$$G_0^{\ 0} + \Lambda g_0^{\ 0} + \sigma H_0^{\ 0} = -8\pi G(\langle T_0^{\ 0} \rangle_{reg} + f(m^2)) \ . \tag{3.126}$$

The anomalous finite term $f(m^2)$ may well be an artifact of the non-covariant procedure used to arrive at Eq. (3.126) and should probably be dropped, so that one ends up with the same result as was obtained through regularization. One encounters similar anomalous terms when attempting to recast the subtracted pressure terms in manifestly covariant form (see Ref. 36 for further discussion).

The analogous calculations have been carried out for the

scalar field with conformal coupling [i.e. $(\nabla_\mu \nabla^\mu + m^2 + \frac{1}{6} R)\phi = 0$] in the anisotropically expanding metric of Eq. (3.78) and in the closed, flat, and open Robertson-Walker metrics by Fulling, Parker, and Hu.[37] For the massless field in the Robertson-Walker metrics they found that the vacuum expectation value of the formal energy-momentum tensor T_{oo} is (there is a misprint in their Eq. (5.29), i.e. their a^2 should read a^3)

$$\langle T_{oo}\rangle = \begin{cases} [4\pi^2 a^3(t)]^{-1}\int_0^\infty dq q^2 \omega_q \text{ if spatial curvature} \leq 0, & (3.127) \\ [4\pi^2 a^3(t)]^{-1} \sum_{q=1}^{\infty} q^2 \omega_q \text{ if spatial curvature} > 0. & (3.128) \end{cases}$$

Here

$$\omega_q = q/a(t) \;. \qquad (3.129)$$

In this case there is no particle production and the vacuum is easily found by defining positive frequency solutions of the wave equation as those solutions which would become the standard positive frequency solutions if a(t) were to become static during any interval. We know that if $\langle T_\mu{}^\mu\rangle$ vanishes (see Eq. (3.104) and the discussion following it) then $\langle T_{oo}\rangle$ must vanish in the cases of negative or vanishing spatial curvature, and that $\langle T_{oo}\rangle$ must be proportional to $a^{-4}(t)$ in the spatially closed universe. One can justify dropping the expression in Eq. (3.127) from the energy-momentum tensor because when expressed in terms of the physical momentum variable $p = q/a(t)$ it is identical with the Minkowski space zero-point energy. However, in the closed case, one must use a more subtle argument to find the constant of proportionality in Eq. (3.104) relating $\langle T_{oo}\rangle$ to $a^{-4}(t)$.

Such an approach has been given by Ford.[99] Following the method used by Casimir[98] in finding the vacuum energy of the electromagnetic field between two conducting plates, one introduces a high-energy cut-off into the divergent sum in Eq. (3.128), replacing it by

$$\begin{aligned} \langle T_{oo}(t)\rangle_{\alpha} &= [4\pi^2 a^4(t)]^{-1} \sum_{q=1}^{\infty} q^3 e^{-\alpha q} \\ &= \frac{3}{2\pi^2\alpha^4} + \frac{1}{480\pi^2 a^4(t)} + O(\alpha) , \end{aligned} \tag{3.130}$$

where the terms $O(\alpha)$ vanish in the limit that the cut-off is removed ($\alpha \to 0$). One then requires that the difference of the regularized tensor $\langle T_{oo}(t)\rangle_{reg} = \rho_{reg}(t)$ at any two different times satisfy

$$\begin{aligned} \rho_{reg}(t_2) - \rho_{reg}(t_1) &= \lim_{\alpha\to 0} \left\{ \langle T_{oo}(t_2)\rangle_{\alpha} - \langle T_{oo}(t_1)\rangle_{\alpha} \right\} \\ &= \frac{1}{480\pi^2 a^4(t)} . \end{aligned} \tag{3.131}$$

Then by requiring that $\rho_{reg}(t)$ must vanish in the limit that $a(t) \to \infty$, since in that case one approaches flat spacetime, one obtains the result

$$\rho_{reg}(t) = \frac{1}{480\pi^2 a^4(t)} . \tag{3.132}$$

for the vacuum energy-density of the massless conformal scalar field in the closed Robertson-Walker universe. This vacuum energy-density is at all times many orders of magnitude smaller than the energy-density of the observed black-body radiation.

Possible Avoidance of the Cosmological Singularity

A number of theorems have been proved by Penrose and Hawking which imply the inevitability of singularities in gravitational collapse and cosmology, under a number of rather general assumptions which involve the classical Einstein equation with vanishing cosmological constant,

$$G_{\mu\nu} = -8\pi G T_{\mu\nu} \ . \tag{3.133}$$

See Hawking and Ellis[114] for a discussion of these theorems and references to the literature on them. The theorems which apply to the cosmological singularity generally involve an "energy-condition" on $T_{\mu\nu}$ which appears quite reasonable from the point of view of ordinary matter. For example, one such condition is the strong energy condition that

$$T_{\mu\nu}W^{\mu}W^{\nu} - \frac{1}{2}W_{\lambda}W^{\lambda}T \geq 0 \tag{3.134}$$

for any timelike vector W^{μ}. In the Robertson-Walker metrics for example, with $T_o{}^o = \rho$ and $T_i{}^j = -P\delta_i{}^j$, this condition becomes

$$\rho + P \geq 0 \quad \text{and} \quad \rho + 3P \geq 0 \ , \tag{3.135}$$

which can only be violated for small P if the energy density becomes negative, or for positive ρ if P becomes sufficiently negative that $P < -\frac{1}{3}\rho$. For the minimally coupled scalar field with $T_{\mu\nu}$ given by Eq. (3.113) one has (with ϕ regarded as a classical field)

$$\rho + 3P = 2(\partial_o\phi)^2 - m^2\phi^2 \ , \tag{3.136}$$

which can become negative when $m \neq 0$. Hawking and Ellis (Ref. 14, p. 96) argue that this could only lead to avoidance of the singularity when the radius of curvature of spacetime becomes less than m^{-1}, or of order 10^{-12} cm for π-mesons, which from the classical point of view is effectively a singular region.

To proceed further at that scale one should use the quantized scalar field (quantization of gravity is probably not necessary at that scale). Parker and Fulling[29] have constructed an explicit model with the quantized, minimally coupled scalar field coupled to the gravitational field through Eq. (3.111), i.e.,

$$G_{\mu\nu} = -8\pi G\langle T_{\mu\nu}\rangle \ , \tag{3.137}$$

in which a closed collapsing Friedmann universe, which according to classical theory should collapse to a state of infinite density, avoids the singularity by reaching a minimum radius of order m^{-1} and passes over into an expanding Friedmann universe with a maximum radius of about 10^{30} cm. The state vector they used contained approximately 10^{40} particles "condensed" in the lowest mode in a state which is a coherent superposition of states of definite particle number (as occurs in models of superfluidity and superconductivity). Therefore their results were insensitive to the method of regularization of $\langle T_{\mu\nu}\rangle$, which mainly affects the higher modes. The "bounce" was produced by large negative pressure fluctuations connected with the violation of the dominant energy condition discussed in the paragraph following Eq. (3.110). Although $\rho = \langle T_o^{\ o}\rangle$ was positive, the pressure $P = -\langle T_i^{\ i}\rangle$ (no sum) became sufficiently negative to violate Eq. (3.135). The fluctuations occured on a time scale of order m^{-1}, so that their effect became significant only when the radius of the universe was of the order of m^{-1}. A bounce occured if the pressure was experiencing a negative fluctuation at that time (i.e. with a probability of roughly 50%). If the state vector used were to result from a transition due to interactions occuring when $a(t)$ approached m^{-1}, then the model might be of physical interest. Jacobs[115] has attempted to extend the model in a promising way by using the composite particle picture of hadrons in which the "bounce" is assured as a result of the numerous hadron resonances successively making transitions to states in which negative pressure fluctuations occur as the singularity is approached.

Such quantum effects are not limited to the spin-0 field. Epstein, Glaser, and Jaffe[116] have proved a very general theorem to the effect that in any local field theory in Minkowski space there exist state vectors such that the energy-density ρ is negative in some region. It seems very likely that this theorem will also hold for $\langle T_{oo} \rangle_{reg}$ in local field theories in curved spacetime. Their theorem applies to interacting fields as well as free fields. Of course, the theorem says nothing about how likely such a state would be in a realistic system.

It is also possible that the vacuum energy-momentum of a quantized field may violate the various energy conditions. The vacuum energy in Eq. (3.132) is positive, but in general there is no obvious reason why it must always be so. In fact, the vacuum energy-density of the electromagnetic field between two conducting plates is negative and equal to the constant value (Casimir, Ref. 98, Ford, Ref. 99 and references given there),

$$\rho = - \frac{\pi^2}{720R^4} , \qquad (3.138)$$

where R is the distance between the plates.

Unless the use of a local energy-momentum tensor and the corresponding regularization techniques (including those used in Minkowski space) are on the wrong track, then it is clear that there exist a variety of effects involving quantized matter fields which can violate the various energy conditions, and which hold out the possibility of avoiding the formation of singularities. Gowdy[147] has suggested that particle creation effects may produce topology changes (also see Dyson[148]).

Within the context of the classical Einstein equation (3.133), Bekenstein[117,118] has worked with the unquantized conformally invariant scalar field of zero mass. The energy-momentum tensor of this field violates the strong energy condition. Bekenstein has constructed interesting exact solutions of the coupled Einstein-conformal scalar field equations which never develop a singularity, both for the free field, and when interactions with other fields are present. The dimensions at which the "bounce" occurs in these models depends on the values assigned to various constants appearing in the solutions, and can evidently occur at macroscopic dimensions.

It would carry us too far from our main topic to discuss other attempts to avoid singularities through the introduction of terms quadratic in the Riemann tensor into Einstein's equation (Gurovich,[119] Nariai,[120] Nariai and Tomita[121]), through the introduction of torsion (Trautman[122]), through hadron interactions (Bahcall and Frautschi[123]), through the effects of quantized gravity (Wheeler,[124] Misner,[125] Ryan,[126] Liang[127]), or through other effects involving quantized matter (Lund[128]).

4. PARTICLE CREATION BY BLACK HOLES

Let me now sketch the calculation of the spectrum of a spherically symmetric black hole. I will follow Hawking's original method[15,16] because of its clear and unambiguous physical basis. Some of the intermediate calculations may be rather lengthy, but the physical input and output are unambiguous. An alternate approach is given by the elegant path integral calculation of Hartle and Hawking.[64] (In this section $G = \hbar = c = 1$.)

In order to avoid possible ambiguities about boundary conditions in the distant past, consider a spherically symmetric body of mass M collapsing to form a black hole. The space-time is pictured below in incoming Eddington-Finkelstein coordinates (Fig. 3) and in a Penrose diagram (Fig. 4).[129] In the former diagram the horizontal planes correspond to spatial sections in the equatorial plane, $\theta = \pi/2$, with the Schwarzschild radial coordinate measuring distance from the origin. Radially incoming light rays (but not outgoing ones) are plotted along 45° lines.

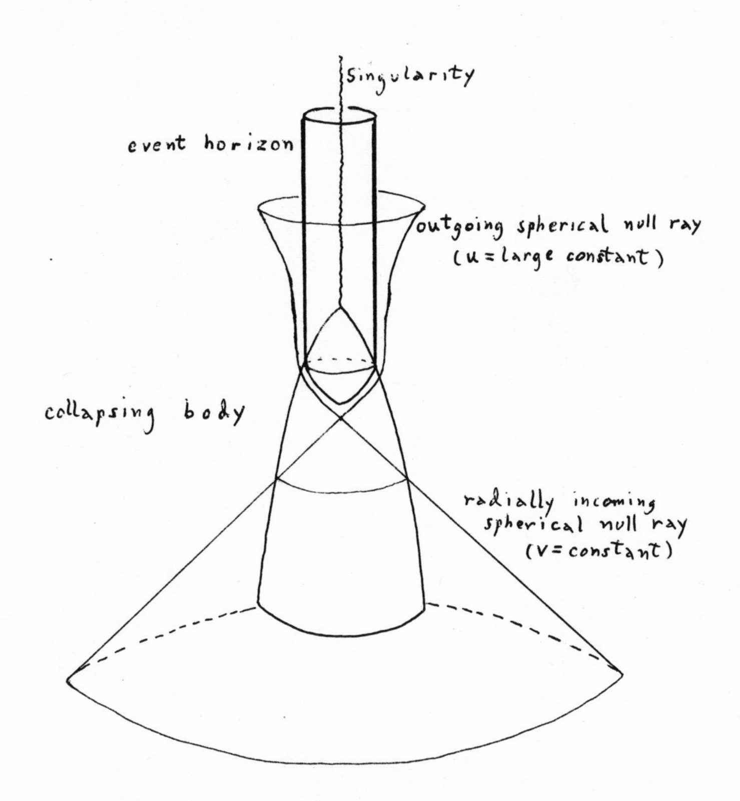

Fig. 3. Eddington-Finkelstein diagram of collapse

In the Penrose diagram both incoming and outgoing light rays are plotted at 45° angles and both angular coordinates are suppressed. Each point represents a spherical shell of area $4\pi r^2$. Distances are also conformally distorted to bring ∞ to a finite distance. Future and past null infinity are denoted by $\mathscr{I}^+$ and $\mathscr{I}^-$, respectively.

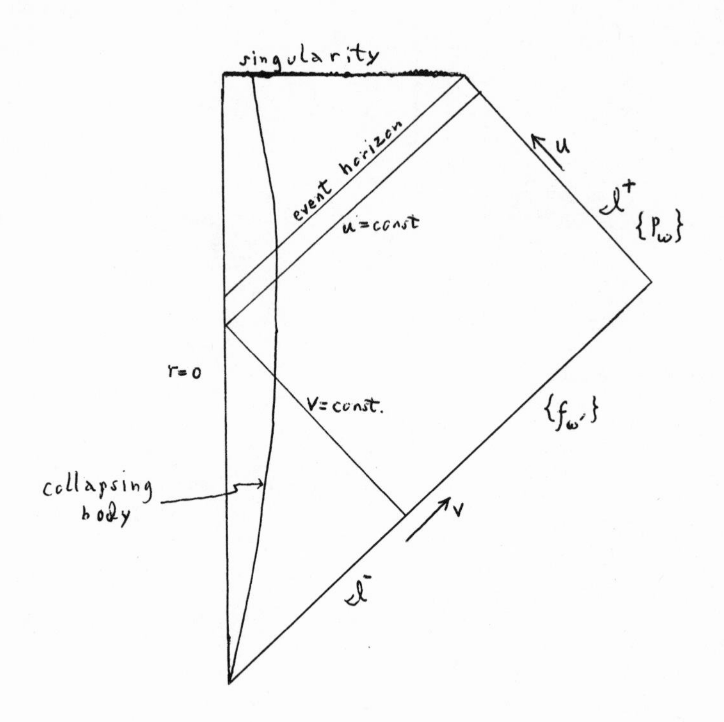

Fig. 4. Penrose diagram of collapse

We will consider a neutral scalar field obeying Eq. (2.1) with m = 0, and quantized in accordance with Eqs. (2.1)-(2.8). We will be interested in what is observed at infinity at very late times. This will depend on the characteristics of the Schwarzschild geometry outside the event horizon, and not on the detailed form of the collapse or the nature of the collapsing body. Therefore, let us first consider the form of the solutions in the Schwarzschild space-time r > 2M. The metric is

$$ds^2 = (1 - \frac{2M}{r})dt^2 - (1 - \frac{2M}{r})^{-1} dr^2 - r^2 d\Omega^2 , \qquad (4.1)$$

where $d\Omega^2 = d\theta^2 + \sin^2\theta d\phi^2$. One finds the following separated solutions g of Eq. (2.1) (the scalar curvature is zero):

$$g \propto r^{-1} R_{\omega\ell}(r) Y_{\ell m}(\Omega)\, e^{-i\omega t} \qquad (4.2)$$

with the radial function $R_{\omega\ell}(r)$ satisfying

$$\frac{d^2}{dr^{*2}} R_{\omega\ell} + \left\{\omega^2 - [\ell(\ell+1)r^{-2} + 2Mr^{-3}][1 - 2Mr^{-1}]\right\} R_{\omega\ell} = 0, \qquad (4.3)$$

with

$$r^* = r + 2M \ln|r(2M)^{-1} - 1| . \qquad (4.4)$$

Note that $r^* \to -\infty$ as $r \to 2M$ and $r^* \to \infty$ as $r \to \infty$. Equation (4.3) is analogous to the Schrödinger equation, with ω^2 playing the role of energy and $[\ell(\ell+1)r^{-2} + 2Mr^{-3}][1 - 2Mr^{-1}]$ the potential. Since the potential vanishes sufficiently fast as $r \to 2M$ ($r^* \to -\infty$) and $r \to \infty$, the solutions in those regions are superpositions of $\exp(\pm i\omega r^*)$. Hence the solutions g of Eq. (4.2) will be superpositions of terms proportional to $r^{-1} \exp(-i\omega u) Y_{\ell m}$ and $r^{-1} \exp(-i\omega v) Y_{\ell m}$ in the regions $r^* \to \pm\infty$, where

$$u = t - r^* , \qquad v = t + r^* \qquad (4.5)$$

are null coordinates. Outgoing spherical null waves are characterized by u = const., and incoming spherical null waves by v = const. In terms of u and v the Schwarzschild metric is

$$ds^2 = (1 - 2Mr^{-1}) du dv - r^2 d\Omega^2 . \qquad (4.6)$$

The effect of the potential in Eq. (4.3) is to scatter incoming waves or outgoing waves so that they become superpositions. Thus, if one sends a purely incoming wave $r^{-1} \exp(-i\omega v) Y_{\ell m}$ from infinity, part of it will reach the horizon and be absorbed and a part will be reflected back out to infinity. The absorption probability (or transmission coefficient) for such spherical waves, $\Gamma_{\omega\ell}$, is also equal to the probability for an outgoing wave $r^{-1}\exp(-i\omega u)Y_{\ell m}$ emitted from just outside the horizon ($r^* = -\infty$) to reach infinity, as follows from the reciprocity property of the transmission coefficient in one-dimensional scattering.

Let us now consider the quantized field ϕ in the space-time of the collapsing body. In order to specify the initial state we must specify a complete set of positive and negative frequency solutions of the field equation at early times. (Because the results at very late times depend only on the Schwarzschild metric, for which $R = 0$, the value of γ in Eq. (2.1) is not important, although R does not vanish in the collapsing body itself. We will take $\gamma = 0$, or $\nabla_\mu \nabla^\mu \phi = 0$.) Let us denote the complete set by

$$f_{\omega'\ell m}(r,\theta,\phi,t) \text{ (positive frequency sol. at early times)}$$

$$f^*_{\omega'\ell m}(r,\theta,\phi,t) \text{ (negative frequency sol. at early times)} \quad (4.7)$$

A wave packet formed from the $f_{\omega'\ell m}$ over a range of ω' values will be far outside the collapsing body at sufficiently early times, and will be a superposition of incoming waves $r^{-1}\exp(-i\omega' v)Y_{\ell m}$. We will not need to know the precise form of the $f_{\omega'\ell m}$ in the interior geometry of the collapsing body.

The field in the entire space-time can be written as

$$\phi = \sum_{\ell m} \int d\omega' \, (a_{\omega'\ell m} f_{\omega'\ell m} + a^\dagger_{\omega'\ell m} f^*_{\omega'\ell m}) \, . \quad (4.8)$$

Here the normalization of the $f_{\omega'\ell m}$ is

$$(f_{\omega_1'\ell_1 m_1}, f_{\omega_2'\ell_2 m_2}) = \delta(\omega_1'-\omega_2')\delta_{\ell_1\ell_2}\,\delta_{m_1 m_2} \quad (4.9)$$

where the conserved scalar product is defined in Eq. (2.2). Then it follows from the canonical commutation relations of Eq. (2.5) that

$$\left.\begin{aligned} [a_{\omega_1'\ell_1 m_1}, a^{\dagger}_{\omega_2'\ell_2 m_2}] &= \delta(\omega_1'-\omega_2')\delta_{\ell_1\ell_2}\delta_{m_1 m_2} \\ [a_{\omega_1'\ell_1 m_1}, a^{\dagger}_{\omega_2'\ell_2 m_2}] &= 0\ . \end{aligned}\right\} \quad (4.10)$$

The $a_{\omega_1'\ell_1 m_1}$ are annihilation operators corresponding to incoming field quanta or particles at early times.

The specification of the state $|0\rangle$ of the field is that there be no such particles,

$$a_{\omega'\ell m}|0\rangle = 0 \qquad \text{for all } \omega'\ell m\ . \quad (4.11)$$

To find out what is observed at infinity at late times we must again decompose the field into positive and negative frequency components, but at late times. Let the $p_{\omega\ell m}$ be defined such that a superposition of them over a range of ω forms a wave packet localized at large r^* at late times, which is a superposition of outgoing positive frequency waves $r^{-1}\exp(-i\omega u)Y_{\ell m}$. The $p_{\omega\ell m}$ are positive frequency solutions at late times and large r^*. They and the $p^*_{\omega\ell m}$ form a complete set for expanding any solution of the field equation which is purely outgoing at late times. However, the most general solution of the wave equation has a part which is incoming at the horizon at late times. Therefore we must introduce a set of solutions $q_{\omega\ell m}$ such that a superposition of them over a range of ω forms a wave packet which is localized near the horizon at late times and has zero Cauchy data on future null infinity $\mathscr{I}^+$. It is physically clear that the precise form of the basis functions $q_{\omega\ell m}$ should not affect what is observed at infinity at late times. That depends only on the $p_{\omega\ell m}$. Normalization is chosen such that

$$\left.\begin{aligned} (p_{\omega_1\ell_1 m_1}, p_{\omega_2\ell_2 m_2}) &= \delta(\omega_1-\omega_2)\delta_{\ell_1\ell_2}\delta_{m_1 m_2} \\ (q_{\omega_1\ell_1 m_1}, q_{\omega_2\ell_2 m_2}) &= \delta(\omega_1-\omega_2)\delta_{\ell_1\ell_2}\delta_{m_1 m_2} \end{aligned}\right\} \quad (4.12)$$

Since the $p_{\omega\ell m}$ and $q_{\omega\ell m}$ are in disjoint regions at late times, one will also have

$$(q_{\omega_1 \ell_1 m_1}, p_{\omega_2 \ell_2 m_2}) = 0 \ . \tag{4.13}$$

One automatically also has relations like $(q,q^*) = 0$, $(q,p^*) = 0$, $(p,p^*) = 0$.

The field ϕ can be written as

$$\phi = \sum_{\ell m} \int d\omega \left\{ b_{\omega \ell m}\, p_{\omega \ell m} + c_{\omega \ell m}\, q_{\omega \ell m} + b^{\dagger}_{\omega \ell m}\, p^*_{\omega \ell m} + c^{\dagger}_{\omega \ell m}\, q^*_{\omega \ell m} \right\} , \tag{4.14}$$

with

$$\left. \begin{array}{l} [b_{\omega_1}, b^{\dagger}_{\omega_2}] = \delta(\omega_1 - \omega_2),\ [b_{\omega_1}, b_{\omega_2}] = 0,\ [b_{\omega_1}, c_{\omega_2}] = 0 \\ [c_{\omega_1}, c^{\dagger}_{\omega_2}] = \delta(\omega_1 - \omega_2),\ [c_{\omega_1}, c_{\omega_2}] = 0,\ [b_{\omega_1}, c^{\dagger}_{\omega_2}] = 0 \end{array} \right\} . \tag{4.15}$$

The $b_{\omega \ell m}$ are annihilation operators for particles outgoing at late times at infinity. We want to find

$$\langle N_{\omega \ell m} \rangle = \langle 0 | b^{\dagger}_{\omega \ell m}\, b_{\omega \ell m} | 0 \rangle , \tag{4.16}$$

the expectation value of the number of particles outgoing at infinity at late times. The physical meaning of the $a_{\omega' \ell m}$ and $b_{\omega \ell m}$ operators is clear because the decomposition into positive and negative frequency parts in the essentially flat space at infinity is unambiguous. The meaning of the $c_{\omega \ell m}$ depends on the definition of positive frequency near the horizon. We need not go into that here, because the observations at infinity do not depend on it.

Because the $f_{\omega' \ell m}$ and $f^*_{\omega' \ell m}$ are a complete set for expanding any solution of the field equation, one can write

$$p_{\omega \ell m} = \int d\omega' \, (\alpha_{\omega \ell m\, \omega' \ell m}\, f_{\omega' \ell m} + \beta_{\omega \ell m\, \omega' \ell m}\, f^*_{\omega' \ell, -m}) , \tag{4.17}$$

where α and β are complex numbers, independent of the coordinates (the subscript $-m$ on f^* is necessary because we have normalized the $Y_{\ell m}$ such that $Y^*_{\ell m} = Y_{\ell,-m}$). From Eq. (4.14) one has

$$b_{\omega\ell m} = (\phi, p_{\omega\ell m}) . \tag{4.18}$$

Expressing ϕ and $p_{\omega\ell m}$ in terms of f and f^* by means of Eqs. (4.8) and (4.17), and making use of the orthonormality relation yields

$$b_{\omega\ell m} = \int d\omega' \, (\alpha^*_{\omega\ell m\,\omega'\ell m} \, a_{\omega'\ell m} - \beta_{\omega\ell m\,\omega'\ell m} \, a^\dagger_{\omega'\ell,-m}) . \tag{4.19}$$

It follows, using Eq. (4.11), that

$$\langle N_{\omega\ell m}\rangle = \langle 0|b^\dagger_{\omega\ell m} \, b_{\omega\ell m}|0\rangle = \int d\omega' \, |\beta_{\omega\ell m\,\omega'\ell m}|^2 . \tag{4.20}$$

The quantity $\beta_{\omega\omega'}$ is found from Eq. (4.17) as

$$\beta_{\omega\ell m\,\omega'\ell m} = (p_{\omega\ell m}, f^*_{\omega'\ell,-m}) .$$

Similarly one has (4.21)

$$\alpha_{\omega\ell m\,\omega'\ell m} = (p_{\omega\ell m}, f_{\omega'\ell m}) .$$

All these relations are analogous to the ones found in Eqs. (2.28) through (2.36). The p_ω are positive frequency at late times at ∞ while the $f_{\omega'}$ are positive frequency at early times at infinity. If there were no mixing of positive and negative frequency solutions caused by the formation of the black hole, then $\beta_{\omega\omega'}$ would vanish and no particles would be observed at infinity at late times.

In order to evaluate the conserved scalar products in Eq. (4.21) consider a wave packet formed from the p_ω which reaches infinity at late times. It will be a superposition at late times of positive frequency outgoing waves $r^{-1}\exp(-i\omega u)Y_{\ell m}$. At very early times this wave packet must have originated at infinity (since no wave packets get out through the future event horizon and there is no past event horizon). At early times it is therefore a superposition of positive and possibly also negative frequency incoming waves, $r^{-1}\exp(-i\omega' v)Y_{\ell m}$ and $r^{-1}\exp(i\omega' v)Y_{\ell m}$, respectively. The coefficients $\beta_{\omega\omega'}$ relevant to the particles observed at late times can be read off from this superposition by means of Eq. (4.21) evaluated at early times (since the $\beta_{\omega\omega'}$ are independent of coordinates). Because the scattering from the static geometry is not the cause of the particle creation, let me neglect it, effectively replacing Eq. (4.3) in the exterior region by

$$\frac{d^2}{dr^{*2}} R_{\omega\ell} + \omega^2 R_{\omega\ell} = 0 \ . \tag{4.22}$$

This simplifies our considerations and at the same time yields the main result. The effect of scattering will be put in at the end through the absorption probabilities $\Gamma_{\omega\ell}$ discussed previously.

An example of a spherical wave like that under consideration which reaches infinity at late times is shown in Fig. 3. Such a wave starts at infinity at early times traveling along an incoming path v = const., and passes through the collapsing body just before formation of the horizon. After passing r = 0 it moves along an outgoing path u = const., which remains close to the horizon for a long time before peeling off to infinity at late times. One can deduce a number of things from this picture. First, because the wave $r^{-1} \exp(-i\omega u) Y_{\ell m}$ reaches infinity at late times with a finite frequency ω and there is a very large red shift, the incoming waves $r^{-1} \exp(\pm i\omega' v) Y_{\ell m}$ which contribute to it at early times must have a very high frequency ω'. Hence the coefficients $\beta_{\omega\omega'}$ relevant to the particles observed at late times are those with very large ω', i.e. we are interested in the asymptotic form of $\beta_{\omega\omega'}$ for large ω'.

Second, because the above wave still has a high frequency when it reaches the collapsing body and as it moves through the interior geometry (remember we are considering the equation $\nabla^\mu \nabla_\mu \phi = 0$ which has no interaction with the matter, but only with the metric), we can trace the propagation of the wave through the interior geometry by geometrical optics.

The third point concerns the justification for not including interactions of ϕ with the collapsing matter. When the wave packet starts at early times it contains no quanta or particles of the field. As it propagates along, there appears a finite probability for observing a particle having the wave function under consideration. It is plausible that this probability builds gradually after the wave enters the collapsing body, especially after formation of the horizon, and that most of the increase in the probability of finding a quantum in the wave occurs after the wave has left the matter behind and is traveling outward near the horizon for a very long time before it moves off to infinity. Thus, as the wave travels through the collapsing matter there is very little probability of finding a quantum of the field in it to be scattered by the collapsing matter. The probability does not suddenly increase at the moment when the horizon first forms because in a local inertial frame in that region nothing special or discontinuous is occuring. Thus, the free field description should be accurate, especially in the limit when the packet reaches infinity at late times after remaining near the horizon for a very long time. This is a heuristic argument which can be made more precise.

Thus we may use geometrical optics to trace the wave packet components which have the form $r^{-1} \exp(-i\omega u)Y_{\ell m}$ at lates times, backward to early times where they can each be expressed as a superposition of waves $r^{-1} \exp(\pm i\omega' v)Y_{\ell m}$ with large values of ω'. As we trace the wave $r^{-1} \exp(-i\omega u)Y_{\ell m}$ backward in time along the path of the wave packet (a null ray) its phase remains constant at the value ωu, where u is the value of the u coordinate of the packet at late times (see Fig. 4). At early times the packet moves along a path v = const., but since the phase is constant it has the form $r^{-1} \exp(-i\omega u(v))Y_{\ell m}$. Here u is the constant value of the u coordinate which the wave moves along after passing r = 0, and v is the constant value of the v coordinate which the wave moves along before reaching r = 0. The problem thus reduces to that of finding the function u(v).

Hawking[15,16] has given an elegantly simple derivation of that function. Let $v = v_0$ be the coordinate of the particular incoming null ray which reaches the horizon just as it is formed (and is doomed to remain at r = 2M until freed as a result of the loss of mass through the radiation process). Refer to Fig. 5.

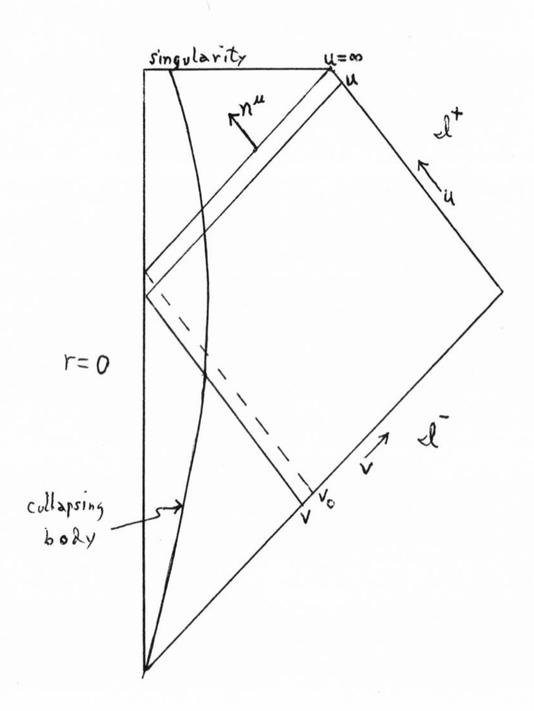

Fig. 5. Ray tracing to find u(v).

At a point on the horizon outside the collapsing body let n^u be the given null vector which points radially inward there. Choose ε such that $-\varepsilon n^u$ connects the horizon to the ray at u, where u is assumed to be large. To find the value of ε as a function of u to within a constant factor, do the following. This functions depends only on the local geometry. Therefore, imagine the exterior Schwarzschild geometry continued to cover the entire space time, including a past horizon. Parallel transport the vector $-\varepsilon n^u$ in along a null generator of the future horizon to the point at which the past and future horizons intersect. It will still connect the future horizon to the same null ray at u. Choose an affine parameter λ on the past horizon such that λ vanishes at the intersection of the two horizons and λ increases toward the future. As is well known, such an affine parameter is related to u on the past horizon by

$$\lambda = -C\, e^{-\kappa u} \tag{4.23}$$

with $C > 0$, and where

$$\kappa = (4M)^{-1} \tag{4.24}$$

is the surface gravity at the horizon. Eq. (4.23) also holds for rotating and charged black holes with appropriate values of κ. Choose C such that $n^\mu = dx^\mu/d\lambda$ in a local coordinate patch at the point where the horizons intersect. Then, in a local inertial frame in which $d^2x^\mu/d\lambda^2$ vanishes because λ is affine, so that n^μ is constant near $\lambda = 0$, integrate $n^\mu d\lambda$ along the past horizon from $\lambda = 0$ to $\lambda(u)$ (the value of λ where the null ray intersects the past horizon). One finds $x^\mu(\lambda) - x^\mu(0) = \lambda n^\mu$. But since $-\varepsilon n^\mu$ connects the horizon to the ray at u, it is also true that $x^\mu(\lambda) - x^\mu(0) = -\varepsilon n^\mu$. Hence

$$\varepsilon = -\lambda(u) = C\, e^{-\kappa u} . \tag{4.25}$$

Now that ε has been found in terms of u, parallel transport the vector $-\varepsilon n^\mu$ back to the original point. Since the vector is now in the exterior region, one can return to the original geometry with the collapsing body and no past horizon. The value of ε is still given by Eq. (4.25).

Now to find the relation between the u of the ray at late times and its value of v at early times parallel transport the vector $-\varepsilon n^\mu$ in the original space-time along the null generator of the past horizon and then along the path of the radially incoming null ray at v_o out to early times and large distances from the matter. The vector $-\varepsilon n^\mu$ will still connect the same two null rays, namely it will point from the ray at v_o to the one at v (which is less than v_o). Thus in u,v coordinates

$$v - v_o = -\varepsilon n^v .$$

But at infinity the space is nearly flat so that n^v is a constant D (with D > 0). Hence

$$v - v_o = -\varepsilon D = -CDe^{-\kappa u} . \tag{4.26}$$

Thus the desired function is

$$u = -4M \ln[(v_o - v)/(CD)] . \tag{4.27}$$

The final result will not depend upon the value of CD.

The spherical wave $p_{\omega\ell m}$ has the following form at late times for r > 2M (recall that I am neglecting the potential in the radial equation for simplicity):

$$p_{\omega\ell m} = N\omega^{-\frac{1}{2}} r^{-1} \exp(-i\omega u) Y_{\ell m} , \tag{4.28}$$

where the factor of $\omega^{-\frac{1}{2}}$ is necessary so that (p_ω, p_ω) is independent of ω, and where N is a real normalization constant which we need not evaluate ($N = 2^{-3/2} \pi^{-1}$). According to Eq. (4.27), at early times outside the collapsing body $p_{\omega\ell m}$ must have the form

$$p_{\omega\ell m} = \begin{cases} N\omega^{-\frac{1}{2}} r^{-1} \exp[i4M\omega \ln (\frac{v_o - v}{CD})] Y_{\ell m} & \text{for } v < v_o \\ 0 & \text{for } v > v_o . \end{cases} \tag{4.29}$$

The function $p_{\omega\ell m}$ vanishes at early times for $v > v_o$ because any null ray which is incoming along $v > v_o$ will enter the horizon of the black holes and not reach infinity at late times. The functions $f_{\omega'\ell m}$ at early times outside the collapsing body are given by

$$f_{\omega'\ell m} = N(\omega')^{-\frac{1}{2}} r^{-1} \exp(-i\omega' v) Y_{\ell m} \tag{4.30}$$

where N is the same normalization constant as in Eq. (4.28). By using Eq. (4.22) instead of Eq. (4.3) we have avoided complications involving backscattering which are not relevant to understanding the cause of the particle creation or the black-body part of the spectrum. It is the asymptotic forms of $\alpha_{\omega\omega'}$ and $\beta_{\omega\omega'}$ for large ω' which are of interest for finding the rate of particle production at late times. Those asymptotic forms for large ω' are found from Eq. (4.17) using Eq. (4.29) and Eq. (4.30) for p and f at early times. By taking Fourier transforms with respect to v one finds

$$N(\omega')^{-\frac{1}{2}} r^{-1} Y_{\ell m} \alpha_{\omega\ell m\omega'\ell m} = \frac{1}{2\pi} \int_{-\infty}^{\infty} dv\, e^{i\omega' v} p_{\omega\ell m}$$

or

$$\alpha_{\omega\ell mi\omega'\ell m} = \frac{1}{2\pi} \int_{-\infty}^{v_o} dv (\omega'/\omega)^{\frac{1}{2}} e^{i\omega' v} \exp[i4M\omega \ln (\frac{v_o - v}{CD})] \tag{4.31}$$

and

$$\beta_{\omega\ell m\omega'\ell m} = \frac{1}{2\pi} \int_{-\infty}^{v_o} dv (\omega'/\omega)^{\frac{1}{2}} e^{-i\omega' v} \exp[i4M\omega \ln (\frac{v_o - v}{CD})] . \tag{4.32}$$

By evaluating these integrals[15,16] one finds that

$$|\alpha_{\omega\omega'}|^2 = \exp(8\pi M\omega)\,|\beta_{\omega\omega'}|^2 . \tag{4.33}$$

From $(p_\omega, p_\omega) = 1$ and Eq. (4.17) it follows that

$$1 = \int d\omega \,(|\alpha_{\omega\omega'}|^2 - |\beta_{\omega\omega'}|^2) . \tag{4.34}$$

It follows that the average number of particles in the mode $\omega\ell m$ outgoing at infinity at late times is

$$\langle N_{\omega\ell m}\rangle = \int d\omega' \,|\beta_{\omega\omega'}|^2 = [\exp(8\pi M\omega) - 1]^{-1} . \tag{4.34}$$

This is a black-body spectrum. The effect of the scattering of the radiation by the exterior geometry is to reflect a fraction back into the black hole. [Mathematically, this means that if p_ω is the part of the wave which propagates by geometrical optics, then $(p_\omega, p_\omega) = \Gamma_\omega$ and hence Γ_ω appears as a factor on the left side of Eq. (4.34).] As mentioned earlier, the probability for an outgoing wave in mode $\omega\ell m$ starting from just outside the horizon to reach infinity is the same as the absorption probability $\Gamma_{\omega\ell}$ for an incoming wave starting from infinity to be absorbed by the black hole. Thus, expression (4.34) is replaced by

$$\langle N_{\omega\ell m}\rangle = \Gamma_{\omega\ell}\,[\exp(8\pi M\omega) - 1]^{-1} , \tag{4.35}$$

which is the same as the average number emitted from a thermal body with absorptivity the same as that of the black hole, and with temperature

$$T = (8\pi k_B M)^{-1} \tag{4.36}$$

where k_B is Boltzmann's constant. In cgs units

$$T = \frac{\hbar c^3}{8\pi k_B G M} = (1.2 \times 10^{26}\ {}^\circ\mathrm{K})\,\left(\frac{1\ \mathrm{gm}}{M}\right) . \tag{4.37}$$

For a solar mass black hole this yields $T = 6 \times 10^{-8}\ {}^\circ\mathrm{K}$, while for $M \approx 10^{14}$ gm one finds $T \approx 10^{12}\ {}^\circ\mathrm{K}$.

To find the average number of particles emitted to infinity per frequency interval ω to $\omega + d\omega$ in spherical mode ℓ,m, the simplest procedure is to impose periodic boundary conditions in a very large spherical shell of radius R surrounding the black hole.[17] Then $\langle N_{\omega\ell m}\rangle$ is the number of outgoing particles contained within the volume of the sphere at any given late time in the mode $\omega\ell m$ (there is a steady flux). Since the radial solutions at large distances go as $\exp(\pm i\omega r)$, the periodic boundary conditions require $\omega = 2\pi n R^{-1}$, with n an integer, and the number of states in the range ω to $\omega + d\omega$ with given values of ℓ,m is $\Delta n = (2\pi)^{-1} R d\omega$. The time for the set of particles originally within the radius R to pass out through the shell is R. Hence, the number of particles per unit time passing out through a shell of large radius at late times, with ω in the range ω to $\omega + d\omega$, and in spherical mode ℓm, is

$$(2\pi)^{-1} \langle N_{\omega\ell m}\rangle \, d\omega \, . \tag{4.38}$$

The power or luminosity L of the black hole is therefore

$$L = (2\pi)^{-1} \sum_{\ell=0}^{\infty} (2\ell+1) \int_0^{\infty} d\omega \, \omega \, \Gamma_{\omega\ell} [\exp(8\pi M\omega) - 1]^{-1} \, . \tag{4.39}$$

Similar equations hold for neutrino, photon, and graviton emission, and for rotating and charged black holes. Page[17] has evaluated these expressions and finds that for an uncharged, nonrotating black hole of mass $M \gg 10^{17}$ gm the luminosity is

$$L = 2.0 \times 10^{-4} \, \hbar c^6 G^{-2} M^{-2} = (3.4 \times 10^{46}) \left(\frac{M}{1 \text{ gm}}\right)^{-2} \text{erg sec}^{-1} \tag{4.40}$$

in the form of neutrinos, photons and gravitons (81% in the four kinds of neutrinos, 17% in photons, and 2% in gravitons). For 5×10^{14} gm $\ll M \ll 10^{17}$ gm one must also include the production of relativistic electrons, which results in the replacement of 3.4 by 6.3 in Eq. (4.40). For smaller mass black holes that numerical factor increases as more species of particles are emitted. The corresponding results for rotating black holes are given in Ref. 24.

The lifetime may be calculated on the basis of energy conservation,

$$\frac{dM}{dt} = - L(M) \ , \tag{4.41}$$

where L is given by the above results in the quasi-static approximation. Assuming that the numerical factor in Eq. (4.40) does not decrease as the mass M approaches zero, one finds that primordial black holes with an initial mass of less than 4×10^{14} gm should have evaporated by now.

The above derivation shows that the average number of created particles in mode $\omega \ell m$ has a thermal distribution. One can also calculate the probability for observing a given number of particles in mode $\omega \ell m$. Wald[132] first considered this problem quantum mechanically, and showed that the probability distribution is thermal. Hawking[133] and Parker[134] showed this in somewhat different ways. Bekenstein,[135] using an information theory approach, also calculated this probability distribution. Gibbons and Perry[136] have shown that a black hole can remain in thermal equilibrium with a heat bath even in the presence of particle interactions. Further aspects of particle production by black holes are discussed in Refs. 131, 137-151, 164-167.

5. MISCELLANEOUS TOPICS

Particle Creation by Singularities

In the space-time of a body collapsing to form a black hole, and in a statically (or adiabatically) bounded expansion of the universe, the initial conditions specifying the state of the quantized field had an unambiguous physical interpretation. When a space-time has a singularity, such as a region of infinite curvature, which can influence observations in the asymptotic region (i.e. a "naked" singularity), then it is necessary to specify boundary conditions near the singularity. There seems to be no clear physical reason for choosing one particular boundary condition over the other possibilities.

For space-times having a singularity which can only influence events in the asymptotic region after some specific time, one can make unambiguous predictions as to what is observed outside the future light cone of the singularity. As the curvature increases before formation of the singularity, the rate of particle creation is independent of any boundary conditions on the singularity. One can thus try to answer the question of whether enough particle production will occur so that the mass present is effectively radiated away before formation of the singularity.

For example, Davies[144] considered a 2-dimensional model of a body of mass M and charge $Q > M$ collapsing. An event horizon does not form in such a collapse. By examining the regularized energy-momentum tensor of a massless scalar field, he finds that the particle creation rate grows without limit as the body collapses (in that model the decrease in mass is assumed to be quasi-static and the back-reaction is not taken into account explicitly). Davies concludes that the matter will be completely radiated away before the naked singularity has had time to form.

There are very few known classical, asymptotically flat solutions of the Einstein equations in which a naked singularity forms from non-singular initial conditions,[156] and these generally depend on very special choices of initial conditions and/or unusual equations of state. Nevertheless, in such models one can unambiguously calculate the flux of created particles observed in the asymptotic region as a result of the formation of the singularity. In one class of such models, the geodesics of pressureless collapsing dust particles converge to form a naked singularity.[157,158] Ford and Parker[159] have considered several models of that type, and find that the particle creation rate does not increase without limit as the geodesics of the collapsing dust converge to form a singularity. One interesting example of that type which they considered is a singularity which forms at the Schwarzschild radius of a non-rotating collapsing dust cloud, so that the singularity is also a surface of infinite red-shift (see Fig. 6).

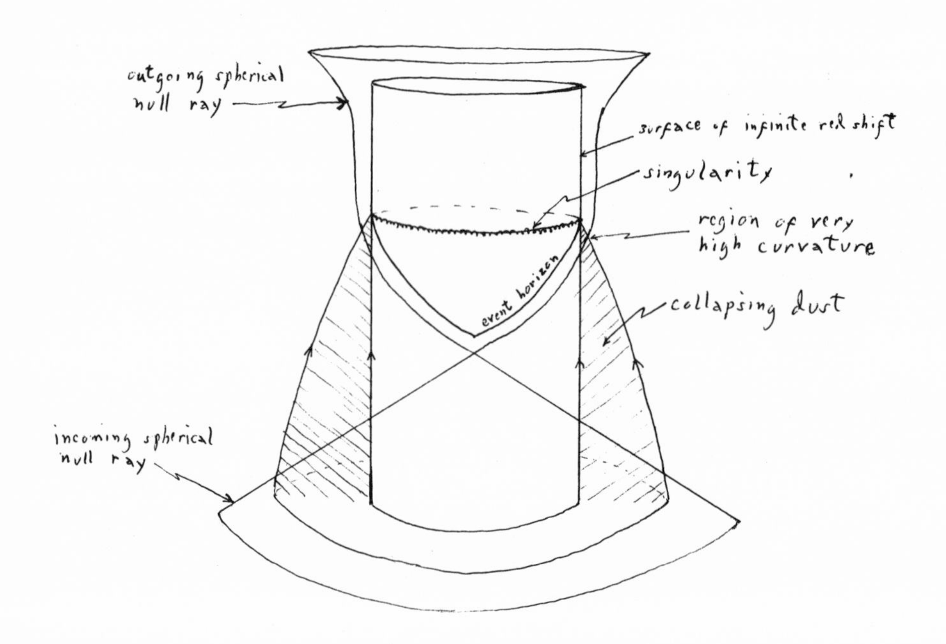

Fig. 6. Eddington-Finkelstein diagram of collapse to singularity

In this example, the inner surface of the dust cloud is at rest at the Schwarzschild radius of the cloud. Particles on that surface remain at rest, as in the corresponding Newtonian example. The particles in the successive shells of the cloud are given the correct initial conditions so that all the shells converge at the position of the inner surface of the cloud simultaneously, forming a singularity at which the density and curvature are infinite. Inside the inner radius of the dust is flat space-time, and beyond the outer radius is Schwarzschild space-time. One might expect the high curvature in the collapsing dust just before formation of the singularity to affect the spectrum of created particles observed at infinity at late times. However, they find that the spectrum is thermal and has the same temperature, $(8\pi k_B M)^{-1}$, as for a black hole of the same mass.

The reason is the following. Consider null rays which pass through the cloud just before formation of the singularity, and reach $\mathscr{I}^+$ at late times (i.e. large values of u) with a given finite frequency ω. Because the singularity forms on a surface of infinite red-shift, the proper frequency of the rays as they pass through the region of high curvature in the dust cloud is very high, and grows without limit as one considers rays passing closer to the singularity. One can show that in the cloud the proper wavelength of rays with successively large values of u approaches zero more rapidly than the characteristic length associated with the Riemann tensor, so that the rays which reach infinity at late times can be propagated by geometrical optics. Then the spectrum observed at late times can be calculated exactly as if the dust cloud had collapsed to a black hole. One then obtains a thermal spectrum of temperature $(8\pi k_B M)^{-1}$. (The $\Gamma_{\omega\ell}$ in the spectrum [see Eq. (4.35)] will not be affected because the scattering from the space-time occurs mainly when the proper frequency has been red-shifted to lower values after the ray has passed through the dust.) These conclusions have also been confirmed by a detailed calculation involving the matching of coordinates at the boundaries of the dust cloud. In other models of the same type, but with matter having a positive bounded pressure, one would expect the curvature in the collapsing body to grow at a slower rate than in the present model as the singularity is approached, so that geometrical optics remains valid, and the same thermal spectrum is obtained.

One can conclude that in these models particle creation is not intense enough to prevent formation of the singularity. Furthermore, according to the above result, one cannot distinguish between a model of the above type and a black hole of the same mass by means of asymptotic observations made at late times. This can be viewed as a further aspect of the "no hair" theorem according to which a black hole is entirely characterized by its mass, angular momentum, and charge.[129] From that viewpoint, the singular surface of infinite red shift is to be thought of as the event horizon of a black hole with a singularity on it. The thermal spectrum observed at late times will be unchanged if one imagines the singularity to remain "frozen" at the Schwarzschild radius after its formation. Some static or stationary solutions of the Einstein equations having singular infinite red-shift surfaces [160,161] can perhaps be thought of as the end state of a singular collapse analogous to the one considered above. In that case, one expects the singular infinite red-shift surface to radiate, with the effective local temperature of the surface given by $\kappa/2\pi$ where κ is the surface gravity.

Conservation of Energy-Momentum and Particle Creation

One might at first expect that the process of particle creation from vacuum must violate the local conservation law

$$\nabla_\nu T^{\mu\nu} = 0 \ . \tag{5.1}$$

However, the formal energy-momentum tensor formed from the quantized field, as well as the regularized expressions for $T_{\mu\nu}$ do obey Eq. (5.1) at all times, even when particle creation is occuring. Eq. (5.1) constitutes four equations involving 10 quantities, and thus is not sufficient to require that if $T_{\mu\nu}$ vanishes on an initial Cauchy hypersurface then it vanishes everywhere. One can, however, impose various "energy conditions" on the components of $T_{\mu\nu}$ which appear reasonable for ordinary matter, and require that no particle creation can occur.[103] However, these energy conditions are violated by the expectation value of $T_{\mu\nu}$ during the particle creation process.[104] That has already been discussed in the paragraph following Eq. (3.110).

When the space-time possesses an infinitesimal isometry, i.e. when the metric is form-invariant under infinitesimal transformations

$$x'^{\mu} = x^{\mu} + \varepsilon\xi^{\mu}(x) \tag{5.2}$$

with $|\varepsilon|$ small, then it is possible to extend the local conservation of energy-momentum, Eq. (5.1), to a global conservation law (Eq. (5.5) below). It is of interest to investigate the relationship between the global conservation law and particle creation by gravitational fields. As is well known,[14,129] if the transformation (5.2) is an isometry then the vector field $\xi^{\mu}(x)$, known as a Killing vector field, satisfies the equation

$$\nabla_{\mu}\xi_{\nu} + \nabla_{\nu}\xi_{\mu} = 0 \ . \tag{5.3}$$

From Eqs. (5.2), (5.3), and the symmetry of $T_{\mu\nu}$ one has

$$\nabla_{\nu}(\xi_{\mu}T^{\mu\nu}) = (-g)^{-\frac{1}{2}}\,\partial_{\nu}[(-g)^{\frac{1}{2}}\,\xi_{\mu}T^{\mu\nu}] = 0 \ . \tag{5.4}$$

It follows in the usual way from the divergence theorem that if S is a complete 3-dimensional Cauchy hypersurface, then the quantity

$$\int_S \xi_{\mu}T^{\mu\nu}(-g)^{\frac{1}{2}}\,dS_{\nu} \tag{5.5}$$

is conserved under continuous deformations and displacements of S. Here $T^{\mu\nu}$, for example, can be taken to be a finite (regularized) matrix element or expectation value of the energy-momentum tensor of a quantized field in curved space-time.

In the Schwarzschild metric of Eq. (4.1) there exists a Killing vector field with components

$$\xi_t = 1 - 2Mr^{-1} , \qquad \xi_r = \xi_\theta = \xi_\phi = 0 , \tag{5.6}$$

as well as a rotational Killing vector field. For $r > 2M$, the vector field (5.6) is timelike and future directed. For $r < 2M$, that vector field becomes spacelike owing to the change in sign of g_{oo}. In the cosmological metric of Eq. (3.1),

$$\xi_t = 0, \quad \xi_x = a^2(t)v_x, \quad \xi_y = a^2(t)v_y, \quad \xi_z = a^2(t)v_z, \tag{5.7}$$

(with v_x, v_y, v_z any constants) is a Killing vector field. It is globally spacelike. In the coordinates of Eq. (3.1), the de Sitter universe corresponds to

$$a(t) = \exp(Ht) . \tag{5.8}$$

Here H is a positive constant (the Hubble constant) equal to $(\Lambda/3)^{\frac{1}{2}}$, where Λ is the cosmological constant [see Eq. (3.106)]. In that case, there is an additional independent Killing vector field with components

$$\xi_t = 1, \quad \xi_j = He^{2Ht}\, x^j \qquad (j = 1,2,3) . \tag{5.9}$$

[One could equally well replace x^j by $(x^j - x^j_o)$ with x^j_o constant in Eq. (5.9), but that only yields a linear combination of (5.7) and (5.9).] The vector ξ_μ is timelike when

$$1 - H^2\, e^{2Ht}\, r^2 > 0 , \tag{5.10}$$

where $r = [\sum_i (x^i)^2]^{\frac{1}{2}}$, or equivalently, when the proper distance from the spatial origin satisfies

$$e^{Ht}\, r < H^{-1} \equiv d , \tag{5.11}$$

where d is the proper radius of the event horizon of any of the geodesics, $\vec{x}$ = constant [see Eq. (3.107)]. Thus, the Killing vector field of Eq. (5.9) is timelike inside the event horizon of an observer at the spatial origin, is null on that horizon, and is spacelike outside of it.

Returning to a general space-time with a Killing vector field ξ_μ, suppose that the state vector can be chosen such that $T^{\mu\nu}(x)$ is non-zero only in a narrow world-tube (i.e. $g_{\mu\nu}$ does not change much on a cross section of the tube) centered on the world-line $x^\mu = X^\mu(\lambda)$ of a classical particle, where λ is a suitable affine parameter (e.g. proper time if $m \neq 0$). Eventually the wave packet describing the particle will spread, but the above description should be good for a large range of λ. For each value of λ the path cuts one of a non-intersecting family of complete space-like Cauchy hypersurfaces S. Label that surface $S(\lambda)$. The conserved quantity of Eq. (5.5) can then be written

$$\xi_\mu(\lambda)p^\mu(\lambda) = \text{conserved} , \tag{5.12}$$

where $\xi_\mu(\lambda) = \xi_\mu(X^\sigma(\lambda))$ and

$$p^\mu(\lambda) \equiv \int_{S(\lambda)} T^{\mu\nu}(-g)^{\frac{1}{2}} \, dS_\nu . \tag{5.13}$$

In a locally inertial coordinate system at the event $x^\mu = X^\mu(\lambda)$, the quantity $p^\mu(\lambda)$ is the four-momentum of the particle represented by the localized wave packet. If $\xi_\mu(\lambda)$ is timelike and future directed, then one can Lorentz transform the locally inertial coordinate system at $x^\mu = X^\mu(\lambda)$ so that $\xi_\mu(\lambda)$ has components with $\xi_o(\lambda)$ positive and $\xi_i(\lambda)$ vanishing $(i = 1,2,3)$. In that local inertial frame, $\xi_\mu(\lambda)p^\mu(\lambda)$ is proportional to the energy of the particle, with a positive proportionality constant. Similarly, if $\xi_\mu(\lambda)$ is spacelike then one can find a local inertial frame in which $\xi_o(\lambda)$ vanishes. In that frame $\xi_\mu(\lambda)p^\mu(\lambda)$ is proportional to the component of the three-momentum along the spatial direction determined by $\vec{\xi}(\lambda)$.

Consider a situation (Fig. 7) in which $T^{\mu\nu}$ is zero on some initial Cauchy hypersurface S and becomes non-zero on a pair of world-tubes originating near some event (i.e. creation of a localized pair by the gravitational field).

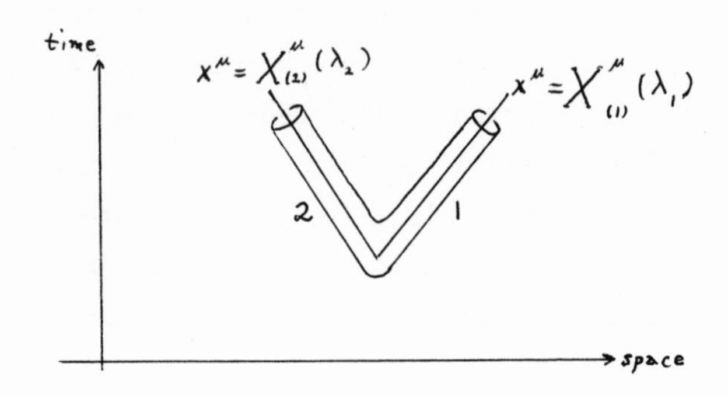

Fig. 7. Localized pair creation

In a space-time which is creating pairs, one would expect there to be a finite probability of observing such a process. The space-like Cauchy hypersurface S can be displaced to intersect the two world tubes at arbitrary points of spacelike separation (it cannot be made to intersect only one world tube because S must remain spacelike). The conservation law of Eq. (5.5) therefore yields

$$\xi_\mu(\lambda_1)p^\mu_{(1)}(\lambda_1) + \xi_\mu(\lambda_2)p^\mu_{(2)}(\lambda_2) = 0 \ , \qquad (5.14)$$

where the quantities with subscripts 1 and 2 are defined along the respective world tubes numbered 1 and 2 in exactly the same way as the quantities in Eq. (5.12) and Eq. (5.13) were defined. The events $x^\mu = X^\mu_{(1)}(\lambda_1)$ and $x^\mu = X^\mu_{(2)}(\lambda_2)$, at which the quantities in Eq. (5.14) are evaluated, have a spacelike separation.

If we suppose that $\xi_\mu(\lambda_1)p^\mu_{(1)}(\lambda_1)$ in Eq. (5.14) is positive, it follows that $\xi_\mu(\lambda_2)p^\mu_{(2)}(\lambda_2)$ is negative. Since S can be held fixed at a point on one world line and displaced along the other, it follows that $\xi_\mu(\lambda_1)p^\mu_{(1)}(\lambda_1)$ and $\xi_\mu(\lambda_2)p^\mu_{(2)}(\lambda_2)$ are separately conserved after the two world lines have become distinct. Suppose ξ^μ is time-like and future directed in the region where the pair creation occurs [as in Eq. (5.6) for $r > 2M$, or Eq. (5.9) within the event horizon of an observer at $\vec{x} = 0$]. Then according to the discussion following Eq. (5.13), particle 2 will have negative energy in some local inertial frame. However, particle 2 can only be *detected* with positive energy in any local inertial frame. One can conclude that pair production by the gravitational field in such a region (where ξ_μ is timelike and future directed) can occur only if particle 2 is directed toward a region where ξ_μ becomes spacelike. In that second region, the fact that $\xi_\mu(\lambda_2)p_{(2)}(\lambda_2)$ is negative only implies that a component of the three-momentum in a local inertial frame is negative. Furthermore, this quantum mechanical tunneling of particle 2 into the second region, where ξ_μ is spacelike, must occur in a sufficiently short proper distance that if particle 2 is detected (i.e. localized) before reaching the second region, then the disturbance caused by the measurement process is large enough (according to the uncertainty principle) to account for the positive energy of the detected particle.

For example, consider the Schwarzschild metric with the Killing vector field of Eq. (5.6). Suppose that the pair is created at $r > 2M$ with a radial orientation. Particle 2 "tunnels" through the event horizon into the black hole, where its energy in local inertial frames is positive. If the pair is created near the event horizon, so that $s \equiv r - 2M$ is small with respect to $2M$, the proper distance of the point at r from the event horizon is

$$d = \int_{2M}^{r} (1 - 2Mr^{-1})^{-\frac{1}{2}}\, dr \simeq 2(2M)^{\frac{1}{2}} s^{\frac{1}{2}} . \tag{5.15}$$

If one tries to localize particle 2 before it enters the event horizon, then the position-momentum uncertainty relation implies that the measurement will disturb the radial component of momentum by at least (in units with $\hbar = c = G = 1$)

$$\Delta p \sim d^{-1} . \tag{5.16}$$

Therefore, we expect that the negative value of the energy $E_{(2)}$ of particle 2 near r in a local inertial frame instantaneously at rest in the Schwarzschild coordinates satisfies $|E_{(2)}| \lesssim d^{-1}$. (We are considering massless particles.) It follows that the energy $E_{(1)}$ of particle 1 near r also satisfies $E_{(1)} \lesssim d^{-1}$. As $r \to 2M$ or $s \to 0$ the upper limit approaches infinity, but what is of interest is the energy E_∞ of particle 1 when it reaches infinity. The gravitational red-shift then gives

$$E_\infty = (1 - 2Mr^{-1})^{\frac{1}{2}} E_{(1)} \lesssim (2M)^{-\frac{1}{2}} s^{\frac{1}{2}} d^{-1} ,$$

or

$$E_\infty \lesssim (4M)^{-1} . \tag{5.17}$$

If we assume that particles are actually created, then the range of observed frequencies $\omega = E_\infty$ will be given by Eq. (5.17). This is consistent with the previous quantum field theory results, according to which the probability of observing E_∞ outside the range of Eq. (5.17) is small. One can also use the above reasoning as the basis for a heuristic estimate of the luminosity of the black hole. Namely, replace Eq. (1.17) by Eq. (5.17) and proceed exactly as in Eqs. (1.18)-(1.23), but in units with G = 1, and with M replaced by 4M. (The same arguments holds if one has an ergoregion in which ξ_μ becomes spacelike without an event horizon. Particle creation and its classical analogue have been discussed in such a case by Ashtekar and Magnon[162] and Friedman.[163])

When particle 2 enters the black hole it decreases the total mass. If M is the original mass of the black hole, then after particle 1 has been dispersed to the asymptotic region and particle 2 has passed through the event horizon, the new mass energy of the black hole will be

$$M' = M + \xi_\mu p^\mu_{(2)} < M \ . \tag{5.18}$$

Thus, the loss of mass of the black hole is equal to the energy $\xi_\mu p^\mu_{(1)}$ radiated to infinity.

Next consider a pair created in de Sitter space-time near $\vec{x} = 0$, with $\xi_\mu(\lambda_2)p^\mu_{(2)}(\lambda_2) < 0$, where ξ_μ is the Killing vector field of Eq. (5.9). As in the previous example, suppose particle 2 "tunnels" through to the region where ξ_μ is spacelike on the opposite side of the event horizon of the observer at $\vec{x} = 0$, a proper distance $d = H^{-1}$ away. For a massless particle this means that observers on geodesics $\vec{x}$ = const. have a proper time interval t to detect particle 2 before it reaches that event horizon, where t is found by integrating $ds^2 = 0$. Thus,

$$\int_0^t a^{-1}(t')dt' = \int_0^{H^{-1}\exp(-Ht)} dx \ . \tag{5.19}$$

or

$$t = H^{-1} \ln 2 \ . \tag{5.20}$$

The position-momentum or the time-energy uncertainty relation then imply that one has a minimum disturbance ΔE in the energy of particle 2, as measured by observers on the geodesics $\vec{x}$ = const., of order

$$\Delta E \sim t^{-1} = H \ln 2 \sim H \ . \tag{5.21}$$

[If we had used ξ_μ with x^i in (5.9) replaced by $x^i - x^i_o$ with $x^i_o \neq 0$, then the proper distance and t-interval for particle 2 to reach the region where ξ_μ is spacelike would have been smaller, leading to a larger ΔE. Thus, the use of the Killing vector field ξ^μ which has components $\xi_o = 1$, $\xi_i = 0$ at the position where the pair is created, yields the smallest value of E, and thus, the most stringent limit on the allowed energy range of each created particle.] Therefore, the negative energy $E_{(2)}$ near $\vec{x} = 0$ must satisfy $|E_{(2)}| \lesssim H$, and similarly for particle 1 one has (with $E_{(1)} = \omega$) the condition

$$\omega \lesssim H \ . \tag{5.22}$$

This is consistent with the temperature of the black-body radiation which Gibbons and Hawking[49] claim an observer on a geodesic $\vec{x}$ = const. can detect in the de Sitter universe. The wavelength of such particles is of the order of the proper radius of the event horizon. (For particles created at a coordinate distance r from the detector, the observed energy will be reduced by a red shift factor $[1 - (r/r_E)^2]^{\frac{1}{2}}$.)

In the Einstein universe, which is a closed Robertson-Walker universe of constant radius, one might expect a similar phenomenon to occur. Such a universe has a global future-directed timelike Killing vector field. Nevertheless, one can imagine a pair created from vacuum such that $\xi_\mu p^\mu_{(1)} > 0$ and $\xi_\mu p^\mu_{(2)} < 0$ in Eq. (5.14), with particle 1 and 2 each traveling half way around the universe in opposite directions and meeting on the other side where they annihilate into vacuum (Fig. 8).

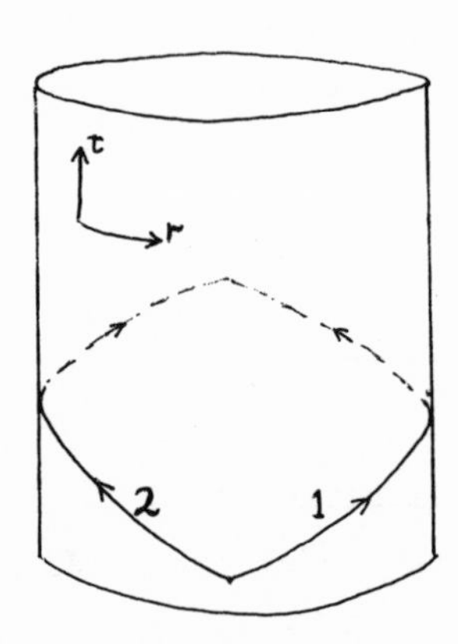

Fig. 8. Static Einstein Universe (one spatial dimension)

If a is the static radius of the Einstein universe, and we are dealing with massless particles (e.g. photons), then the time interval Δt (with respect to the preferred geodesics) for such a process to occur is of order a. That is the maximum time interval available to measure the negative energy of particle 2 before annihilation. Therefore, the time-energy uncertainty relation requires that the magnitude of the energy $|\omega|$ of particle 1 or 2 must be in the range

$$|\omega| \lesssim a^{-1} , \tag{5.23}$$

corresponding to wavelengths which are of the order of the radius of the universe. Such a phenomenon could not occur in an open universe, and is a non-local effect depending on the topology. Such a process involving real particles may well be forbidden. For example, if particle 1 is detected before the annihilation event, then there will be an arbitrarily long time-interval available to detect particle 2 with negative energy, which presumably is a forbidden situation. In the de Sitter example such a question does not arise because particle 2 will reach a region where ξ_μ is spacelike whether or not particle 1 is detected.

Regardless of the possibility of the process described in the previous paragraph involving real particles, Fig. 8 suggests a virtual process which would be expected to occur in a closed universe (provided the expansion is not too rapid to allow the members of the pair to meet again, as when event horizons exist). In such a process, particles 1 and 2 would be virtual. Consider the static Einstein universe. The energy density associated with such a process, assuming it occured with probability of order 1 for all energies $\omega \lesssim a^{-1}$ allowed by the uncertainty principle, would be

$$\rho \sim \int_0^{a^{-1}} d\omega\, \omega^2 \omega \sim a^{-4} , \tag{5.24}$$

where the density of states ω^2 has been inserted in Eq. (5.24). Thus, the contribution of these processes to the vacuum energy is finite. Such processes, in which the particles circumnavigate the space-time are not possible in an open universe with simple topology. If one assumes that the infinite contribution of the vacuum diagrams which also occur in flat space-time should be subtracted from the vacuum energy, then one arrives at Eq. (5.24) as the gravitationally active vacuum energy density in the closed Einstein universe. This is consistent with the result arrived at by the cut-off method, Eq. (3.132), and gives an understanding of why one has an additional contribution to the vacuum energy in a closed universe as compared with an open universe.

In a model like the de Sitter universe, in which event horizons are present and the photons cannot meet for a second time on the opposite side of the universe, one would not expect such a contribution to the vacuum energy. This would resolve the paradox raised by Dowker and Critchley concerning vacuum energy in the de Sitter universe (i.e. one cannot simply conformally transform the Einstein universe's vacuum energy to the de Sitter universe). However, in such a universe related phenomena appear, as discussed previously, in which the event horizon plays a role analogous to the closed topology. In the light of these considerations, it is perhaps not too surprising that the energy density of the scalar de Sitter radiation in Eq. (3.110) is the same as the scalar vacuum energy of Eq. (3.105) in an Einstein universe having the same radius as that of the de Sitter event horizon.

Finally, in a non-stationary metric such as an expanding universe in which there is no time-like Killing vector field even locally, one cannot extend the local conservation law of Eq. (5.1) to a global energy conservation in the form of Eq. (5.14). In such a situation, pair production from vacuum in which both members have positive energy can occur as a result of excitation of the quantum oscillators of the matter or radiation field by the variation of the metric in time-like directions. When a space-like Killing vector field exists, as in Eq. (5.7), then Eq. (5.14) leads to conservation (with a red shift factor included) of linear momentum. We already considered such pair creation in Section 3.

In these lectures, various types of particle creation processes and vacuum phenomena involving strong gravitational fields have been discussed. I hope that the reader has been guided toward an understanding of this beautiful and rapidly developing field.

References

* Work supported in part by the National Science Foundation (PHY 72-05161A02).

1. L. Parker, Ph.D. Thesis, Harvard University, 1966.

2. L. Parker, Phys. Rev. Lett. 21, 562 (1968).

3. L. Parker, Phys. Rev. 183, 1057 (1969).

4. L. Parker, Phys. Rev. D3, 346 (1971).

5. R.U. Sexl and H.K. Urbantke, Acta Phys. Austriaca 26, 339 (1967).

6. R.U. Sexl and H.K. Urbantke, Phys. Rev. 179, 1247 (1969).

7. Ya. B. Zeldovich, Pisma v. Zh. ETF 12, 443 (1970) [JETP Lett. 12, 307].

8. Ya. B. Zeldovich and A.A. Starobinsky, Zh. ETF 61, 2161 (1971) [JETP 34, 1159].

9. For the classical analogue of this pair creation see E. Schrödinger, Physica 6, 899 (1939), and Ref. 10. Schrodinger was aware of the connection with pair creation.

10. E. Schrödinger, Proc. Roy. Irish Acad. 46, 25 (1940).

11. V.N. Lukash and A.A. Starobinsky, Zh. ETF 66, 1515 (1974) [JETP 32, 742 (1975)].

12. L. Parker in Proceedings of the Second Latin American Symposium on Relativity and Gravitation, Dec. 1975 (to be published).

13. L. Parker, Nature 261, 20 (1976).

14. S. Weinberg, Gravitation and Cosmology (Wiley & Sons, New York) (1972).

15. S.W. Hawking, Nature 248, 30 (1974).

16. S.W. Hawking, Commun. Math. Phys. 43, 199 (1975).

17. D.N. Page, Phys. Rev. D13, 198 (1976). Further references are given in Section 4 below.

18. Ya. B. Zeldovich, Pisma v. Zh. ETF 14, 270 (1971) [JETP Lett. 14, 180].

19. Ya. B. Zeldovich, Zh ETF 62, 2076 (1972) [JETP 35, 1085].

20. C.W. Misner, Bull. Am. Phys. Soc. 17, 472 (1972).

21. W.G. Unruh, Phys. Rev. D10, 3194 (1974).

22. L. Ford, Phys. Rev. D12, 2963 (1975).

23. B.S. DeWitt, Phys. Reports 19, 295 (1975).

24. D.N. Page, "Emission Rates From a Rotating Black Hole", preprint OAP-452.

25. G.F. Chapline, Phys. Rev. D12, 2949 (1975).

26. B.J. Carr, "Primordial Black Hole Mass Spectrum", preprint OAP-389, 1975.

27. S.W. Hawking, Mon. Nat. R. Astr. Soc. 152, 75 (1971).

28. B.J. Carr and S.W. Hawking, Mon. Nat. R. Astr. Soc. 168, 399 (1974).

29. L. Parker and S.A. Fulling, Phys. Rev. D7, 2357 (1973). Other references on avoidance of the cosmological singularity are given in Section 3.

30. W. Heisenberg and H. Euler, Zs. f. Phys. 98, 714 (1936).

31. J. Schwinger, Phys. Rev. 82, 664 (1951).

32. Ya. B. Zeldovich, in Magic Without Magic, ed. by J.R. Klauder (W.H. Freeman and Co., San Francisco, 1972).

33. A.G. Doroshkevitch, V.N. Lukash and I.D. Novikov, Zh. Eksp. Teor. Fiz. 64, 1457 (1973) [JETP 37, 739 (1973)].

34. C. Møller, Royaumont Conf. Proc. (Centre Nat. de la Recherche Sci., Paris, 1962) pp. 21-29.

35. L. Parker and S.A. Fulling, Phys. Rev. D9, 341 (1974).

36. S.A. Fulling and L. Parker, Ann. Phys. (N.Y.) 87, 176 (1974).

37. S.A. Fulling, L. Parker and B.L. Hu, Phys. Rev. D10, 3905 (1974).

38. S.M. Christensen, Ph.D. Thesis, University of Texas at Austin, 1975.

39. S.M. Christensen, Phys. Rev. D (in press).

40. P.J.E. Peebles, Physical Cosmology (Princeton University Press, New Jersey, 1971).

41. E.R. Harrison, A. Rev. Astr. Astrophys. 11, 155 (1973).

42. J.D. Bekenstein, Nuovo Cim. Lett. 4, 7371 (1972).

43. J.D. Bekenstein, Phys. Rev. D7, 2333 (1973).

44. J.D. Bekenstein, Phys. Rev. D9, 3292 (1974).

45. S.W. Hawking, Phys. Rev. Lett. 26, 1344 (1971).

46. W.G. Unruh, Phys. Rev. D (in press).

47. S.A. Fulling, Ph.D. Thesis, Princeton University, 1972.

48. S.A. Fulling, Phys. Rev. D7, 2850 (1973).

49. G.W. Gibbons and S.W. Hawking, "Cosmological Event Horizons, Thermodynamics, and Particle Creation", to be published.

50. R. Figari, R. Höegh-Krohn and C.R. Nappi, Commun. Math. Phys. 44, 265 (1975).

51. T. Imamura, Phys. Rev. 118, 1430 (1960).

52. Examples dealing with production of spin $\frac{1}{2}$ particles are Refs. 1-4, 17, 21, and 53, 54, 55 below. Examples dealing with the Maxwell field include Refs. 1, 3, 10, 17, 56, and 78. References on graviton production are given in Section 3.

53. W.G. Unruh, Phys. Rev. Lett. 31, 1265 (1973).

54. V.M. Frolov, S.G. Mamayev and V.M. Mostepanenko, Phys. Lett. 55A, 389 (1976).

55. J. Audretsch, Nuovo Cim. B17, 248 (1973).

56. L. Parker, Phys. Rev. D5, 2905 (1972).

57. See for example L. Parker, Phys. Rev. D7, 976 (1973), Appendix B.

58. H.K. Urbantke, Nuovo Cim. 63B, 203 (1969).

59. For further discussion of canonical quantization in curved space-time see Ref. 47, and Ref. 60 below. One of the earliest discussions is in Ref. 34.

60. A. Ashtekar and A. Magnon, Proc. Roy. Soc. Lond. A346, 375 (1975).

61. N.N. Bogoliubov, Zh. ETF 34, 58 (1958) [JETP 7, 51 (1958)].

62. W. Rindler, Am. J. Phys. 34, 1174 (1966).

63. P.C.W. Davies and S.A. Fulling, "Radiation From a Moving Mirror and Black Hole Evaporation" (in preparation).

64. J.B. Hartle and S.W. Hawking, Phys. Rev. D13, 2188 (1976).

65. A.S. Lapedes, "Thermal Particle Production In Two Taub Nut Type Spacetimes", preprint.

66. W. Israel, Phys. Lett. (in press).

67. S.A. Fulling, "Alternative Vacuum States in Static Space-Times with Horizons", preprint.

68. N. Woodhouse, Phys. Rev. Lett. 36, 999 (1976).

69. A.A. Grib and S.G. Mamaev, Yad. Fiz. 10, 1276 (1969) [Sov. J. Nucl. Phys. 10, 722 (1970)].

70. A.A. Grib and S.G. Mamaev, Yad. Fiz. 14, 800 (1971) [Sov. J. Nucl. Phys. 14, 450 (1972)].

71. B.K. Berger, Ph.D. Thesis, University of Maryland, 1972.

72. B.K. Berger, Ann. Phys. (N.Y.) 83, 458 (1974).

73. B.K. Berger, Phys. Rev. D11, 2770 (1975).

74. B.K. Berger, Phys. Rev. D12, 368 (1975).

75. M. Castagnino, A. Verbeure and R.A. Weder, Phys. Lett. A8, 99 (1974).

76. M. Castagnino, A. Verbeure and R.A. Weder, Nuovo Cimento 26B, 396 (1975).

77. C.W. Misner, Phys. Rev. D8, 3271 (1973).

78. B. Mashoon, Phys. Rev. D8, 4297 (1973).

79. P. Chitre and J.B. Hartle, (in preparation).

80. R. Penrose, in Relativity, Groups and Topology, ed. by C. DeWitt and B.S. DeWitt (Gordon and Breach Publishers, Inc., New York, 1964).

81. S. Deser, M.J. Duff and C.J. Isham, Nucl. Phys. B (in press).

82. P.C.W. Davies, S.A. Fulling and W.G. Unruh, Phys. Rev. D13, 2720 (1976).

83. P.C.W. Davies and S.A. Fulling, "Quantum Vacuum Energy in Two-Dimensional Space-times" (to be published).

84. S.M. Christensen and S.A. Fulling, "Trace Anomalies and the Hawking Effect", preprint.

85. L.P. Grishchuk, Zh. ETF 67, 825 (1974) [JETP 40, 409 (1975)].

86. L.P. Grishchuk, Nuovo Cim. Lett. 12, 60 (1975).

87. R. Gowdy, Phys. Rev. Lett. 27, 826 (1971).

88. D.J. Raine and C.P. Winlove, Phys. Rev. D12, 946 (1975).

89. P.S. Epstein, Proc. Nat. Acad. Sci. (U.S.) 16, 627 (1930).

90. C. Eckart, Phys. Rev. 35, 1303 (1930).

91. R. Kulsrud, Phys. Rev. 106, 205 (1957).

92. L. Parker, Phys. Rev. Lett. 28, 705 (1972).

93. B.L. Hu, Phys. Rev. D8, 1048 (1973).

94. B.L. Hu, Phys. Rev. D9, 3263 (1974).

95. B.L. Hu, S.A. Fulling and L. Parker, Phys. Rev. D8, 2377 (1973).

96. The idea for the present proof is due to Sidney Coleman (private communication, 1972).

97. C.G. Callan, S. Coleman and R. Jackiw, Ann. Phys. (N.Y.) 59, 42 (1970) and S. Coleman and R. Jackiw, ibid. 67, 552 (1971).

98. H.B.G. Casimir, Kon. Ned. Akad. Wet. Proc. 51, 793 (1948).

99. L. Ford, Phys. Rev. D11, 3370 (1975).

100. L. Ford, "Quantum Vacuum Energy in a Closed Universe" (submitted for publication).

101. J.S. Dowker and R. Critchley, J. Phys. (London) A9, 535 (1976).

102. J.S. Dowker and R. Critchley, Phys. Rev. D13, 224; 3224 (1976).

103. S.W. Hawking, Commun. Math. Phys. 18, 301 (1970).

104. S. Bonazzola and F. Pacini, Phys. Rev. 148, 1269 (1966).

105. R. Ruffini and S. Bonazzola, Phys. Rev. 187, 1767 (1969).

106. B.S. DeWitt, Phys. Rev. 162, 1195, 1239 (1967).

107. R. Utiyama and B.S. DeWitt, J. Math. Phys. 3, 608 (1962).

108. S. Deser and P. van Nieuwenhuizen, Phys. Rev. Lett. 32, 245 (1974).

109. S. Deser and P. van Nieuwenhuizen, Phys. Rev. D10, 401 (1974).

110. G. t'Hooft, Nucl. Phys. B62, 444 (1973).

111. G. t'Hooft and M. Veltman, Ann. Inst. H. Poincaré (in press).

112. L. Halpern, Ark. Fys. 34, 539 (1967).

113. E. Streeruwitz, Phys. Rev. D11, 3378 (1975).

114. S.W. Hawking and G.F.R. Ellis, Large Scale Structure of Space-Time (Cambridge University Press, London, 1973).

115. K.C. Jacobs, Ann. New York Acad. Sci. 262, 462 (1974); and private communication.

116. H. Epstein, V. Glaser and A. Jaffe, Nuovo Cim. 31, 1016 (1965).

117. J.D. Bekenstein, Ann. Phys. (N.Y.) 82, 535 (1974).

118. J.D. Bekenstein, Phys. Rev. D11, 2072 (1975).

119. V. Ts. Gurovich, Dokl. Akad. Nauk. SSSR 195, 1300 (1970) [Sov. Phys. Dokl. 15, 1105 (1971)].

120. H. Nariai, Prog. Theor. Phys. (Kyoto) 46, 433 (1971).

121. H. Nariai and K. Tomita, Prog. Theor. Phys. (Kyoto) 46, 776 (1971).

122. A. Trautman, Nature Phys. Sci. 242, 7 (1973). For a pair creation mechanism involving torsion see Ref. 168.

123. J.N. Bahcall and S. Frautschi, Ap. J. 170, L81 (1971).

124. J.A. Wheeler in Relativity, Groups, and Topology, ed. by B.S. DeWitt and C. DeWitt (Gordon & Breach, New York, 1964).

125. C.W. Misner, Phys. Rev. 186, 1319, 1328 (1969).

126. M. Ryan, Hamiltonian Cosmology (Springer, Berlin, 1972).

127. E.P.T. Liang, Phys. Rev. D5, 2458 (1972).

128. F. Lund, Phys. Rev. D8, 3253 (1973).

129. C.W. Misner, K.S. Thorne and J.A. Wheeler, Gravitation (W.H. Freeman and Co., San Francisco, 1973).

130. P. Candelas and D.J. Raine, Phys. Rev. D12, 965 (1975).

131. H.C. Ohanian, "On Particle Creation in Gravitational Collapse", preprint.

132. R. Wald, Commun. Math. Phys. 45, 9 (1975).

133. S.W. Hawking, "Fundamental Breakdown of Physics in Gravitational Collapse", preprint.

134. L. Parker, Phys. Rev. D12, 1519 (1975).

135. J.D. Bekenstein, Phys. Rev. D12, 3077 (1975).

136. G.W. Gibbons and M.J. Perry, Phys. Rev. Lett. 36, 985 (1976).

137. W.H. Press and S.A. Teukolsky, Nature 238, 211 (1972).

138. A.A. Starobinsky, Zh. ETF 64, 48 (1973) [JETP 37, 28 (1973)].

139. A.A. Starobinsky and S.M. Churilov, Zh. ETF 65, 3 (1973) [JETP 38, 1 (1974)].

140. W.G. Unruh, Phys. Rev. Lett. 31, 1265 (1973).

141. G.W. Gibbons, Commun. Math. Phys. 44, 245 (1975).

142. D.G. Boulware, Phys. Rev. D11, 1404 (1975).

143. D.G. Boulware, Phys. Rev. D13, 2169 (1976).

144. P.C.W. Davies, Proc. Roy. Soc. A (in press).

145. U.H. Gerlach, Phys. Rev. D (in press).

146. T. Damour and R. Ruffini, Phys. Rev. D (in press).

147. R.H. Gowdy, "Future Cauchy Horizons, Topology Change, and Exploding Black Holes", preprint.

148. F.J. Dyson, preprint.

149. V.P. Frolov, Usp. Fiz. Nauk 118, 473 (1976).

150. V.P. Frolov, I.V. Volovich and V.A. Zagrebnov, "Quantum Particle Creation (Hawking Effect) By Non-Stationary Black Holes", preprint-60.

151. I.D. Novikov, "Particle Creation and Vacuum Polarization By Black Holes", preprint-268.

152. S.L. Adler, J. Lieberman, Y.J. Ng and H-S. Tsao, Phys. Rev. D14, 359 (1976).

153. S.L. Adler, Phys. Rev. D14, 379 (1976).

154. H. Nariai and K. Tanabe, Prog. of Theor. Phys. 55, 1116 (1976).

155. H. Nariai, "Propagators for a Scalar Field in Some Bianchi-type I Universes", preprint RRK76-7.

156. For a recent survey see G.T. Horowitz, Princeton University Senior Thesis.

157. P. Yodzis, H.J. Seifert and H. Müller zum Hagen, Commun. Math. Phys. 34, 135 (1973).

158. H. Müller zum Hagen, P. Yodzis and H.J. Seifert, Commun. Math. Phys. 37, 29 (1974).

159. L. Ford and L. Parker (in preparation).

160. C. Duncan, P. Esposito and L. Witten, "On a Static Axially Symmetric Solution of the Einstein Equations," Phys. Rev. D (to be published).

161. A. Ashtekar and A. Magnon, C. R. Acad. Sc., Paris, 280A, 741, 833 (1975).

162. J. Friedman, "Ergoregion Instabilities" (to be published).

163. S.L. Adler, J. Lieberman and Y.J. Ng, "Regularization of the Stress-Energy Tensor for Vector and Scalar Particles Propagating in a General Background Metric," Preprint COO-2220-83.

164. R. Wald, Phys. Rev. D13, 3176 (1976).

165. J.D. Bekenstein and A. Meisels, "Einstein A and B Coefficients for a Black Hole," preprint.

166. D.G. Boulware, Phys. Rev. D12, 350 (1975).

167. S.W. Hawking, Phys. Rev. D13, 191 (1976).

168. G.D. Kerlick, Phys. Rev. D12, 3004 (1975).

)-(-SPACE AND NULL INFINITY

M. Ko and E.T. Newman
Department of Physics and Astronomy
K.P. Tod
Department of Mathematics
University of Pittsburgh
Pittsburgh, Pennsylvania

Contents

Introduction

Our object in this article is to provide a review of one approach to asymptotically flat space-times, and to show how this approach leads to the introduction of an associated four complex dimensional manifold, $\mathcal{H}$-space, with remarkable properties.

The article falls into two parts.

In the first, we provide an introduction to the mathematical techniques found useful in the study of asymptotically flat space-times, starting with the description of tensor calculus on a two-surface. This finds application to a calculus of null directions in a space-time manifold, the two-surface being a section of the light cone at each point.

A review of the spin-coefficient formalism follows, with its application to asymptotically flat space-times, in a coordinate system based on outgoing null hypersurfaces. The role of the asymptotic shear, σ^o, of these hypersurfaces in determining the asymptotic solution is remarked and the question is raised of finding a family of null hypersurfaces with zero σ^o. After complexifying the $\mathcal{I}^+$ of space-time, a four complex parameter family of such "good cuts" can be found and this is the manifold of $\mathcal{H}$-space. An integral expression is given for a metric on this manifold and we state, without proof, that the metric is Riemannian.

The second part of the article sets out a new analysis of $\mathcal{H}$-space which differs from earlier treatments in that the integral expression for the metric is largely avoided. This is accomplished by the use of the light-cone calculus of the first part and results in a considerable simplification in the formulae. The metric and curvature tensors of $\mathcal{H}$-space are obtained from the solution of the good cut equation simply by algebraic and differential operations.

The connection with two other approaches to $\mathcal{H}$-space is investigated.

We show how the non-linear graviton of Penrose may be constructed from a solution of the good cut equation, with σ^o playing the role of infinitesimal transition function, and we show how the two forms of the $\mathcal{H}$-space metric obtained by Plebanski arise in our formalism.

Part I: Mathematical Preliminaries

E.T. Newman
Department of Physics and Astronomy
K.P. Tod
Department of Mathematics
University of Pittsburgh

A: Tensor Calculus On A Two-Surface

We recall how two dimensional tensor calculus can be phrased in terms of scalars with the property of spin weight and introduce the operators, ð, (pronounced edth) and $\bar{ð}$, which correspond to covariant differentiation.[1]

On any real, two dimensional manifold, S, we can introduce isothermal coordinates, x and y, so that the metric is conformally flat:

$$ds^2 = \frac{1}{(p(x,y))^2}(dx^2 + dy^2)$$

or, introducing a complex coordinate, $\zeta = x + iy$,

$$ds^2 = \frac{1}{(P(\zeta,\bar{\zeta}))^2} d\zeta d\bar{\zeta} \ . \tag{A.1}$$

In a later section we will complexify S by allowing x and y to be complex; then x-iy is independent of x + iy and will be denoted by $\tilde{\zeta}$. Throughout these notes, the appearance of a barred quantity will recall the possibility of complexification and the replacement of the barred quantity by an independent tilded quantity.

Two vectors spanning the tangent space to S are

$$a = \frac{1}{\sqrt{2}} P \frac{\partial}{\partial x} \qquad b = \frac{1}{\sqrt{2}} P \frac{\partial}{\partial y} \ , \qquad a \cdot a = b \cdot b = \frac{1}{2} \ ,$$

which we take in the linear combinations

$$m = a - ib \qquad \bar{m} = a + ib$$
$$= \sqrt{2}\, P \frac{\partial}{\partial \zeta} \qquad = \sqrt{2}\, P \frac{\partial}{\partial \bar{\zeta}} \tag{A.2}$$

observing the normalization

$$m \cdot m = \bar{m} \cdot \bar{m} = 0; \quad m \cdot \bar{m} = 1.$$

Coordinate transformations preserving the form (A.1) of the metric are just

$$\zeta \rightarrow \zeta' = f(\zeta); \qquad \bar{\zeta} \rightarrow \bar{\zeta}' = \bar{f}(\bar{\zeta})$$

under which

$$m \rightarrow m' = e^{i\psi} m; \qquad \bar{m} \rightarrow \bar{m}' = e^{-i\psi} \bar{m} \tag{A.3}$$

where $e^{2i\psi} = \dfrac{\bar{f}_{,\bar{\zeta}}}{f_{,\zeta}}$.

From the two-dimensionality of the tangent space to S, it follows that only totally symmetric, trace-free tensor fields "count", since any traces can be subtracted off with the use of the metric, and any skew parts with the use of the skew tensor $\varepsilon_{ab} = m_{[a}\bar{m}_{b]}$.

If $\eta_{ab\ldots d}$ is such a real, totally symmetric, trace-free tensor field, it has a unique expression as

$$\eta_{a\ldots d} = \eta \bar{m}_a \ldots \bar{m}_d + \bar{\eta} m_a \ldots m_d \tag{A.4}$$

where

$$\eta = \eta_{a\ldots d}\, m^a \ldots m^d$$
$$\bar{\eta} = \eta_{a\ldots d}\, \bar{m}^a \ldots \bar{m}^d$$

Further, under (A.3), the scalar η transforms as

$$\eta \rightarrow \eta' = e^{is\psi}\, \eta \tag{A.5a}$$

$$\bar{\eta} \rightarrow \bar{\eta}' = e^{-is\psi}\, \bar{\eta} \tag{A.5b}$$

where s is the number of indices of $\eta_{a\ldots.d}$.

A scalar with the transformation law (A.5a) under (A.3) is said to be of spin weight[1] s (similarly, (A.5b) characterizes spin weight -s) and we see that such a spin weighted scalar uniquely determines a real, totally symmetric, trace-free tensor field tangent to S.

We define the operators $\eth$ and $\bar{\eth}$ for a quantity with positive spin weight s by projecting the covariant derivative, with respect to the metric (A.1), of the corresponding tensor in the m and $\bar{m}$ directions; thus[1,2]:

$$\eth\eta = \sqrt{2}\, \eta_{a\ldots b;c}\, m^a \ldots m^b m^c \tag{A.6a}$$

$$\bar{\eth}\eta = \sqrt{2}\, \eta_{a\ldots b;c}\, m^a \ldots m^b \bar{m}^c \tag{A.6b}$$

and similarly for negative spin weight:

$$\eth\xi = \sqrt{2}\, \xi_{a\ldots b;c}\, \bar{m}^a \ldots \bar{m}^b m^c \tag{A.6c}$$

$$\bar{\eth}\xi = \sqrt{2}\, \xi_{a\ldots b;c}\, \bar{m}^a \ldots \bar{m}^b \bar{m}^c \tag{A.6d}$$

The following properties of $\eth$ and $\bar{\eth}$ are immediate.

i) $\eth$ and $\bar{\eth}$ are linear and satisfy the Leibnitz rule, from the corresponding properties of the covariant derivative. Also $\overline{\eth\eta} = \bar{\eth}\bar{\eta}$.

ii) If η has spin weight s, then $\eth\eta$ has spin weight $s+1$ and $\bar{\eth}\eta$ has spin weight $s-1$.

iii) The Ricci identity for the covariant derivative becomes

$$(\bar{\eth}\eth - \eth\bar{\eth})\eta = -2sK\eta \tag{A.7}$$

on a spin weight s quantity η, where K is the Gaussian curvature of S,

$$K = \eth\bar{\eth}\, \ln P .$$

On an $s = 0$ quantity, $\eth\bar{\eth} = \bar{\eth}\eth$ is the Laplacian.

iv) From (A.2), coordinate expressions for $\eth$ and $\bar{\eth}$ may be obtained:

$$\eth\eta = 2P^{1-s} \frac{\partial}{\partial\zeta} (P^s\eta) \tag{A.8}$$

$$\bar{\eth}\eta = 2P^{1+s} \frac{\partial}{\partial\bar{\zeta}} (P^{-s}\eta)$$

for η of spin weight s.

Indeed, one could take these as definitions and derive the other properties.

Henceforth, we shall restrict S to be a metric sphere so that

$$P = P_o = \frac{1}{2}(1+\zeta\bar{\zeta})$$

and

$$\zeta = e^{i\phi} \cot \frac{\theta}{2}$$

which is the usual stereographic coordinate. This would give $\eth$ and $\bar{\eth}$ in terms of θ and ϕ as

$$\eth\eta = -(\sin\theta)^{s} \left(\frac{\partial}{\partial\theta} + \frac{i}{\sin\theta} \frac{\partial}{\partial\phi} \right) ((\sin\theta)^{-s}\eta)$$

$$\bar{\eth}\eta = -(\sin\theta)^{-s} \left(\frac{\partial}{\partial\theta} - \frac{i}{\sin\theta} \frac{\partial}{\partial\phi} \right) ((\sin\theta)^{s}\eta) \quad .$$

The spherical harmonics, $Y_{\ell m}(\zeta,\bar{\zeta})$, are the appropriately normalized, regular solutions of

$$\eth\bar{\eth}\, Y_{\ell m} = -\ell(\ell + 1)\, Y_{\ell m}$$

regarded as spin weight zero quantities.

One can define spin weight s spherical harmonics by[1,2]

$$\begin{aligned} {}_sY_{\ell m} &= \frac{(\ell - s)!}{(\ell + s)!}\, \eth^{s}\, Y_{\ell m} & 0 \le s \le \ell \\ &= (-1)^{s} \frac{(\ell - s)!}{(\ell + s)!}\, \bar{\eth}^{-s} Y_{\ell m} & -\ell \le s \le 0 \\ &= 0 & |s| > \ell \end{aligned}$$

These form a complete set of spin weight s functions and are eigenfunctions of $\eth\bar{\eth}$. In the same way, one can define ${}_sY_{\ell m}$ for s and ℓ half-odd-integral.

The raising and lowering properties of $\eth$ and $\bar{\eth}$ on the ${}_sY_{\ell m}$ may be conveniently summarized by a diagram in which ${}_sY_{\ell m}$ is represented by the point (ℓ,s) in the plane (suppressing m, see figure 1). Then $\eth$ moves us one point in the positive s direction and $\bar{\eth}$ moves us one point in the negative s direction. Any application of $\eth$ and $\bar{\eth}$ which takes us beyond the lines $s = \pm\ell$ necessarily gives zero, and this property enables us to solve simple differential equations. For instance, if η is a spin weight zero quantity, so that it is a linear combination of points on the ℓ-axis, and satisfies

$$\eth^{2}\eta = 0 \tag{A.9}$$

then η can only contain spherical harmonics corresponding to points which two applications of $\eth$ carry across the line $s = \ell$. Thus the general solution of (A.9) is

$$\eta = a\, Y_{oo}(\zeta,\bar{\zeta}) + b_m\, Y_{1m}(\zeta,\bar{\zeta}) \tag{A.10}$$

for four constants a, b_m.

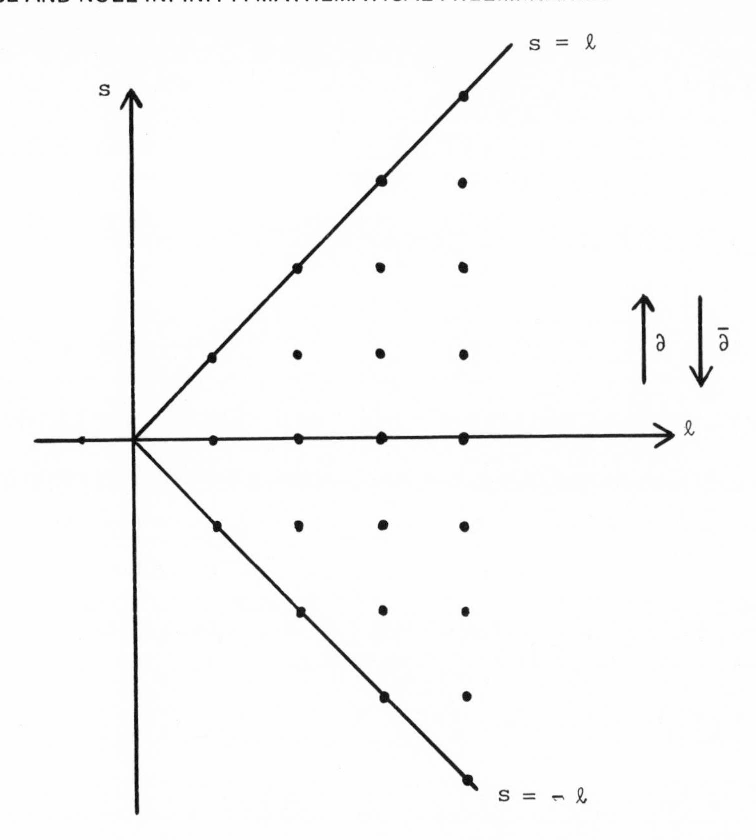

Figure 1: The grid of ${}_sY_{\ell m}$, m suppressed, showing the action of ð and $\bar{ð}$.

B: The Light Cone Calculus

In this section we use the ð operator of the previous section to develop a calculus of null vector fields on a space-time manifold.

Initially restricting to the tangent space of one point, we may suppose we have a preferred unit timelike vector, t^a, and use this to take a section, S, of the light cone. That is, we consider all null vectors, ℓ^a, scaled so that

$$\ell_a t^a = \frac{1}{\sqrt{2}} \quad . \tag{B.1}$$

This section of the light cone receives the metric of a two-sphere from the Minkowski metric of the tangent space, and we can take a stereographic coordinate, ζ, on S. We remark that ζ can be fixed uniquely by choosing an orthonormal tetrad with time-like leg t^a. We then label each of the normalized null vectors, ℓ^a, by the point where it meets S, giving a hybrid object $\ell^a(\zeta,\bar{\zeta})$, which is a null vector for each ζ and sweeps out the entire (normalized) light cone as ζ varies.

In terms of an orthonormal tetrad, the components of ℓ^a may be taken to be

$$\frac{1}{2\sqrt{2}\,P_o}\left(1+\zeta\bar{\zeta},\ \zeta+\bar{\zeta},\ \frac{\zeta-\bar{\zeta}}{i},\ -1+\zeta\bar{\zeta}\right)$$

and we observe that these are Y_{oo} and Y_{1m}, so that from (A.9) and (A.10),

$$\eth^2 \ell^a = 0 \quad . \tag{B.2}$$

That is, the components of ℓ^a in any frame, regarded as functions on S, will satisfy (B.2).

Dropping the restriction to a single point, we suppose we have a time-like vector field t^a and a null vector field $\ell^a(x^b,\zeta,\bar{\zeta})$ satisfying (B.1) and (B.2) at every point, x^b. (We may suppose ζ to be defined everywhere either by decree, or by a global choice of orthonormal tetrad).

From this field of null vectors, we proceed to construct a tetrad by

$$\begin{aligned} m^a &= \eth\ell^a \\ \bar{m}^a &= \bar{\eth}\ell^a \\ n^a &= \ell^a + \eth\bar{\eth}\ell^a \end{aligned} \tag{B.3}$$

from which it follows that

$$t^a = \frac{1}{\sqrt{2}}(\ell^a + n^a) \quad .$$

Then (B.1), (B.2) and the null-ness of ℓ^a imply that the space-time metric is

$$g^{ab} = \ell^a n^b + \ell^b n^a - m^a\bar{m}^b - m^b\bar{m}^a \tag{B.4}$$

so in particular

$$\eth g^{ab} = 0 = \bar{\eth} g^{ab} \quad .$$

For future use we observe that if we make a Lorentz transformation:

$$
\begin{aligned}
L^a &= \alpha \ell^a + \beta m^a \\
M^a &= \gamma \ell^a + \delta m^a \\
\bar{M}^a &= \beta n^a + \alpha \bar{m}^a \\
N^a &= \delta n^a + \gamma \bar{m}^a
\end{aligned}
\tag{B.5}
$$

where $\alpha, \beta, \gamma, \delta$ are allowed to be functions of x^a, ζ and $\bar{\zeta}$ subject only to $\alpha\delta - \beta\gamma = 1$, then again

$$g^{ab} = L^a N^b + L^b N^a - M^a \bar{M}^b - M^b \bar{M}^a$$

although now, of course, L^a has no simple behaviour. In general, $\eth^2 L^a \neq 0$.

This calculus raises the following questions:

i) how would Einstein's equations appear in terms of the hybrid object, $\ell^a(x^b, \zeta, \bar{\zeta})$?

ii) how far would they simplify if $\ell_a = \ell_{,a}$ for some function $\ell(x^b, \zeta, \bar{\zeta})$?

iii) could we choose α, β, γ and δ in (B.5) to simplify them further?

We shall find partial answers to these questions in the second part of this article.

C: The Spin Coefficient Formalism and the Introduction of $\mathcal{H}$-Space

The starting point of the spin coefficient formalism is a null tetrad $(\ell^a, m^a, \bar{m}^a, n^a)$, now a function only of position, with the metric given by

$$g_{ab} = 2\ell_{(a} n_{b)} - 2m_{(a} \bar{m}_{b)} \quad .$$

The spin coefficients[3,4] are the tetrad components of the covariant derivatives of the tetrad vectors and thus are linear combinations of the Ricci rotation coefficients. However, there is an economy of notation since the twenty-four real Ricci rotation coefficients are combined into twelve complex spin coefficients, and it becomes feasible to introduce a separate letter for each one. A similar economy is achieved with the Riemann tensor, whose ten real components (in vacuo) are combined into five complex tetrad components.

Introducing four differential operators, the tetrad components of the covariant derivative operator, the spin coefficient equations form a set of first order equations which can be written out explicitly, that is, without the use of the summation convention.

There are three groups of equations: the definitions of the spin coefficients, the Ricci identities and the Bianchi identities.

For future use, we recall the definitions of some of the spin coefficients:

$$\rho = m^a \bar{m}^b \ell_{a;b} \quad ; \quad \bar{\rho} = \bar{m}^a m^b \ell_{a;b}$$

$$\sigma = m^a m^b \ell_{a;b} \quad ; \quad \bar{\sigma} = \bar{m}^a \bar{m}^b \ell_{a;b}$$

(when ℓ^a is tangent to a geodesic congruence, these are the optical scalars of the congruence and have geometrical significance; in particular σ, the shear, provides a measure of the distortion of the congruence)[4].

A typical Weyl tensor component is

$$\psi_o = -C_{abcd}\ell^a m^b \ell^c m^d \quad ; \qquad \bar{\psi}_o = -C_{abcd}\ell^a \bar{m}^b \ell^c \bar{m}^d \ .$$

Once again, we remark the possibility of using this formalism in a complexified space-time by freeing barred from unbarred quantities; thus $\bar{\rho}$ and $\bar{\psi}_o$ become $\tilde{\rho}$ and $\tilde{\psi}_o$, independent of ρ and ψ_o and so on[5].

To apply the formalism to asymptotically flat, vacuum space-times a choice of coordinate system and tetrad is made which simplifies the equations sufficiently that asymptotic solutions can be found[6]. The conformal factor is chosen to make the sections of I^+ metrically spherical, and the generators of I^+ are labelled by a stereographic coordinate ζ. Introducing the vector n^a, tangent to the generators, and choosing a specific section as origin, a one parameter family of sections labelled by the coordinate u is defined by going equal affine distances up each generator. Each of these sections then determines a null hypersurface in the interior of the manifold (at least near I^+) by taking, at each point of the section, the ray from the interior which meets it orthogonally. The fourth coordinate, r, is then affine distance along the generators of the constant u hypersurfaces.

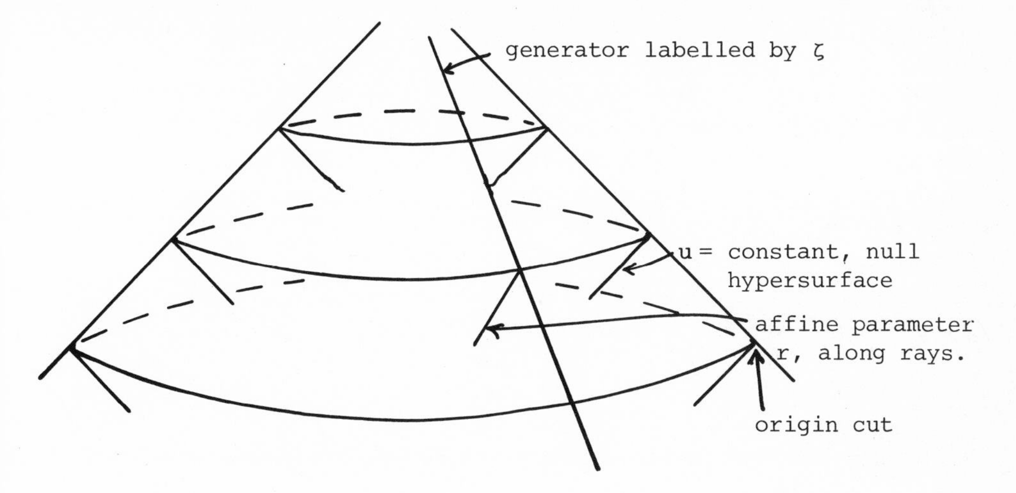

Figure 2: $\mathcal{I}^+$ showing the $(u,r,\zeta,\bar{\zeta})$ coordinate system.

To fix the tetrad, ℓ^a is taken tangent to the outgoing rays in the constant u hypersurface and the vectors m^a, $\bar{m}^a$ and n^a are parallelely propagated along ℓ^a.

The condition of asymptotic flatness is imposed on the Weyl tensor as*

$$\psi_o = \frac{\psi_o^o}{r^5} + O(r^{-6}) \quad .$$

For the other Weyl tensor components, the radial dependence is found to be

$$\psi_1 = \frac{\psi_1^o}{r^4} + O(r^{-5})$$

* This condition can be deduced from the existence of $\mathcal{I}^+$ but is compatible with more general structures at null infinity[7].

$$\psi_2 = \frac{\psi_2^o}{r^3} + O(r^{-4})$$

$$\psi_3 = \frac{\psi_3^o}{r^2} + O(r^{-3})$$

$$\psi_4 = \frac{\psi_4^o}{r} + O(r^{-2})$$

which expresses the Peeling Theorem[8]. We note also the results

$$\rho = -\frac{1}{2}\ell^a{}_{;a} = -\frac{1}{r} + O(r^{-3})$$

$$\sigma = \frac{\sigma^o(u,\zeta,\bar{\zeta})}{r^2} + O(r^{-4}) \quad .$$

The asymptotic solution is determined by the freely specifiable function $\sigma^o(u,\zeta,\zeta)$, the asymptotic shear, together with a certain number of freely specifiable functions of ζ and $\bar{\zeta}$ only. In particular[6]

$$\psi_4^o = -\ddot{\bar{\sigma}}^o$$

$$\psi_3^o = \eth\dot{\bar{\sigma}}^o$$

$$\psi_2^o - \bar{\psi}_2^o = \bar{\eth}^2\sigma^o - \eth^2\bar{\sigma}^o + \bar{\sigma}^o\dot{\sigma}^o - \sigma^o\dot{\bar{\sigma}}^o$$

from which we see that the asymptotic shear determines the gravitational radiation field.

The BMS group, in this context, may be defined in terms of its infinitesimal generators, the asymptotic killing vectors[11]. These are vectors, ξ_a, such that $\xi_{(a;b)}$ vanishes on $\mathcal{I}^+$ in the rescaled manifold, or falls off to zero sufficiently fast in the physical manifold. In particular, the supertranslations correspond to changing the initial choice of origin cut for the u coordinate on $\mathcal{I}^+$, (see figure 3). This will change the tetrad, and in particular will give a new asymptotic shear. If the new origin cut is given in the old u-coordinate as

$$u = \alpha(\zeta,\bar{\zeta})$$

then the asymptotic shear of the new origin cut is found to be

$$\sigma^{o\prime}(\zeta,\bar{\zeta}) = \sigma^o(\alpha(\zeta,\bar{\zeta}),\zeta,\bar{\zeta}) - \eth^2\alpha(\zeta,\bar{\zeta}) \quad .$$

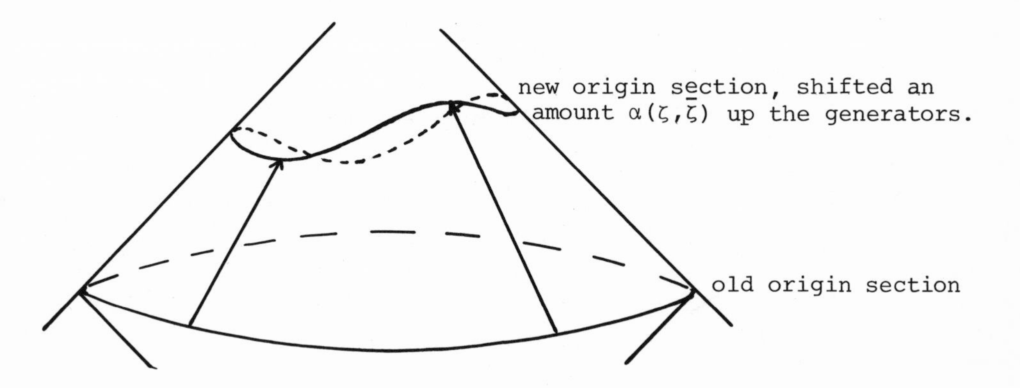

Figure 3: Sketch of $\mathcal{I}^+$, showing the effect of a supertranslation $\alpha(\zeta,\bar{\zeta})$ on the origin section.

The question then arises of whether $\alpha(\zeta,\bar{\zeta})$ can be chosen such that the new asymptotic shear vanishes; in other words, do there exist asymptotically shear-free cuts?

If $u = Z(\zeta,\bar{\zeta})$ were to be such a cut then

$$\eth^2 Z(\zeta,\bar{\zeta}) = \sigma^o(Z(\zeta,\bar{\zeta}),\zeta,\bar{\zeta})$$

and it is intuitively clear that, in general, this equation will have no solutions, because σ^o, being complex, corresponds to two real functions while Z is just one.

It is at this point that we introduce the long heralded complexification! That is, we wish to allow complex supertranslations, and so we complexify $\mathcal{I}^+$ by allowing u to be complex and freeing $\bar{\zeta}$ from ζ, denoting it by $\tilde{\zeta}$. It is customary to represent $\mathcal{I}^+_c$, complexified $\mathcal{I}^+$, by a cube with the three real axes playing the rôle of three complex axes[9]. There is a "diagonal" with u real and $\tilde{\zeta} = \bar{\zeta}$ corresponding to the original real $\mathcal{I}^+$, and a complex supertranslation is represented by a surface in the cube (see figure 4).

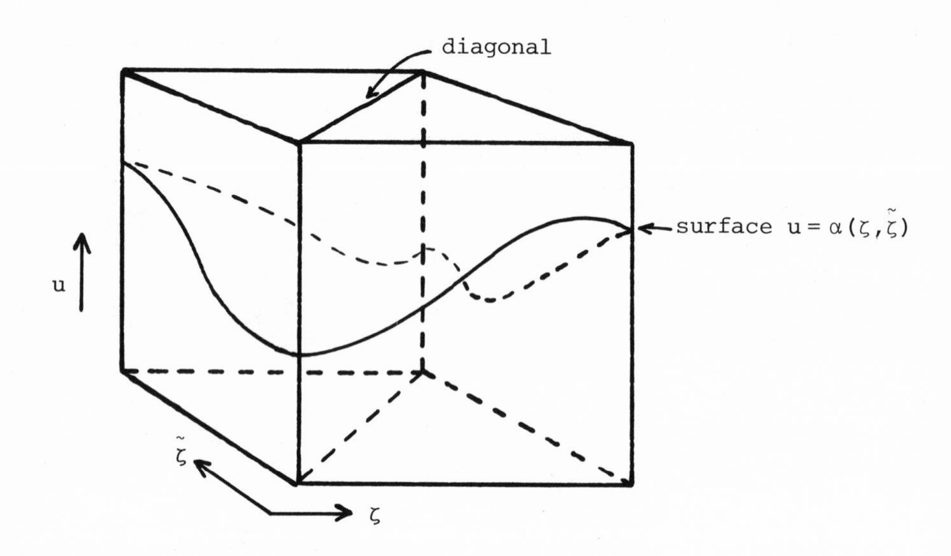

Figure 4: I_c^+ with the surface representing a complex supertranslation.

The equation for an asymptotically shear-free cut (briefly, a "good cut") is now

$$\eth^2 Z(\zeta,\tilde{\zeta}) = \sigma^o(Z(\zeta,\tilde{\zeta}),\zeta,\tilde{\zeta}) \tag{C.1}$$

which we refer to as the good cut equation.

If the asymptotically flat space-time under consideration were Minkowski space, equation (C.1) would have zero on the right, becoming

$$\eth^2 Z = 0 \tag{C.2}$$

which we recall from (A.9), (A.10) and (B.2) has a four complex parameter family of regular solutions which can be written

$$Z = z^a \ell_a(\zeta, \tilde{\zeta})$$

where the z^a are four complex parameters. In complexified Minkowski space, however, there is a four complex parameter family of completely shear-free outgoing null hypersurfaces, which therefore automatically give good cuts, and these are the light cones of points. We thus have a correspondence between the points of Minkowski space and the solutions of the good cut equation. Furthermore, the metric separation of two points of Minkowski space can be calculated from the corresponding good cuts as follows: if the points have (Cartesian) coordinates z^a and $z^a + dz^a$, with corresponding cuts Z and Z + dZ then

$$dZ = dz^a \ell_a = dz^a Z_{,a} \tag{C.3}$$

and it can be checked that

$$\eta_{ab} dz^a dz^b = \frac{1}{8\pi i} \left[\oint \frac{d\Omega}{(dZ)^2} \right]^{-1} \tag{C.4}$$

where $d\Omega = \dfrac{2d\zeta \wedge d\tilde{\zeta}}{i(1 + \zeta\tilde{\zeta})^2}$ is the usual area element on the (complex) unit sphere and the integral is carried out over a sphere contour.

Recapitulating, the solutions of (C.2) form a four complex dimensional manifold on which it is possible to define a (flat) Riemannian metric. This is the situation which generalizes to give $\mathcal{H}$-space.

Penrose and Atiyah[10] have recently confirmed the long-standing conjecture that, for small enough σ^o, equation (C.1) admits a four complex parameter family of regular solutions, which will be the points of $\mathcal{H}$-space. Then, remarkably, the same definition (C.4) with dZ as in (C.3) results in a quadratic (i.e. not Finslerian, as might at first appear) metric on $\mathcal{H}$-space. That is

$$g_{ab} dz^a dz^b = \frac{1}{8\pi i} \left[\oint \frac{d\Omega}{(dZ)^2} \right]^{-1}$$

where $dZ = dz^a Z_{,a}$.

Further properties of $\mathcal{H}$-space may be obtained from laborious calculations with this integral expression. In particular, for the Riemann and Ricci tensors one finds

$$R_{ab} = 0$$

$$R_{abcd} = i\overset{*}{R}_{abcd} .$$

Thus $\mathcal{H}$-space satisfies Einstein's vacuum equations and the Riemann tensor is anti-self dual.

In the second part of this article, these and other properties of $\mathcal{H}$-space are obtained by a different method, drawing on the Light Cone Calculus of section B, which avoids the manipulation of integral expressions.

REFERENCES FOR PART I

1. R.W. Lind, J. Messmer and E.T. Newman, J. Math. Phys. 13 1884 (1972).

2. E.T. Newman and R. Penrose, J. Math. Phys. 7, 863 (1966). J.N. Goldberg et al., J. Math. Phys. 8 2155 (1967).

3. E.T. Newman and R. Penrose, J. Math. Phys. 3 566 (1962).

4. F.A.E. Pirani in "Lectures on General Relativity" Brandeis Summer School, Prentice Hall (N.Y.) 1965.

5. As an example of the spin coefficient formalism applied to complex spaces see C.W. Fette, A.L. Janis, E.T. Newman, J. Math. Phys. 17 660 (1976).

6. A.R. Exton, E.T. Newman and R. Penrose, J. Math. Phys. 10 1566 (1969).

7. R. Penrose, Proc. Roy. Soc. Land. A284 159 (1965).

8. R. Penrose in Battelle Pencontres, Benjamin (N.Y.) 1968.

9. R. Penrose and M.A.H. MacCallum, Phys. Rep. 6C 242 (1973). E.J. Flaherty, "Hermitian and Kahlerian Geometry in Relativity", Lecture Notes in Physics #46, Springer Verlag 1976.

10. R. Penrose, private conversation.

11. L. Tamburino and J. Winicour, Phys. Rev. 150 1039 (1966).

Part II: $\mathcal{H}$-Space - A New Approach

M. Ko and E.T. Newman*
Dept. of Physics and Astronomy
K.P. Tod
Dept. of Mathematics
University of Pittsburgh

1. Introduction

In this part of the article we describe an alternative approach to the subject of $\mathcal{H}$-space and present some new results. This approach is, we believe, both more transparent and more powerful than the earlier approach based on contour integral theorems.

In Sec. II we will review one of the earlier approaches, i.e. the origin of the "good cut" equation and its relation (by means of contour integrals over the solutions) to the metric and curvature properties of $\mathcal{H}$-space. In Sec. III we will show that the metric of $\mathcal{H}$-space can be explicitly exhibited in terms of algebraic and differential operations applied to the solution of the "good cut" equation, with no recourse to the integral operations of Sec. II. In Secs. IV and V we show how to express the covariant derivative and curvature tensor in terms of the good cut equation solutions. In Sec. VI we discuss the relationship of our approach to $\mathcal{H}$-space with that of Penrose[3,5] via deformed twistor space. In particular we show how the deformation function (overlap function) is related to the shear, σ. Finally in Sec. VII we associate our approach to that of Plebanski[7,8] and show the meaning (from our point of view) of the two canonical Plebanski coordinate systems.

2. Review

In the case of general asymptotically flat solutions to the vacuum Einstein (or Einstein-Maxwell) equations, one cannot find many (or frequently any) null surfaces which are asymptotically shear-free - i.e. which most resemble ordinary null cones in Minkowski space. If however the space is analytic in the neighborhood of null

* This research was supported in part by a grant from the National Science Foundation.

infinity one can in fact find complex null surfaces in the analytically extended physical space, which are asymptotically shear-free. Using a Bondi coordinate system $(u,\zeta,\tilde{\zeta})$ to describe null infinity and furthermore with $\sigma^o(u,\zeta,\tilde{\zeta})$, the Bondi shear*, analytic in the three slots, the equation which describes the intersection of null infinity with the asymptotically shear-free complex null surfaces is $u = Z(\zeta,\tilde{\zeta})$, which satisfies

$$\eth^2 Z = \sigma^o(Z,\zeta,\tilde{\zeta}) \ . \tag{2.1}$$

Eq. (2.1) is referred to as the good cut equation and its solutions, $u = Z(\zeta,\tilde{\zeta})$ as good cut functions.

Though for a considerable length of time it appeared intuitively clear that the regular** solutions of (2.1) formed a four-complex parameter set, i.e. that the solution set formed a four complex dimensional manifold ($\mathcal{H}$-space), it was not until recently that Penrose and Atiyah*** using the Kodaira deformation theorems proved (for sufficiently small σ^o) the existence and four dimensional nature of $\mathcal{H}$. We can hence write the general good cut function as

$$u = Z(z^a,\zeta,\tilde{\zeta}) \tag{2.2}$$

where the z^a are four complex parameters (or coordinates) which label the good cut functions (or points in $\mathcal{H}$).

From the properties of the good cut function, one can prove[1,2] the following rather remarkable theorem.

Defining $dZ = Z,_a dz^a$, $Z,_a \equiv \dfrac{\partial Z}{\partial z^a}$

and

$$ds^2 = \left\{\frac{1}{8\pi i}\oint \frac{d\Omega}{(dZ)^2}\right\}^{-1} \tag{2.3}$$

where $d\Omega = \dfrac{2}{i}\dfrac{d\zeta \wedge d\tilde{\zeta}}{(1+\zeta\tilde{\zeta})^2}$ is the area element on the complex unit sphere

*In the real space $\tilde{\zeta} = \bar{\zeta}$, the complex conjugate of ζ and u is real. The Bondi shear is thus usually given as a function of three real variables. We are considering the analytic extension of it to the complex null infinity $C\mathcal{I}^+$. ζ and $\tilde{\zeta}$ are defined on the extended complex plane and u on the complex plane. $C\mathcal{I}^+$ thus has the topology of $S^2 \times S^2 \times R^2$.

**By regular we mean that $Z(\zeta,\tilde{\zeta})$ have an expansion in spherical harmonics.

***Private communication.

and the integral is taken over the real unit sphere (or, in order to avoid poles, over an appropriate deformation of the real unit sphere into the complex sphere), then

$$ds^2 = g_{ab}(z^c)dz^a dz^b \qquad (2.4)$$

where g_{ab} is a complex non-degenerate symmetric tensor field on $\mathcal{H}$, i.e. the good cut function induces, via (2.3), a complex metric on $\mathcal{H}$.

By repeated differentiation of the integral (2.3) one can (after considerable labor) construct[2] integral expressions for the affine connection and curvature tensor of the metric (2.4). One can then show the further remarkable properties of $\mathcal{H}$-space, namely that

$$R_{ab} = 0 \qquad (2.5)$$

$$R_{abcd} = i\, R^{*}_{abcd} \qquad (2.6)$$

i.e. the metric (2.3) and (2.4) satisfies the vacuum Einstein equations and the Riemann (Weyl) tensor is anti-self dual.

Penrose[3,5] begins his construction of $\mathcal{H}$-space with flat projective twistor space (CP^3). By deforming the complex structure of a portion of CP^3, and appealing to the Kodaira deformation theorems he arrives at $\mathcal{H}$-space.

3. The Metric

We begin this section with a discussion of the good cut function $u = Z(z^a, \zeta, \tilde{\zeta})$. The point of view being adopted here (and in the remainder of the paper) is that we are not attempting to solve the good cut equation but rather that Z is a known function of the six arguments z^a, ζ and $\tilde{\zeta}$ (which satisfies (2.1)) and that we can differentiate and algebraically manipulate it freely. In particular we wish to show that Z is a master function from which all properties of $\mathcal{H}$ can be obtained simply by differential and algebraic operations.

For any fixed but arbitrary values of ζ and $\tilde{\zeta}$, Z can be considered as a scalar function on $\mathcal{H}$ (or more precisely, as a cross section of the complex sphere bundle over $\mathcal{H}$) and one can ask for properties of the gradient of Z, namely $Z_a \equiv \partial Z/\partial z^a$. Properties of Z_a can be investigated by realizing it satisfies the (important) equation*

*By σ we mean the shear evaluated on $C\mathcal{I}^+$ in the conformally rescaled manifold. This is the same as the asymptotic shear σ^o used in (2.1).

$$\eth^2 Z_a = \dot{\sigma} Z_a, \tag{3.1}$$

$$\dot{\sigma}(Z(z^a,\zeta,\tilde{\zeta}),\zeta,\tilde{\zeta}) \equiv \frac{\partial\sigma(Z,\zeta,\tilde{\zeta})}{\partial Z}$$

which is obtained simply by taking the gradient of (2.1).

Z_a can always be put into the form

$$Z_a = V(\ell_a + Wm_a) \tag{3.2}$$

where V is a particular solution to (3.1) i.e.

$$\eth^2 V = \dot{\sigma} V \tag{3.3}$$

and

$$m_a = \eth \ell_a \tag{3.4a}$$

$$\eth^2 \ell_a = \eth m_a = 0 \tag{3.4b}$$

$$1 + \eth W = V^{-2}\ . \tag{3.4c}$$

This is simply shown by direct substitution of (3.2) into (3.1), using (3.3) and (3.4). Further we have

$$\eth Z_a = \eth V \ell_a + (V^{-1} + W\eth V) m_a\ . \tag{3.5}$$

We have shown that the gradients of Z and $\eth Z$ are two independent vector fields (for each value of ζ and $\tilde{\zeta}$) which are expressible as linear combinations of the two vector fields ℓ_a and m_a which have simple ζ and $\tilde{\zeta}$ behavior from (3.4b). One can in fact invert (3.2) and (3.5) and obtain

$$\ell_a = (V^{-1} + W\eth V) Z_a - WV\eth Z_a \tag{3.6}$$

$$m_a \equiv \eth \ell_a = -\eth V Z_a + V\eth Z_a. \tag{3.7}$$

If one is given (for each fixed value of ζ and $\tilde{\zeta}$) a regular vector field $\ell_a(z^a,\zeta,\tilde{\zeta})$ which satisfies equation (3.4b)*, three additional linearly independent fields can be constructed by

*ℓ_a consists of four independent linear combinations of $\ell = 0$ and $\ell = 1$ spherical harmonics.

$$m_a = \eth \ell_a$$

$$\tilde{m}_a = \tilde{\eth} \ell_a \qquad (3.8)$$

$$n_a = \ell_a + \eth\tilde{\eth} \ell_a \ .$$

The symmetric non-degenerate tensor field formed from these four vectors

$$g_{ab} = \ell_a n_b + n_a \ell_b - m_a \tilde{m}_b - \tilde{m}_a m_b \qquad (3.9)$$

is easily shown to satisfy $\eth g_{ab} = 0$ and $\tilde{\eth} g_{ab} = 0$ and thus be independent of ζ and $\tilde{\zeta}$. If we now use (3.6) with (3.8) in (3.9) we will have produced a symmetric tensor field on $\mathcal{H}$ which is defined as the metric tensor on $\mathcal{H}$. It can be shown to be identical to the metric given by (2.3).

When these operations are performed and the results simplified (a rather lengthy but straightforward computation) the metric becomes

$$g_{ab} = 2\mathcal{D} Z_{(a} \eth Z_{b)} + 2\mathcal{E} Z_{(a} Z_{b)} + 2\mathcal{F} \eth Z_{(a} \eth Z_{b)} + 2Z_{(a} \eth\tilde{\eth} Z_{b)} - 2\eth Z_{(a} \tilde{\eth} Z_{b)} \qquad (3.10)$$

with

$$\mathcal{D} = -\eth \mathcal{F} \qquad (3.11)$$

$$\mathcal{E} = 1 - \mathcal{F}\dot{\sigma} + \frac{1}{2} \eth^2 \mathcal{F}$$

and

$$\mathcal{F} \equiv V^2 [\tilde{\eth} W + W^2] \ . \qquad (3.12)$$

It should be noted that though we are close to our goal of expressing the metric of $\mathcal{H}$-space in terms of Z and its derivatives we are not quite there yet. V which was a particular solution to (3.1) is expressible as a linear combination of the general solution Z_a, i.e.

$$V = Z_a v^a \qquad (3.13)$$

where v^a is a vector field on $\mathcal{H}$. W, from (3.4c), must depend on the choice of V or v^a and thus it would appear that $\mathcal{F}$ and hence g_{ab} were dependent on v^a. This however turns out not to be the case - one can show by direct calculation that

$$\frac{\partial \mathcal{F}}{\partial v^a} = 0 \tag{3.14}$$

and hence that g_{ab} does not depend on the particular V chosen. We now show how $\mathcal{F}$ can be obtained in an alternate manner that does not depend on using a V.

From (3.10) one can derive the scalar products of the four fields Z_a, $\eth Z_a$, $\tilde{\eth} Z_a$ and $\eth\tilde{\eth} Z_a$, namely

$$\begin{array}{lll} Z^a Z_a = 0 & \eth Z^a \eth Z_a = 0 & \tilde{\eth} Z^a \tilde{\eth} Z_a = -2\mathcal{F} \\ Z^a \eth Z_a = 0 & \eth Z^a \tilde{\eth} Z_a = -1 & \tilde{\eth} Z^a \eth\tilde{\eth} Z_a = -\eth\mathcal{F} \\ Z^a \tilde{\eth} Z_a = 0 & \eth Z^a \eth\tilde{\eth} Z_a = 0 & \eth\tilde{\eth} Z^a \eth\tilde{\eth} Z_a = -2\mathcal{E} \\ Z^a \eth\tilde{\eth} Z_a = 1 \ . & & \end{array} \tag{3.15}$$

As the fields change with changing ζ and $\tilde{\zeta}$ these relations remain unchanged. In fact only three are essential ($Z^a Z_a = 0$, $\eth Z^a \tilde{\eth} Z_a = -1$ and $\tilde{\eth} Z^a \tilde{\eth} Z_a = -2\mathcal{F}$) all others being obtainable from these by differentiation. Note further that

$$\tilde{\eth} Z^a \tilde{\eth} Z_a = \tilde{\eth}(Z^a \tilde{\eth} Z_a) - \tilde{\eth}^2 Z_a Z^a = -2\mathcal{F}$$

or

$$\tilde{\eth}^2 Z_a Z^a = 2\mathcal{F} \ . \tag{3.16}$$

This now permits us to obtain $\mathcal{F}$ directly from Z in the following fashion. $\tilde{\eth}^2 Z_a$ (known from differentiating $Z(z^a,\zeta,\tilde{\zeta})$) is a vector field and hence can be expanded in the basis Z_a, $\eth Z_a$, $\tilde{\eth} Z_a$, $\eth\tilde{\eth} Z_a$ (also known) in the form

$$\tilde{\eth}^2 Z_a = \Lambda_1 Z_a + \Lambda_2 \eth Z_a + \Lambda_3 \tilde{\eth} Z_a + \Lambda_4 \eth\tilde{\eth} Z_a \ . \tag{3.17}$$

Since each of the Λ_i can be determined algebraically from (3.17) and we see from (3.15), (3.16) and (3.17) that

$$\Lambda_4 = 2\mathcal{F} \ , \tag{3.18}$$

then $\mathcal{F}, \mathcal{E}, \mathcal{D}$ and the metric g_{ab} are determined from knowledge of $Z(z^a,\zeta,\tilde{\zeta})$ and its derivatives.

One can prove, either from (3.16) or (3.12) by repeated $\eth$ operations on $\mathcal{F}$, using (3.1) that

$$\eth^3 F - 4\eth F\dot{\sigma} - 2F\eth\dot{\sigma} = -2\tilde{\eth}\dot{\sigma} \, . \tag{3.19}$$

[For later use, in Sec. VII we give

$$\Lambda_1 = 3F - \dot{\sigma}F^2 + \frac{1}{2} F\eth^2 F - \frac{1}{4}(\eth F)^2 - \frac{1}{2}\eth\tilde{\eth}F \tag{3.20}$$

$$\Lambda_2 = \tilde{\eth}F \tag{3.21}$$

$$\Lambda_3 = -\eth F \;] \tag{3.22}$$

Also for later use we give another set of four linearly independent vector fields

$$L_a = Z_a \, , \qquad M_a = \eth Z_a$$

$$\tilde{M}_a = \frac{1}{2}\eth F Z_a - F\eth Z_a + \tilde{\eth} Z_a \tag{3.23}$$

$$N_a = (1 + \frac{1}{2}\eth^2 F - \dot{\sigma}F) Z_a - \frac{1}{2}\eth F \eth Z_a + \eth\tilde{\eth} Z_a$$

with the relations

$$L_a N^a = -M_a \tilde{M}^a = 1$$

$$L_a L^a = L_a M^a = L_a \tilde{M}^a = N_a M^a = N_a \tilde{M}^a = N_a N^a = M_a M^a = \tilde{M}_a \tilde{M}^a = 0 \tag{3.24}$$

so that

$$g_{ab} = L_a N_b + N_a L_b - M_a \tilde{M}_b - \tilde{M}_a M_b \, .$$

It should be emphasized that $\eth^2 L_a \neq 0$ and that this set of vectors is essentially different in their ζ and $\tilde{\zeta}$ behavior from the $\ell_a, m_a, \tilde{m}_a$ and n_a.

4. The Covariant Derivative

We have seen in the previous section that directly from $Z(z^a, \zeta, \tilde{\zeta})$ we can obtain a metric $g_{ab}(z^a)$ and express it solely in terms of "sidewards" derivatives ($\eth$ and $\tilde{\eth}$ operations) on the null vector field $Z_a(z^a, \zeta, \tilde{\zeta})$. ($Z_a$ for any fixed ζ and $\tilde{\zeta}$ is the null field but for fixed z^a and varying ζ and $\tilde{\zeta}$ it defines the null cone at z^a.)

We will now obtain an expression for the covariant derivative of Z_a. Since the sidewards derivatives yield the other independent vectors, we will have obtained the covariant derivatives of the four independent fields Z_a, $\eth Z_a$, $\tilde{\eth} Z_a$, $\tilde{\eth}\eth Z_a$ and thus the general expression.

Using the notation $Z_{ab} \equiv Z_{a;b}$ we obtain from (3.1) that

$$\eth^2 Z_{ab} = \dot{\sigma} Z_{ab} + \ddot{\sigma} Z_a Z_b \ . \tag{4.1}$$

The general solution to (4.1) is

$$Z_{ab} = Z^p_{ab} + \gamma^c_{ab} Z_c \tag{4.2}$$

where Z^p_{ab} is a particular solution and $\gamma^c_{ab} Z_c$ (with $\gamma^c_{ab}(z^d)$) is the general solution to the homogeneous equation. We will choose a specific particular solution and show that the γ^c_{ab} must be zero for the Z_{ab} to be $Z_{a;b}$.

Let

$$Z^p_{ab} = \alpha Z_a Z_b + 2\beta Z_{(a} \eth Z_{b)} + \gamma \eth Z_a \eth Z_b \tag{4.3}$$

and substitute into (4.1). After simplification we obtain three differential equations for the α, β and γ, the first two of which are

$$\eth^2 K = \dot{\sigma} K, \quad H = -\eth K \tag{4.4}$$

with

$$K = 3\beta + \eth\gamma, \quad H = 2\alpha + \eth\beta + \gamma\dot{\sigma}. \tag{4.5}$$

Noting that K is spin-weight $s = -1$ we can prove that $K = H = 0$. Writing $K = Vk$ where V is, as in Sec. III, an $s = 0$ solution to $\eth^2 V = \dot{\sigma} V$, (4.4) becomes

$$\eth(V^2 \eth k) = 0$$

or $\quad V^2 \eth k = c = \text{constant}$

or $\quad \eth k = c(1 + \eth W)$

or $\quad \eth(k - cW) = c.$

Since $k - cW$ is $s = -1$ then $c = 0$. Hence

$$\eth k = 0$$

which implies $k = 0$ and thus $K = H = 0$. We thus have from (4.5)

$$\beta = -\frac{1}{3}\,\eth\gamma \tag{4.6}$$

$$\alpha = \frac{1}{6}\,\eth^2\gamma - \frac{1}{2}\,\gamma\dot{\sigma} \tag{4.7}$$

The last equation, which relates γ to $\ddot{\sigma}$, is

$$\frac{1}{6}\,\eth^4\gamma - \frac{5}{3}\,\eth(\dot{\sigma}\eth\gamma) - \frac{1}{2}\,\gamma\eth^2\dot{\sigma} + \frac{3}{2}\,\gamma\dot{\sigma}^2 = \ddot{\sigma} \quad . \tag{4.8}$$

Eq. (4.3) with (4.6), (4.7) and (4.8) becomes our particular solution. Note that if our particular solution is the covariant derivative of Z_a then γ can be associated with F by

$$\gamma = F_{,a}Z^a .^* \tag{4.9}$$

This is seen by taking the covariant derivative of $2F = -\tilde{\eth}Z_a\tilde{\eth}Z^a$

$$F_{,a} = -\tilde{\eth}Z_{ab}\tilde{\eth}Z^b$$

$$F_{,a}Z^a = -\tilde{\eth}Z_{ab}\tilde{\eth}Z^bZ^a$$

$$= -\tilde{\eth}(Z_{ab}Z^a\tilde{\eth}Z^b) + Z_{ab}\tilde{\eth}Z^a\tilde{\eth}Z^b + Z_{ab}\tilde{\eth}^2Z^bZ^a$$

and since, from (4.3), $Z_{ab}Z^a = 0$ then

$$F_{,a}Z^a = Z_{ab}\tilde{\eth}Z^a\tilde{\eth}Z^b = \gamma .$$

If we now take the expression for the metric (3.10) and demand that

$$g_{ab;c} = 0$$

and substitute for Z_{ab} and its sideward derivatives, (4.2) and (4.3) we obtain an algebraic statement that $\gamma^c_{ab} = 0$ and that hence

$$Z_{a;b} = Z_{ab} = \alpha\, Z_aZ_b + 2\beta Z_{(a}\eth Z_{b)} + \gamma\eth Z_a\eth Z_b \tag{4.10}$$

*Once it is recognized that γ is related to F one can show that (4.8) is not an independent equation and that it can be derived by differentiating (3.19).

with

$$\gamma = F,_a z^a$$

$$\beta = -\frac{1}{3}\eth\gamma$$

$$\alpha = \frac{1}{6}\eth^2\gamma - \frac{1}{2}\gamma\dot{\sigma}\,.$$

(Note that the Penrose bracket expression

$$\{Z_a, Z_b\} \equiv Z_a \eth Z_b - Z_b \eth Z_a = 2L_{[a}M_{b]} = 2\ell_{[a}m_{b]} \tag{4.11}$$

can be shown, after a brief calculation using (4.10), to be covariantly constant, i.e.

$$\{Z_a, Z_b\}_{;c} \equiv 0 \tag{4.12}$$

and that hence self dual bivectors or equivalently unprimed spinors can be parallely transported over $\mathcal{H}$.)

If we consider the four vectors $L_a, M_a, \tilde{M}_a$, and N_a of (3.23) (for each value of ζ and $\tilde{\zeta}$) as a null tetrad system for use in the spin coefficient formalism[9*] one can easily, from (4.10) and its sidewards derivates, calculate all the spin coefficients with the result that all the untilded spin coefficients vanish and the only nonvanishing tilded spin coefficients are

$$\begin{aligned} \underline{\tilde{\sigma}} &= \gamma \\ \underline{\tilde{\tau}} &= \underline{\tilde{\beta}} = -\beta \\ \underline{\tilde{\mu}} &= \underline{\tilde{\gamma}} = \alpha \\ \underline{\tilde{\nu}} &= \eth\alpha + 2\dot{\sigma}\beta \end{aligned} \tag{4.13}$$

where the spin coefficients are underlined to avoid confusion with our α, β, γ.

*Though the spin coefficient formalism was originally developed for use in 4 dimensional real space-time, it goes over almost unchanged to four dimensional complex space-time (e.g. $\mathcal{H}$-space), the only essential change is that barred quantities which were formerly the complex conjugate of the unbarred quantities are now independent quantities. To denote this independence the bar is replaced by a tilde (e.g. $\bar{m}_a \to \tilde{m}_a$).

If the vanishing of the unprimed spin coefficients is used in the field equations[9] of the spin coefficient formalism one immediately obtains that

$$R_{ab} = 0 \tag{4.14}$$

and

$$R_{abcd} - i\overset{*}{R}_{abcd} = 0 \; . \tag{4.15}$$

In the following section we will show how the curvature properties including (4.14) and (4.15) can be calculated directly from (4.10).

5. The Curvature Tensor

In order to calculate the curvature tensor directly from $Z(z^a,\zeta,\tilde{\zeta})$ one begins by taking the covariant derivative of (4.10)

$$\begin{aligned} Z_{abc} \equiv Z_{a;bc} &= \alpha(Z_{ac}Z_b + Z_aZ_{bc}) + 2\beta(Z_{(a|c|}\eth Z_{b)} + Z_{(a}\eth Z_{b)c}) \\ &+ \gamma(\eth Z_{ac}\eth Z_b + \eth Z_a\eth Z_{bc}) + \alpha_{,c}Z_aZ_b + 2\beta_{,c}Z_{(a}\eth Z_{b)} \\ &+ \gamma_{,c}\eth Z_a\eth Z_b \; . \end{aligned} \tag{5.1}$$

By (4.10) and $\eth$ applied to (4.10), (5.1) is expressible in terms of $Z_{,a}$, its derivatives and F and its derivatives. Now since

$$\begin{aligned} Z_{a[bc]} &= \frac{1}{2} R^d_{\ abc} Z_d \\ \eth Z_{a[bc]} &= \frac{1}{2} R^d_{\ abc} \eth Z_d \\ \eth\tilde{\eth} Z_{a[bc]} &= \frac{1}{2} R^d_{\ abc} \eth\tilde{\eth} Z_d \\ \tilde{\eth} Z_{a[bc]} &= \frac{1}{2} R^d_{\ abc} \tilde{\eth} Z_d \end{aligned} \tag{5.2}$$

we have, using (3.10)

$$\begin{aligned} \frac{1}{2} R_{eabc} &= (2EZ_e + \mathcal{D}\eth Z_e + \eth\tilde{\eth} Z_e) Z_{a[bc]} \\ &+ (\mathcal{D}Z_e + 2F\eth Z_e - \tilde{\eth} Z_e)\eth Z_{a[bc]} \\ &+ Z_e\eth\tilde{\eth} Z_{a[bc]} - \eth Z_e\tilde{\eth} Z_{a[bc]} \; . \end{aligned} \tag{5.3}$$

To express (5.3) in a more convenient form, one starts with the following identities on F, which are derived from differentiation on (3.16),

$$F_{,d}M^d = \frac{1}{3}\,\eth(F_{,d}L^d)$$

$$F_{,d}N^d = \frac{1}{3}\,\eth(F_{,d}\tilde{M}^d) + \frac{\gamma}{3} \tag{5.4}$$

$$F_{,d}\tilde{M}^d = -\tilde{\eth}(F_{,d}L^d) - 3\beta F - \frac{3}{2}\,\gamma\eth F$$

$$F_{,d}N^d = -\tilde{\eth}(F_{,d}M^d) + 2\alpha F + \frac{\beta}{2}\,\eth F - \gamma E$$

where $L^d, M^d, \tilde{M}^d, N^d$ are the tetrad vectors in (3.23). By taking appropriate directional directives of (5.4), one can show that the Riemann tensor can be written as follows,

$$\begin{aligned}
R_{eabc} = &-4\gamma_{,d}L^d\{M_{[e}N_{a]}M_{[b}N_{c]}\} \\
&-4\gamma_{,d}M^d\{M_{[e}N_{a]}(\tilde{M}_{[b}M_{c]} + N_{[b}L_{c]} + (\tilde{M}_{[e}M_{a]} + N_{[e}L_{a]})M_{[b}N_{c]}\} \\
&-4\alpha_{,d}L^d\{M_{[e}N_{a]}L_{[b}\tilde{M}_{c]} + L_{[e}\tilde{M}_{a]}M_{[b}N_{c]} + \\
&\qquad + (\tilde{M}_{[e}M_{a]} + N_{[e}L_{a]})(\tilde{M}_{[b}M_{c]} + N_{[b}L_{c]})\} \\
&-4(\eth\alpha + 2\dot{\sigma}\beta)_{,d}L^d\{L_{[e}\tilde{M}_{a]}(\tilde{M}_{[b}M_{c]} + N_{[b}L_{c]} + \\
&\qquad + (\tilde{M}_{[e}M_{a]} + N_{[e}L_{a]})L_{[b}\tilde{M}_{c]}\} \\
&-4(\eth\alpha + 2\dot{\sigma}\beta)_{,d}M^d\{L_{[e}\tilde{M}_{a]}L_{[b}\tilde{M}_{c]}\} \quad .
\end{aligned} \tag{5.5}$$

It follows from (5.5) since each term is anti-self dual, that R_{eabc} is left-flat (i.e. 4.15 holds) and from the tetrad scalar products that the Ricci tensor vanishes. The tetrad components of the Weyl tensor[13], which are

$$\tilde{\psi}_0 = \gamma_{,d}L^d$$

$$\tilde{\psi}_1 = \gamma_{,d}M^d = -\beta_{,d}L^d$$

$$\tilde{\psi}_2 = \alpha_{,d}L^d = -\beta_{,d}M^d \tag{5.6}$$

$$\tilde{\psi}_3 = (\eth\alpha + 2\dot{\sigma}\beta)_{,d}L^d = \alpha_{,d}M^d$$

$$\tilde{\psi}_4 = (\eth\alpha + 2\dot{\sigma}\beta)_{,d}M^d,$$

satisfy the following identities,

$$\eth\tilde{\psi}_0 = 4\tilde{\psi}_1$$

$$\eth\tilde{\psi}_1 = 3\tilde{\psi}_2 + \dot{\sigma}\tilde{\psi}_0$$

$$\eth\tilde{\psi}_2 = 2\tilde{\psi}_3 + 2\dot{\sigma}\tilde{\psi}_1 \tag{5.7}$$

$$\eth\tilde{\psi}_3 = \tilde{\psi}_4 + 3\dot{\sigma}\tilde{\psi}_2$$

$$\eth\tilde{\psi}_4 = 4\dot{\sigma}\tilde{\psi}_3 .$$

Consequently a knowledge of $\tilde{\psi}_0 = \gamma_{,d}L^d$ determines the other components by sidewards differentiation.

6. The Non-Linear Graviton

In this section, our purpose is to discuss the relationship of our approach to ℋ-space with that of Penrose via the Non-Linear Graviton[3,5]. Specifically, we demonstrate how the deformed twistor space representing a non-linear graviton can be constructed from a complete solution of the good cut equation, and that the metric defined by Penrose's technique will agree with the metric constructed in Section III. The deformed twistor space of a non-linear graviton is a deformation of part of CP^3*. It has the structure of a fibre bundle with fibres C^2 and projection π over the Riemann sphere, CP^1. To give the structure of the bundle, we give a coordinate $\tilde{\zeta}$ on the sphere (allowing $\tilde{\zeta} = \infty$), a covering $\{U_i\}_{i\varepsilon I}$ of the sphere and fibre coordinates (p_i, q_i) on $\pi^{-1}(U_i)$. Then the bundle is determined by the transition functions (p_{ij}, q_{ij}) defined on $\pi^{-1}(U_i \cap U_j)$ by

*This account of the non-linear graviton will be somewhat cursory and we refer to the original paper for the details[3,5].

$$p_i = p_{ij}(p_j, q_j)$$

$$q_i = q_{ij}(p_j, q_j) \quad \text{not summed.}$$

There is extra structure in the shape of a four parameter family of holomorphic curves which are the analogues of straight lines in the undeformed CP3 and which are interpreted as the points of $\mathcal{H}$. The metric of $\mathcal{H}$ is then defined in terms of these "lines" in a way to be seen later. We may represent the holomorphic curves in $\pi^{-1}(U_i)$ by

$$p_i = p_i(z^a, \tilde{\zeta})$$

$$q_i = q_i(z^a, \tilde{\zeta}) \quad . \tag{6.1}$$

where z^a are the four complex parameters, which correspond to coordinates in $\mathcal{H}$.

These structures from our point of view are obtained from a complete solution of the good cut equation via the construction of asymptotic twistor space, $P\mathcal{T}$ [9,10]. We recall[11,12] the representation of $C\mathcal{I}^+$ as a cube where the diagonal represents the real section or $\mathcal{I}^+$. (See Fig. 1).

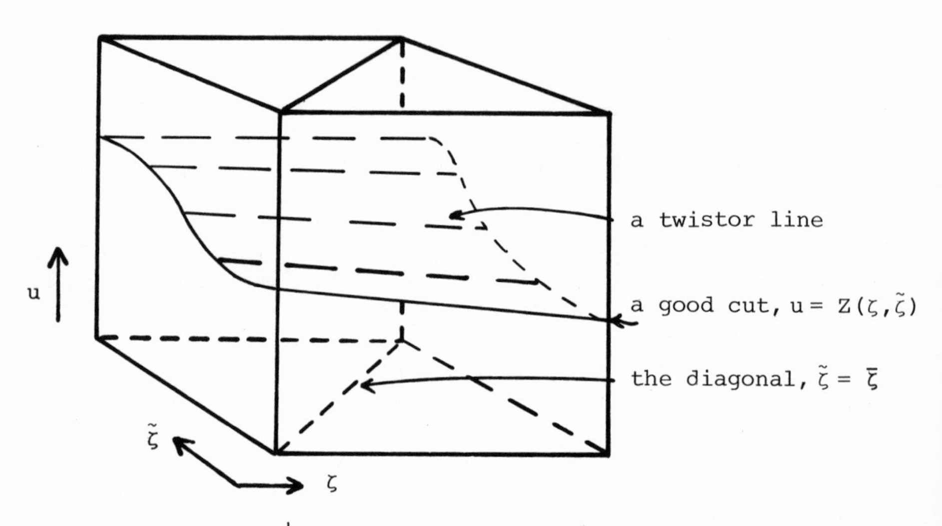

Figure 1: $C\mathcal{I}^+$ showing a good cut ruled by twistors.

In this picture a good cut is a surface given by

$$u = Z(\zeta,\tilde{\zeta})$$

satisfying

$$\eth^2 Z = \sigma(Z,\zeta,\tilde{\zeta}) . \tag{6.2}$$

For fixed $\tilde{\zeta}$, the solutions to equation (6.2) are lines, so that the good cut is ruled by these preferred lines which are projective twistors, the elements of $P\mathcal{T}$.

Taking $\tilde{\zeta}$ as a coordinate on the Riemann sphere, the bundle structure of $P\mathcal{T}$ is clear, in that two coordinates are required to specify a twistor lying in a particular surface of constant $\tilde{\zeta}$.

If a particular line is given by

$$\tilde{\zeta} = \tilde{\zeta}_1$$

$$u = Z(\zeta),$$

so that $\eth^2 Z(\zeta) = \sigma(Z(\zeta),\zeta,\tilde{\zeta}_1)$, then we define the quantities

$$p(\zeta) = Z - \zeta\eth Z$$

$$q(\zeta) = \tilde{\zeta}_1 Z + \eth Z . \tag{6.3}$$

These will satisfy

$$Z = \frac{p + \zeta q}{1 + \zeta\tilde{\zeta}_1} \tag{6.4}$$

and

$$p_{,\zeta} + \zeta q_{,\zeta} = 0. \tag{6.5}$$

Now for our fibre coordinates we may fix ζ equal to some value ζ_1 and take

$$p_1 = p(\zeta_1)$$

$$q_1 = q(\zeta_1)$$

so that an element of $P\mathcal{T}$ appears as a triple (p_1,q_1,ζ_1). If instead we had fixed ζ equal to ζ_2 we would obtain coordinates

$$p_2 = p(\zeta_2)$$

$$q_2 = q(\zeta_2)$$

and, at least, in principle, we could determine the transition functions (p_{12}, q_{12}) from (6.3).

We will come in a moment to the question of the covering $\{U_i\}$ of the sphere; in the meantime we recall that there is to be a four parameter family of holomorphic curves in $P\mathcal{T}$ representing the points of $\mathcal{H}$. These will come from the good cut function $Z(z^a;\zeta,\tilde{\zeta})$ via the definitions

$$p(z^a;\zeta,\tilde{\zeta}) = Z(z^a;\zeta,\tilde{\zeta}) - \zeta\eth Z(z^a;\zeta,\tilde{\zeta})$$

$$q(z^a;\zeta,\tilde{\zeta}) = \tilde{\zeta}Z(z^a;\zeta,\tilde{\zeta}) + \eth Z(z^a;\zeta,\tilde{\zeta}) \quad . \tag{6.6a}$$

For if we fix ζ equal to ζ_1, we obtain the curves as in (6.1) by

$$p_1 = p(z^a;\zeta_1,\tilde{\zeta}) = p_1(z^a;\tilde{\zeta})$$

$$q_1 = q(z^a;\zeta_1,\tilde{\zeta}) = q_1(z^a;\zeta). \tag{6.6b}$$

Now for ζ fixed at ζ_1, as $\tilde{\zeta}$ varies the quantities defined in (6.6) will eventually encounter singularities when we get too far from the real slice in $C\mathcal{I}^+$. Consequently for a particular ζ_1, (6.6) is only non-singular for a neighborhood, say U_1, of $\bar{\zeta}_1$ in the $\tilde{\zeta}$ sphere, and it is this property which we use to determine the covering $\{U_i\}$ of the sphere. This situation is shown diagramatically in Fig. 2, where the singular region of $C\mathcal{I}^+$ is shown shaded.

While we cannot in general solve for the transition functions, we can find an infinitesimal transition function when ζ_1 and ζ_2 differ only infinitesimally.

We suppose

$$\zeta_2 = \zeta_1 + \varepsilon$$

then

$$p_2 = p(\zeta_2)$$

$$= p(\zeta_1) + \varepsilon \left.\frac{\partial p}{\partial \zeta}\right|_{\zeta_1} + O(\varepsilon^2)$$

or

$$p_2 = p_1 + \varepsilon \left.\frac{\partial p}{\partial \zeta}\right|_{\zeta_1} + O(\varepsilon^2)$$

and

$$q_2 = q_1 + \varepsilon \left.\frac{\partial q}{\partial \zeta}\right|_{\zeta_1} + O(\varepsilon^2) \; .$$

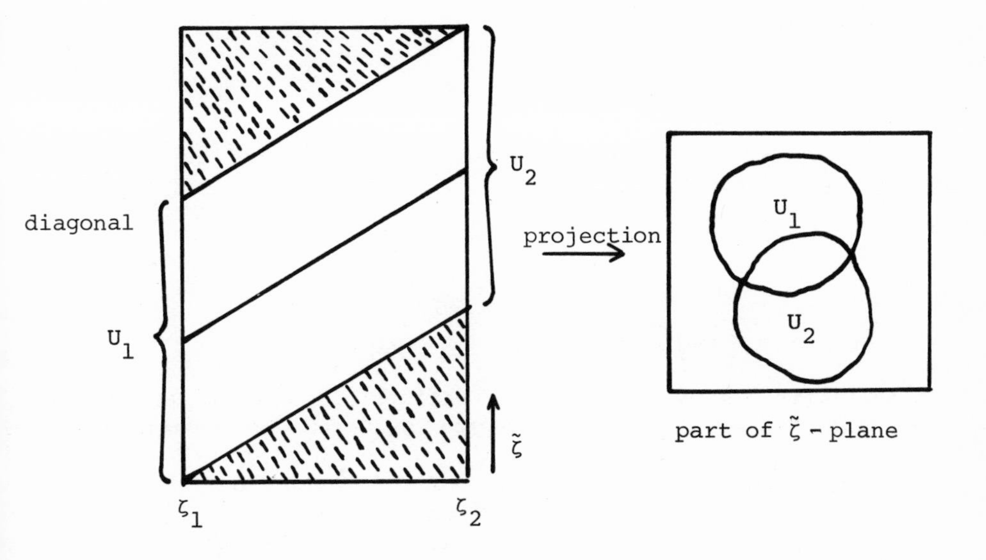

Figure 2: $C\mathcal{I}^+$ from above, with singular regions shaded, showing typical patches U_1 and U_2.

From (6.3) and (6.5) we find

$$\frac{\partial p}{\partial \zeta} = \frac{-\zeta}{1+\zeta\tilde{\zeta}}\,\sigma(Z,\zeta,\tilde{\zeta})$$

$$\frac{\partial q}{\partial \zeta} = \frac{1}{1+\zeta\tilde{\zeta}}\,\sigma(Z,\zeta,\tilde{\zeta})$$

whence, replacing Z in σ by (6.4),

$$p_2 = p_1 - \frac{\varepsilon\zeta_1}{1+\zeta_1\tilde{\zeta}_1}\,\sigma\left(\frac{p_1+\zeta_1 q_1}{1+\zeta_1\tilde{\zeta}_1},\zeta_1,\tilde{\zeta}_1\right) + 0\,(\varepsilon^2)$$

$$q_2 = q_1 + \frac{\varepsilon}{1+\zeta_1\tilde{\zeta}_1}\,\sigma\left(\frac{p_1+\zeta_1 q_1}{1+\zeta_1\tilde{\zeta}_1},\zeta_1,\tilde{\zeta}_1\right) + 0\,(\varepsilon^2) \qquad (6.7)$$

so that asymptotic shear, σ, appears in the infinitesimal transition functions.

We may cast these equations into a simpler form if we introduce a function $h(u,\zeta,\tilde{\zeta})$ such that

$$\frac{d}{du}\,h(u,\zeta,\tilde{\zeta}) = \sigma(u,\zeta,\tilde{\zeta})$$

and write by means of (6.4)

$$h(Z;\zeta,\tilde{\zeta}) = H(p,q;\zeta,\tilde{\zeta})$$

for then (6.7) becomes

$$p_2 = p_1 - \varepsilon\,\frac{\partial H}{\partial q_1} + 0(\varepsilon^2)$$

$$q_2 = q_1 + \varepsilon\,\frac{\partial H}{\partial p_1} + 0(\varepsilon^2)$$

and the infinitesimal transition functions appear as a canonical transformation with the time-integral of the shear as Hamiltonian.

Turning now to a consideration of the metrics defined in the good-cut formalism and for the non-linear graviton, we first remark that the conformal metrics agree. This is because two nearby heavenly points are null separated in the good cut picture[2] if the corresponding infinitesimally separated good cuts intersect along a twistor line and this is precisely the condition that the corresponding holomorphic curves in the deformed twistor-space[3] meet.

For the full metric, we must recall Penrose's definition of the $\mathcal{H}$-space metric. If P, Q and R are three infinitesimally separated points of $\mathcal{H}$ with P and Q null separated and P and R null separated, Penrose gives a construction for the scalar produce of the vectors PQ and PR in terms of the corresponding holomorphic curves. By linearity, this is sufficient to define the metric. If the points P,Q,R, have coordinates z^a, $z^a + dz^a$ and $z^a + \delta z^a$ respectively then the corresponding holomorphic curves may be written:

$$
\begin{aligned}
\text{P:}\quad & p = p(z^a;\tilde{\zeta}) \\
& q = q(z^a;\tilde{\zeta}) \\
\text{Q:}\quad & p = p(z^a;\tilde{\zeta}) + dz^a p_a(z^b;\tilde{\zeta}) + 0(dz^2) \\
& q = q(z^a;\tilde{\zeta}) + dz^a q_a(z^b;\tilde{\zeta}) + 0(dz^2) \\
\text{R:}\quad & p = p(z^a;\tilde{\zeta}) + \delta z^a p_a(z^b;\tilde{\zeta}) + 0(\delta z^2) \\
& q = q(z^a;\tilde{\zeta}) + \delta z^a q_a(z^b;\tilde{\zeta}) + 0(\delta z^2).
\end{aligned}
$$

The fact that PQ is a null vector means, as remarked above, that the line corresponding to Q meets the line corresponding to P at some value, $\tilde{\zeta}_1$ say, of $\tilde{\zeta}$ i.e.

$$
\begin{aligned}
& dz^a p_a(z^b;\tilde{\zeta}_1) = 0 \\
& dz^a q_a(z^b;\tilde{\zeta}_1) = 0
\end{aligned}
\tag{6.8}
$$

Similarly, because PR is null, there will be a $\tilde{\zeta}_2$ at which

$$
\begin{aligned}
& \delta z^a p_a(z^b;\tilde{\zeta}_2) = 0 \\
& \delta z^a q_a(z^b;\tilde{\zeta}_2) = 0\ .
\end{aligned}
$$

This situation is shown in Figure 3.

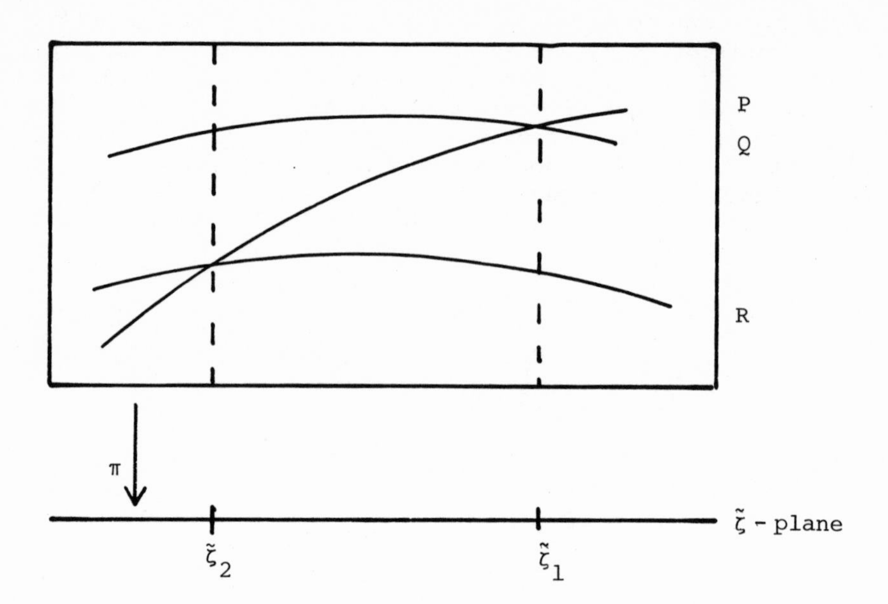

Figure 3: Asymptotic twistor space, showing the holomorphic curves corresponding to the points P,Q and R of ℋ-space.

The curves for P and Q intersect so that the points P and Q are null separated in ℋ-space, and likewise P and R.

Penrose's expression for the metric is then

$$g_{ab}dz^a\delta z^b = \frac{(\tilde{\zeta}_2 - \tilde{\zeta}_1)}{(\tilde{\zeta} - \tilde{\zeta}_1)(\tilde{\zeta} - \tilde{\zeta}_2)} (p_a(z^c,\tilde{\zeta})q_b(z^c,\tilde{\zeta})$$

$$- p_b(z^c,\tilde{\zeta})q_a(z^c,\tilde{\zeta}))dz^a\delta z^b \quad .$$

Substituting for p_a and q_a from (6.6) gives

$$g_{ab}dz^a\delta z^b = \frac{(\tilde{\zeta}_2 - \tilde{\zeta}_1)(1+\zeta\tilde{\zeta})}{(\tilde{\zeta} - \tilde{\zeta}_1)(\tilde{\zeta} - \tilde{\zeta}_2)}(Z_a \eth Z_b - Z_b \eth Z_a)dz^a\delta z^b \tag{6.9}$$

We observe that this is independent of ζ, or in our terminology, is well defined as a function of $\tilde{\zeta}$, because of the good cut equation (3.1) written in the form

$$\eth^2 Z_a = ((1+\zeta\tilde{\zeta})\eth Z_a)_{,\zeta} = \dot{\sigma} Z_a \quad .$$

Then, following Penrose, we remark that as a function of $\tilde{\zeta}$, the right hand side of (6.9) has no poles and so must be a constant, i.e. must be independent of $\tilde{\zeta}$. Consequently we may evaluate it at any value of $\tilde{\zeta}$ and we choose $\tilde{\zeta}_1$. Recalling (6.8), we write

$$\begin{aligned} dz^a Z_a &= (\tilde{\zeta} - \tilde{\zeta}_1)(dz^a Z_a)_{,\tilde{\zeta}} + O((\tilde{\zeta} - \tilde{\zeta}_1)^2) \\ &= \frac{(\tilde{\zeta} - \tilde{\zeta}_1)}{1 + \zeta\tilde{\zeta}}\tilde{\eth}(dz^a Z_a) + O((\tilde{\zeta} - \tilde{\zeta}_1)^2) \end{aligned} \tag{6.10}$$

and

$$dz^a \eth Z_a = \frac{(\tilde{\zeta} - \tilde{\zeta}_1)}{1+\zeta\tilde{\zeta}}\tilde{\eth}(dz^a \eth Z_a) + O((\tilde{\zeta} - \tilde{\zeta}_1)^2)$$

Substituting this in (6.9) and setting $\tilde{\zeta} = \tilde{\zeta}_1$ gives

$$g_{ab}dz^a\delta z^b = -(\tilde{\eth} Z_a \eth Z_b - \tilde{\eth}\eth Z_a Z_b)\Big|_{\tilde{\zeta}_1} dz^a\delta z^b \ .$$

Now we recall from (3.10) the metric obtained from the good cut formalism and remembering that this also is independent of $\tilde{\zeta}$ we evaluate it at $\tilde{\zeta}_1$ to find

$$g_{ab}dz^a\delta z^b = (\eth\tilde{\eth} Z_a Z_b - \tilde{\eth} Z_a \eth Z_b)\Big|_{\tilde{\zeta}_1} dz^a\delta z^b$$

and the two metrics agree.

7. The Relation to the Approach of Plebanski

In a recent paper J. Plebanski[7], beginning with (4.14) and (4.15), showed that there are two different types of coordinate systems which can be naturally introduced in an ℋ-space and that to each type of coordinate system there existed a scalar quantity satisfying a wave-like equation, knowledge of which determines the ℋ-space metric.

It is the purpose of this section to show the relation of Plebanski's work with ours.

Again assuming that we know $Z(z^a,\zeta,\tilde{\zeta})$ and hence $\eth Z$, $\tilde{\eth}Z$ and $\eth\tilde{\eth}Z$, we can consider this, for each value of ζ and $\tilde{\zeta}$, as knowing four scalar functions on $\mathcal{H}$ which can be chosen as natural coordinates. In particular, at fixed but arbitrary $\zeta = \zeta_0$ and $\tilde{\zeta} = \tilde{\zeta}_0$, we introduce

$$p = Z(z^a,\zeta_0,\tilde{\zeta}_0)$$

$$q = -\eth Z$$

$$x = \eth\tilde{\eth}Z + Z$$

$$y = \tilde{\eth}Z \tag{7.1}$$

as preferred coordinates. (Note that if a second direction is chosen i.e. $\zeta = \zeta_1$ and $\tilde{\zeta} = \tilde{\zeta}_1$ there is a second set of (p,q,x,y) and that knowledge of $Z(p,q,x,y,\zeta,\tilde{\zeta})$ is equivalent to knowing the coordinate transformation to the new set.) From (7.1) we have

$$\delta^p_a = Z_a(z^a,\zeta_0,\tilde{\zeta}_0)\,, \quad \text{or} \quad dp = Z,_a dz^a$$

$$\delta^q_a = -\eth Z_a \qquad dq = \eth Z,_a dz^a$$

$$\delta^x_a = \eth\tilde{\eth}Z_a \qquad dx = \eth\, Z,_a dz^a$$

$$\delta^y_a = \tilde{\eth}Z_a \qquad dy = \tilde{\eth}Z,_a dz^a \tag{7.2}$$

which, from (3.10) leads to

$$g_{ab}dz^a dz^b = 2(E-1)dp^2 - 2\mathcal{D}dpdq + 2\mathcal{F}dq^2 + 2dpdx + 2dqdy \tag{7.3}$$

or

$$g_{ab} = \left(\begin{array}{cccc} \partial^2 F - 2F\dot{\sigma} & \eth F & 1 & 0 \\ \eth F & 2F & 0 & 1 \\ \hline 1 & 0 & 0 & 0 \\ 0 & 1 & 0 & 0 \end{array}\right) \tag{7.4}$$

(with $\mathcal{D},E,\mathcal{F}$ evaluated at $\zeta = \zeta_0$, $\tilde{\zeta} = \tilde{\zeta}_0$) for the metric in this coordinate system. This, we claim, is the second Plebanski form[7]. In order to prove this we must show that there exists a function $\theta(p,q,x,y)$ such that

$$\frac{1}{2}\,\eth F = \theta_{xy} \tag{7.5}$$

$$-F = \theta_{xx} \tag{7.6}$$

$$F\dot{\sigma} - \frac{1}{2}\,\eth^2 F = \theta_{yy} \tag{7.7}$$

and

$$\theta_{xx}\theta_{yy} - \theta_{xy}^2 + \theta_{xp} + \theta_{yq} = 0\ . \tag{7.8}$$

From (3.17) (with (7.2)), (3.18), (3.20), (3.21), (3.22) we have at $\zeta = \zeta_0$, $\tilde{\zeta} = \tilde{\zeta}_0$

$$\tilde{\eth}^2 Z_{,x} = 2F \tag{7.9}$$

$$\tilde{\eth}^2 Z_{,y} = -\eth F \tag{7.10}$$

$$\tilde{\eth}^2 Z_{,q} = -\tilde{\eth} F \tag{7.11}$$

$$\tilde{\eth}^2 Z_{,p} = F - \dot{\sigma}F^2 + \frac{1}{2}\,F\eth^2 F - \frac{1}{4}\,(\eth F)^2 - \frac{1}{2}\,\eth\tilde{\eth}F \tag{7.12}$$

In addition by applying $\eth$ to (3.17) we obtain

$$\eth\tilde{\eth}^2 Z_a = (\eth\Lambda_1 + \Lambda_2\dot{\sigma} + \Lambda_4\tilde{\eth}\dot{\sigma})Z_a + (\Lambda_1 + \eth\Lambda_2 - 2\Lambda_4)\eth Z_a$$
$$+(-\eth^2 F + 2F\dot{\sigma})\tilde{\eth}Z_a + \eth F\eth\tilde{\eth}Z_a \tag{7.13}$$

from which it follows that

$$\eth\tilde{\eth}^2 Z_{,x} = \eth F \tag{7.14}$$

$$\eth\tilde{\eth}^2 Z_{,y} = 2F\dot{\sigma} - \eth^2 F \tag{7.15}$$

$$\eth\tilde{\eth}^2 Z_{,q} = -(\Lambda_1 + \partial\Lambda_2 - 2\Lambda_4) \tag{7.16}$$

$$\eth\tilde{\eth}^2 Z_{,p} = (\eth\Lambda_1 + \Lambda_2\dot{\sigma} + \Lambda_4\tilde{\eth}\dot{\sigma} - \partial F) \tag{7.17}$$

By comparing (7.14) and (7.10) it is seen that we can introduce a $\theta(p,q,x,y)$ such that

$$\theta_x = -\frac{1}{2}\tilde{\eth}^2 Z$$

$$\theta_y = \frac{1}{2}\eth\tilde{\eth}^2 Z . \tag{7.18}$$

From (7.18), with (7.9) - (7.12) and (7.14) - (7.17), it is straightforward to show that (7.5) - (7.8) are identically satisfied. This completes the proof that (7.3) is of the second Plebanski type.

From (7.5) - (7.7), using the identities

$$\frac{\partial}{\partial x} = L^a \frac{\partial}{\partial z^a}$$

$$\frac{\partial}{\partial y} = -M^a \frac{\partial}{\partial z^a} \tag{7.19}$$

and those derived from (5.4), one can show that Plebanski's expression for the tetrad components of the Weyl tensor agree with (5.6) if the following identification is made on Plebanski's one-forms,

$$e^1 = dp \leftrightarrow L_a dz^a$$

$$e^4 = -dq \leftrightarrow M_a dz^a$$

$$e^2 \leftrightarrow N_a dz^a$$

$$e^3 \leftrightarrow -\tilde{M}_a dz^a .$$

Although it is Plebanski's second form of the metric which arises naturally from our approach, we may also obtain his first form by a different choice of coordinates and tetrad. Two of the coordinates will be p and q as in (7.1) at fixed values $\zeta_0, \tilde{\zeta}_0$ of ζ and $\tilde{\zeta}$, while for the other two we take

$$r = Z(z^a, \zeta_0, \tilde{\zeta}_1)$$

$$s = -\eth Z(z^a, \zeta_0, \tilde{\zeta}_1)$$

that is, Z and $\eth Z$ evaluated at the same value of ζ but a different fixed $\tilde{\zeta}$. Initially we impose no restriction on $\tilde{\zeta}_1$ (except that it be different from $\tilde{\zeta}_0$) but we shall later find that a specific choice, relative to $\tilde{\zeta}_0$, simplifies our expressions.

A useful notation for the coordinates is to omit the $(\zeta,\tilde{\zeta})$ arguments so that Z and $\eth Z$ are thought of as evaluated at $(\zeta_0,\tilde{\zeta}_0)$ and introduce $\hat{Z}$ and $\eth\hat{Z}$ for the same thing at $(\zeta_0,\tilde{\zeta}_1)$. Thus

$$p = Z \qquad r = \hat{Z}$$

$$q = -\eth Z \qquad s = -\eth\hat{Z} .$$

For the tetrad we take

$$L_a = Z_a = \delta^p_a \qquad \text{or} \qquad dp = L_a dZ^a$$

$$M_a = \eth Z_a = -\delta^q_a \qquad dq = \eth Z_a dZ^a \tag{7.20a}$$

as in (7.2) and (3.23) and for the other two, take linear combinations

$$\tilde{M}_a = A\hat{Z}_a - B\eth\hat{Z}_a = A\delta^r_a + B\delta^s_a , \qquad Adr + Bds = \tilde{M}_a dZ^a$$

$$N_a = C\hat{Z}_a - D\eth\hat{Z}_a = C\delta^r_a + D\delta^s_a , \qquad Cdr + Dds = N_a dZ^a \tag{7.20b}$$

where the coefficients A,B,C and D are fixed by the usual normalization requirements:

$$L^a\tilde{M}_a = M^a N_a = 0$$

$$L^a N_a = -M^a\tilde{M}_a = 1 .$$

(The fact that $\tilde{M}^a$ and N^a are both null follows from (3.15). At this stage N_a and $\tilde{M}_a$ are not as in (3.23).)

We find

$$A = \frac{Z^a\eth\hat{Z}_a}{\Delta} \qquad B = \frac{Z^a\hat{Z}_a}{\Delta}$$

$$C = \frac{\eth Z^a\eth\hat{Z}_a}{\Delta} \qquad D = \frac{\eth Z^a\hat{Z}_a}{\Delta} \tag{7.21}$$

where $\Delta = -(AD - BC)^{-1}$

$$= (Z^a\hat{Z}_a)(\eth Z^b\eth\hat{Z}_b) - (Z^a\eth\hat{Z}_a)(\eth Z^b\hat{Z}_b)$$

$$= 4Z^{[a}\eth Z^{b]}\,\hat{Z}_{[a}\eth\hat{Z}_{b]} \tag{7.22}$$

$$= \{Z^a,Z^b\}\,\{\hat{Z}_a,\hat{Z}_b\}$$

so that Δ is independent of position in $\mathcal{H}$ by (4.12).

For the metric we now have

$$g_{ab}dz^a dz^b = 2(L_a N_b - M_a \tilde{M}_b) dz^a dz^b$$

$$= 2(Cdpdr + Ddpds + Adqdr + Bdqds) \qquad (7.23)$$

which has the general appearance of Plebanski's first form, if we can exhibit A,B,C and D as partial derivatives of some other function.

To this end, we expand the vector $\tilde{\eth} Z_a$ in our basis

$$\tilde{\eth} Z_a = \pi_1 Z_a - \pi_2 \eth Z_a + \pi_3 \hat{Z}_a - \pi_4 \eth \hat{Z}_a \quad . \qquad (7.24)$$

Contracting this with Z^a and using (3.15) and (7.21) gives

$$\pi_3 B - \pi_4 A = 0 \qquad (7.25)$$

and contracting with $\eth Z^a$ similarly gives

$$-1 = \pi_3 D\Delta - \pi_4 C\Delta \ . \qquad (7.26)$$

Solving (7.25) and (7.26) for π_3 and π_4 we find

$$B = \pi_4$$

$$A = \pi_3$$

so that

$$A = (\tilde{\eth} Z)_{,r}$$

$$B = (\tilde{\eth} Z)_{,s} \quad .$$

In a similar way we expand $\eth\tilde{\eth} Z_a$. $\tilde{\eth}\hat{Z}_a$ and $\eth\tilde{\eth}\hat{Z}_a$ in our basis to obtain

$$A = (\tilde{\eth} Z)_{,r} = (\eth\tilde{\eth}\hat{Z})_{,q}$$

$$B = (\tilde{\eth} Z)_{,s} = (\tilde{\eth}\hat{Z})_{,q}$$

$$C = (\eth\tilde{\eth}Z)_{,r} = (\eth\tilde{\eth}\hat{Z})_{,p}$$

$$D = (\tilde{\eth}\tilde{\eth}Z)_{,s} = (\tilde{\eth}\hat{Z})_{,p} \tag{7.27}$$

and these are just the integrability conditions for the existence of a single function $\Omega(p,q,r,s)$ with the properties

$$A = \Omega_{,qr}$$

$$B = \Omega_{,qs}$$

$$C = \Omega_{,pr}$$

$$D = \Omega_{,ps} \quad . \tag{7.28}$$

The metric now takes the form

$$g_{ab}dz^a dz^b = 2(\Omega_{,pr}dpdr + \Omega_{,ps}dpds + \Omega_{,qr}dqdr$$
$$+ \Omega_{,qs}dqds) \tag{7.29}$$

subject to equation (7.22) which becomes

$$\Omega_{,pr}\Omega_{,qs} - \Omega_{,ps}\Omega_{,qr} = \frac{1}{\Delta} \quad . \tag{7.30}$$

Recalling that Δ is a constant on $\mathcal{H}$, we see that we have essentially Plebanski's first form of the heavenly metric.

We may now make the simplifying choice of $\tilde{\zeta}_1$ relative to $\tilde{\zeta}_0$ which was mentioned above.

Comparing (7.27) and (7.28) we see that it is tempting to identify

$$\Omega_{,p} = \eth\tilde{\eth}Z + Z$$

$$\Omega_{,q} = \tilde{\eth}Z \quad . \tag{7.31}$$

but this requires another integrability condition, namely

$$(\partial\tilde{\partial}Z + Z)_{,q} = (\tilde{\partial}Z)_{,p}$$

or

$$\tilde{M}^a(\partial\tilde{\partial}Z_a + Z_a) = N^a\tilde{\partial}Z_a \quad . \tag{7.32}$$

Using (7.20) we may rearrange (7.32) as

$$\{\hat{Z}^a, \hat{Z}^b\}\ \tilde{\eth}\{Z_a, Z_b\} = 0 \tag{7.33}$$

and it is this which will be used to fix the choice of $\tilde{\zeta}_1$. Specifically we choose $\tilde{\zeta}_1$ so that

$$\{\hat{Z}_a, \hat{Z}_b\} = \tilde{\eth}^2\{Z_a, Z_b\} \ . \tag{7.34}$$

It is a straightforward calculation from (3.15) that

$$\tilde{\eth}\{Z_a, Z_b\}\ \tilde{\eth}^2\{Z^a, Z^b\} = 0$$

and

$$\{Z_a, Z_b\}\ \tilde{\eth}^2\{Z^a, Z^b\} = 1 \tag{7.35}$$

from which it follows that the right hand side of (7.34) is simple, so that it is possible to find a $\tilde{\zeta}_1$ satisfying (7.34). Further this $\tilde{\zeta}_1$ will satisfy (7.33) and finally, from (7.32), with this value of $\tilde{\zeta}_1$ we will have Δ equal to 1. Now with $\Delta = 1$ the N_a and $\tilde{M}_a$ of (7.20b) are as in (3.23).

With this value for Δ, we have exactly Plebanski's first form for the metric and furthermore, equations (7.31) show how to make the transition to the second form by introducing the new coordinates:

$$x = \Omega_{,p} = \eth\tilde{\eth}Z + Z$$

$$y = \Omega_{,q} = \tilde{\eth}Z$$

which were used in the earlier parts of this section.

ACKNOWLEDGEMENTS

We wish to thank Professor John R. Porter of the Mathematics Department, University of Pittsburgh, for many useful discussions and conversations.

REFERENCES

1. E.T. Newman, General Relativity and Gravitation; Proceedings of the Seventh International Conference (GR7) Tel-Aviv University, June 23-28, 1974, John Wiley and Sons, N.Y. 1975.

2. R. Hansen, E.T. Newman, R. Penrose, in preparation (intended to appear).

3. R. Penrose, GRG, 7, 31 (1976).

4. E.T. Newman, GRG, 7, 107 (1976).

5. R. Penrose, "The Nonlinear Graviton", 1975 Gravity Research Foundation Prize Essay.

6. C.W. Fette, A.I. Janis, E.T. Newman, J. Math. Phys. 17, 660 (1976).

7. J.F. Plenbaski, J. Math. Phys. 16, 2395 (1975).

8. J.F. Plebanski, S. Hacyan, J. Math. Phys. 16, 2403 (1975).

9. M. Ko, E.T. Newman, R. Penrose, "Kahler Structure of Asymptotic Twistor Space", in preparation.

10. R. Penrose, "Twistor Theory, Its Aims and Achievements", in Quantum Gravity (eds. C.J. Isham, R. Penrose, D.W. Sciama) Oxford University Press, (1975).

11. E.J. Flaherty, "Hermitian and Kahlerian Geometry in Relativity" Lecture Notes in Physics #46, Springer Verlag (1976).

12. R. Penrose and M. MacCallum, Phys. Rep. 6C, No. 4, (1973).

13. E. Newman, R. Penrose, J. Math. Phys. 3, 566 (1962).

PHYSICS IN CONE SPACE

BRIAN DAVID BRAMSON

MERTON COLLEGE

OXFORD, U.K.

1 INTRODUCTION

In front of me stands a huge potted plant. Incoming radiation from the sun falls on its leaves and is converted to electrical and chemical energy. The plant grows and changes its shape. In particular, it changes its trace-free quadrupole moment and, consequently, radiates gravitational waves.

Newton's theory of gravitation does not provide us with a means of discussing gravitational radiation but it does enable us to define the multipole moments of a given system of matter. If the matter density be ρ, then to take moments about some point X in space-time we must perform integrals over the spatial three-surface V through X. Generically, these may be written

$$M_w(X) = \int dV\, w\, \rho \ , \tag{1.1}$$

where w is some tensorial weighting factor and where indices have been suppressed. If X is chosen as the origin of Cartesian coordinates y_i on V, then w takes the forms 1 (for the mass), y_i (for the dipole moment), $y_i y_j - 1/3\, y^2\, \delta_{ij}$ (for the trace-free quadrupole moment) and so on. Quite generally, restricting to the trace-free moments yields weighting factors w which are trace-free, symmetric tensor fields on Euclidean three-space and which, consequently, satisfy Laplace's equation

$$\nabla^2 w = 0 \ . \tag{1.2}$$

In view of this, equation (1.1) may be rewritten by Gauss' theorem as an integral over any two-surface S lying in V and surrounding the matter. Defining the Newtonian potential ϕ by Poisson's equation

$$\nabla^2 \phi = \rho \ , \tag{1.3}$$

together with the boundary condition

$$\phi \to 0 \quad \text{at spatial infinity,}$$

yields

$$M_w(X) = \int dS.(w\nabla\phi - \phi\nabla w) \ . \tag{1.4}$$

Thus, in Newton's theory, the multipole moments emerge as three-dimensional tensor fields on four-dimensional Galilean space-time: and, if we impose the trace-free condition, they may be expressed as two-dimensional surface integrals.

The question, of course, arises as to whether a similar analysis is feasible within the framework of general relativity. In other words, can we define the momentum, angular momentum, quadrupole moment, charge, electromagnetic dipole moment and so on for a reasonably isolated system of matter? Let us, therefore, try to generalise the Newtonian expression (1.4). There are, however, several features of this expression which we should discuss and clarify before launching into our programme.

First of all, it involves integrating the gravitational field potential, the matter density having disappeared on use of the field equation (1.3). So, we may expect the generalisation to involve integrating certain components of the conformal curvature tensor (if S lies *outside* the matter) and of the electromagnetic field.

Secondly, the value of the integral (1.4) is not very sensitive to the choice of integration surface. That is to say, as long as the surface lies in the given constant time hyperplane through X and as long as it surrounds all of the matter, it makes no difference whether it is chosen to be close in or far out: for, in Newton's theory the gravitational field does not contribute to the total mass and other moments. In general relativity, however, we may expect a different state of affairs. The gravitational field itself carries energy and momentum; so, we had better ensure that the surface of integration lies at infinity (whatever that means) in order to be able to include all of the gravitational effects.

Thirdly, the integrand in (1.4) involves a weighting factor w with tensor indices. This means that we are adding up tensors at different points in Euclidean space. In curved space-time, however, this is the one thing which we are prohibited from doing owing to the absence of a global parallelism; that is unless the curvature is sufficiently weak on the surface of integration. Amazingly, it turns out that if the surface is taken at null infinity, then a parallelism may be defined even in the presence of gravitational waves.

The problems involving the choice of integration surface and the method of integrating tensorial quantities arise, in fact, in all classical theories based upon a *non*-Abelian gauge group [1]: in general relativity this group is precisely the Lorentz group acting in the tangent space at each point of the manifold.

Let us consider the question of the integration surface. We have already agreed that this should lie at infinity. Spacelike infinity is a possible candidate but if we seek to discuss the evolution of some system and ask, for example, how its rest mass changes as waves are radiated, such a choice does not prove useful. This, as we have heard from Professor Geroch, is because the mass defined at spacelike infinity includes the energy contained in the radiation, incoming or outgoing, and therefore remains constant. So instead, we shall choose

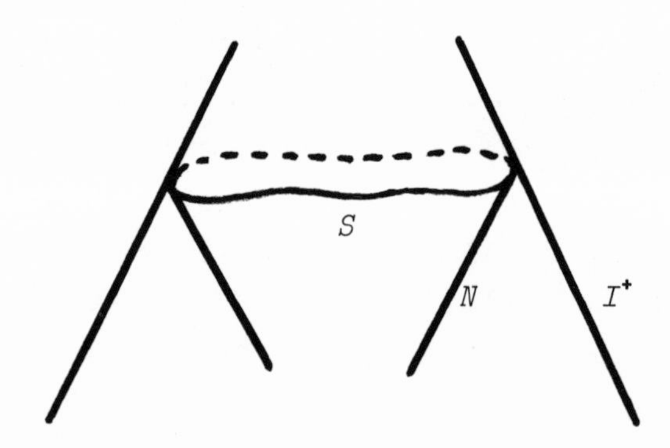

FIGURE 1. Each smooth cut S is the intersection of a unique null hypersurface N with I^+.

smooth two-surfaces at null infinity (defined later) and, if we wish to examine the evolution of the system in terms of the waves which it *radiates*, we had better make it future null infinity $\mathcal{I}^+$. Such two-surfaces are variously called *cuts* or *cross-sections* (figure 1). Each cut defines a unique outgoing null hypersurface (cone) generated by the null rays which leave it orthogonally and enter the interior of the space-time [2]. The set of all smooth cross-sections of $\mathcal{I}^+$ or, equivalently, the set of all outgoing null hypersurfaces forms an infinite dimensional manifold called *cone space* C^+ and will be discussed more fully later. The point is that if the moments of the system are to be defined by performing two-surface integrals at null infinity, then they become fields on cone space rather than on the original space-time manifold.

Let us next turn to the second problem, namely the question of integrating vectors or tensors over surfaces. To be specific, suppose we have a field of one-forms T_a (covariant vector field) which we wish to integrate over some cut S of $\mathcal{I}^+$ to define the total four-momentum of the system. The basic idea is to seek a canonically aligned field of tetrads δ^a_{a} defined on $\mathcal{I}^+$ and unique up to rigid Lorentz rotation. (Note that I am using Penrose's abstract index convention here [3] in a crucial way. The italic index a on δ^a_{a} indicates a vector and on T_a it indicates a one-form. On the other hand an upright index a or $\underline{a}$ takes values 0, 1, 2 and 3. So δ^a_{a} indicates a set of four basis vectors.) Then the canonical components $T_{\mathrm{a}} = T_a \delta^a_{\mathrm{a}}$ are to be integrated over S. The resulting quantity

$$P_{\mathrm{a}} = \int dS\, T_{\mathrm{a}} \tag{1.5}$$

is then defined uniquely up to Lorentz rotation of the tetrads and is called an *asymptotic vector* (strictly a one-form).

Let us suppose that our quest for canonically aligned frames is successful and ask the question "In what space do the integrated quantities like P_{a} sit? Are the multipole moments of the system geometric objects defined on the original space-time manifold as in Newton's theory? If not, do they lie in some other manifold and, if so, what is the physical meaning of this new manifold?"

To begin with, let me repeat that by performing the integrals (1.5) over each cut of I^{+}, P_{a} becomes a field on cone space. If the original space-time is flat, two simplifications emerge. (In practice, of course, the vanishing of the gravitational field will mean that T_{a} will be zero; so imagine for the moment that we are dealing with linear spin-two theory on Minkowski space.) First of all, the frames δ^{a}_{a} may be aligned by parallel transport and chosen uniquely up to Lorentz rotation. Secondly, cone space itself contains Minkowski space-time as a canonical four-dimensional submanifold. This arises trivially by considering those cross-sections of I^{+} which come from the light cones of points of the space-time (figure 2). These are called *good* or *shear-*

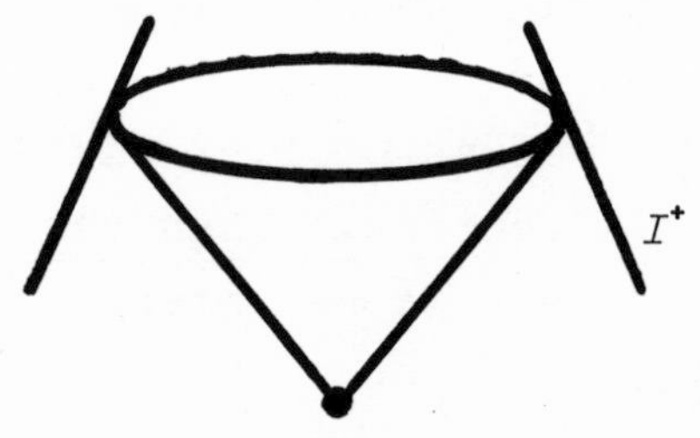

FIGURE 2. To each point of Minkowski space-time there corresponds a shear-free cut of I^{+}.

free cross-sections. If we now restrict the integral (1.5) to these special sections, then P_a becomes a field on the original Minkowski manifold and, at each point, lies in the cotangent space. (More precisely, these are the components of such an object in a standard orthonormal frame.)

Before generalising to curved space let me elaborate a little on the shear-free condition mentioned here and which is of crucial importance when we come to take into account the gravitational bending of light. Consider a point source at rest in flat space and emitting light in all directions. In space-time this defines a null congruence consisting of all the future light cones emanating from some timelike geodesic but, for the moment, consider a description of the emitted beam in the three-space orthogonal to the timelike direction defined by the source. Then a small plane circular object held at right-angles to the beam will cast a circular shadow on a screen also held at right-angles to the beam a small distance d from the object. If, however, we interpose an astigmatic lens between source and object (figure 3),

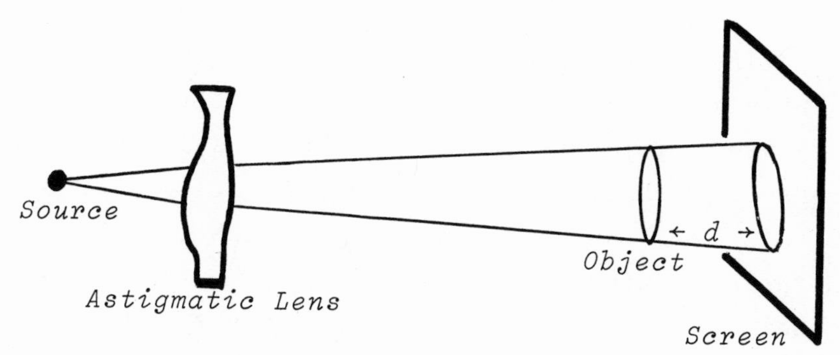

FIGURE 3. A space picture: a circular object casts an elliptical shadow owing to the presence of an astigmatic lens.

then the shadow on the screen will no longer be circular but rather, to first order in d, elliptical. This is an application of a theorem by Ehlers and Sachs [4]. The eccentricity of the ellipse and the orientation of its major axis is determined by the value at the object of a complex field σ which measures the *shear* of the beam. In particular, if we attempt to produce back the rays which have been refracted by the lens, then we shall find that they do not appear to have come from a point source.

Now, the passage of a beam of light through a region of gravitation can very usefully be discussed as if the gravity were a sequence of lenses [5]. In particular, pure conformal curvature ($R_{ab}=0$) gives rise to astigmatic focusing locally and is therefore a source of shear. The null cone of some point of a conformally curved manifold will be generated by rays which, though shear-free near the vertex, will possess shear in the asymptotic region. Of course, the light cones of points of Minkowski space are everywhere shear-free; which is how we were able to set up the correspondence in figure 2 between the points of flat space-time and shear-free cross-sections of $\mathcal{I}^+$. (By the shear of a cross-section is meant the shear at null infinity of the corresponding null hypersurface.)

When the space-time is asymptotically flat (defined later) and stationary an extraordinary state of affairs emerges: for it turns out that we can still find shear-free cuts of $\mathcal{I}^+$ and the space M^+ of such cuts is still isomorphic to Minkowski space-time, even though the original manifold M is curved. The effect of the stationary curvature is to separate the manifolds M and M^+. In the absence of gravitation, these were trivially isomorphic but we now have two for the price of one (three if we introduce I^- but, for the moment, ignore this).

The question naturally arises as to the significance of the new manifold M^+. First of all, it is contained in cone space; so that the integrated quantities, like the momentum, when restricted to M^+, will be defined on flat space even though the original space-time is curved. In particular, it turns out that there exists a skew-symmetric asymptotic tensor field M^{ab}, defined on M^+ and constructed from surface integrals, which transforms under translations as a moment of momentum: in other words, if O and X belong to M^+ with relative position vector X^a, then

$$M^{ab}(X) = M^{ab}(O) - (X^a P^b - X^b P^a) \; . \qquad (1.6)$$

That such a law, usually associated with flat space theories, should arise in general relativity is both remarkable and auspicious. It provides us with a means of defining the intrinsic spin of the system together with a picture of its centre of mass as a timelike geodesic in M^+. In fact, by formulating physical laws in cone space rather than in the original manifold, many other features of flat space theories persist.

It is my purpose here to try to persuade you that for a wide class of physical situations it is cone space and not the original, curved, Riemannian manifold which should be used to define the physical space-time. Such situations will arise whenever the experimentalists are forced to observe from the asymptotic region; and this will occur either through expediency (consider observations of a distant star) or else through inevitability, because of the very nature of things, (for example, when the system in question has atomic or nuclear dimensions). From this point of view M^+ may be considered as an

$M^+ = \{$  $| \sigma = 0 \text{ on } I^+ \}$

FIGURE 4. A null hypersurface, shear-free at infinity, defines an image point which lies in Minkowski space-time M^+ (stationary M).

asymptotic observation space. After all, each member of M^+ is a null hypersurface generated by rays which at infinity appear to have come from some point: this is the meaning of the shear-free condition there. In the language of elementary optics this is precisely what we mean by an *image point* and so M^+ may be regarded as a space-time of such image points (figure 4).

In the non-stationary case and particularly when there is outgoing gravitational radiation we cannot use the shear-free condition to produce a Minkowski space of cuts of I^+. This is because of the intimate connection between the time rate of change of the shear and the outward flow of energy (see later). Professor Newman has a very cunning way around this problem, described in his lectures, which entails finding complex cuts of the complexified I^+. The space $\mathcal{H}$ which he ends up with is Riemannian with four complex dimensions and empty. In the stationary case it is just the complexification of M^+ but, more generally, it possesses a "one-handed" curvature. It may well turn out that in the presence of radiation $\mathcal{H}$ space is the manifold with which we should be dealing. For the present, however, I propose to examine the consequences of using real cuts of I^+ exclusively.

We shall see that it is still possible to find Minkowski subspaces of cone space, even in the presence of radiation, although there are infinitely many of them.

Initially, the strategy will be to pick any one of them, to formulate physical laws in it and then to ask whether the laws depend upon the choice of space and, if so, whether or not they are physically meaningful. It may well be that we shall be forced to impose conditions on the outgoing gravitational radiation field in the remote past [6] in order to select out a unique space. The details of this, however, are not yet finalised.

To complete this introduction let me state that the fundamental idea is to construct physical laws in cone space because it is directly connected with what the asymptotic observer sees. The original Riemannian manifold will be used as a ladder in order to generate the dynamics but will be discarded as soon as possible. I hope that this does not sound too callous.

2 SPINOR EQUATIONS

Although we shall rely heavily on the use of the spinor formalism in general relativity, because of its elegance and economy, it would be be beyond the scope of this article to include a comprehensive discussion of spinors. I shall instead sketch some of their basic properties and refer the reader elsewhere [3,4,7] should more background be required.

Abstract Indices, Frame Components and Connection Symbols

The index convention employed will be the one described in [3]. A smooth vector field v^a defined on a manifold will mean, at each point, a certain geometric entity rather than a set of components; and it is best thought of as a derivative operator $v^a\nabla_a$ acting on scalar fields. The *abstract index* 'a' merely serves to label the type of creature with which we are concerned and is included at all to enable us to perform easily certain operations like symmetrising and contraction. Given a basis δ^{a}_{a} consisting of four independent vectors δ^{a}_{0}, δ^{a}_{1}, δ^{a}_{2} and δ^{a}_{3}, we may of course assign to v^{a} a set of components, and hence scalar fields, v^{a} by means of the expansion

$$v^{a} = v^{\mathrm{a}}\,\delta^{a}_{\mathrm{a}} \;. \tag{2.1}$$

Similarly a one form ξ_{a} may be expanded in terms of the dual basis,

$$\xi_{a} = \xi_{\mathrm{a}}\,\delta^{\mathrm{a}}_{a} \tag{2.2}$$

and where

$$\delta^{a}_{\mathrm{b}}\,\delta^{\mathrm{c}}_{a} = \delta^{\mathrm{c}}_{\mathrm{b}} \quad \text{, the Kronecker delta.} \tag{2.3}$$

Now suppose that we are given a derivative operator ∇_{a} whose action on tensor fields satisfies Leibniz and linearity and which acts on scalar fields as a gradient. Then the components $v^{\mathrm{b}}{}_{;\mathrm{a}}$ of $\nabla_{a}v^{b}$ are given by the expression

$$v^{\mathrm{b}}{}_{;\mathrm{a}} = \delta^{a}_{\mathrm{a}}\,\delta^{\mathrm{b}}_{b}\,\nabla_{a}\,v^{b} = \nabla_{\mathrm{a}}\,v^{\mathrm{b}} + \Gamma_{\mathrm{a}}{}^{\mathrm{b}}{}_{\mathrm{c}}\,v^{\mathrm{c}} \quad , \tag{2.4}$$

and where the connection symbols $\Gamma_{\mathrm{a}}{}^{\mathrm{b}}{}_{\mathrm{c}}$ are defined by

$$\Gamma_{\mathrm{a}}{}^{\mathrm{b}}{}_{\mathrm{c}} = \delta^{\mathrm{b}}_{b}\,\nabla_{\mathrm{a}}\,\delta^{b}_{\mathrm{c}} \;. \tag{2.5}$$

If the manifold has a metric g_{ab}, the derivative operator is determined uniquely by the additional conditions that it preserves the metric and is torsion-free. Henceforth, we shall be dealing with such a manifold M, with Lorentz signature, and such a derivative operator.

The historical development of general relativity has emphasized the use of basis vectors which are provided naturally in terms of a system of coordinates. In this case the 64 components of $\Gamma_{\mathrm{a}}{}^{\mathrm{b}}{}_{\mathrm{c}}$ reduce to 40 Christoffel symbols. An alternative system, often more useful, is found by choosing basis vectors which are orthonormal. In this case the components of the connection reduce to 24 *Ricci rotation coefficients* and the local Lorentz invariance of the theory [1,8,9], regarded as a gauge theory, is explicitly exhibited.

At each point of the manifold the Lorentz group acts on vectors in the tangent space so as to preserve the metric inner product:

$$g_{ab}\, u^a v^b = g_{ab}\, L^a{}_c\, L^b{}_d\, u^c v^d \quad . \tag{2.6}$$

In an orthonormal frame, the components of the metric are given by

$$g_{\mathrm{ab}} = diag.(1\ -1\ -1\ -1) \tag{2.7}$$

and the matrix $L^{\mathrm{a}}{}_{\mathrm{b}}$, which represents the group element $L^a{}_b$, therefore satisfies

$$g_{\mathrm{ab}}\, L^{\mathrm{a}}{}_{\mathrm{c}}\, L^{\mathrm{b}}{}_{\mathrm{d}} = g_{\mathrm{cd}} \ ; \tag{2.8}$$

which means to say that it belongs to O(1,3).

Spinor Calculus

We may construct *spin space*, which is a two-dimensional complex vector space, at each point of the manifold by considering the covering group of the Lorentz group. This acts on spinors according to

$$\alpha^A \rightarrow \Lambda^A{}_B\, \alpha^B \tag{2.9}$$

in such a way as to preserve the inner product

$$\alpha_A\, \beta^A = \varepsilon_{AB}\, \alpha^A \beta^B = \varepsilon_{AB}\, \Lambda^A{}_C\, \Lambda^B{}_D\, \alpha^C \beta^D \ . \tag{2.10}$$

This is analogous to equation (2.6) but where ε_{AB} is, for the moment, any skew-symmetric object lying in the dual spin space tensored with itself. In a normalised spin basis $\delta_{\mathrm{A}}{}^A$, satisfying $\delta_{0A}\delta_1{}^A = 1$, the components of

ε_{AB} are given by

$$\varepsilon_{AB} = \begin{pmatrix} 0 & 1 \\ -1 & 0 \end{pmatrix} ; \tag{2.11}$$

and the matrix $\Lambda^{A}{}_{B}$, which represents the group element $\Lambda^{A}{}_{B}$, satisfies

$$\varepsilon_{AB}\, \Lambda^{A}{}_{C}\, \Lambda^{B}{}_{D} = \varepsilon_{CD} ; \tag{2.12}$$

which means to say that it belongs to SL(2,C). Note that since spin space is two-dimensional, any skew-symmetric two index spinor will be proportional to ε_{AB}. In other words

$$\left.\begin{aligned} &\gamma_{AB} = \gamma_{[AB]} = \tfrac{1}{2}\gamma_{C}{}^{C}\, \varepsilon_{AB} , \\ &\text{where} \\ &\gamma_{C}{}^{C} = \varepsilon^{CD}\, \gamma_{CD} \quad \text{and} \quad \varepsilon_{AB}\, \varepsilon^{CB} = \delta_{A}{}^{C} . \end{aligned}\right\} \tag{2.13}$$

With each spinor α^{A} is associated a complex conjugate spinor $\bar{\alpha}^{A'}$ (though $\varepsilon_{A'B'}$ is written without the bar). This enables contact with the geometry of the manifold, for the tensor product of spin space with itself is precisely the space of four-vectors at each point. This is written

$$v^{a} = v^{AA'} . \tag{2.14}$$

With this identification we may define a symmetric tensor $k_{ab} = \varepsilon_{AB}\, \varepsilon_{A'B'}$ which has the property, from (2.10) and its complex conjugate, that $k_{ab}u^{a}v^{b}$ is invariant under Lorentz rotations of the vectors u^{a} and v^{a}. k_{ab} is therefore conformal to the metric g_{ab}. So, we shall scale ε_{AB} until $k_{ab}=g_{ab}$. We may now write

$$g_{ab} = \varepsilon_{AB}\, \varepsilon_{A'B'} . \tag{2.15}$$

There is still a phase freedom in ε_{AB} but it turns out to be unimportant for our purposes (see later) and we shall fix it once and for all.

In view of equation (2.15) and the skew-symmetry of ε_{AB}, it follows that the real vector

$$l^a = \alpha^A \bar{\alpha}^{A'} \tag{2.16}$$

is null. The spinor α^A is not represented completely by l^a because of the freedom to change its phase. However, the skew tensor

$$P^{ab} = \alpha^A \alpha^B \varepsilon^{A'B'} + \varepsilon^{AB} \bar{\alpha}^{A'} \bar{\alpha}^{B'} \tag{2.17}$$

is sensitive to the phase of α^A and defines a two-dimensional vector space spanned by l^a and a spacelike vector orthogonal to l^a:

$$P^{ab} = l^a s^b - l^b s^a \quad , \tag{2.18}$$

where

$$s^a = \alpha^A \bar{\beta}^{A'} + \beta^A \bar{\alpha}^{A'} \quad ,$$

for any β^A satisfying $\alpha_A \beta^A = 1$. The spinor α^A is then completely represented by the null vector l^a together with a two-plane element known as the *flag plane*. Everything we shall do will, in fact, be independent of the phases of our spinors, and hence of the orientation of the flag plane, but the flag will be used in diagrams merely to indicate the presence of a spinor.

We have noted the geometric significance of a one-index spinor. It turns out that many higher valence spinors have important tensor translations. Fortunately, by virtue of the expressions (2.13), we need only consider symmetric spinors. Of these, we shall need the fact that a two-index spinor ϕ_{AB} defines a general skew-symmetric second rank tensor according to the scheme

$$F_{ab} = \phi_{AB} \varepsilon_{A'B'} + \varepsilon_{AB} \bar{\phi}_{A'B'} \quad ; \tag{2.19}$$

and a four-index spinor C_{ABCD} (also symmetric) defines a tensor with the symmetries and trace-free properties of the Weyl conformal tensor

$$C_{abcd} = C_{ABCD}\,\varepsilon_{A'B'}\,\varepsilon_{C'D'} + \varepsilon_{AB}\,\varepsilon_{CD}\,\bar{C}_{A'B'C'D'} \quad . \qquad (2.20)$$

Spinor Equations

Given smooth spinor fields on the manifold we may define the action of the derivative operator ∇_a, introduced previously, by demanding that not only should it satisfy Leibniz and linearity, act on scalar fields as a gradient, be torsion-free and preserve the metric, but that it should also annihilate the spinors ε_{AB} and $\varepsilon_{A'B'}$. In accordance with the identification (2.14) this operator may now be written as $\nabla_{AA'}$.

Let us, accordingly, consider a smooth spinor field $\alpha^A(x)$ and ask for the simplest, first order, linear, differential equations which it may satisfy. The derivative of this field may be split into its irreducible parts by symmetrising and antisymmetrising on the unprimed indices, thus:

$$\nabla^{A}{}_{A'}\,\alpha^B = \nabla^{(A}{}_{A'}\,\alpha^{B)} + \tfrac{1}{2}\,\varepsilon^{AB}\,\nabla_{CA'}\,\alpha^C \quad , \qquad (2.21)$$

where we have exploited the equations (2.13). By setting each piece of (2.21) equal to zero, we obtain two fundamental equations. On the one hand we get the *twistor equation* for spinors of valence one [10]

$$\nabla^{(A}{}_{A'}\,\omega^{B)} = 0 \quad , \qquad (2.22)$$

while, on the other, the neutrino equation

$$\nabla^{A}{}_{A'}\,\phi_A = 0 \quad . \qquad (2.23)$$

Both of these equations go over naturally to higher

valence. If we take symmetric spinors with n indices, then the obvious generalisations are

$$\nabla^{(A}{}_{A'}\,\omega^{BC....K)} = 0 \tag{2.24}$$

and

$$\nabla^{A}{}_{A'}\,\phi_{AB....J} = 0 \quad . \tag{2.25}$$

(Given a general space-time, however, solutions to (2.24) will not exist globally while solutions to the zero rest mass field equations (2.25) will be constrained algebraically by the Weyl curvature spinor [3] for spin $n/2 \geqslant 3/2$. This is a consequence of the spinor equivalent of the Ricci identities.)

By analogy with the neutrino equation (2.23), equation (2.25) gives us the spinor form of Maxwell's source-free equations on setting n=2: this follows from (2.19). Furthermore, putting n=4 we get the spinor equivalent of the empty space Bianchi identities; which follows from (2.20). Thus the spinor formalism provides us with a language in which the equations governing the behaviour of the neutrino, the photon and the graviton are special cases of the basic equation (2.25) and so a remarkable unification is achieved.

The two sets of equations (2.24) and (2.25) have the property that they are invariant under the conformal rescalings of the metric,

$$\left.\begin{aligned} \hat{g}_{ab} &= \Omega^2 g_{ab} \\ \hat{\varepsilon}_{AB} &= \Omega\,\varepsilon_{AB} \end{aligned}\right\}, \tag{2.26}$$

where Ω is a real, smooth, positive, scalar field on the manifold. Under this rescaling the derivative operator transforms according to a law of the form

$$\hat{\nabla}_{AA'}\,\alpha_B = \nabla_{AA'}\,\alpha_B + T_{AA'B}{}^{C}\,\alpha_C \quad , \tag{2.27}$$

for some spinor-tensor $T_{AA'B}{}^{C}$. Demanding that $\hat{\nabla}_a$ act on scalars as a gradient and be torsion-free yields the relation

$$T_A{}^{A'}{}_B{}^C = \bar{T}_{D'(A}{}^{D'A'}\delta_{B)}{}^C + \tfrac{1}{2}\varepsilon_{AB}T_D{}^{A'DC} \quad , \tag{2.28}$$

while the condition that it preserves $\hat{\varepsilon}_{AB}$ is

$$T_{AA'D}{}^D = -T_{AA'} = -\Omega^{-1}\nabla_{AA'}\Omega \quad . \tag{2.29}$$

Thus it follows [3] that

$$\left.\begin{aligned} &\hat{\nabla}_{AA'}\alpha_B = \nabla_{AA'}\alpha_B - T_{BA'}\alpha_A \\ &\text{and similarly} \\ &\hat{\nabla}_{AA'}\beta^B = \nabla_{AA'}\beta^B + \delta_A{}^B T_{CA'}\beta^C \end{aligned}\right\} . \tag{2.30}$$

From the relations (2.30) it follows that the twistor equations (2.24) are invariant under conformal rescalings of the metric if we scale ω according to

$$\hat{\omega}^{AB....J} = \omega^{AB...J} \quad . \tag{2.31}$$

For n=1 this is certainly right and proper. ω^A defines a null vector field $\omega^A\bar{\omega}^{A'}$, which satisfies the conformal Killing equation [11], and as such is defined independently of any metric.

In the same way it may be shown that the zero rest mass field equations (2.25) are conformally invariant on scaling ϕ according to

$$\hat{\phi}_{AB....J} = \Omega^{-1}\phi_{AB....J} \quad . \tag{2.32}$$

For n=2, ϕ_{AB} defines the electromagnetic field tensor F_{ab} according to equation (2.19). F_{ab} may readily be seen to have zero conformal weight and this is understandable on the grounds that, being a two-form, it also is defined independently of the metric. For n=4, however, the matter is a little delicate. By analogy with equation (2.20), ϕ_{ABCD} defines a tensor K_{abcd} according to

$$K_{abcd} = \phi_{ABCD}\,\varepsilon_{A'B'}\,\varepsilon_{C'D'} + \varepsilon_{AB}\,\varepsilon_{CD}\,\bar{\phi}_{A'B'C'D'} \quad . \qquad (2.33)$$

This implies that

$$\hat{K}_{abcd} = \Omega\, K_{abcd} \quad , \qquad (2.34)$$

which means to say that it has the wrong conformal weight to be the Weyl conformal tensor. The Weyl tensor, by virtue of its relation to the geometry, scales according to

$$\hat{C}_{abcd} = \Omega^2\, C_{abcd} \quad . \qquad (2.35)$$

The resolution of this paradox involves the use of two four-index spinors. Henceforth, we shall write

$$\psi_{ABCD} = \phi_{ABCD} \quad ,$$

scale it according to (2.32) and call it the *spin-two field*. The spinor C_{ABCD}, on the other hand, has zero conformal weight and will be called the *curvature spinor*. The two are essentially identical with respect to the physical metric but, more generally, they are related by the expression

$$\psi_{ABCD} = \phi\, C_{ABCD} \quad , \qquad (2.36)$$

where ϕ is a scalar field which scales according to

$$\hat{\phi} = \Omega^{-1}\,\phi \quad .$$

It is possible to construct a conformally invariant formulation of general relativity, using g_{ab} and ϕ [12], and in which the physical metric is selected from all the conformally related metrics by the requirement that ϕ be constant: units may then be chosen to make this constant unity.

Twistors

Let us now turn to the twistor equation (2.22) whose general solution in flat space-time is given by [10]

$$\omega^A(X) = \omega^A(O) - i\, x^{AA'}\, \pi_{A'} \quad , \tag{2.37}$$

where x^a is the position vector from the point O to the point X and where $\omega^A(O)$ and $\pi_{A'}$ are arbitrary constant spinors. We say that the solution (2.37) defines a *twistor* Z^α of valence $\begin{bmatrix}1\\0\end{bmatrix}$ which, with respect to the origin O, may be represented by the pair of spinors $\omega^A(O)$ and $\pi_{A'}$; and we write

$$Z^\alpha \underset{O}{\leftrightarrow} [\omega^A(O)\ ,\ \pi_{A'}] \quad . \tag{2.38}$$

The set of solutions of the form (2.37) defines a complex, linear, vector space called *twistor space*. Since each spin space has two complex dimensions, it follows that twistor space is C^4.

Of course, the representation (2.38) is not unique and if we choose X as the new origin, then

$$Z^\alpha \underset{X}{\leftrightarrow} [\omega^A(X)\ ,\ \pi_{A'}] \quad . \tag{2.39}$$

The difference between $\omega^A(X)$ and $\omega^A(O)$ amounts, essentially to a moment of $\pi_{A'}$.

The twistor itself, together with its complex conjugate,

$$\bar{Z}_\alpha \leftrightarrow [\bar{\pi}_A\ ,\ \bar{\omega}^{A'}(X)] \quad , \tag{2.40}$$

may be used to describe a massless system in special relativity [11]. To see this define

$$\omega_{AB} = i\, \omega_{(A}\, \bar{\pi}_{B)} \tag{2.41}$$

and

$$M_{ab} = \omega_{AB}\,\varepsilon_{A'B'} + \bar{\omega}_{A'B'}\,\varepsilon_{AB} \quad . \tag{2.42}$$

This last expression is a skew-symmetric tensor field on Minkowski space and which, according to (2.37), satisfies

$$M_{ab}(X) = M_{ab}(O) - (x_a P_b - x_b P_a) \tag{2.43}$$

where

$$P_a = \bar{\pi}_A \pi_{A'} \quad . \tag{2.44}$$

The last two equations describe the behaviour, under translations, of the relativistic angular momentum of a system of mass zero. Furthermore, if we calculate its Pauli-Lubanski spin vector,

$$S^a = \tfrac{1}{2}\,\varepsilon^{abcd}\,P_b\,M_{cd} \quad , \tag{2.45}$$

we find that it is parallel to the momentum vector:

$$S^a = s\,P^a \quad , \tag{2.46}$$

where the helicity s is given by

$$2s = Z^\alpha \bar{Z}_\alpha \quad . \tag{2.47}$$

The fact that the spin and momentum vectors are parallel is not an automatic consequence of the Poincaré algebra but arises because we have chosen only those massless systems given by a twistor description. As far as is known, these are the only physical ones.

We may also use the twistor Z^α to define a kind of complex centre of mass of the system just described. This arises by considering that set of points in complex Minkowski space-time for which the principal (or angular momentum) part ω^A of the twistor vanishes. Solving equation (2.37) with zero left hand side and allowing x^a to be complex gives

$$x^{AA'} = (\chi_{B'}\,\pi^{B'})^{-1} i\omega^{A}(O)\,\chi^{A'} + \lambda^{A}\,\pi^{A'} \quad , \tag{2.48}$$

for any fixed $\chi^{A'}$ not proportional to $\pi^{A'}$ and for arbitrary λ^{A}. This, therefore, defines a complex, two-dimensional plane in complex Minkowski space-time, all of whose tangent vectors are null, and represents the twistor up to overall scale factor. It contains real points if and only if Z^{α} is null, which means to say that s vanishes, and then it contains a real, null geodesic:

$$x^{AA'} = (i\,\bar{\omega}^{B'}\pi_{B'})^{-1}\,\omega^{A}\,\bar{\omega}^{A'} + \lambda\,\bar{\pi}^{A}\,\pi^{A'} \quad , \tag{2.49}$$

for real λ.

The complex two planes, which arise naturally here, are, of course, the flat space analogues of the null strings described in the lectures of Professor Plebanski.

Ambiguity of Phase

To end this section let us consider the question, raised earlier, of the phase ambiguity in ε_{AB}. If we choose a new ε_{AB} defined by

$$\hat{\varepsilon}_{AB} = exp\,2i\theta\;\varepsilon_{AB} \quad , \tag{2.50}$$

for real θ, then, by analogy with the discussion for conformal rescalings, we shall require a new derivative operator $\hat{\nabla}_{AA'}$ such that

$$\left.\begin{aligned} &\hat{\nabla}_{AA'}\,\alpha_{B} = (\nabla_{AA'} - i\,\nabla_{AA'}\theta)\,\alpha_{B} \\ &\text{and} \\ &\hat{\nabla}_{AA'}\,\alpha^{B} = (\nabla_{AA'} + i\,\nabla_{AA'}\theta)\,\alpha^{B} \quad . \end{aligned}\right\} \tag{2.51}$$

We shall require that any scalar or tensor quantities constructed from spinors, as for example in equations (2.15) to (2.20), be invariant under the phase transformation (2.50). To this end we shall demand that *all* spinors should transform according to the following rule. Let S be a spinor with upstairs indices p unprimed and q primed and downstairs indices r unprimed and s primed. Then set

$$\hat{S} = exp\, i(-p+q+r-s)\theta\, S \quad . \tag{2.52}$$

This will guarantee that $\nabla_{AA'}S$ picks up the same phase factor by virtue of (2.51). In particular, the spin coefficients and field variables, introduced later, will be invariant and we shall, henceforth, ignore the phase freedom in ε_{AB}.

3 ASYMPTOTICALLY FLAT SPACE-TIMES

The potted plant, introduced in the first section, has the property that the gravitational field which it produces gets progressively weaker as we walk further and further away from it. We say that the associated space-time is *asymptotically flat*. A precise definition of asymptotic flatness has been given in the article by Geroch in this volume and in [3, 13] but for completeness it is included here.

The idea of bringing infinity in by rescaling the metric was suggested by Penrose following the important pioneering work on isolated, radiating systems by Bondi and colleagues [14] and Sachs [15, 16]. The idea has its roots in projective geometry by means of which lines which were parallel in the Euclidean two-plane are made to meet at infinity. Again, certain ordinary differential equations are analysed at infinity by making the substitution $y=1/x$ and examining the behaviour of solutions near $y=0$.

For typographical reasons, the notation to be used

here is that the original space-time metric will be written $\tilde{g}_{ab}$ whereas the rescaled metric will be written g_{ab}. Similarly, the original manifold without boundary will be written $\tilde{M}$ while the extended manifold is written M.

Basic Definitions and the Structure of Null Infinity

A space-time $\tilde{M}$ is said to be *asymptotically simple* if extendible to a manifold M containing $\tilde{M}$ and with boundary $I = M - \tilde{M}$ such that the following conditions are satisfied.

(A) On M there exists a smooth scalar field $\Omega \geqslant 0$ and a smooth metric

$$g_{ab} = \Omega^2 \tilde{g}_{ab} . \tag{3.1}$$

(B) On the boundary I

$$\Omega = 0 \quad \text{but} \quad \nabla_a \Omega \neq 0 \quad . \tag{3.2}$$

(C) Every null geodesic has two end points on I.

The last condition is much too strong for our purposes. Not only does it exclude the Schwarzschild solution, because null rays can disappear down the black hole, but it also excludes, for example, spherically symmetric stars with radii between twice and three times their masses, owing to the presence of closed null orbits. Of course, outside of the embarrassing region, such a space-time looks just like one which satisfies conditions (A), (B) and (C); and, indeed, if we fill in the region $r \leqslant 3m$ in a suitable manner with matter, then the new manifold will be asymptotically simple.

Accordingly, $\tilde{M}$ is said to be *weakly* asymptotically simple if it possesses a subset which is isometric to an open neighbourhood of the I of some asymptotically simple space-time. If, in addition, the cosmological constant vanishes then M is said to be asymptotically *flat*. Henceforth, we shall be dealing exclusively with such space-times.

Of crucial importance will be the underlying philosophy that, as far as the geometry and physics are concerned, the boundary I is to be treated as if it lay in the finite region. In particular, the conformal factor, the rescaled metric, curvature tensor, spin-one and spin-two fields are all to be perfectly well behaved on I; which means differentiable as many times as we need.

Let us, for definiteness, consider a dynamical system comprised of interacting matter sufficiently localized that its energy-momentum tensor is purely electromagnetic within some open neighbourhood of I. Then the intrinsic structure of I is determined by the regularity of the rescaled Ricci tensor there and is obtained from the expression [13]

$$P_{ab} = \tilde{P}_{ab} - \Omega^{-1} \nabla_a \nabla_b \Omega + \tfrac{1}{2} \Omega^{-2} g_{ab} \nabla_c \Omega \nabla^c \Omega , \qquad (3.3)$$

where

$$P_{ab} = -\tfrac{1}{2}(R_{ab} - 1/6\, R\, g_{ab}) .$$

Near I we use Einstein's field equations

$$\tilde{P}_{ab} = \tilde{\phi}_{AB} \bar{\tilde{\phi}}_{A'B'} , \qquad (3.4)$$

the right hand side being the Maxwell stress tensor, and make the substitution

$$\phi_{AB} = \Omega^{-1} \tilde{\phi}_{AB} , \qquad (3.5)$$

which defines the rescaled spin-one field and is a special case of equation (2.32). The rescaled Ricci tensor is thus obtained from the expression

$$\Omega^2 P_{ab} = \Omega^4 \phi_{AB} \bar{\phi}_{A'B'} - \Omega \nabla_a \nabla_b \Omega + \tfrac{1}{2} g_{ab} \nabla_c \Omega \nabla^c \Omega ; \qquad (3.6)$$

and setting $\Omega=0$ yields immediately that I is a null hypersurface,

$$\nabla_c \Omega \, \nabla^c \Omega = 0 \,, \tag{3.7}$$

and hence generated by null geodesics. In fact, this information is carried entirely by the trace of equation (3.6). The trace-free part yields, on I,

$$\nabla_a \nabla_b \Omega = \tfrac{1}{4} g_{ab} \Box\Omega \tag{3.8}$$

which means that the generators of I, with tangent vector $\nabla^a \Omega$, are shear-free.

As discussed in the article by Geroch, I, which is now called *null infinity*, splits into the union $I^+ \cup I^-$ of two disjoint subsets depending on whether we are considering the future or past end points of null geodesics (see (C) below equation (3.2)). In passing, let me mention the sky at night as an example (figure 5). An observer in Minkowski space looks out along his past light cone. At each instant of time v, measured by his personal clock, he sees the stars sitting on the celestial sphere which is a cross-section of I^-; and an event, say the occultation of an artificial satellite, will be labelled by three coordinates v, θ and ϕ. Thus the sky (I^-) has topology $R^1 \times S^2$ and I shall assume that I^- and

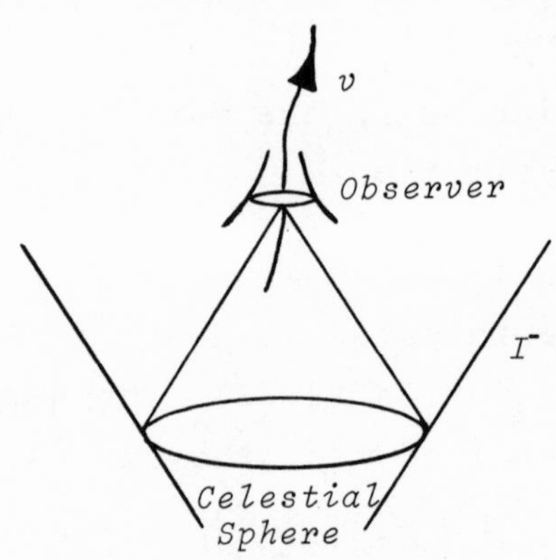

FIGURE 5. The sky (I) has topology $R^1 \times S^2$.

$\mathcal{I}^+$ will have this topology quite generally.

In practice, we shall sometimes need to use both future and past null infinity; for example, if we want to discuss the scattering of starlight around the sun within the framework of field theory. On the other hand, if we wish to establish relations between the evolution of some given dynamical system and the radiation fields it produces, then the use of $\mathcal{I}^+$ alone will suffice. Let us, for the present, restrict our attention to the latter case though everything we do will be capable of immediate generalisation.

Coordinates for Null Infinity

The normal to $\mathcal{I}^+$ is given by (see figure 6)

$$n_a = -\nabla_a \Omega \Big|_{\Omega=0} , \tag{3.9}$$

with the sign chosen so that the vector n^a (with index raised with the four-metric), which is tangent to the generators, is future pointing. Now the intrinsic three-geometry of $\mathcal{I}^+$ is degenerate since n^a is null. The line element is two-dimensional and, hence, conformally flat or, equivalently, conformally spherical. So, labelling the generators of $\mathcal{I}^+$ (points on the S^2) by angular coordinates θ and ϕ or, more conveniently, by stereographic coordinates ζ and $\bar{\zeta}$, where

$$\zeta = \exp i\phi \cot \tfrac{1}{2}\theta , \tag{3.10}$$

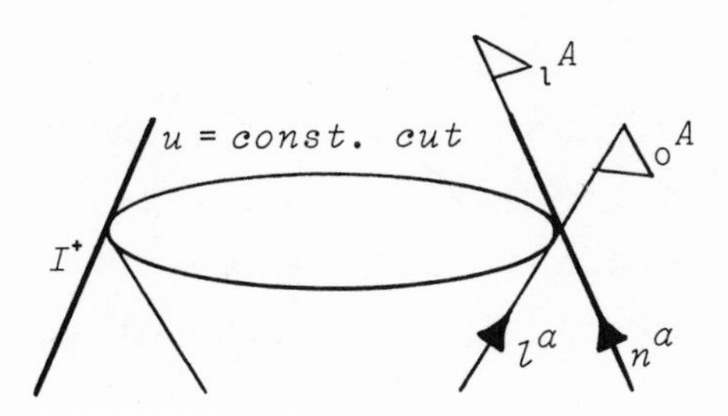

FIGURE 6. The spinor dyad o^A, ι^A on $\mathcal{I}^+$.

the line element takes the form

$$dl^2 = F^2\, d\zeta\, d\bar{\zeta} \quad , \tag{3.11}$$

where F is a scalar field on I^+. (Strictly speaking, F is a *weighted scalar* since it depends upon the choice of conformal factor.) The freedom in ζ is that of a fractional linear transformation

$$\zeta \rightarrow (c\zeta + d)^{-1}\,(a\zeta + b) \tag{3.12}$$

and will appear again later when we come to recognize the SL(2,C) freedom in frame alignment on I^+.

We may extend ζ and $\bar{\zeta}$ to a system of coordinates for I^+ by defining a monotonically increasing parameter u up each generator and scaled such that

$$n^a \nabla_a u = 1 \quad . \tag{3.13}$$

This is defined up to $u \rightarrow u + f(\zeta,\bar{\zeta})$ (see also [17]) and so, to tie down the freedom, we shall choose some particular cross-section S of I^+ and put

$$u = 0 \quad \text{on } S \quad . \tag{3.14}$$

We now have a one-parameter family of cross-sections $u = \text{constant}$ and hence, associated with them, a one-parameter family of outgoing null hypersurfaces as explained in the first section. Labelling these hypersurfaces also by $u = \text{constant}$ we may define a field of normals

$$l_a = \nabla_a u \quad , \tag{3.15}$$

at least in some neighbourhood of I^+, and in this neighbourhood the generators of the family give rise to a twist-free, null, geodesic congruence with tangent vector l^a. Note that on I^+, by virtue of equation (3.13),

$$l_a n^a = 1 \quad . \tag{3.16}$$

For completeness, though we shall not need it here, we may extend u, ζ and $\bar{\zeta}$ to a coordinate system for M at least near I^+. This is achieved first by using u, ζ and $\bar{\zeta}$ to label the members of the l congruence and secondly by defining r to be an affine parameter along each null ray such that

$$\left.\begin{aligned} l^a \nabla_a r &= 1 , \\ r &= 0 \quad \text{on } I^+ . \end{aligned}\right\} \tag{3.17}$$

Of course, this radial parameter is adapted naturally to the rescaled metric and is roughly the inverse of the usual luminosity parameter $\tilde{r}$. Note that, by virtue of equations (3.9), (3.16) and (3.6), it follows that

$$\Omega = -r + O(r^3) \ . \tag{3.18}$$

Note, further, that on I^+ the four-metric is given by

$$ds^2 = 2\, du\, dr - F^2(u,\zeta,\bar{\zeta})\, d\zeta\, d\bar{\zeta} \quad , \tag{3.19}$$

where F is defined in equation (3.11). This follows by considering an infinitesimal element of the form

$$dx^a = n^a\, du + l^a\, dr + dy^a \quad , \tag{3.20}$$

where dy^a lies in one of the $u = \text{constant}$ cuts of I^+ and is hence orthogonal to n^a and l^a.

Now, with the null vectors l^a and n^a on I^+ we may associate a pair of spinors:

$$\delta_0{}^A = o^A \quad : \quad \delta_1{}^A = \iota^A \quad : \quad o_A \iota^A = 1 \ , \tag{3.21}$$

by using equation (2.16). The spinors are defined up to phase and, for convenience, because of (3.16), have been normalised to unity. They give rise to a null tetrad of vectors according to the scheme [18]

$$\left.\begin{aligned} l^a &= o^A \bar{o}^{A'} \quad : \quad n^a = \iota^A \bar{\iota}^{A'} \quad : \\ m^a &= o^A \bar{\iota}^{A'} \quad : \quad \bar{m}^a = \iota^A \bar{o}^{A'} \quad . \end{aligned}\right. \tag{3.22}$$

In turn, these define the derivative operators

$$D = l^a \nabla_a = \partial/\partial r \quad : \quad \Delta = n^a \nabla_a = \partial/\partial u$$
$$\delta = m^a \nabla_a \quad \text{and} \quad \bar{\delta} = \bar{m}^a \nabla_a \ , \tag{3.23}$$

the last two operators acting tangent to the u = constant cross-sections. In fact we can choose ζ and $\bar{\zeta}$ (or adjust the phases of the spinors) such that $\delta = 2^{\frac{1}{2}} F^{-1} \partial/\partial\zeta$.

The Field Variables and their Evolution

It is now time to turn to the field variables on I^+ which are to be used to describe the system. Basically, these are the various components of the Maxwell and spin-two fields with respect to the spin basis just introduced and are written

$$\phi_0 = \phi_{AB}\, o^A o^B \ , \quad \phi_1 = \phi_{AB}\, o^A \iota^B \ , \quad \phi_2 = \phi_{AB}\, \iota^A \iota^B \tag{3.24}$$

and

$$\psi_0 = \psi_{ABCD}\, o^A o^B o^C o^D \ , \quad \psi_1 = \psi_{ABCD}\, o^A o^B o^C \iota^D \ ,$$
$$\psi_2 = \psi_{ABCD}\, o^A o^B \iota^C \iota^D \ , \quad \psi_3 = \psi_{ABCD}\, o^A \iota^B \iota^C \iota^D \ , \tag{3.25}$$
$$\psi_4 = \psi_{ABCD}\, \iota^A \iota^B \iota^C \iota^D \ .$$

These are precisely the quantities which enter into the *peeling-off* theorems [3]. In particular, the *null data* ϕ_2 and ψ_4 give information about the outgoing electromagnetic and gravitational radiation fields, being the coefficients of $1/\tilde{r}$ in asymptotic expansions of the unrescaled fields. ϕ_1 contains information about the charge and about the intermediate-zone electromagnetic radiation field and ϕ_0 describes the Maxwell dipole moment and the near-zone radiation. ψ_3 like ψ_4 describes gravitational radiation while ψ_2 encodes the momentum of the system together with the intermediate-zone radiation. Finally, ψ_1 and ψ_0 give the angular momentum and quadrupole moment, respectively, and, further, contain the near-zone radiation field. There do exist other field components, namely the various derivatives of ϕ_0 and ψ_0 off $\mathcal{I}^+$. They would be required if we decided to examine the higher multipole moments of the system but, for simplicity, they will be ignored here.

The evolution of the system will be determined by Maxwell's equations

$$\nabla^A{}_{A'}\phi_{AB} = 0 \tag{3.26}$$

on $\mathcal{I}^+$ and by the spin-two equations with source. The latter may be obtained by starting with the unrescaled Bianchi identities [3]

$$\tilde{\nabla}^D{}_{A'}\tilde{C}_{ABCD} = \tilde{\nabla}_{(C}{}^{B'}\tilde{P}_{AB)A'B'}\ , \tag{3.27}$$

with $\tilde{P}_{ab}$ determined by equation (3.4), and by using the definition of the rescaled spin-two field

$$\psi_{ABCD} = \Omega^{-1}\tilde{C}_{ABCD} = \Omega^{-1}C_{ABCD} \tag{3.28}$$

(see discussion in section two). This gives

$$\nabla^D{}_{A'}\psi_{ABCD} = 3\,\iota_{(A}\phi_{BC)}\bar{\phi}_{A'}{}^{B'}\bar{\iota}_{B'} \tag{3.29}$$

on $\mathcal{I}^+$.

When we take components, equations (3.26) and (3.29) lead to partial differential equations for the quantities ϕ_r and ψ_r. Since we are working with a particular basis $\delta_{\underline{A}}{}^A = (o^A,\ \iota^A)$ we shall require the connection symbols, by analogy with equation (2.5). There are twelve complex *spin coefficients* (compare 24 Ricci rotation coefficients)

$$\Gamma_{\underline{AA'BC}} = \delta_{\underline{B}}{}^B\nabla_{\underline{AA'}}\delta_{\underline{C}B} = \tag{3.30}$$

AA' \ BC	00	01 10	11
$D=00'$	κ	ε	π
$\bar{\delta}=10'$	ρ	α	λ
$\delta=01'$	σ	β	μ
$\Delta=11'$	τ	γ	ν

Note that we have assumed tacitly that the spinor basis is continued off I^+ in a smooth manner. We shall, however, require the values of the spin coefficients only *on* I^+. Many, in fact, vanish and most of the rest can be made to vanish by choosing the conformal factor appropriately.

First of all, define o^A off I^+ such that $l^a = o^A \bar{o}^{A'}$. By virtue of equation (3.15) and the nullity of l^a, it follows that l^a is parallelly transported along the l congruence. So, adjust the phase of o^A to make $Do^A = 0$. Hence

$$\kappa = 0 = \varepsilon \; . \tag{3.31}$$

The freedom in o^A is now a phase $exp\; i\theta(u,\zeta,\bar{\zeta})$ and in ι^A the inverse phase. Secondly, since $\nabla_{[a} l_{b]} = 0$ it follows that

$$\rho = \bar{\rho} = -\tfrac{1}{2} \nabla_a l^a, \tag{3.32}$$

(the l congruence, being surface-forming, is twist-free and has expansion -2ρ), and that

$$\tau = \bar{\alpha} + \beta \; .$$

Thirdly, from equations (3.8) and (3.9) it follows that

$$\nu = 0 \quad , \tag{3.33}$$

(the generators of I^+ are null geodesics), that

$$\lambda = 0 \quad , \tag{3.34}$$

(the generators are shear-free), that

$$\mu = \bar{\mu} = \tfrac{1}{3} \nabla_a n^a = -(\gamma + \bar{\gamma}) \tag{3.35}$$

and that

$$\bar{\alpha} + \beta = 0 \; . \tag{3.36}$$

If we now choose the conformal factor to make the generators of I^+ non-expanding, so that the quantity F

in the line element (3.11) is independent of u, it follows that

$$\mu = 0 \tag{3.37}$$

and, by adjusting the phase of ι^A, we can make

$$\gamma = 0\ . \tag{3.38}$$

The freedom in o^A and ι^A now amounts to the phase transformations

$$o^A \to \exp i\theta(\zeta,\bar{\zeta})\ o^A \ : \quad \iota^A \to \exp -i\theta(\zeta,\bar{\zeta})\ \iota^A\ .$$

Quite generally, any field component η which transforms according to the rule

$$\eta \to \exp 2is\theta\ \eta \tag{3.39}$$

under phase change of spinor basis is said to have *spin-weight* s. It is then very useful to introduce the differential operator $\eth$ (edth) [6] according to

$$\eth\eta = (\delta - 2s\beta)\eta \tag{3.40}$$

(see article by Newman & Tod in this volume for elaboration). β $(=-\bar{\alpha})$ is, essentially, the Christoffel symbol appropriate to the intrinsic two-geometry of the cross-sections u = constant.

We are now in a position to deduce the evolution of the field components from equations (3.26) and (3.29). Choosing a conformal factor for which $\mathcal{I}^+$ is non-expanding and putting $^\circ = \Delta = \partial/\partial u$, the relevant expressions are

$$\dot{\phi}_0 = \eth\phi_1 + \sigma\phi_2\ , \tag{3.41}$$

$$\dot{\phi}_1 = \eth\phi_2\ : \tag{3.42}$$

$$\dot{\psi}_0 = \eth\psi_1 + 3(\sigma\psi_2 + \bar{\phi}_2\phi_0)\ , \tag{3.43}$$

$$\dot{\psi}_1 = \eth\psi_2 + 2(\sigma\psi_3 + \bar{\phi}_2\phi_1)\ , \tag{3.44}$$

$$\dot{\psi}_2 = \eth\psi_3 + \sigma\psi_4 + \bar{\phi}_2\phi_2 \tag{3.45}$$

and

$$\dot{\psi}_3 = \eth\psi_4 \ . \tag{3.46}$$

In point of fact we have used only half of the possible equations. The other half involve derivatives of the field variables off $\mathcal{I}^+$ and, for our present purposes, they will not be required.

The evolution equations (3.41) through (3.46) were first obtained in reference [19] by using the Newman-Penrose formalism with respect to the unrescaled metric and by making expansions in powers of $1/\tilde{r}$. Notice that the only spin coefficients that enter are β and σ . β itself depends merely on the intrinsic geometry of $\mathcal{I}^+$ and can be made to vanish by choosing a conformal factor to flatten $\mathcal{I}^+$ completely. $\mathcal{I}^+$ is then regarded topologically as the Riemann sphere and has the line element

$$dl^2 = 2\,d\zeta\,d\bar{\zeta} \ ,$$

with the factor 2 for convenience, and the phase of m^a can then be chosen in such a way that

$$\eth = \delta = \partial/\partial\zeta \ . \tag{3.47}$$

Thus σ, which measures the shear on $\mathcal{I}^+$ of the null hypersurfaces $u = \text{constant}$, is the last remnant of the fact that we started with a curved space-time theory.

The link between the spin-two field and the geometry is found by using the Bianchi identities on $\mathcal{I}^+$ together with the Ricci identities in spinor form [4]. For simplicity we shall choose a gauge to flatten $\mathcal{I}^+$. The Bianchi identities there, as distinct from the spin-two equations, are found by removing the tildes from equation (3.27) and using (3.28). (Note that the curvature spinor C_{ABCD} actually vanishes on $\mathcal{I}^+$.) The result is

$$\iota^D \psi_{ABCD} = \bar{o}^{A'} \nabla^{B'}_{(C} P_{AB)A'B'} \tag{3.48}$$

and we may equally well replace P_{ab} by the trace-free tensor

$$\Phi_{ab} = -\tfrac{1}{2}(R_{ab} - \tfrac{1}{4} R\, g_{ab}) \ . \tag{3.49}$$

It follows that on I^+

$$\psi_4 = \Delta\, \Phi_{110'0'} - \bar{\delta}\, \Phi_{110'1'} - \bar{\sigma}\, \Phi_{111'1'} \ . \tag{3.50}$$

Next, to obtain the relevant components of Φ_{ab} we evaluate the time derivatives of the spin coefficients σ, β (=0) and μ (=0) and use the Ricci identities. These give

$$\dot{\sigma} = \Delta(o^A\, \delta\, o_A) = o^A[\Delta, \delta]\, o_A = -\Phi_{001'1'} \ , \tag{3.51}$$

where use has been made of the relations

$$\Delta o^A = 0 = \Delta \iota^A = \delta \iota^A \ .$$

Similarly,

$$\Phi_{011'1'} = -\dot{\beta} = 0 = -\dot{\mu} = \Phi_{111'1'} \ . \tag{3.52}$$

Hence, combining the last three equations and using the reality of the Ricci tensor, we obtain a relation between the spin-two field and the shear on I^+:

$$\psi_4 = -\ddot{\bar{\sigma}} \ . \tag{3.53}$$

Similarly [19, 39], it may be shown that

$$\psi_3 = -\eth\, \dot{\bar{\sigma}} \tag{3.54}$$

and that

$$\psi_2 - \bar{\psi}_2 = \bar{\eth}^2 \sigma - \eth^2 \bar{\sigma} + \bar{\sigma}\, \dot{\sigma} + \sigma\, \dot{\bar{\sigma}} \ . \tag{3.55}$$

Equations (3.53) to (3.55) hold true not only when I^+ is flattened but also if it is chosen to be metrically spherical. They are not true if we restrict I^+ merely to be expansion-free. We shall see, later, that the flat and spherical metrics form an equivalence class and, henceforth, we shall restrict our attention to these special gauges.

To end this section, it is worthwhile listing the data which may be given freely on $\mathcal{I}^+$ as far as the evolution equations are concerned. Basically, there are two functions $\phi_2(u,\zeta,\bar{\zeta})$ and $\sigma(u,\zeta,\bar{\zeta})$ on the whole of $\mathcal{I}^+$, which tell us about the outgoing radiation fields, together with initial data ϕ_0, ϕ_1, ψ_0, ψ_1 and $\mathrm{Re}.\,\psi_2$ given on one particular cross-section $u = u_0$. The initial data, which are functions of u_0, ζ and $\bar{\zeta}$, amount to the knowledge of certain multipole moments specified after equation (3.25). Giving the radiation fields for all time may seem a strange way of setting up the problem *ab initio* but it is characteristic of using null as opposed to spacelike hypersurfaces. The point is that an equation like

$$\ddot{\phi}_0 = \eth^2\phi_2 + \dot{\sigma}\,\phi_2 + \sigma\,\dot{\phi}_2 , \qquad (3.56)$$

obtained from (3.41) and (3.42), may be read from left to right or from right to left. In other words, knowledge of the radiation determines the evolution of dipole moment or, alternatively, the behaviour of the dipole determines the radiation field.

In conclusion, we should note that as far as the *physics* is concerned the quantities σ and ϕ_2 may not be given freely. We shall see, later, that the condition that only a finite amount of energy be radiated (not unreasonable) leads to the demand that both $\mathring{\sigma}$ and ϕ_2 be square integrable.

4 WEIGHTED SCALARS AT NULL INFINITY

At this stage let us restate the aim which was discussed in the introduction. We want to define spinor fields in cone space and this, roughly speaking, will be achieved as follows. Each point P in cone space defines and is defined by a unique cross-section S of $\mathcal{I}^+$. Given some field component ϕ on $\mathcal{I}^+$ we shall integrate it over S weighted with some appropriate spinorial weighting factor w. This procedure defines a spinor Φ, say, at P. Repeating the performance for all (smooth) cross-sections gives rise to a field on cone space

$$\Phi_w(\triangledown) = \int_S dS\, w\, \phi \quad . \tag{4.1}$$

However, we must ensure that this field is well defined: in other words that the value of each integral is independent of the choice of conformal factor. It turns out that the field components not only have spin weights but also conformal (or boost) weights; and the idea is to choose w in such a way that the differential form involved in (4.1) has total weight zero. This is very similar to the intertwining theory described in reference [20].

Suppose that we are given a conformal factor Ω together with a cross-section S. We have seen in the previous section that this leads to a unique normal n_a for $\mathcal{I}^+$, a unique retarded time parameter u with $u=0$ on S, a unique one-parameter family of cuts $u = \text{constant}$ and, finally, a unique null vector l^a (see figure 6). The relevant equations are (3.9), (3.13), (3.14) and (3.15). Furthermore, associated with l^a and n^a are the normalised spinors o^A and ι^A up to phase.

Now consider a new conformal factor $\hat{\Omega}$ determined by

$$\hat{\Omega} = [\,\Theta(u,\zeta,\bar{\zeta}) + O(r)]\,\Omega \,. \tag{4.2}$$

On I^+ this leads to the rescalings

$$\begin{aligned} \hat{g}_{ab} &= \Theta^2 g_{ab}\,, & \hat{g}^{ab} &= \Theta^{-2} g^{ab}\,, \\ \hat{\varepsilon}_{AB} &= \Theta\,\varepsilon_{AB}\,, & \hat{\varepsilon}^{AB} &= \Theta^{-1}\,\varepsilon^{AB} \end{aligned} \tag{4.3}$$

and, further,

$$\hat{n}_a = \Theta\, n_a \quad \text{and} \quad \hat{n}^a = \Theta^{-1} n^a \,. \tag{4.4}$$

Hence, we obtain a new u parameter and a new family of cuts. Demanding that S be one of these, since we want to check the scaling behaviour of expressions like (4.1) for fixed S, we set

$$\hat{u} = G(u,\zeta,\bar{\zeta}) \quad \text{such that} \quad 0 = G(0,\zeta,\bar{\zeta}) \tag{4.5}$$

and where, from (4.4),

$$\dot{G} = \Theta \,. \tag{4.6}$$

On the fixed cut $u = 0$, it then follows that

$$\hat{l}_a = \Theta\, l_a \quad \text{and} \quad \hat{l}^a = \Theta^{-1}\, l^a \,. \tag{4.7}$$

Hence, on the cut in question, the spinor dyad transforms according to

$$\begin{aligned} \hat{o}_A &= k\, o_A\,, & \hat{o}^A &= \bar{k}^{-1} o^A\,, \\ \hat{\iota}_A &= \bar{k}\,\iota_A\,, & \hat{\iota}^A &= k^{-1}\iota^A\,, \end{aligned} \tag{4.8}$$

where

$$\Theta = k\bar{k} \,. \tag{4.9}$$

(In allowing k to be complex we have incorporated the phase or spin-weight transformation mentioned in the previous section.) Off the cut S the transformation of o_A is more complicated, since the family of cuts is itself transformed, and is given by

$$\hat{o}_A = k\, o_A - \bar{k}^{-1}\,\delta G\,\iota_A\,, \tag{4.10}$$

but we will seldom require the use of this.

Any quantity η on S which transforms according to the law

$$\hat{\eta} = k^p \bar{k}^q \eta \qquad (4.11)$$

under the combined conformal, spin and boost transformations (4.3) and (4.8) will be said to have *weight* $\{p,q\}$. (This is an extension of work by Geroch, Held and Penrose [21] to include conformal rescalings of the metric on I^+.) The *boost-weight* of η is $\frac{1}{2}(p+q)$ while its *spin-weight* s is $\frac{1}{2}(p-q)$. A simple example is provided by the electromagnetic null datum ϕ_2. From equation (3.24) and the scaling law (4.8) together with

$$\hat{\phi}_{AB} = \Theta^{-1} \phi_{AB} , \qquad (4.12)$$

it follows that ϕ_2 has weight $\{-3,-1\}$.

Of course, not all quantities on I^+ possess well defined weights. The spin coefficients β and γ, for example, transform according to

$$\hat{\beta} = \bar{k}^{-2} (\beta + \delta\, log\, k)$$
and
$$\hat{\gamma} = k^{-1} \bar{k}^{-1} (\gamma - \Delta\, log\, \bar{k}) , \qquad (4.13)$$

the proof of which involves the behaviour (2.30) of the derivative operator under conformal rescalings. By comparison, the derivative operators δ and Δ, when applied to η, behave according to

$$\hat{\delta}\, \hat{\eta} = k^p \bar{k}^{q-2} (\delta + p\, \delta\, log\, k + q\, \delta\, log\, \bar{k})\, \eta$$
and
$$\hat{\Delta}\, \hat{\eta} = k^{p-1} \bar{k}^{q-1} (\Delta + p\, \Delta\, log\, k + q\, \Delta\, log\, \bar{k})\, \eta , \qquad (4.14)$$

although the second of these holds good only if η transforms according to (4.11) all over I^+ under the transformation (4.5) which preserves S. (The complication arises because of the way (4.10) o_A transforms off the given cut.)

Even though δ and Δ do not have well defined weights, it is easily seen, by using the connection symbols β and γ as correction terms, that we can construct new operators $\eth$ (introduced earlier) and $\text{þ}'$ (thorn prime) [20] whose weights are well defined under the appropriate circumstances. The first of these is given by

$$\eth\eta = \delta\eta - (p-q)\beta\eta \quad , \tag{4.15}$$

and has weight $\{0,-2\}$ as long as it acts on quantities with $q=0$. The second is defined by

$$\text{þ}'\eta = (\Delta + p\bar{\gamma} + q\gamma)\eta \tag{4.16}$$

and has weight $\{-1,-1\}$ as long as η has weight $\{p,q\}$ all over I^+.

Occasionally, we shall want to apply $\eth$ to an object on S with definite spin-weight but with no definite boost-weight. In this connection there is a useful result concerning the weight of $\eth^n$ but which works only if we restrict to conformal rescalings which take us from one conformal factor to another within the same equivalence class. The equivalence relation is given by

$$\hat{\Omega} \sim \Omega \Leftrightarrow \hat{\Omega} = \Theta\Omega \quad \text{with} \quad \eth^2\Theta^{-1} = 0 \text{ on } I^+ \tag{4.17}$$

and the result is then [6]:

$\eth^n$ has weight $\{0,-n-1\}$ when acting on quantities with $q=n-1$.

To show all this, useful identities are

$$\hat{\eth} = \bar{k}^{-2}k^{2s}\eth k^{-2s}$$

and

$$(\Theta^{-2}\eth)^n = \Theta^{-n-1}\eth^n\Theta^{1-n} \tag{4.18}$$

for Θ restricted by (4.17) and (4.9).

If we start with a flat metric on I^+ then, using equation (3.47), the other metrics in its class are given by

$$dl^2 = \Theta^2 \, d\zeta \, d\bar{\zeta}$$

where (4.19)

$$\Theta = (A + \bar{B}\zeta + B\bar{\zeta} + C\zeta\bar{\zeta}) .$$

Furthermore, if we restrict $\mathscr{I}^+$ to be expansion-free, so that A, B and C are constant (A and C real, B complex), then the metrics (4.19) comprise all the flat and spherical metrics. The conformal factors Ω which give rise to them are said to form a *Bondi class* and $\mathscr{I}^+$ is then said to have a *Bondi scaling*. The associated $u = \text{constant}$ cuts are often said to form a Bondi family.

An important quantity is the shear σ of a given cross-section of $\mathscr{I}^+$. This has weight {1,-3} but the weights of $\dot{\sigma}$ {0,-4} and $\ddot{\sigma}$ {-1,-5}, which enter into the relations (3.53) and (3.54), are defined only if we restrict to the Bondi class of conformal factors. More generally, these equations should be written

$$\psi_4 = \text{þ}' N$$

and (4.20)

$$\psi_3 = \eth N ,$$

where the *news function* N has weight {-4,0} and in a Bondi gauge is given by

$$N = -\dot{\bar{\sigma}} \quad . \tag{4.21}$$

In a general gauge, $\dot{\sigma}$ gives information not only of the radiation but also of the coordinate system employed. As an example, consider a system which is radiation-free so that $\dot{\sigma} = 0$ in any given Bondi system. If we make a conformal rescaling (4.3) but with Θ independent of u, so that $\mathscr{I}^+$ remains expansion-free, then, using (4.10), it follows that the shear transforms according to

$$\hat{\sigma} = k\,\bar{k}^{-3}(\sigma - \hat{u}\,\eth^2\Theta^{-1}) \quad . \tag{4.22}$$

Hence, unless (4.17) is satisfied, which gives another Bondi gauge, $\mathring{\sigma}$ is now non-zero.

Table 1 presents the weights of certain important quantities on $\mathcal{I}^+$. For completeness, I have included the representation of a vector in cone space as well as the twistorially aligned frame on $\mathcal{I}^+$: both of these are introduced later.

TABLE 1. Weighted quantities on $\mathcal{I}^+$.

QUANTITY		WEIGHT	REMARKS
Surface element	dS	$\{2,2\}$	
Representation of vector in C^+	V	$\{1,1\}$	
Spinor along generator of $\mathcal{I}^+$	ι^A	$\{-1,0\}$	
	ι_A	$\{0,1\}$	
Spinor along generator of outgoing null cone	o^A	$\{0,-1\}$	*only on the fixed cut* S
	o_A	$\{1,0\}$	
Twistorially aligned frame	$Z_{\mathrm{A}}{}^A$	$\{0,0\}$	
Weighting factors	o_{A}	$\{1,0\}$	
	$\bar{o}_{\mathrm{A}'}$	$\{0,1\}$	
Corrected derivatives	$ð$	$\{0,-2\}$	*only if* $q=0$
tangent to S	$\bar{ð}$	$\{-2,0\}$	*only if* $p=0$
Corrected derivative up generator of $\mathcal{I}^+$	$Þ'$	$\{-1,-1\}$	*restricted for example to null data*
Shear on $\mathcal{I}^+$ *of outgoing cone*	σ	$\{1,-3\}$	
News function	N	$\{-4,0\}$	$=-\dot{\bar{\sigma}}$ *if Bondi*
Components of Maxwell field	ϕ_2	$\{-3,-1\}$	
	ϕ_1	$\{-2,-2\}$	
	ϕ_0	$\{-1,-3\}$	*compare* [22]
Components of spin-two field	ψ_4	$\{-5,-1\}$	*homogeneity of*
	ψ_3	$\{-4,-2\}$	*twistor*
	ψ_2	$\{-3,-3\}$	*functions*
	ψ_1	$\{-2,-4\}$	$\phi_r(W_\alpha,X_\alpha)$
	ψ_0	$\{-1,-5\}$	

5 FRAME ALIGNMENT AT NULL INFINITY

In this section we shall show how to obtain reference frames on I^+ which are unique up to rigid Lorentz transformation. From such frame fields we shall be able to construct the spinorial weighting factors w to be used in expressions like (4.1) and this will enable us to define spinor fields in cone space. We shall begin by considering the situation which arises when the original manifold is flat.

Frame Alignment in Flat Space

Consider a constant vector field $\tilde{v}^a$ on Minkowski space-time $\tilde{M}$. It satisfies the equation

$$\tilde{\nabla}_a \tilde{v}^b = 0 \tag{5.1}$$

and maps the space-time to itself by translation. In particular, a point O gets mapped to a point P, where $\tilde{v}^a$ is the displacement vector from O to P, and points on the light cone of O are mapped to corresponding points on the light cone of P.

Figure 7 depicts the behaviour of this vector field when we extend it to the manifold M with boundary and, particularly, to I^+. Under the conformal rescaling

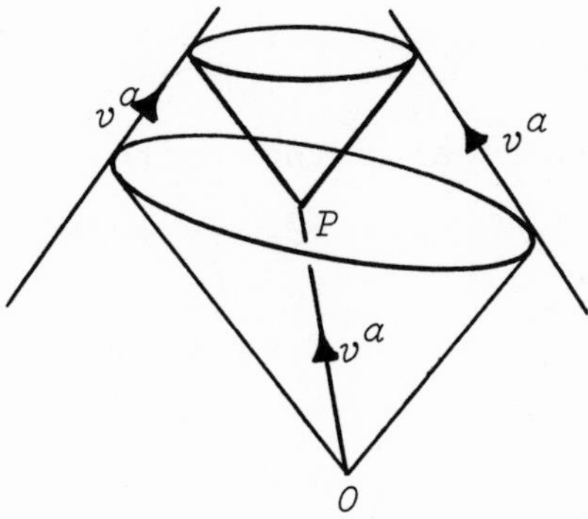

FIGURE 7. The behaviour of a translation at null infinity for flat $\tilde{M}$.

$$g_{ab} = \Omega^2 \tilde{g}_{ab} \qquad (5.2)$$

we shall set

$$v^a = \tilde{v}^a \quad , \qquad (5.3)$$

for a vector field thought of as a derivative operator is defined quite independently of any metric. Note that the norm of v^a is given by

$$g_{ab} v^a v^b = \Omega^2 \tilde{g}_{ab} v^a v^b \qquad (5.4)$$

which vanishes on I^+. The meaning of this is clear from figure 7 since v^a maps the end points of the rays generating the null cone of O to the corresponding end points of the null cone of P. In fact, if the null vectors l^a and n^a are defined as in section 3, then

$$v^a = l_b v^b n^a \quad \text{on } I^+. \qquad (5.5)$$

This relation is independent of the choice of conformal factor and accompanying slicing of I^+. The translation v^a is thus represented on I^+ by the scalar $l_b v^b$ which has weight {1,1} under conformal rescalings.

Similarly, if we take a tetrad of vectors δ^a_{a}, orthonormal with respect to $\tilde{g}_{ab}$ and aligned in $\tilde{M}$ by parallel transport, then on I^+ each of its legs will be tangent to the generators:

$$\delta^a_{\mathrm{a}} = l_{\mathrm{a}} n^a \quad . \qquad (5.6)$$

The linear independence of the vectors δ^a_{a} is lost at each point of I^+ but now emerges as the functional independence of the four weighted scalars $l_{\mathrm{a}}(u,\zeta,\bar{\zeta})$. To illustrate this crucial point, choose Ω so that the metric of I^+ is spherical: then the l_{a} consist simply of linear combinations of the first four spherical harmonics Y_{00}, Y_{1m} [23]. Like the δ^a_{a}, the l_{a} are defined uniquely up to rigid Lorentz transformation and provide an example of the type of weighting factor w which we seek.

Analogous discussions proceed for the behaviour of spin frames. Let $\tilde{z}_{\mathrm{A}}{}^{A}$ be a parallelly aligned spinor dyad defined on Minkowski space-time,

$$\tilde{\nabla}_{AA'}\tilde{z}_{\mathrm{A}}{}^{A} = 0 \ , \tag{5.7}$$

and normalised to unity,

$$\tilde{\varepsilon}_{AB}\,\tilde{z}_{\mathrm{A}}{}^{A}\,\tilde{z}_{\mathrm{B}}{}^{B} = \varepsilon_{\mathrm{AB}} = \begin{pmatrix} 0 & 1 \\ -1 & 0 \end{pmatrix} . \tag{5.8}$$

Analogous to equation (5.3), we shall scale the dyad according to

$$z_{\mathrm{A}}{}^{A} = \tilde{z}_{\mathrm{A}}{}^{A} \tag{5.9}$$

and extend it to the whole of M. The normalisation condition then becomes

$$\varepsilon_{AB}\,z_{\mathrm{A}}{}^{A}\,z_{\mathrm{B}}{}^{B} = \Omega\,\varepsilon_{\mathrm{AB}} \tag{5.10}$$

and since this vanishes on I^{+} it follows that the legs of the dyad are proportional there. Furthermore, setting $v^{a} = z_{0}{}^{A}\bar{z}_{0'}{}^{A'}$ and $z_{1}{}^{A}\bar{z}_{1'}{}^{A'}$ in turn and noting (5.5), it follows that

$$z_{\mathrm{A}}{}^{A} = o_{\mathrm{A}}\,\iota^{A} \quad \text{on } I^{+} , \tag{5.11}$$

where ι^{A} is defined in equation (3.22) and where the scalars o_{A} have weight {1,0}. The freedom in the o_{A} is precisely the freedom in the $z_{\mathrm{A}}{}^{A}$ with which we started, namely the group SL(2,C) acting rigidly; and the quantity o_{A} is the desired spinorial weighting factor.

Given the spin frame $z_{\mathrm{A}}{}^{A}$ together with its complex conjugate we may construct a tetrad of vectors δ^{a}_{a} by taking linear combinations of the null vectors $z_{\mathrm{A}}{}^{A}\,\bar{z}_{\mathrm{A'}}{}^{A'}$, namely

$$\delta^{a}_{a} = \sigma_{a}{}^{AA'} z_{A}{}^{A} \bar{z}_{A'}{}^{A'} \quad . \tag{5.12}$$

Thus

$$l_{a} = \sigma_{a}{}^{AA'} o_{A} \bar{o}_{A'} \quad . \tag{5.13}$$

A convenient choice for the coefficients $\sigma_{a}{}^{AA'}$, to ensure that the δ^{a}_{a} are orthonormal with respect to the unrescaled metric, is to take them to be the unit and Pauli matrices with factor $2^{-\frac{1}{2}}$ [4]. These, of course, are the Infeld-van der Waerden connecting symbols.

Conformally Invariant Equations of Alignment

Let us examine a little more closely the equations (5.7) and (5.8) for spin frame alignment in Minkowski space-time. Equation (5.7) implies trivially that the twistor equations

$$\tilde{\nabla}^{(B}{}_{B'} \tilde{z}_{A}{}^{A)} = 0 \tag{5.14}$$

are satisfied; but the surprising fact is that the only solutions of (5.14) consistent with the normalisation condition (5.8) are the constant solutions, namely the solutions of (5.7). In other words

$$(5.7) \cap (5.8) \Leftrightarrow (5.14) \cap (5.8) \quad .$$

This follows by using (2.37) to write down the general solutions to (5.14):

$$\tilde{z}_{A}{}^{A}(x) = \tilde{z}_{A}{}^{A}(O) - i x^{AA'} \Pi_{AA'} \quad , \tag{5.15}$$

where $\Pi_{0A'}$ and $\Pi_{1A'}$ are constant spinors. Then, imposing the normalisation condition for all values of x implies that the $\Pi_{AA'}$ have to vanish.

The advantage of using equation (5.14) rather than (5.7) to align the spinor dyads lies in its conformal invariance, for we now have equations of alignment in M given in terms of the rescaled metric. In particular, the dyads satisfy the following three conditions on $\mathcal{I}^+$.

$$\left.\begin{array}{ll} \text{(A)} & \nabla^{(B}{}_{B'} Z_{\mathrm{A}}{}^{A)} = 0 \ , \\ \text{(B)} & Z_{\mathrm{A}}{}^{A} /\!/ \iota^{A} \\ \text{and} \quad \text{(C)} & D(\varepsilon_{AB} Z_{\mathrm{A}}{}^{A} Z_{\mathrm{B}}{}^{B}) = -\varepsilon_{\mathrm{AB}} \ , \end{array}\right\} \tag{5.16}$$

the last of these arising from the condition (5.10). In fact, the solutions of (5.16) have precisely the form (5.11). Condition (A) leads to the relations

$$\eth o_{\mathrm{A}} = 0 = Þ' o_{\mathrm{A}} \tag{5.17}$$

while (C) taken with (A) says that

$$o_{\mathrm{A}} \bar{\eth} o_{\mathrm{B}} - o_{\mathrm{B}} \bar{\eth} o_{\mathrm{A}} = \varepsilon_{\mathrm{AB}} \ . \tag{5.18}$$

When $\tilde{M}$ is curved there will be, in general, no global solutions to equation (5.14). Nevertheless, for an asymptotically flat space-time we may impose the conditions (5.16) on $\mathcal{I}^+$. It turns out [24] that the relations obtained are precisely the relations (5.17) and (5.18) and the solutions (5.11) are, therefore, the same as in flat space. The key equations are

$$\left.\begin{array}{l} o_{A} D Z_{\mathrm{A}}{}^{A} = 0 \ , \\ \iota_{A} D Z_{\mathrm{A}}{}^{A} + o_{A} \bar{\delta} Z_{\mathrm{A}}{}^{A} = 0 \ , \\ o_{A} \delta Z_{\mathrm{A}}{}^{A} = 0 \\ \text{and} \\ \iota_{A} \delta Z_{\mathrm{A}}{}^{A} + o_{A} \Delta Z_{\mathrm{A}}{}^{A} = 0 \ . \end{array}\right\} \tag{5.19}$$

The point is that, having imposed condition (B) on $\mathcal{I}^+$, the shear σ does not appear. (In actual fact, in the presence of gravitational radiation the *only* solutions to (A) on $\mathcal{I}^+$ are those which satisfy (B).)

It is now straightforward to show that the line element (3.11) on $\mathcal{I}^+$ may be rewritten as

$$dl^2 = 2\, o_A\, d\, o^A\, \bar{o}_{A'} d\, \bar{o}^{A'} \quad . \tag{5.20}$$

To see this, consider an infinitesimal element on $\mathcal{I}^+$,

$$dx^a = du\, n^a + \bar{\varepsilon} m^a + \varepsilon \bar{m}^a \quad , \tag{5.21}$$

(where m^a and $\bar{m}^a$ were introduced in section 3) with length $-2\varepsilon\bar{\varepsilon}$. The use of (5.17) and (5.18) then gives

$$o_A\, d\, o^A = \varepsilon\, o_A\, \bar{\delta}\, o^A = \varepsilon \quad .$$

Furthermore, the surface element of the cross-sections of $\mathcal{I}^+$ may be expressed as

$$dS = i\, o_A\, d\, o^A \wedge \bar{o}_{A'}\, d\, \bar{o}^{A'} \tag{5.22}$$

and note that this has the correct weight {2,2}.

Asymptotic Spin Space

Let us now consider the implications of the results of this section. For flat $\tilde{M}$ each parallelly propagated spinor field $\tilde{\omega}^A$ satisfies the equations

$$\left.\begin{array}{l} \nabla^{(A}{}_{A'}\, \omega^{B)} = 0 \\ \text{with} \\ \omega^A \parallel \iota^A \end{array}\right\} \text{ on } \mathcal{I}^+ \quad ; \tag{5.23}$$

and there exists a natural isomorphism between the spin space of $\tilde{M}$ and the space of solutions to the equations (5.23). When $\tilde{M}$ is asymptotically flat we may still use (5.23) to *define* a spin space over cone space.

DEFINITION. Asymptotic spin space $\mathcal{S}$ is the space of solutions on $\mathcal{I}^+$ to the equations (5.23). By restricting the solutions to each cross-section of $\mathcal{I}^+$, $\mathcal{S}$ then defines a spin space at each point of cone space but with trivial spin bundle: (trivial because we may compare spinors at different points in cone space). Henceforth we shall identify each local spin space with asymptotic spin space.

Equation (5.11) which solves (5.16) gives us a normalised basis for $\mathcal{S}$ and we may, therefore, make the expansion

$$\omega^A = \omega^{\mathrm{A}} z_{\mathrm{A}}{}^{A} , \tag{5.24}$$

the components ω^{A} being constants with freedom SL(2,C). Furthermore, we may raise and lower the index A using the symbol $\varepsilon_{\mathrm{AB}}$.

DEFINITION. The space T of *asymptotic vectors* is the tensor product of $\mathcal{S}$ with its complex conjugate $\bar{\mathcal{S}}$.

T is spanned by the tetrad δ^a_{a}, defined in equation (5.12), and a member v^a of T may be given components by analogy with (5.24). In particular, this involves the weight {1,1} quantity

$$V = v^{\mathrm{AA'}} o_{\mathrm{A}} \bar{o}_{\mathrm{A'}} = v^{\mathrm{a}} l_{\mathrm{a}} \tag{5.25}$$

(compare equation (5.5)). Again, we may raise and lower the index a by using the Minkowski metric g_{ab}.

For flat $\tilde{M}$, the space T may be identified with the tangent space to the original manifold and, in particular, with the group of translations. When $\tilde{M}$ is no longer flat the identification with its tangent space is lost but T still defines the translations on I^+. We are thus led to consider motions in cone space and these will be discussed in the following section.

For completeness, we shall exhibit an explicit solution to the equations for frame alignment. We shall need to exploit the toplogy S^2 of the cross-sections of

$\mathcal{I}^+$; and so consider a conformal factor to make $\mathcal{I}^+$ metrically spherical. In that case the line element becomes

$$dl^2 = (1 + \zeta\bar{\zeta})^{-2}\, 2\, d\zeta\, d\bar{\zeta} \tag{5.26}$$

and, using equations (3.47) and (4.18), it follows that we can arrange things such that

$$\eth = (1 + \zeta\bar{\zeta})^{1-s}\, \partial/\partial\zeta\, (1 + \zeta\bar{\zeta})^{s} \quad . \tag{5.27}$$

Hence (5.17) yields

$$o_A = (1 + \zeta\bar{\zeta})^{-\frac{1}{2}}\, \xi_A(\bar{\zeta}) \quad , \tag{5.28}$$

where ξ_A is analytic on the sphere. However, for the function V of equation (5.25) to be *globally defined*, in particular as $\zeta \to \infty$ ($\theta = 0$), we require that ξ_A have the form

$$\xi_A = (c\bar{\zeta} + d\ ,\ a\bar{\zeta} + b) \tag{5.29}$$

where, from the normalisation condition (5.18),

$$ad - bc = 1 \quad . \tag{5.30}$$

The freedom in o_A, like the freedom in $\bar{\zeta}$ itself, is thus SL(2,C) and a convenient choice is given by

$$o_A = (1 + \zeta\bar{\zeta})^{-\frac{1}{2}}\, (1\ ,\ \bar{\zeta}) \tag{5.31}$$

Note finally that if we choose $\mathcal{I}^+$ to be flat or spherical, o_A satisfies

$$\bar{\eth}^2 o_A = 0$$

in addition to (5.17). Consequently, using equation (5.25),

$$\eth^2 V = 0 = \bar{\eth}^2 V = \mathring{V} \tag{5.32}$$

and this characterises asymptotic vector space.

6 CONE SPACE

Basic Definitions and Affine Structure

We have waited this long before commencing a detailed discussion of cone space since we needed the results of the last two sections. Cone space C^+ is the space of smooth cross-sections of future null infinity or, equivalently, the space of outgoing null hypersurfaces well behaved in some neighbourhood of I^+.

$$C^+ = \{\bigtriangledown\} . \tag{6.1}$$

Since it is our intention to describe properties of a given dynamical system by performing integrals over the cuts of I^+, we are forced to focus attention on cone space rather than on the original manifold. It may be asked "Why not consider just those cross-sections which arise from points of M?" The problem is that, owing to the gravitational bending of light, there is no guarantee that, given a point in M, a corresponding cross-section will exist at all; and, if it does, there is no reason for it to be globally defined. For the present, then, we shall consider all smooth cuts of I^+.

First of all, cone space is an infinite dimensional manifold. A convenient system of labelling may be found by choosing a conformal factor to make I^+ spherical. Next, take any cross-section with smooth shear and call it the origin O of C^+. Any other point P in C^+ (figure 8)

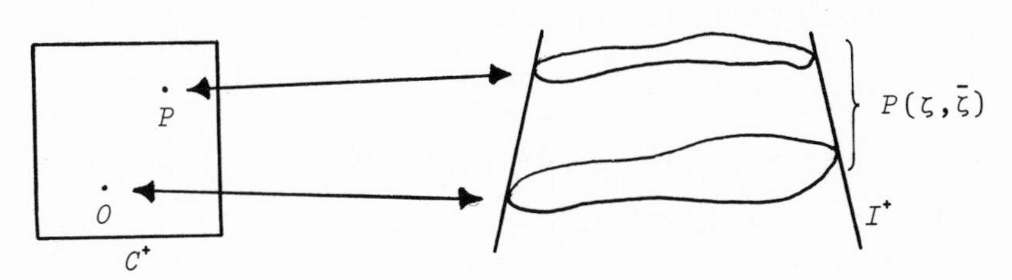

FIGURE 8. Cone space may be labelled by the set of smooth functions on the sphere.

defines a second cross-section of I^+ and may be labelled relative to O by means of a smooth real function $P(\zeta,\bar{\zeta})$ on the sphere. On each generator of I^+ this function tells us the affine distance u we must walk to go from the first cut to the second. In this way C^+ is put into (1-1) correspondence with the set of smooth functions on the sphere. There are many different ways of setting up a topology for cone space. One way is to make use of the positive definite metric

$$\int dS\,(P_1-P_2)^2 \tag{6.1}$$

though this depends upon the choice of conformal factor. The dependence of all this on the degree of smoothness of the functions need not concern us here but, for definiteness, we shall take them to be analytic, (i.e. real analytic functions on the sphere or complex analytic separately in ζ and $\bar{\zeta}$).

To turn cone space from a topological space to a manifold we must exhibit a system of coordinates and, to this end, we may use the coefficients p_{lm} which arise in the expansion of the function in spherical harmonics:

$$P(\zeta,\bar{\zeta}) = p_{lm}\,Y_{lm}(\zeta,\bar{\zeta}) \; . \tag{6.2}$$

Let us next consider the tangent space TC^+ to cone space. We proceed by analogy with figure 7 and equation (5.5). A vector v at a point P of cone space may be represented on the cut of I^+ corresponding to P by a vector field parallel to n^a:

$$v \leftrightarrow V(\zeta,\bar{\zeta})\,n^a \quad . \tag{6.3}$$

The vector field is independent of the choice of conformal factor so that, since n^a has weight $\{-1,-1\}$, V has weight $\{1,1\}$. If ε is small then εv defines a point Q close to P and, in the given u system, the cut corresponding to Q may be obtained from the one corresponding to P by walking a distance $\varepsilon V(\zeta,\bar{\zeta})$ up each generator. (The restriction to a spherical $\mathcal{I}^+$ is here relaxed.) The weighted function V thus represents the vector v in the tangent space to cone space.

Like Minkowski space, cone space possesses a natural affine structure [25]. The tangent bundle is trivial because we may compare vectors at different points. The vector v at P may be used to define a point Q a finite distance away and is then the unique relative position vector $\vec{PQ}$. Further, we have the important addition law

$$\vec{OP} + \vec{PQ} = \vec{OQ} \ .$$

The easiest way to see all this is to choose any conformal factor for which $\mathcal{I}^+$ is non-expanding. The metric of $\mathcal{I}^+$ is then independent of u which, in turn, becomes an affine parameter since n^a is now parallelly transported up each generator. Two vectors v_1 and v_2 at different points will be equal if their representations $V_1(\zeta,\bar{\zeta})$ and $V_2(\zeta,\bar{\zeta})$ are equal. Two points O and P have relative position vector p represented by the function $P(\zeta,\bar{\zeta})$ which, on each generator of $\mathcal{I}^+$, determines the affine distance apart of the corresponding cross-sections. (Figure 8 exemplifies this.)

In the previous section we showed how a constant vector field in Minkowski space gave rise to a vector field on $\mathcal{I}^+$ by means of equation (5.5). In the same way a constant vector field on cone space is given by a vector field on $\mathcal{I}^+$ according to the correspondence

$$v(\bigtriangledown) \leftrightarrow V(u,\zeta,\bar{\zeta}) \ , \qquad (6.4)$$

where V has weight $\{1,1\}$ and satisfies

$$\text{Þ}' V = 0 \ . \qquad (6.5)$$

This means, of course, that if $\mathcal{I}^+$ is non-expanding, V is independent of u. Such a vector field defines a motion of cone space called a *supertranslation*. The asymptotic vectors introduced in the previous section are special cases of supertranslations and are called *translations*. They form a subspace of the supertranslations and, in a Bondi scaling, satisfy equations (5.32).

The Invariant Norm and Minkowski Submanifolds

We shall now define a norm on the tangent space to cone space and hence on the space of supertranslations. For each $v \in TC^+$ at P, represented by the weighted function $V(\zeta,\bar{\zeta})$, define $|v|$ by Newman's integral [26]

$$|v|^{-2} = (4\pi)^{-1} \int dS\, V^{-2} \quad , \tag{6.6}$$

with the integral performed over the cut corresponding to P and having topology S^2. (Strictly speaking, a *real* sphere integration works only for V positive or negative definite: more generally, homologous deformations of the contour into the complex have to be considered.) Note that the differential form involved in the norm (6.6) has weight {0,0}, which is why it was chosen, so that, unlike the expression (6.1), it is quite independent of the choice of conformal factor. Note, secondly, that

$$|\lambda v|^2 = \lambda^2 |v|^2 \quad , \quad \text{for real } \lambda, \tag{6.7}$$

so that $|v|^2$ is homogeneous of degree two in v; which is right and proper. Note, thirdly, that (6.6) does give the correct expression when restricted to vectors in Minkowski space or to asymptotic vectors. To see this, use equations (5.22) and (5.25) and consider

$$|v|^2 = 4\pi \left(\int \frac{o_A do^A \wedge \bar{o}_{A'} d\bar{o}^{A'}}{(v^{AA'} o_A \bar{o}_{A'})^2} \right)^{-1} \tag{6.8}$$

with result $v^a v_a$. The easiest way to establish this is to choose $o_A = (1,\bar{\zeta})$, thus flattening $\mathcal{I}^+$. Next, free $\bar{\zeta}$ from ζ and write it as η. The denominator in (6.8) is now analytic separately in ζ and η. Then perform an open integral in η between two fixed end points and follow it with an S^1 in ζ. The resulting contour is, in fact, homologous to the required S^2.

However, (6.6) does not give rise to a Riemannian metric on the whole of TC^+. To see this, take $\mathcal{I}^+$ to be spherical and, as counter example, choose a two-dimensional subspace of vectors defined by

$$V = a + b\cos^2\theta \quad . \tag{6.9}$$

(To avoid complications with the contour, impose the restriction $a(a+b)>0$.) With regard to the expansion (6.2) this subspace is spanned by the spherical harmonics Y_{00} and Y_{20}. The result of (6.6) is now

$$|v|^2 = 2a^2[(1 + b/a)^{-1} + (b/a)^{-\frac{1}{2}}\tan^{-1}(b/a)^{\frac{1}{2}}]^{-1} \tag{6.10}$$

The point is that $|v|^2$ is a non-quadratic function of the components a and b.

We have seen that, according to the chosen norm, cone space is a non-Riemannian manifold. Nevertheless, it does possess Riemannian and, in fact, Minkowskian submanifolds. This follows by choosing any origin O in cone space and by using T to construct the four-parameter family of cuts M_O^+ obtained from O by translation. In a gauge for which $\mathcal{I}^+$ is divergence-free we may label these cuts by using (5.25) thus:

$$M_O^+ = \{\leftrightarrow\,|\,u = x^{AA'}o_A\bar{o}_{A'}\} \quad , \tag{6.11}$$

where we have put u=0 at O. Use of equation (6.8) shows that M_O^+ is trivially a Minkowski space and the coordinates $x^{AA'}$ appearing in (6.11) are standard Minkowski coordinates, being the components of the position vector from the origin O to the point in question. There are, of course, infinitely many of these spaces, the freedom being the supertranslations factored out by the translations. Naturally, if the original manifold is flat, then a unique space of cuts is singled out according to figure 2. For the moment, we shall postpone the question as to whether a unique space can be chosen in general.

Although the norm (6.6) has nasty properties, being non-Riemannian, we may use it to define the Bondi-Metzner-Sachs group. When first introduced [14, 15] the BMS group emerged as the group of coordinate transformations acting on $\tilde{M}$ which preserved certain asymptotic properties of the components of the metric tensor. Later [27, 13], it was considered as a group of transformations of I^+ onto itself. I wish to adopt a slightly different viewpoint here and consider it instead as the group of motions of cone space. Of crucial importance will be the affine structure of cone space: in particular, that to each pair of points P and Q in C^+ we may associate a unique position vector $\vec{PQ}$ which is the supertranslation required to get from P to Q. The definition runs directly analogous to that of the Poincaré group as the group of motions of Minkowski space-time.

DEFINITION. The BMS group is the group of mappings of cone space onto itself which acts linearly on the space of position vectors and which preserves the norm of each position vector.

It follows that the subgroup of BMS which leaves invariant some point O in C^+ is the Lorentz group (compare [25]) and it turns out that BMS is the semi-direct product of the Lorentz group with the Abelian group of supertranslations. To see this, take I^+

divergence-free and label the points in C^+ by using the supertranslations from some origin O. Let P have position vector p relative to O. Under the action of an element of BMS let the images of O and P have position vectors $f(O)$ and $f(p)$ with respect to O. Then from the condition of linearity,

$$f(p) - f(O) = Lp \quad , \tag{6.12}$$

where L acts linearly on p. Furthermore,

$$|Lp| = |p| \quad \text{for all } p \tag{6.13}$$

and, in particular, when p is a translation. Hence, the group of possible L s is a subgroup of the Lorentz group. To show that it is the whole of the Lorentz group we need to define the action of the Lorentz group on the supertranslations. If p is a translation, it is represented, according to equation (5.25), by the weighted function

$$P(\zeta,\bar{\zeta}) = p(o_A , \bar{o}_{A'}) = p^{AA'} o_A \bar{o}_{A'} \quad . \tag{6.14}$$

Under a Lorentz rotation of the vector p we have

$$p' (o_A , \bar{o}_{A'}) = p(o^*_A , \bar{o}^*_{A'}) \ , \tag{6.15}$$

where

$$o^*_A = \Lambda^B{}_A o_B \quad . \tag{6.16}$$

Note that

$$o^*_A \, d\, o^{*A} = o_A \, d\, o^A \quad . \tag{6.17}$$

If p is a supertranslation, then it is represented by a function $p(o_A , \bar{o}_{A'})$ homogeneous of degree one in each of its arguments. (Remember that the ratio of the components of o_A may be used to label the generators of I^+.) The action of the Lorentz group on p is now *defined* by equations (6.15) and (6.16); and using (6.8), but with p' replacing v, and (6.17) , it follows that $|p'| = |p|$.

Hence we have shown that the BMS group consists of mappings of the form

$$f(p) = Lp + f(O) \quad , \tag{6.18}$$

which is a Lorentz rotation followed by a supertranslation.

We can pick out the Poincaré group by taking that subgroup of BMS which preserves a given Minkowski space M^+_O but, of course, it will not be unique. The possible choice of a canonical Poincaré group, like the choice of a canonical Minkowski space, will again be postponed.

In section 5 there was discussed a special equivalence class of conformal factors for which the metric of I^+ is either spherical or flat. In such a Bondi gauge a translation V is characterised by equation (5.32). It follows that we can obtain a geometric meaning for the u = constant slicings associated with these special gauges by noting that $V = 1$ solves (5.32). For, if u=0 corresponds to the point O in C^+, then the u = constant cuts define a geodesic in M^+_O. Furthermore, by using the norm (6.6) this geodesic is timelike or null depending upon whether I^+ is spherical or flat.

7 SPINOR FIELDS IN CONE SPACE

We are now well placed, armed with the results of the preceding sections, to define fields on cone space in order to describe the state and behaviour of a given dynamical system. For each point in cone space we shall perform two-dimensional integrals over the corresponding cross-section of I^+. The quantities entering into the integrations will be the weighted scalars listed in table 1; namely, the element of two-surface dS, the various components of the spin-one and spin-two fields, ϕ_r and

ψ_r, the shear σ and news function N and finally the weighting factors o_A, $\bar{o}_{A'}$ and l_a. The differential forms involved will be constructed to have weights {0,0} and so guarantee the invariance of the integrals under conformal rescalings of the metric on I^+.

Charge, Momentum and the Evolution Equations

The simplest example of the procedure just described is the electric charge of the system

$$q = -1/2\pi \int dS\, \phi_1 \tag{7.1}$$

which is a constant scalar field on cone space. The constancy follows from equation (7.7), derived later, and expresses the fact that there are no massless, charged fields.

A more complicated example is the Bondi-Sachs expression [14, 16] for the four-momentum of the system,

$$P^a(\bigtriangledown) = -\int dS\, l^a\, (\psi_2 - \sigma N) \quad , \tag{7.2}$$

which is an asymptotic vector field on cone space. (Strictly speaking, we have the components of such a field in a standard orthonormal frame.)

Of course, we shall want to know how the momentum evolves in time and, more generally, given two points O and A in C^+ we shall want to compare $P^a(O)$ with $P^a(A)$. This is feasible because of the triviality of the vector bundle. So, before introducing more fields onto cone space we shall develop the evolution equations (3.41) through (3.46).

For simplicity of calculation (but for no other reason) we shall choose a Bondi conformal factor and set up a family of cuts u = constant for which u=0 corresponds to the point O in C^+. In this system let the position vector $\vec{OA}$ of a neighbouring point A be represented by the infinitesimal supertranslation $a(\zeta,\bar{\zeta})$. With the same conformal factor we shall consider a new family of cuts u' = constant given by

$$u' = u - a(\zeta,\bar{\zeta}) \quad , \tag{7.3}$$

with $u'=0$ corresponding to A. On $\mathcal{I}^+$ the preservation of equation (3.15) generates a transformation of l_a given by

$$l'_a = l_a - \nabla_a a \tag{7.4}$$

and this in turn means that

$$\left.\begin{array}{l} o'_A = o_A + \eth a\, \iota_A \\ \text{together with} \\ \iota'_A = \iota_A \quad . \end{array}\right\} \tag{7.5}$$

At each point of $\mathcal{I}^+$ the equations (7.5) generate transformations of the field variables and of the shear. However, given the field components on the section $u=0$, we need the transformed components on the *new* section $u = a(\zeta,\bar{\zeta})$. Using the defining equations (3.24) and (3.25) together with the evolution equations (3.41) to (3.46), we get, to first order in a,

$$\phi'_0 = \phi_0 + 2\eth a\, \phi_1 + a(\eth\phi_1 + \sigma\phi_2) \quad , \tag{7.6}$$

$$\phi'_1 = \phi_1 + \eth a\, \phi_2 + a\eth\phi_2 \quad , \tag{7.7}$$

$$\phi'_2 = \phi_2 + a\,\dot{\phi}_2 \quad : \tag{7.8}$$

$$\psi'_0 = \psi_0 + 4\eth a\, \psi_1 + a(\eth\psi_1 + 3\sigma\psi_2 + 3\bar{\phi}_2\phi_0) \quad , \tag{7.9}$$

$$\psi'_1 = \psi_1 + 3\eth a\, \psi_2 + a(\eth\psi_2 + 2\sigma\psi_3 + 2\bar{\phi}_2\phi_1) \quad , \tag{7.10}$$

$$\psi'_2 = \psi_2 + 2\eth a\, \psi_3 + a(\eth\psi_3 + \sigma\psi_4 + \bar{\phi}_2\phi_2) \quad , \tag{7.11}$$

$$\psi'_3 = \psi_3 + \eth a\, \psi_4 + a\eth\psi_4 \quad , \tag{7.12}$$

and

$$\psi'_4 = \psi_4 + a\,\dot{\psi}_4 \quad , \tag{7.13}$$

with ψ_3 and ψ_4 given by equations (3.53) and (3.54).

Finally,

$$\sigma' = \sigma - \eth^2 a + a\dot{\sigma} \quad . \tag{7.14}$$

Applying all this to the Bondi-Sachs momentum, which, in view of the correction term $-\sigma N$ and equation (3.55), is real, now gives

$$\Big[P_a\Big]_0^A = -\int dS\, l_a\, a\, (|\dot{\sigma}|^2 + |\phi_2|^2) \quad , \tag{7.15}$$

to first order in a. If we now restrict attention to any one of the Minkowski submanifolds, M_0^+, of cone space discussed in the previous section, it follows that P^a becomes a vector field in flat space. Furthermore, using the expression (6.11) we find that the derivative of P^a is given by

$$\nabla_a P_b = -\int dS\, l_a\, l_b\, (|\dot{\sigma}|^2 + |\phi_2|^2) \quad , \tag{7.16}$$

which is a trace-free, symmetric tensor in M_0^+. If t^a is any future pointing, timelike asymptotic vector $(t^a l_a > 0)$ then, assuming that P^a is likewise timelike and future pointing,

$$t^a \nabla_a m^2 \leqslant 0 \tag{7.17}$$

where

$$m = (P^a P_a)^{\frac{1}{2}} \quad . \tag{7.18}$$

(Even if we do not make reasonable assumptions about P^a, (7.17) will still hold if we replace t^a by P^a.) Equation (7.17) says that as electromagnetic and gravitational waves leave the system so its rest mass decreases, which is by no means surprising. If we assume that the radiation dies off sufficiently fast as $u \to \pm\infty$, then the total amount of momentum leaving the system is given by

$$\Big[P_a\Big]_\infty^{-\infty} = \Pi_a = \int_{-\infty}^{\infty} du \int dS\, l_a\, (|N|^2 + |\phi_2|^2) \quad ; \tag{7.19}$$

and this will be of crucial importance when we come to provide a quantum description of the radiation process.

In and Out Fields

Associated with the outgoing waves we may define a pair of *out fields* in C^+ ; namely, a spin-one field

$$\phi^{out}_{AB} = (2\pi)^{-1} \int dS \, o_A \, o_B \, \dot{\phi}_2 \tag{7.20}$$

and a spin-two field

$$\psi^{out}_{ABCD} = (2\pi)^{-1} \int dS \, o_A \, o_B \, o_C \, o_D \, \dot{\psi}_4 \quad : \tag{7.21}$$

and, by virtue of the equations (6.11), (7.8) and (7.13), these satisfy the free field equations

$$\nabla_{AA'} \phi^{AB}_{out} = 0 = \nabla_{AA'} \psi^{ABCD}_{out} \tag{7.22}$$

when restricted to M^+_0. Notice that the free fields depend only on the null data, ϕ_2 and ψ_4, of the original fields on $\mathcal{I}^+$; that is to say, only on the radiation components and not on their Coulomb parts. Consequently, as long as the radiation is suitably smooth, the *out* fields are singularity-free in M^+_0 (and, indeed, in the whole of cone space).

More generally, given null datum ϕ_n for some spin $\frac{1}{2}n$ field on $\mathcal{I}^+$ we may construct the corresponding *out* field in cone space:

$$\phi^{out}_{AB\ldots J} = (2\pi)^{-1} \int dS \, o_A \, o_B \ldots o_J \, \text{þ}' \, \phi_n \tag{7.23}$$

and this is a solution to the zero rest mass field equations in M^+_0. If the original manifold of points is flat and if we choose M^+_0 to consist of the corresponding canonical cuts of $\mathcal{I}^+$ then the expression (7.23) is just the Kirchoff representation of the final value problem

[28] ; which is to say that (7.23) then represents the unique free field on Minkowski space with null datum ϕ_n on I^+.

A similar analysis follows if we replace I^+ by I^-. The *out* fields are then replaced by the *in* fields and we now have the free fields which enter into a discussion of scattering theory. As an example, consider the scattering in flat space of a hypothetical massless scalar field off a given fixed source

$$\Box\phi = J \quad . \tag{7.24}$$

Depending upon whether we specify data on I^- or I^+ the solutions may be given in terms of the retarded or advanced Green's functions.

$$\phi = \phi_{in}(x) + \int d^4y\, G_R(x,y)\, J(y) \tag{7.25}$$

solves the initial value problem while

$$\phi = \phi_{out}(x) + \int d^4y\, G_A(x,y)\, J(y) \tag{7.26}$$

solves the final value problem. Thus, three fields enter the problem. Of course, the interacting or *interpolating* field ϕ may be eliminated to give a relationship between the *in* and *out* fields [29] :

$$\phi_{out} = \phi_{in} + \int d^4y\, G_C(x,y)\, J(y) \quad , \tag{7.27}$$

and where G_C is the commutator function with support on the light cone.

When the original manifold $\tilde{M}$ is asymptotically flat the three fields ϕ_{in}, ϕ and ϕ_{out} are defined on three different manifolds $M^-_{0^-}$, $\tilde{M}$ and $M^+_{0^+}$. An interesting programme would be to use the formalism to describe, say, the classical scattering of light around the sun. Given null datum on I^- we would aim to solve Maxwell's equations in M and finally calculate data on I^+. Use of equation (7.20) and a similar expression would then give us a pair of free fields *in* and *out* on a pair of flat spaces together with a rule (S matrix) for getting from one to the other. An important question would then be "To what extent can we replace an asymptotically flat space-time by a pair of flat spaces together with a set of S matrices governing the scattering of massless fields?"

Potentials and Quantisation

Not only do we need to understand the classical theory of scattering but we shall also want a quantum description in order, for example, to discuss the Hawking process which was explained in the lectures of Professor Parker. Since the *in* and *out* fields are free there is, naturally, no great difficulty in obtaining a consistent quantisation procedure.

We shall begin by considering the spin-one case and introduce the left-handed potentials

$$\alpha^{out}_{AA'} = -(2\pi)^{-1} \int dS \, o_A \, \bar{o}_{A'} \, T^{-1} \eth (T\phi_2) \tag{7.28}$$

and

$$\chi^{out}_{A'B'} = (4\pi)^{-1} \int dS \, \bar{o}_{A'} \, \bar{o}_{B'} \, T^{-2} \eth (T^2 \phi_1) \quad . \tag{7.29}$$

T is any smooth, timelike (nowhere vanishing) supertranslation and, thus, satisfies equation (6.5). It is introduced to ensure that the integrals are well defined. In M^+_0, α^{out}_a satisfies the relations

$$\left.\begin{array}{l} \nabla_A{}^{A'} \alpha^{out}_{BA'} = -\phi^{out}_{AB} \\ \text{and} \\ \nabla^A{}_{A'} \alpha^{out}_{AB'} = 0 \quad ; \end{array}\right\} \tag{7.30}$$

and hence also the Lorenz condition

$$\nabla_a \alpha^a_{out} = 0 \ . \tag{7.31}$$

Note, in particular, that α^{out}_a is a complex potential for ϕ^{out}_{AB} but is not a potential for $\bar{\phi}^{out}_{A'B'}$ and it turns out to correspond to photons of one definite helicity. The usual real potential is obtained by adding in its complex conjugate. The spinor field $\chi^{out}_{A'B'}$ satisfies

$$\left.\begin{aligned} &\Box \chi^{out}_{A'B'} = 0 \\ \text{and} \\ &\nabla_A{}^{B'} \chi^{out}_{B'A'} = -\alpha^{out}_{AA'} \end{aligned}\right\} \tag{7.32}$$

and so is a Hertz potential [31] for α^{out}_a .

The freedom in the choice of T, which appears in the definitions (7.28) and (7.29), amounts to the freedom to make gauge transformations of the second kind,

$$\alpha^{out}_a \to \alpha^{out}_a - \nabla_a \Lambda \ .$$

If we choose T to be a timelike translation,

$$T = t^{AA'} o_A \bar{o}_{A'} \ , \tag{7.33}$$

this puts α^{out}_a into the Coulomb gauge:

$$t^a \alpha^{out}_a = 0 \ . \tag{7.34}$$

(Since we are dealing with free fields there is no inconsistency between the transversality of α^{out}_a and the Lorenz condition.) On integrating by parts and using equations (5.17) and (5.18), it follows that (7.28) may be rewritten

$$\alpha^{out}_a = (2\pi)^{-1} \int dS \, m_a \, \phi_2 \ , \tag{7.35}$$

with the *polarization vector* m_a given by

$$m_{AA'} = T^{-1} o_A o_B t^B{}_{A'} \ . \tag{7.36}$$

Next, choosing any Bondi scaling for I^+, make a

Fourier analysis of ϕ_2 up each generator (compare [32]):

$$\phi_2(u,\zeta,\bar{\zeta}) = (2\pi)^{-\frac{1}{2}} \int_0^\infty \omega\, d\omega [e^{-i\omega u} \alpha_1(\omega,\zeta,\bar{\zeta}) + e^{i\omega u} \beta_1^\dagger(\omega,\zeta,\bar{\zeta})] . \quad (7.37)$$

The coefficients $\alpha_1^\dagger$ and $\beta_1^\dagger$ are going to be the creation operators for momentum state photons of opposite helicity. Roughly speaking, ω measures the energy and ζ and $\bar{\zeta}$ tell us the direction of motion of each photon. Substituting (7.37) into (7.35) and using (6.11) leads to the identification

$$p_a = \omega\, l_a(\zeta,\bar{\zeta}) \quad . \quad (7.38)$$

From this it follows that the differential form

$$d^3p/2p^0 = 1/6\, \varepsilon_{abcd}\, t^a\, dp^b \wedge dp^c \wedge dp^d / p^e t_e \quad ,$$

which is independent of t^a, may be rewritten

$$d^3p/2p^0 = dS \wedge \omega\, d\omega \quad , \quad (7.39)$$

use being made of equation (5.22). Finally, writing

$$a_1(p) = \alpha_1(\omega,\zeta,\bar{\zeta}) \quad \text{and} \quad b_1(p) = \beta_1(\omega,\zeta,\bar{\zeta}) \; , \quad (7.40)$$

it follows that the electromagnetic potential becomes

$$\alpha_a^{out} = (2\pi)^{-3/2} \int d^3p/2p^0\, m_a [e^{-ip.x} a_1(p) + e^{ip.x} b_1^\dagger(p)] . \quad (7.41)$$

The field may thus be quantised according to the standard prescription pertaining to a non-interacting quantum field theory, namely

$$[a_1(p), a_1^\dagger(p')] = 2p^0\, \delta^3(p-p') = [b_1, b_1^\dagger] \quad . \quad (7.42)$$

I would stress, however, that the interpretation of the underlying Fock space is somewhat unconventional and very intriguing. This arises from the fact that the momentum space involved is the one dual to M_0^+ rather than the original curved $\tilde{M}$. Thus, we are forced to conclude that the photons themselves exist in cone space.

A similar analysis goes through in the spin-two case. Here there are four left-handed potentials for the field, the first two being given by

$$\Gamma^{out}_{AA'BC} = -(2\pi)^{-1}\int dS\, o_A \bar{o}_{A'} o_B o_C\, T^{-1}\eth(T\psi_4) \qquad (7.44)$$

and

$$h^{out}_{AA'BB'} = (4\pi)^{-1}\int dS\, o_A \bar{o}_{A'} o_B \bar{o}_{B'}\, T^{-2}\eth(T^2\psi_3) \; . \qquad (7.45)$$

The first of these has the spinorial structure of a set of spin coefficients, or spin currents, while the second defines a complex, second rank, trace-free, symmetric tensor and plays a role similar to that of the perturbation to the metric in linearised theory. The equations which they obey are;

$$\nabla_A{}^{A'}\,\Gamma^{out}_{BA'CD} = -\psi^{out}_{ABCD} \; ,$$
$$\nabla^A{}_{A'}\,\Gamma^{out}_{AB'CD} = 0 \; , \qquad (7.46)$$

$$\nabla_A{}^{A'}\,h^{out}_{BA'CC'} = -\Gamma^{out}_{CC'AB}$$

and (7.47)

$$\nabla^A{}_{A'}\,h^{out}_{AB'CC'} = 0 \; .$$

In particular, in view of the last equation, it follows that h^{out}_{ab} satisfies the de Donder gauge condition.

$$\nabla_a h^{ab}_{out} = 0 \; . \qquad (7.48)$$

As in the Maxwell case, the choice (7.33) for T also makes h^{out}_{ab} totally transverse,

$$t^a h^{out}_{ab} = 0 \; , \qquad (7.49)$$

and it now follows, on using (4.20), that

$$h^{out}_{ab} = (2\pi)^{-1}\int dS\, m_a m_b N \; . \qquad (7.50)$$

The quantisation procedure, described above for the spin-one case, may be applied again here. Making the Fourier expansion

$$N = -\dot{\sigma} = (2\pi)^{-\frac{1}{2}}\int_0^\infty \omega\, d\omega (e^{-i\omega u}\alpha_2 + e^{i\omega u}\beta_2^\dagger) \qquad (7.51)$$

and inserting into equation (7.50), yields

$$h_{ab}^{out} = (2\pi)^{-3/2}\int d^3p/2p^0\, m_a m_b [e^{-ip.x} a_2(p) + e^{ip.x} b_2^\dagger(p)]\,. \qquad (7.52)$$

The commutation relations are, of course,

$$[a_i(p), a_j^\dagger(p')] = 2p^0\,\delta^3(p-p')\delta_{ij} = [b_i, b_j^\dagger]\,, \qquad (7.53)$$

(all others vanishing) which covers both the spin-one and spin-two cases. Again, the gravitons themselves live in cone space and not in the original manifold.

We have demonstrated a method for quantising the *out* fields associated with an asymptotically flat space-time. Let us now check that our identification of the Fock operators a_i and b_i from the expansions (7.37) and (7.51) is consistent with the expression (7.19) for the total momentum radiated away from the given dynamical system. Direct substitution, together with the expressions (7.38) and (7.39), does indeed give the standard result

$$\Pi_a = \sum_{i=1}^{2}\int d^3p/2p^0\, p_a [a_i^\dagger(p)\, a_i(p) + b_i^\dagger(p)\, b_i(p)]\,, \qquad (7.54)$$

where the operators have been normal ordered, so that the quantity inside the square brackets is the particle number operator for momentum p_a.

To see how this formalism should work in practice, let us return to the question of the gravitational scattering of test photons off some given classical source. For simplicity, suppose that I^- is a Cauchy surface so that given the null datum there we may, in principle, solve Maxwell's equations on M and obtain the null datum on I^+. (Solving the equations is the hard part.) For arbitrary choices of $M^-_{O^-}$ and $M^+_{O^+}$, with $O^- \in C^-$ and $O^+ \in C^+$, we may construct the *in* and *out* fields ϕ^{in}_{AB} and ϕ^{out}_{AB}, each one up to supertranslation. More importantly, we may define the corresponding creation and annihilation operators according to the decomposition (7.37) and the identifications (7.40). It follows [30] that the scattering of the field will induce a transformation of the type

$$a^{out}(p) = \int d^3k/2k^0 \, [A(p,k)\, a^{in}(k) + B(p,k)\, b^{\dagger in}(k)] \, . \quad (7.55)$$

We must, however, ask whether the physical predictions given by this transformation depend upon the choices of origins O^- and O^+.

So, consider a new Minkowski space $M^+_{S^+}$ obtained by supertranslating the origin O^+. In the original system (6.11) of coordinates this new space may be parametrised according to

$$M_S = \{ ⟗ \,|\, u = x^{AA'} o_A \bar{o}_{A'} + s(o_A, \bar{o}_{A'}) \} \, , \quad (7.56)$$

where s is homogeneous of degree $\{1,1\}$ in its arguments. Substituting (7.56) into (7.37) results in expressions for the supertranslated Fock operators of the form

$$a^{out}_s(p) = exp[-is(\bar{\pi}_A, \pi_{A'})]\, a^{out}(p) \, , \quad (7.57)$$

where

$$p_a = \bar{\pi}_A \pi_{A'} \, .$$

Similar transformations are obtained for the *in* operators but, in general, for different s. The crucial point is that the coefficients A and B occurring in (7.55) acquire only phase factors. Thus, the rate of particle production derived from $|B|^2$, and the scattering amplitude, obtained from $|A|^2$ (at least for $p \neq k$) are supertranslation invariant. For forward scattering ($p = k$) the situation is not so clear. This is because the splitting of the beam in the forward direction into scattered and unscattered does not appear to be supertranslation invariant. However this is no great drawback since, in practice, the forward differential cross-section is not measured by the experimentalist; and, in any case, owing to the long range nature of the forces, it is divergent.

8 STATIONARY SYSTEMS

It is when we come to analyse stationary, isolated systems in general relativity that their similarity to systems in special relativity becomes most striking. I would stress that there is no approximation involved either here or in the previous section. Certainly nowhere have we made a splitting of the coordinate components of the metric into a Minkowski part plus a small perturbation. We do not have to, because the Lorentz invariance of the theory is made manifest by using orthonormal frames and, in such a basis, the metric components are exactly Minkowskian.

The Canonical Minkowski Space

Let me begin by pointing out that, if the system is stationary, then there is a *canonical* Minkowski subspace of cone space. It is defined by that set of cuts of $\mathcal{I}^+$ which are shear-free and which are thus characterised as if they had arisen from the light cones of points of Minkowski space-time. In order to show this, the idea is to use the timelike Killing vector ξ^a on $\tilde{M}$ and extend it to $\mathcal{I}^+$ where, with respect to the rescaled metric, it will be null. From one outgoing null hypersurface ξ^a will generate a one parameter family of null hypersurfaces. On I^+ ξ^a will map cross-sections to cross-sections and hence, being null, will point up the generators. We shall use ξ^a to define a canonical n^a vector on I^+ and hence define a u parameter in the usual way:

$$n^a = \xi^a = \partial/\partial u \ , \tag{8.1}$$

the cuts and hypersurfaces to be labelled by u = constant. I assert, further, that there exists a conformal factor Ω, independent of u, and for which

$$n_a = -\nabla_a \Omega \quad \text{on } I^+ \ . \tag{8.2}$$

Then the metric of I^+ and the shear of the cross-sections will both be u-independent. Since there is no radiation leaving the system it follows, from equation (4.22), that the metric of I^+ is either flat or spherical. Thus, associated with the Killing vector is a Bondi family of cross-sections of I^+ (compare [28]).

Let T be the timelike translation associated with ξ^a on I^+ and which, in the natural gauge, is given by T=1. Then, from equation (7.10), the stationarity of the system implies that

$$3\eth T\,\psi_2 + T\eth\psi_2 = 0 \tag{8.3}$$

with solution

$$\psi_2 = m/T^3 \ , \tag{8.4}$$

where, for globality reasons, m is constant. (The constraint is that $m(\bar{\zeta})$ be defined everywhere on the sphere or, equivalently, that it be expandable in spherical harmonics.) On using equation (3.55), it follows that

$$-\tfrac{1}{2}i(\bar{\eth}^2\sigma - \eth^2\bar{\sigma}) = \text{Im. } m/T^3 \tag{8.5}$$

and writing [28]

$$\sigma = \eth^2(a + ib) \quad , \tag{8.6}$$

where a and b are real with weight {1,1}, we get

$$T\bar{\eth}^2\eth^2 b = \text{Im.}\, m/T^2 \quad .$$

(This follows most easily if we flatten I^+.) Integrating both sides over a section of I^+ now yields

$$\text{Im.}\, m = 0 = b \ . \tag{8.7}$$

To make the shear vanish we need only perform the supertranslation (7.3). The freedom to do this involves the solutions of the equation,

$$\eth^2 a = 0 \ , \tag{8.8}$$

which defines the translations (5.32). The canonical Minkowski space is thus given by

$$M^+ = \{\leftrightarrow \mid u = x^{AA'} o_A \bar{o}_{A'} + a(\zeta, \bar{\zeta})\} \quad . \tag{8.9}$$

Of course, this space is just the space of image points mentioned in the introduction and is the real part of Newman's $\mathcal{H}$ space.

In the rest of this section, we shall investigate the behaviour of certain multipole moments for stationary systems.

Relativistic Angular Momentum

A recent paper [1] was devoted to the definition of the angular momentum of a general, asymptotically flat system. This was achieved by rewriting general relativity as a Yang-Mills theory (without torsion), based on local Lorentz invariance, and by exploiting the connection (Noether's theorem) between invariance groups and conservation laws. The result was a Gauss' theorem for the relativistic angular momentum (or Lorentz generators) of the system including the gravitational contribution. The quantity M^{ab}, so defined, forms a skew-symmetric, asymptotic tensor on cone space and whose spinor equivalent is given by

$$\omega_{AB} = \frac{1}{2}\int dS\, o_A\, o_B\, (\bar{\psi}_1 - 3\,\bar{\sigma}\bar{\eth}\sigma - \sigma\bar{\eth}\bar{\sigma}) \quad , \tag{8.10}$$

(compare [33] and [34]). Note that the sum of the correction terms involving σ and $\bar{\sigma}$ is well defined since $\bar{\eth}$ acts effectively on $\sigma^3\bar{\sigma}$, which has p=0 (see table 1). Note, further, that the expression (8.10) is a one-point function. The chosen cross-section of integration defines not only the "instant" at which we wish to evaluate the angular momentum but also the point in cone space about which we take moments.

When applied to stationary systems, the correction terms in (8.10) vanish (in fact, whether or not we impose the restriction to shear-free cuts). Further, by restricting to M^+, the angular momentum becomes a field on flat space-time; and its behaviour there is governed by the expression

$$\psi'_1 = \psi_1 + 3\eth\alpha\,\psi_2 + \alpha\eth\psi_2 \quad , \tag{8.11}$$

which comes from equation (7.10) and where α is an infinitesimal translation, having the form

$$\alpha = dx^{AA'}\, o_A\, \bar{o}_{A'} \quad . \tag{8.12}$$

Basically, (8.11) says that ψ_1 transforms as a moment of ψ_2. Indeed, on using equations (5.17) and (5.18), we get the following differential equation for ω_{AB} in M^+:

$$\nabla_{CC'}\,\omega_{AB} = \varepsilon_{C(A}\,P_{B)C'} \quad , \tag{8.13}$$

P_a being the momentum of the system defined in (7.2). In particular, ω_{AB} satisfies the twistor equation in M^+:

$$\nabla^{(C}{}_{C'}\,\omega^{AB)} = 0 \quad . \tag{8.14}$$

Equation (8.13) may be swiftly integrated. If the point $X \in M^{+}$ has position vector x^{a} relative to some origin O, then it follows that

$$\omega_{AB}(X) = \omega_{AB}(O) + x_{(A}{}^{C'} P_{B)C'} \tag{8.15}$$

and this, in turn, is the spinor equivalent of

$$M_{ab}(X) = M_{ab}(O) - (x_{a} P_{b} - x_{b} P_{a}) \quad . \tag{8.16}$$

Thus, as far as the Poincaré generators are concerned, the system behaves as if it were specially relativistic. Of course, in the expression for the angular momentum, (8.16) means that we are really taking moments about image points. It is at this stage that one begins to feel that, perhaps, the matter comprising the system should itself be regarded as living in M^{+} rather than in the original curved manifold $\tilde{M}$. Indeed, from the expression (8.16) it follows that its centre of mass certainly lies in M^{+}. This we shall now discuss.

It proves very interesting to widen the discussion a little by considering the complexification $\mathcal{H}$ of M^{+}. M_{ab} is now written

$$M_{ab} = \omega_{AB}\, \varepsilon_{A'B'} + \varepsilon_{AB}\, \tilde{\omega}_{A'B'} \quad , \tag{8.17}$$

where $\tilde{\omega}_{A'B'}$ is the analytic continuation of $\bar{\omega}_{A'B'}$ from M^{+} to $\mathcal{H}$. Analogous to (8.15), $\tilde{\omega}_{A'B'}$ transforms under translations according to

$$\tilde{\omega}_{A'B'}(X) = \bar{\omega}_{A'B'}(O) + x^{C}{}_{(A'} P_{B')C} \quad . \tag{8.18}$$

The expressions (8.15) and (8.18) are now to be realised for real origin O but for complex x^{a}. (Note that $\bar{x}^{a}$ nowhere appears.) Of particular physical interest are two vectors orthogonal to the momentum; namely,

$$D^{a} = M^{ab} P_{b} = \omega^{AB} P_{B}{}^{A'} + \tilde{\omega}^{A'B'} P^{A}{}_{B'} \quad , \tag{8.19}$$

which is the mass × dipole moment , and the Pauli-Lubanski spin vector

$$S^{a} = M^{*ab} P_{b} = -i(\omega^{AB} P_{B}{}^{A'} - P^{A}{}_{B'} \tilde{\omega}^{A'B'}) , \qquad (8.20)$$

which is real. Under translations the former behaves according to

$$D^{a}(X) = D^{a}(O) + (P^{a} P_{b} - \delta^{a}_{b} m^{2}) x^{b} , \qquad (8.21)$$

while the spin vector is invariant. Note, further, that the vector

$$W^{a} = D^{a} + iS^{a} = 2\,\omega^{AB} P_{B}{}^{A'} \qquad (8.22)$$

transforms like D^{a}. (D^{a} is not, in general, real.)

Since ω^{AB} satisfies the twistor equation in $\mathcal{H}$, it may be regarded as the principal part of a twistor of valence $\begin{bmatrix}2\\0\end{bmatrix}$ called the *angular momentum twistor* [11]. Analogous to the discussion in section 2 on valence one twistors, the angular momentum twistor may be represented geometrically, up to scale, as that set of points in $\mathcal{H}$ for which ω^{AB} or, equivalently ($m \neq 0$), W^{a} vanishes. This defines the *complex centre of mass* of the system,

$$x^{a} = 1/m^{2}\, W^{a}(O) + \lambda P^{a} , \qquad (8.23)$$

with λ an arbitrary complex parameter. The geodesic defined by equation (8.23) may be compared with the real centre of mass of the system which is used more conventionally. This is obtained by demanding that the mass-dipole moment D^{a} vanish and is given by

$$x^{a} = 1/m^{2}\, D^{a}(O) + \lambda P^{a} , \text{ with } \lambda \text{ real.} \qquad (8.24)$$

Thus it follows that the complex centre of mass line may be obtained from its real counterpart by means of the imaginary displacement $iS^{a}/m^{2} + \lambda P^{a}$, with λ pure imaginary.

Electromagnetic Dipole Moment and Gyromagnetic Ratio

The analysis may now be repeated to construct the complex centre of charge of the system. This entails the definition of the electromagnetic dipole spinor

$$d_{AB} = 3/8\pi \int dS\, o_A\, o_B\, T^{-1}\, \bar{\phi}_0 \quad . \tag{8.25}$$

As in the definition of the electromagnetic potential, the quantity T is needed here to guarantee the conformal invariance of the expression. However, since the system is stationary we may choose T canonically to be the timelike translation (7.33) but with t^a the unit vector parallel to P^a.

The behaviour of the dipole moment as a field in M^+ is governed by the expression

$$\phi'_0 = \phi_0 + 2\,\eth\alpha\,\phi_1 + \alpha\,\eth\phi_1 \quad , \tag{8.26}$$

which comes from equation (7.6) and where α is given by (8.12). Stationarity implies, further, that

$$2\,\eth T\,\phi_1 + T\,\eth\phi_1 = 0 \quad . \tag{8.27}$$

Thus it follows that the quantity

$$\chi = 3\phi_0/2T \tag{8.28}$$

transforms according to

$$\chi' = \chi + 3\,\eth\alpha(\phi_1/T) + \alpha\,\eth(\phi_1/T) \quad , \tag{8.29}$$

which may be compared with the transformation law (8.11) for ψ_1. (The comparison is valid since they have the same weights.) It follows that the behaviour of d_{AB} under translations of origin is similar to that of ω_{AB}. Hence,

$$d_{AB}(X) = d_{AB}(O) + x_{(A}{}^{A'}\, q_{B)A'} \quad , \tag{8.30}$$

where

$$q_a = -(2\pi)^{-1} \int dS\, l_a\, T^{-1}\, \bar{\phi}_1 \quad . \tag{8.31}$$

On solving (8.27) and using the definition (7.1), we obtain

$$\phi_1 = -\tfrac{1}{2}q/T^2 \quad , \tag{8.31}$$

by virtue of the norm (6.6). Hence

$$q_a = q\, t_a \quad . \tag{8.32}$$

By analogy with (8.22) we may now construct the complex vector

$$\varepsilon^a + i\,\mu^a = 2\, d^{AB}\, t_B{}^{A'} \quad , \tag{8.33}$$

which is orthogonal to t^a. We identify ε^a with the electric dipole moment of the system and μ^a, which is translation invariant, with its magnetic moment. The complex centre of charge of the system is obtained by making d_{AB} vanish [35] and is, therefore, given by the geodesic

$$x^a = 1/q\,(\varepsilon^a + i\mu^a) + \lambda\, t^a \quad , \lambda \text{ complex } (q \neq 0) \tag{8.34}$$

which bears comparison with the expression (8.23).

Suppose, then, that we have a charged, massive, spinning chunk of matter which is stationary with the property that its complex centres of mass and charge coincide. It then follows that

$$(\varepsilon^a + i\mu^a)/q = (D^a + iS^a)/m^2 \quad , \tag{8.35}$$

with moments taken about any origin X. Consequently, the real centres of charge and mass will coincide and, furthermore, the spin and magnetic moment of the system are related by the expression

$$\mu^a/q = S^a/m^2 \quad . \tag{8.36}$$

In other words, the system has a gyromagnetic ratio given by the Dirac value g=2. This generalises a result that was already known [35] in the absence of gravity. In general relativity, the Dirac value has also been derived

[36] for Kerr-Newman black holes [37].

The use of the complex centres of mass and charge is not quite as esoteric as it sounds. Indeed, recent work by Tod and Perjés [38] seems to indicate that they have important dynamical significance. In the case $g=2$ it is possible to deduce the Papapetrou and Michel-Bargmann-Telegdi equations for the motion of classical, charged, massive, spinning particles through gravitational or electromagnetic shock waves. (The method employed involves decomposition of the massive particle into a pair of massless constituents (twistors) at the point where the complex world line encounters the shock. The twistors are then refracted through the shock by means of a canonical transformation [11] and eventually recombine to form the scattered particle.)

Gravitational Quadrupole Moment

Let us end this section by defining the gravitational quadrupole moment for a general stationary system. The spinor equivalent of this is given in M^+ by

$$Q_{ABCD} = -5/9 \int dS\, o_A\, o_B\, o_C\, o_D\, T^{-1}\, \bar{\psi}_0\ . \tag{8.37}$$

The quantity

$$Q_{ab} = 2\, Q_{ABCD}\, t^C{}_{A'}\, t^D{}_{B'} \tag{8.38}$$

is a complex, trace-free, symmetric tensor which is transverse to t^a and whose real and imaginary parts are, respectively, the mass and spin quadrupole moments. An analysis similar to the equations (8.26) through (8.32) shows that Q_{ABCD} satisfies

$$\nabla_{EE'}\, Q_{ABCD} = 5/3\, \varepsilon_{E(A}\, \omega_{BC}\, t_{D)E'} \tag{8.39}$$

in M^+ and, in particular, the twistor equation.

The choice of normalisation factor in the definition (8.37) is determined by analogy with Newton's theory for which

$$\nabla_i \nabla_j Q^{ij} = 10\, m \quad . \tag{8.40}$$

In fact, because of the natural 3+1 splitting of M^+ by P^a and the associated fact that the moments, like $D^a + iS^a$, $\varepsilon^a + i\mu^a$ and Q^{ab}, all sit in the canonical three-space, the system looks very much like a Newtonian system apart from the existence of spin or magnetic type moments.

9 SUMMARY AND CONCLUSION

Before entering into a discussion of the meaning of all the preceding, which is partly a question of attaching words to mathematics, let us, first, summarise the principal results.

1. The boundary of the extension of an asymptotically flat space-time, which is Einstein-Maxwell sufficiently far out, is a shear-free null hypersurface: (from the regularity of the Ricci tensor there). Further, being the set of end points of null geodesics, it splits into the union of two disjoint pieces, future and past null infinity $I^+ \cup I^-$.

2. The evolution of the system is determined by Maxwell's equations on I^+ and by the spin-two equations. (Of course, they do not tell us everything. Basically, we lose the traces of the multipole moments.) These are expressed, most easily, by choosing a conformal factor to flatten I^+, so that the only spin coefficient entering the equations is the shear σ of the cross-sections of I^+. The relationship between the spin-two field on I^+ and the shear is then determined by the Bianchi identities.

3. With each field component on I^+ may be associated a pair of integers $\{p,q\}$ which describe its

behaviour under simultaneous rescalings of the metric on $\mathcal{I}^+$ and boost and spin transformations of spinor dyad.

4. Use of the twistor equation enables spin frames to be aligned on $\mathcal{I}^+$ uniquely up to rigid SL(2,C). The space spanned by these frames is asymptotic spin space $\mathcal{S}$. The asymptotic vectors or translations are formed by taking the tensor product of $\mathcal{S}$ with its complex conjugate.

5. Cone space C^+ is an infinite dimensional affine space consisting of all smooth cross-sections of $\mathcal{I}^+$. Each vector (supertranslation) in TC^+ is assigned a norm which is constructed to be independent of the choice of conformal factor but which gives rise to a non-Riemannian structure for C^+. However, the four-dimensional submanifold M^+_O, generated by translation away from some arbitrary origin O in C^+, is Minkowski space with respect to the given norm.

6. The BMS group is the group of affine mappings on cone space which preserves the given norm.

7. By integrating the field components over the cross-sections of $\mathcal{I}^+$ and using the twistorially aligned frames as weighting factors (spin-weighted spherical harmonics), we can construct spinor fields on cone space: that is to say, fields in C^+ which, at each point, lie in asymptotic spin space (or its tensor products), the spin bundle being trivial. By restricting to any Minkowski subspace of C^+, the behaviour of the system is then determined by equations in flat space.

8. The Bondi-Sachs four-momentum is an asymptotic vector field on C^+ which in M^+_O is divergence-free and hypersurface-orthogonal. The rest mass of the system decreases with the radiation of electromagnetic and

gravitational waves.

9. The *out* fields of the system may be constructed from the null data on I^+. Thus, associated purely with the outgoing radiation, we now have spinor fields in C^+ which solve the zero rest mass field equations in M_0^+. A hierarchy of Hertz potentials may also be constructed and, in the spin-two case, this gives rise to a transverse, trace-free symmetric tensor h_{ab}^{out} satisfying the de Donder gauge condition in M_0^+. Using I^-, similar procedures generate the *in* fields.

10. The *in* and *out* fields may be quantised in the conventional manner and this is consistent with the Bondi-Sachs momentum. Within the framework of scattering theory, both the differential cross-section (away from the forward direction) and the rate of particle production are supertranslation invariant.

11. For stationary systems cone space possesses a unique Minkowski subspace M^+, defined by the shear-free cross-sections of I^+. In M^+ the relativistic angular momentum of the system transforms as a moment of momentum. The system then behaves as if it were specially relativistic and lying in the space M^+ of image points. In particular, its centre of mass is a timelike geodesic there.

12. Treating the electromagnetic dipole moment of a stationary system analogous to its angular momentum leads to the conclusion that its centre of charge also lies in M^+. If the complex centres of mass and charge coincide, then the system's spin and magnetic moment vectors are proportional with gyromagnetic ratio given by the Dirac value.

There are several aspects of the programme just presented which seem capable of development. The problem of producing higher multipole moments is one. If the behaviour of the momentum, angular momentum and quadrupole moment is anything to go by, then one might expect a whole hierarchy of moments $\{\omega^{(n)}_{AB\ldots J} | n = 0,1,2\ldots\}$, (with $2n$ indices and symmetric), which for stationary systems satisfy

$$\left.\begin{aligned} &\nabla_K{}^{K'}\,\omega^{(n)}_{AB\ldots J} \sim \varepsilon_{K(A}\,t_B{}^{K'}\,\omega^{(n-1)}_{C\ldots J)} \\ &\text{with} \\ &\omega^{(0)} = m \quad . \end{aligned}\right\} \tag{9.1}$$

In particular, each spinor of the hierarchy has vanishing symmetrised derivative and so defines a twistor of valence $\begin{pmatrix} 2n \\ 0 \end{pmatrix}$. The use of the angular momentum twistor has played an important role, not only in reference [38] but also within the framework of a scheme for classifying elementary particles [40, 41]. Whether the twistors of higher valence are relevant, here, presents an exciting possibility.

There arises, of course, the question of multipole moments for radiating systems. If we use the definition (8.25), but where T is an *arbitrary* timelike translation, then it is straightforward to show that in any M_0^+,

$$\left.\begin{aligned} &t^a\,\nabla_a\,\bar{d}_{A'B'} = 3/2\,\chi^{(e)}_{A'B'} \\ &\text{where} \\ &\chi^{(e)}_{A'B'} = (4\pi)^{-1}\int dS\,\bar{o}_{A'}\bar{o}_{B'}\,[\,T^{-2}\,\eth'(T^2\phi_1) + \sigma\phi_2] \end{aligned}\right\} \tag{8.2}$$

is the Maxwell Hertz potential (7.29) with a correction term. This extra term does not affect the equations (7.32) and (7.30) and so the potential gives rise to the *out* field in the usual way. Thus, we have a picture of an oscillating dipole in cone space generating a radiation field.

A similar argument shows that the time rate of change of the quadrupole moment (8.37) is proportional to the lowest Hertz potential for the spin-two *out* field. However, if we calculate the potential h^{out}_{ab} it turns out not to be transverse (though the expression for the field is, in no way, affected). To remedy the situation requires the modification of (8.37) by the addition of correction terms and the uniqueness of these is still under consideration.

The freedom of choice of the quantity $T = t^a l_a$ in the expressions for the multipole moments is a little puzzling although, when we calculate the potentials, it is merely a question of gauge. For stationary systems, t^a may, of course, be chosen canonically by means of P^a. If, however, we consider transitions between states which are, in some sense, stationary in the remote past and future, then in general a canonical t^a does not present itself; for if t^a is aligned with P^a initially, it will not, necessarily, be so finally. One suggestion (due to Newman) is to relax the constraint that t^a be fixed and align it with P^a for each point in cone space. This is being considered currently.

A related problem concerns systems which are radiation-free rather than stationary. Imagine superposing a pair of Maxwell dipoles, d_1^a and d_2^a, moving relative to each other but sufficiently far apart that radiation may be neglected. Each dipole is defined with respect to its own timelike vector, t_1^a or t_2^a, but the dipole moment of the total system will involve the direction t^a of the centre of mass, say. This leads to an unusual law of superposition of the dipoles which is not yet understood.

The most crucial question concerns the validity of using an *arbitrary* Minkowski subspace of cone space. We have seen that the ambiguity is unimportant in the theory of scattering. Again, if we consider the expression (7.19) for the total momentum radiated, then it is clearly supertranslation invariant. The behaviour of the spin vector is a little more delicate. For stationary systems it is supertranslation invariant but this does not appear to be the case if we assume the system merely to be radiation-free. It is possible that by examining the shear as $u \to -\infty$ we can supertranslate to cross-sections which are shear-free in the limit [6] and so pick out a unique Minkowski space; but the use of this is still open to question.

"Why bother to use cone space at all?" it might be asked. "After all, the physics, as far as the asymptotic observer is concerned, is contained completely in the evolution equations (3.41) to (3.46)." To illustrate our viewpoint let us imagine a set of observers with receivers, detectors or cameras (S^2s worth) making measurements of some evolving system from afar. Think of them as lying near to $\mathcal{I}^+$. An observer, labelled by $(\zeta,\bar{\zeta})$, measures some field component at all points along his world line. After a while a meeting is arranged and the information, in the form of functions on $S^2 \times R^1$, is assessed. A great discussion arises as to how best the data is to be interpreted. The majority display an overwhelming psychological need for some sort of space-time description. Someone points out the possibility of obtaining physical laws in flat space. "Integration over spheres," he says, "is the key to the solution of our problems; and I don't care about the freedom to supertranslate." A second person points out that, if we are prepared to put up with a complex space-time, then there is a way of getting a canonical one. "Further," he argues, "interpreting the data as a set of physical laws in $\mathcal{H}$ space is mathematically the most compelling." Yet a third person states that he doesn't much care for space-time in any case and thinks that it should be constructed from something even more fundamental. "The data", he claims, "should be used to produce functions in twistor space."

Needless to say, the discussion is still continuing. It is not completely clear by what criteria we should judge the outcome. Mathematics is the study of patterns and science is the study of patterns in nature. In particular, a scientific law is one which arises out of the quest for simplicity from complexity. So, presumably, the programme which wins is the one which displays the greatest economy coupled with the greatest level of unification. On the other hand, there is something to be said for looking at a problem from differing viewpoints and it may well be that each is valid within its own domain of applicability.

I should like to thank the Science Research Council in conjunction with Oxford University Mathematical Institute for financial support between 1973 and 1975 and Merton College for support since that time. I would also mention the Departments of Physics at the Universities of Pittsburgh and Syracuse for their stimulating hospitality during the Spring of 1975. It is a very great pleasure to thank Professor E.T. Newman for many discussions, particularly on intertwining theory, which led to the streamlining of section 8 of this paper and Professor R. Penrose, F.R.S. for discussions on conformal freedom which led to section 4. Finally, I gratefully acknowledge fruitful conversations with Professors J.N. Goldberg and J. Winicour and with Drs. R.O. Hansen, M. Ludvigsen, P.J. McCarthy, G.A.J. Sparling and M. Walker.

REFERENCES

1. B.D.Bramson, *Proc. R. Soc. Lond.* A341 (*1975*) *463.*
2. B.Aronson, R.W.Lind, J.Messmer & E.T.Newman, *J. Math. Phys.* 12 (*1971*) *2462.*
3. R.Penrose, in *Battelle Rencontres* (ed. C.M.De Witt & J.A.Wheeler): New York, Benjamin (*1968*).
4. F.A.E.Pirani, in *Brandeis Lectures on General Relativity* (ed. S.Deser & K.W.Ford): New York, Prentice Hall (*1965*).
5. R.Penrose, in *Perspectives in Geometry and Relativity*: Bloomington, Indiana University Press (*1966*).
6. E.T.Newman & R.Penrose, *J. Math. Phys.* 7 (*1966*) *863.*
7. R.Penrose, *Ann. Phys.* 10 (*1960*) *171.*
8. D.W.Sciama, in *Recent Developments in General Relativity*: New York, Pergamon (*1962*).
9. T.W.B.Kibble, *J. Math. Phys.* 2 (*1961*) *212.*
10. R.Penrose, *J. Math. Phys.* 8 (*1967*) *345.*
11. M.A.H.MacCallum & R.Penrose, *Phys. Rep.* 6C (*1973*) *242.*
12. B.D.Bramson, *Phys. Lett.* 47A (*1974*) *431.*
13. R.Penrose, in *Group Theory in Non-Linear Problems* (ed. A.O.Barut): D.Reidel (*1974*).
14. H.Bondi, M.G.J. van der Burg & A.W.K.Metzner, *Proc. R. Soc. Lond.* A269 (*1962*) *21.*
15. R.K.Sachs, *Phys. Rev.* 128 (*1962*) *2851.*
16. R.K.Sachs, *Proc. R. Soc. Lond.* A270 (*1962*) *103.*
17. M.K.W.Ko & E.T.Newman, *Gen. Rel. & Grav.* 6 (*1975*) *595.*
18. E.T.Newman & R.Penrose, *J. Math. Phys.* 3 (*1962*) *566.*
19. A.R.Exton, E.T.Newman & R.Penrose, *J. Math. Phys.* 10 (*1969*) *1566.*
20. R.W.Lind, J.Messmer & E.T.Newman, *J. Math. Phys.* 13 (*1972*) *1879.*
21. R.P.Geroch, A.Held & R.Penrose, *J. Math. Phys.* 14 (*1973*) *874.*
22. R.Penrose, *Int. J. Theor. Phys.* 1 (*1968*) *61.*
23. A.Held, E.T.Newman & R.Posadas, *J. Math. Phys.* 11 (*1970*) *3145.*
24. B.D.Bramson, *Proc. R. Soc. Lond.* A341 (*1975*) *451.*

25. P.J.McCarthy, to be published in *Proc. R. Soc. Lond.*

26. E.T.Newman, *Gen. Rel. & Grav.* 7 (*1976*) *107*.

27. R.Penrose, *Phys. Rev. Lett.* 10 (*1963*) *66*.

28. E.T.Newman & R.Penrose, *Proc. R. Soc. Lond.* A305 (*1968*) *175*.

29. D.Lurié, *Particles and Fields*: New York, Interscience (*1968*).

30. S.W.Hawking, in *Quantum Gravity* (ed. C.J.Isham, R.Penrose & D.W.Sciama): Oxford University Press (*1975*).

31. R.Penrose, *Proc. R. Soc. Lond.* A284 (*1965*) *159*.

32. A.Komar, *Phys. Rev.* 134 (*1964*) *1430*.

33. J.Winicour, *J. Math. Phys.* 9 (*1968*) *861*.

34. R.W.Lind, J.Messmer & E.T.Newman, *J. Math. Phys.* 13 (*1972*) *1884*.

35. E.T.Newman & J.Winicour, *J. Math. Phys.* 15 (*1974*) *1113*.

36. G.C.Debney, R.P.Kerr & A.Schild, *J. Math. Phys.* 10 (*1969*) *1842*.

37. K.Chinnapared, W.E.Couch, A.R.Exton, E.T.Newman, A.Prakash & R.J.Torrence, *J. Math. Phys.* 6 (*1965*) *918*.

38. K.P.Tod & Z.Perjés, to be published in *Gen. Rel. & Grav.*

39. M.Walker, private communication.

40. R.Penrose, in *Quantum Theory and the Structure of Time and Space* (ed. L.Castell, M.Drieschner & C.F. von Weizsäcker): Munich, Verlag (*1975*).

41. Z.Perjés, *Phys. Rev.* D11 (*1975*) *2031*.

THE COMPLEX VACUUM METRIC WITH MINIMALLY DEGENERATED CONFORMAL CURVATURE

Jerzy F. Plebanski* and Ivor Robinson**

*Centro de Investigacion y Estudios Avanzados 14-740, Mexico, D.F. Mexico., on leave of absence from the University of Warsaw, Warsaw, Poland
**Institute for Mathematical Sciences, The University of Texas at Dallas, Richardson, Texas.

Abstract: By applying Plebański-Hacyan theorem, the canonical forms of the metric are established for all complex Einstein flat with the minimally (one-sided) algebraically degenerate - conformal curvature. Then Einstein equations are integrated. The solution is expressed in the terms of only one fundamental key function which is determined by a differential equation of the second order and with quadratic non-linearity only, this equation being a generalization of the second heavenly equation.

1. The General Form of the Metric. This paper postulates the null tetrad formalism and uses the apparatus of the differential forms basically in the same notation as [1] (for a more complete description of this formalism see [2]).

The natural extension of this formalism on the case of a complex V_4 together with the pertinent information concerning spinors and the algebraic classification of the conformal curvature of such a space, was given in the first section of [3] and will be applied here.

As is well known, the Goldberg-Sachs theorem [4] plays the key role in the theory of the real algebraically degenerate solutions of the Einstein vacuum equation in real V_4's of the signature (+ + + -). In fact, the proper use of this theorem (as in e.g. [1] or [5]) forms the first step in the search for these solutions. If one considers, however, the Einstein equations from the point of view of their analytic continuation on complex V_4's this theorem has been in a sense misleading. Although Einstein equations like null congruences of non-shearing geodesics, they like even more 2-surfaces, and in particular 2-dimensional totally null geodesic surfaces, which for brevity shall be called null strings. In fact, recently, a theorem was established by Plebański and Hacyan [6], which states that if along a complex V_4 the Einstein vacuum equations are fulfilled, then the heavenly (un-dotted) part of the conformal curvature is algebraically degenerated [i.e., the null tetrad can be selected so that $C^{(5)} = C^{(4)} = 0$], if and only if there exist a congruence of the complex null strings, i.e., the geodesic 2-surfaces with the tangent space spanned by two orthogonal null vectors, these tangent spaces being parallely propagated along the surfaces from the congruence.

Technically, with the metric given in the form:

(1.1) $$ds^2 = 2e^1 \otimes e^2 + 2e^3 \otimes e^4 \,, \quad e^a \in \Lambda^1$$

the theorem can be stated as follows:

(1.2) Assuming $R_{ab} = 0$

$$\left\{\begin{array}{l}\text{The self-dual part}\\ \text{of conf. curvature}\\ \text{is alg. degenerate}\end{array}\right\} \equiv \left\{\begin{array}{l} C^{(5)} = 0 \\ C^{(4)} = 0 \end{array}\right\} \Longleftrightarrow \left\{\begin{array}{l} \Gamma_{422} = 0 \\ \Gamma_{424} = 0 \end{array}\right\} \equiv \left\{\begin{array}{l} e^1 \text{ and } e^3 \text{ are surface}\\ \text{forming, this surface}\\ \text{being the null string.}\end{array}\right\}$$

Now, the objective of this paper is to capitalize on this theorem and to explore -- maintaining as much generality as possible -- the one-sidely degenerated vacuum metrics in complex V_4's . Of course, the usual results of the theory of the algebraically degenerate solutions of $R_{ab} = 0$ in real V_4's of signature (+ + + -) should be -- mutatis mutandi -- contained in the results which one is able to establish on the complex level. The essential virtue of the complex treatment of the solutions of the Einstein equations consists in the possibility of working with the two mutually orthogonal null directions.

Reference [7] points out that the 6-parametric solutions of Einstein-Maxwell equations (basically equivalent to the B. Carter solutions [8]) properly analytically extended to a complex form -- can be then written in the double Kerr-Schild form. Also a more general result, a class of metrics of the type D established by Plebański and Demianski [9] which contains 7 continuous parameters, admits the double KS form (some further general study of the double KS metrics where also the rigorous definition of the concept of DKS conjugation and DKS structure is given, can be found in [10]).

The present work intends to use the existence in a complex V_4 of the orthogonal null directions studying the Einstein equations in as general assumptions as possible -- specifically without postulating a priori the existence of a real cross-section to a V_4 of the signature (+ + + -).

Our first step will consist in establishing the general form of the null tetrad for the one-sidedly degenerate vacuum, with $R_{ab} = 0$ and $C^{(5)} = 0 = C^{(4)}$

For the present purposes, it will be convenient to provide the theorem stated in (1.2) with an explicit cotangent interpretation. Indeed, assuming:

(1.3)
$$\Gamma_{422} = 0 = \Gamma_{424}$$

we easily find from the structure equations $de^a = e^b \wedge \Gamma^a{}_b$ that:

(1.4)
$$de^1 = \alpha \wedge e^1 + \beta \wedge e^3 \qquad de^3 = \gamma \wedge e^1 + \delta \wedge e^3$$
$$\alpha, \beta, \gamma, \delta \in \Lambda^1 .$$

But this is precisely the thesis of the classical Frobenius theorem (see, e.g. [11], or for a more elementary treatment [12]).

Thus (in a singly connected region of V_4) there exist such scalars that:

(1.5)
$$e^1 = f^1{}_u du + f^1{}_v dv \qquad e^3 = f^3{}_u du + f^3{}_v dv$$
$$\det(f^i{}_j) \neq 0$$

In the next step, not losing generality but only executing a tetrad gauge [such as which maintains $\Gamma_{422} = \Gamma_{424} = 0$], we can select the surface forming e^1 and e^3 to assume the form:

(1.6)
$$e^1 = \phi^{-2} du \qquad e^3 = \psi^{-2} dv$$

where $\phi \neq 0 \neq \psi$ can be still chosen as convenient. [Notice then that in the arbitrary coordinates $\{x^\mu\}$ the intersections of the (orthogonal) null surfaces $u(x^\mu)$ = const., $v(x^\mu)$ = const., define a congruence of the 2-dimensional null strings.]

We will use u and v as coordinates together with some two independent variables x' and y' . Thus:

(1.7)
$$e^2 = e^2{}_{x'} dx' + e^2{}_{y'} dy' + e^2{}_u du + e^2{}_v dv ,$$
$$e^4 = e^4{}_{x'} dx' + e^4{}_{y'} dy' + e^4{}_u du + e^4{}_v dv ,$$

However, because in 2 dimensions each vector is proportional to a gradient, there exist the functions x, y, and $\lambda \neq 0 \neq \mu$ such that ($e^2_{x'}$, $e^2_{y'}$)$=\lambda$($x_{x'}$, $x_{y'}$) and ($e^4{}_{x'}$, $e^4{}_{y'}$) $=$ $=\mu$($y_{x'}$, $y_{y'}$) . [With any of λ or μ equal to zero, the e^a's became dependent.] Consequently, (1.7) can be re-written in the form:

(1.8) $$e^2 = \lambda (dx + A du + B dv) \quad , \quad e^4 = \mu (dy + C du + D dv)$$

Finally, using the freedom of choice for ϕ and ψ in (1.6), we can set in (1.8) without losing generality $\lambda = 1 = \mu$.

Therefore, the metric of the complex Einstein-flat space with the conformal curvature of the type [Alg. Deg.] $\otimes$ [Anything] (in the notation of [3]) can be always written in the form of:

(1.9) $$ds^2 = 2e^1 \otimes e^2 + 2e^3 \otimes e^4 \qquad \begin{array}{ll} e^1 = \phi^{-2} du \ , & e^2 = dx + A du + B dv \ , \\ e^3 = \psi^{-2} dv \ , & e^4 = dy + C du + D dv \ , \end{array}$$

and, because of:

(1.10) $$0 \neq e^1 \wedge e^2 \wedge e^3 \wedge e^4 = (\phi\psi)^{-2} du \wedge dx \wedge dv \wedge dy$$

we can use as the convenient local coordinates of the metric:

(1.11) $$\{x^\mu\} = \{u x v y\} .$$

Notice that although the metric depends on both functions, B and C, ($g_{uv} = \phi^{-2}B + \psi^{-2}C$) the choice for the value of the function $\phi^{-2}B - \psi^{-2}C$ remains arbitrary being related to the (residual). freedom of the tetrad gauge. Observe also that the coordinates transformation:

(1.12) $$u = U(u', v') \ , \quad v = V(u', v') \ , \quad \frac{\partial(U, V)}{\partial(u', v')} \neq 0 ,$$

if accompanied by the suitable re-gauge of the tetrad and re-definitions of x and y [again basically the process which leads from (1.7) to (1.11)] maintains the general form of the metric and the tetrad. [See Appendix A_3].

2. Connections and the Canonical Forms of the Metric.

We postulate the tetrad in the form of (1.9) Then the inverse tetrad is given by:

(2.1) $$\begin{array}{ll} \partial_2 = \partial_x \ , & \partial_1 = \phi^2(\partial_u - A\partial_x - C\partial_y) \ , \\ \partial_4 = \partial_y \ , & \partial_3 = \psi^2(\partial_v - B\partial_x - D\partial_y) \ . \end{array}$$

The connection forms $\Gamma_{ab} = \Gamma_{[ab]} = \Gamma_{abc} e^c$ can be computed from:

(2.2) $$de^a = e^b \wedge \Gamma^a{}_b$$

or equivalently from:

(2.3) $$\partial_b \partial_a - \partial_a \partial_b = (\Gamma^s{}_{ab} - \Gamma^s{}_{ba})\partial_s$$

The results of the computation which applies (2.3) are summarized as follows (for the details of the computation see appendix A_1):

(2.4) $$\Gamma_{42} = -\frac{1}{2}(\ln\Phi^{-2})_{,4}e^1 + \frac{1}{2}(\ln\Psi^{-2})_{,2}e^3 ,$$

(2.5) $$\Gamma_{41} = -\Phi^2 A_{,4}e^1 - \frac{1}{2}(\ln\Phi^{-2})_{,4}e^2 + \frac{1}{2}\{(\ln\Psi^{-2})_{,1} - \Phi^2 C_{,4} - \Psi^2 B_{,4}\}e^3 ,$$

(2.6) $$\Gamma_{31} = (\Psi^2 B_{,1} - \Phi^2 A_{,3})e^1 + \frac{1}{2}[-(\ln\Phi^{-2})_{,3} - \Phi^2 C_{,2} + \Psi^2 B_{,2}]e^2 + (\Psi^2 D_{,1} - \Phi^2 C_{,3})e^3 + \frac{1}{2}[(\ln\Psi^{-2})_{,1} - \Phi^2 C_{,4} + \Psi^2 B_{,4}]e^4 ,$$

(2.7) $$\Gamma_{32} = \frac{1}{2}[-(\ln\Phi^{-2})_{,3} + \Phi^2 C_{,2} + \Psi^2 B_{,2}]e^1 + \Psi^2 D_{,2}e^3 + \frac{1}{2}(\ln\Psi^{-2})_{,2}e^4 ,$$

(2.8) $$\pm\Gamma_{12} + \Gamma_{34} = \{\pm\Phi^2 A_{,2} + \frac{1}{2}[(\ln\Psi^{-2})_{,1} + \Phi^2 C_{,4} + \Psi^2 B_{,4}]\}e^1 + \{\pm(\ln\Phi^{-2})_{,2} + \frac{1}{2}(\ln\Psi^{-2})_{,2}\}e^2 + \{\pm\frac{1}{2}[(\ln\Phi^{-2})_{,3} + \Phi^2 C_{,2} + \Psi^2 B_{,2}] + \Psi^2 D_{,4}\}e^3 + \{\pm\frac{1}{2}(\ln\Phi^{-2})_{,4} + (\ln\Psi^{-2})_{,4}\}e^4 .$$

The next step consists in the explicit evaluation of the simplest of the curvature forms:

(2.9) $$\mathcal{A} \equiv d\Gamma_{42} + \Gamma_{42} \wedge (\Gamma_{12} + \Gamma_{34}) = \frac{1}{2}C^{(5)}e^4 \wedge e^2 - \frac{1}{2}R_{44}e^4 \wedge e^1 + [\frac{1}{2}C^{(3)} - \frac{R}{12}]e^3 \wedge e^1 - \frac{1}{2}R_{22}e^3 \wedge e^2 + \frac{1}{2}[C^{(4)} + R_{24}]e^1 \wedge e^2 + \frac{1}{2}[C^{(4)} - R_{24}]e^3 \wedge e^4$$

Because:

(2.10) $$d\Gamma_{42} = -\frac{1}{2}\Phi^2(\Phi^{-2})_{,4a}e^a \wedge e^1 + \frac{1}{2}\Psi^2(\Psi^{-2})_{,2a}e^a \wedge e^3$$

and Γ_{42} is spanned by e^1 and e^3, it is clear that A does not possess any $e^4 \wedge e^2$ contribution, so that we duly have:

(2.11) $$C^{(5)} = 0$$

The theorem from [6] assures now that with all $R_{ab}=0$ equations fulfilled and $\Gamma_{422} = \Gamma_{424} = 0$ necessarily $C^{(4)}=0$. Therefore, reading off from (2.9) the $e^4\wedge e^1$, $e^3\wedge e^2$, $e^1\wedge e^2$ and $e^3\wedge e^4$ components we obtain four conditions:

(2.12)
$$(a)\ -\tfrac{1}{2}R_{44} \equiv -\tfrac{1}{2}\Phi^2\Phi^{-2}{}_{;44} + \tfrac{1}{2}(\ln\Phi^{-2})_{,4}\left[\tfrac{1}{2}(\ln\Phi^{-2})_{,4} + (\ln\Psi^{-2})_{,4}\right] = 0,$$
$$(b)\ -\tfrac{1}{2}R_{22} \equiv -\tfrac{1}{2}\Psi^2\Psi^{-2}{}_{;22} + \tfrac{1}{2}(\ln\Psi^{-2})_{,2}\left[\tfrac{1}{2}(\ln\Psi^{-2})_{,2} + (\ln\Phi^{-2})_{,2}\right] = 0,$$
$$(c)\ \tfrac{1}{2}\left[C^{(4)}+R_{24}\right] \equiv \tfrac{1}{2}\Phi^2\Phi^{-2}{}_{;42} - \tfrac{1}{2}(\ln\Phi^{-2})_{,4}\left[\tfrac{1}{2}(\ln\Psi^{-2})_{,2} + (\ln\Phi^{-2})_{,2}\right] = 0,$$
$$(d)\ \tfrac{1}{2}\left[C^{(4)}-R_{24}\right] \equiv -\tfrac{1}{2}\Psi^2\Psi^{-2}{}_{;24} + \tfrac{1}{2}(\ln\Psi^{-2})_{,2}\left[\tfrac{1}{2}(\ln\Phi^{-2})_{,4} + (\ln\Psi^{-2})_{,4}\right] = 0.$$

For the sake of completness we will add that with $R=0$, the $e^3\wedge e^1$ coefficient in (2.9) gives:

(2.13)
$$\tfrac{1}{2}C^{(3)} = -\tfrac{1}{2}\Phi^2\Phi^{-2}{}_{;43} - \tfrac{1}{2}\Psi^2\Psi^{-2}{}_{;21} +$$
$$+ \tfrac{1}{2}(\ln\Phi^{-2})_{,4}\left\{\Psi^2 D_{,4} + \tfrac{1}{2}\left[(\ln\Phi^{-2})_{,3} + \Phi^2 C_{,2} + \Psi^2 B_{,2}\right]\right\} +$$
$$+ \tfrac{1}{2}(\ln\Psi^{-2})_{,2}\left\{\Phi^2 A_{,2} + \tfrac{1}{2}\left[(\ln\Psi^{-2})_{,1} + \Phi^2 C_{,4} + \Psi^2 B_{,4}\right]\right\}.$$

We will now examine the mechanisms which permit the conditions (2.12) to be fulfilled. First of all, we observe that (2.12) became identities when:

(2.14) Case I: $$\Phi_{,4} \equiv \Phi_{,y} = 0 = \Psi_{,x} \equiv \Psi_{,2}\,.$$

In this case, however, $\Phi = \Phi(x\,u\,v)$ and $\Psi = \Psi(y\,u\,v)$ according to (1.9) can be absorbed [made equal to one] by the proper re-definitions of x and y. Therefore, in the case I we can assume -- without losing generality -- that:

(2.15) Case I: $$\Phi = 1 = \Psi$$

Notice that then:

(2.16) Case I: $$C^{(3)} = 0\,,$$

so that in this case the possible types of the Einstein flat complex V_4 are only [3, 1]⊗[Anything], [4]⊗[Anything] and "Strong hell" [-]⊗[Anything] (for the notation and terminology see [3]).

Case II arises when we are outside of the Case I, i.e., when at least one of the two derivatives, $\Phi_{,y}$ and $\Psi_{,x}$ is different from zero.

We will first examine the sub-case of $\Phi_{,y}=0$, $\Psi_{,x}\neq 0$ [Sub-case $\Phi_{,y}\neq 0$, $\Psi_{,x}=0$ is obtained from this one by replacing x by y, y by x and u by v and v by u.]

Equations (2.12a) and (2.12c) are then fulfilled; moreover, $\Phi = \Phi(xuv)$ can be absorbed (i.e. made equal to one) by the proper re-definition of the coordinate x ($x = x(x'uv)$ such that $\Phi^{-2}(xuv)\frac{\partial x}{\partial x'} = 1$).
Then, with $\Phi = 1$ (2.12$_b$) and (2.12$_d$) became

$$(2.17) \quad (b) \quad (\Psi^{-1})_{xx} = 0 \qquad (d) \quad (\ln \Psi^{-1})_{xy} = 0$$

According (2.17$_d$) $\Psi^{-1} = A(xuv)\cdot B(yuv)$ then (2.17$_b$) demands $A_{xx}=0$ so that $\Psi^{-1} = [\alpha(uv)x + \beta(uv)]\cdot B(yuv)$. Of course, under the present assumptions $\Psi_x \neq 0 \rightarrow \alpha B \neq 0$. A permissible re-definition of x ($x' - \beta/\alpha = x$) permits us to set $\beta = 0$. Thus $\Psi^{-1} = \alpha B\cdot x$; in the last step, we can now absorb αB (i.e. make it equal to one) by the proper re-definition of the coordinate y ($y = y(y'uv)$ such that $\alpha^2(uv)B^2(yuv)\frac{\partial y}{\partial y'} = 1$). Therefore, in the sub-case considered, we can assume without losing generality that:

$$(2.18) \quad \text{II:} \qquad \Phi^{-1} = 1 \qquad \Psi^{-1} = x$$

We will now demonstrate that modulo coordinates transformations (and the corresponding re-adjustments of the null tetrad members) (2.18) exhausts all possibilities within the case II.

Indeed, assume the last remaining possibility within the case II, i.e., assume that $\Phi_y \neq 0 \neq \Psi_x$. Then one easily finds that the equations (2.12) reduce to:

$$(2.19) \quad (a) \quad \partial_y(\Psi/\Phi)^2\partial_y\Phi = 0, \qquad (b) \quad \partial_x(\Phi/\Psi)^2\partial_x\Psi = 0,$$

$$(c) \quad \partial_x(\Psi/\Phi)\partial_y\Phi = 0, \qquad (d) \quad \partial_y(\Phi/\Psi)\partial_x\Psi = 0.$$

Therefore (in the present argument the inessential variables u and v are suppressed):

$$(2.20) \quad \Phi_y(\Psi/\Phi)^2 = A(x) \qquad \Psi_x(\Phi/\Psi)^2 = B(y)$$

$$\Phi_y(\Psi/\Phi) = b(y) \qquad \Psi_x(\Phi/\Psi) = a(x)$$

with all these functions being $\neq 0$. Thus, taking the ratios by sides:

$$(2.21) \quad \Psi/\Phi = A/b, \qquad \Phi/\Psi = B/a,$$

and consequently

$$(2.22) \quad A/a = b/B =: e^{\sigma} \longrightarrow \sigma_x = 0 = \sigma_y.$$

Therefore

(2.23) $$\psi/\phi = e^{\delta} a/b$$

and (2.20) reduce to:

(2.24) $$\phi_y = e^{-\delta} b^2 a^{-1} \quad , \quad \psi_x = e^{\delta} a^2 b^{-1} .$$

Therefore, if:

(2.25) $$a = \sqrt{f_x(x)} \quad , \quad b = \sqrt{g_y(y)} \, ,$$

then we can integrate (2.24) in the form of:

(2.26) $$\phi = \frac{e^{-\delta}}{\sqrt{f_x}} [g + \ell(x)] \quad , \quad \psi = \frac{e^{\delta}}{\sqrt{g_y}} [f + k(y)]$$

This substituted into (2.23) gives as the consistency condition:

(2.27) $$e^{\delta} f(x) - \ell(x) = g(y) - e^{\delta} k(y) =: \rho \rightarrow \rho_x = 0 = \rho_y .$$

Therefore, substituting from (2.27) into (2.26) for $\ell(x)$ and $k(y)$ we obtain:

(2.28) $$\phi = \frac{1}{\sqrt{f_x}} [f + e^{-\delta}(g - \rho)] \quad , \quad \psi = \frac{1}{\sqrt{g_y}} [(g - \rho) + e^{\delta} f]$$

It follows that if we introduce $f = e^{-\delta/2} F(x)$, $g = \rho + e^{\delta/2} G(y)$ and $\Lambda := e^{\delta/4}$ we can write for the most general solution of (2.19) [with $\phi_y \neq 0 \neq \psi_x$]:

(2.29) $$\phi = \Lambda^{-1} \frac{F+G}{\sqrt{F_x}} \qquad \psi = \Lambda \frac{F+G}{\sqrt{G_y}}$$

where F = F(xuv), G = G(yuv) and $\Lambda = \Lambda(uv)$ are arbitrary (of course, subject to the conditions $F_x \neq 0 \neq G_y \, ; \, \Lambda \neq 0$).

Substituting (2.29) into our metric (1.9) we easily obtain:

(2.30) $$ds^2 = 2(F+G)^{-2} \{ \Lambda^2 du \otimes (dF + \ldots) + \Lambda^{-2} dv \otimes (dG + \ldots) \}$$

where $(\cdots)$ denote some terms linear in du and dv . In the next step, we observe that because in the two dimensions any one-form is proportional to a differential, therefore there exists such functions u' , v' and α , β that:

(2.31) $$\Lambda^2 du + \Lambda^{-2} dv = \alpha du'$$
$$\Lambda^2 du - \Lambda^{-2} dv = \beta dv' .$$

By taking the external product by sides we have $-2 du \wedge dv = \alpha\beta du' \wedge dv'$ so that $\alpha \neq 0 \neq \beta$ and u', v' can be used as the new variables in the place of u and v:

(2.32) $$du = \frac{i}{2}\Lambda^{-2}(\alpha\, du' + \beta\, dv')\ , \quad dv = \frac{i}{2}\Lambda^{2}(\alpha\, du' - \beta\, dv')\ .$$

Using this in (2.30) and ordering:

(2.33) $$ds^2 = (F+G)^{-2}\{\alpha\, du' \otimes [d(F+G) + \ldots] + \beta\, dv' \otimes [d(F-G) + \ldots]\}$$

where (...) denote some terms linear in du' and dv'.

In the last step, we introduce the new coordinates:

(2.34) $$x' := -\frac{i}{2}\frac{\alpha}{F+G} \qquad y' := \frac{2\beta}{\alpha^2}(F-G)$$

Then (2.33) assumes the form:

(2.34) $$ds^2 = 2\, du' \otimes (dx' + \ldots) + 2x'^2\, dv' \otimes (dy' + \ldots)\ .$$

Dropping primes, we see that this is precisely the particular form of our metric in the case II, as described by (2.18). This ends our proof of the statement after the last formula.

Therefore, summing up: we demonstrated that without any loss of the generality it is enough to consider the metric (1.9) only in the two cases when:

(2.35) I: $$\Phi^{-1} = 1 \quad , \quad \Psi^{-1} = 1$$

II: $$\Phi^{-1} = 1 \quad , \quad \Psi^{-1} = x$$

Now, according to (2.34) all metrics of type II can be represented in the form

(2.36) $$ds^2 = 2\, du \otimes (dx + \ldots) + 2\, dv \otimes x^2 (dy + \ldots)$$

where (...) denote the general linear combinations of du and dv; this representation of the metric although very simple, implies some assymetry between the variables x and y. For our purposes it is more convenient to execute in (2.36) a coordinate transformation

(2.37) $$x = -\frac{1}{x'+y'}\ , \quad y = x' - y'\ , \quad u = \frac{1}{2}(u'+v')\ , \quad v = \frac{1}{2}(u'-v')$$

which brings (2.36) to the form:

(2.38) $$\text{II}:\quad ds^2 = 2(x+y)^{-2}\{du \otimes (dx + \ldots) \\ dv \otimes (dy + \ldots)\}$$

where (...) denote the general terms in du' and dv' .

Therefore, dropping primes, we can represent our metric according to:

(2.39) II: $$ds^2 = 2(x+y)^{-2}\{du \otimes (dx + P du + R dv) \\ dv \otimes (dy + R du + Q dv)\}$$

where P, Q, and R are the three structural functions. Therefore, each one-sided degenerate vacuum metric can be always represented in the form of

(2.40) $$ds^2 = 2\Phi^{-2}[du \otimes (dx + P du + R dv) + dv \otimes (dy + R du + Q dv)]$$

I: $\Phi = 1$ II: $\Phi = x+y$

We shall consider (2.40) as the canonical form of the one-sidedly degenerate vacuum metric. It has the advantage of being form-invariant with respect to the (involutory) coordinate transformation:

(2.41) $$T: \quad \begin{matrix} x \to y & u \to v \\ y \to x & v \to u \end{matrix}.$$

It should be observed that, because the metric $ds^2 = 2(x+y)^{-2}[du \otimes dx + dv \otimes dy]$ is flat, (2.40) according to the terminology of the reference [10] implies the existence of a simple theorem: all one-sidedly degenerate vacuum metrics are double KS structures with the space $N \subset \Lambda^1$ being spanned by two gradients. This theorem justifies the previous intuitions concerning the distinguished role of the double KS metrics (see [7]) and in particular single KS metrics (see [1]).

We will close this section by providing an invariant description of our types I and II.

The surface element of our null string:

(2.42) $$\Sigma := \tfrac{1}{2}\Sigma_{ab}\, e^a \wedge e^b =: du \wedge dv$$

is characterized by the expansion form:

(2.43) $$\Theta := \Theta_a e^a = \tfrac{1}{2}(u^{;a}{}_{;a} dv - v^{;a}{}_{;a} du),$$

where "$;a$" denotes the covariant (tetradinal) derivative. One easily proves that

(2.44) $$\Sigma_{ab;r}\Sigma^{rc} + \Sigma_{ab}\Theta^c = 0.$$

In the special case of $\Theta = 0$, not merely is Σ covariantly constant on each surface, but one finds that the equations:

(2.45) $$X^a{}_{;r}\Sigma^{rb} = 0$$

have a tetrad of independent solutions. Without difficulty one now recognizes that the types I and II are characterized by $\Theta = 0$ and $\Theta \neq 0$ respectively.

3. Metrics of the Type I. Assuming $\Phi = 1 = \Psi$ and specifying the tetrad gauge so that B = C = :R, and re-defining the structural functions A = :P, D = :Q, we have for the metric and the natural tetrad:

(3.1) I: $$\begin{aligned} ds^2 &= 2du \otimes (dx + \mathcal{P}du + \mathcal{R}dv) + 2dv \otimes (dy + \mathcal{R}du + \mathcal{Q}dv) \\ e^1 &= du, \quad e^2 = dx + \mathcal{P}du + \mathcal{R}dv, \quad \partial_2 = \partial_x, \quad \partial_1 = \partial_u - \mathcal{P}\partial_x - \mathcal{R}\partial_y, \\ e^3 &= dv, \quad e^4 = dy + \mathcal{R}du + \mathcal{Q}dv, \quad \partial_4 = \partial_y, \quad \partial_3 = \partial_v - \mathcal{R}\partial_x - \mathcal{Q}\partial_y. \end{aligned}$$

The connections are now:

(3.2) $$\begin{aligned} \Gamma_{42} &= 0, \quad \Gamma_{31} = (\mathcal{R}_{,1} - \mathcal{P}_{,3})e^1 + (\mathcal{Q}_{,1} - \mathcal{R}_{,3})e^3 \\ \Gamma_{12} + \Gamma_{34} &= (\mathcal{P}_{,2} + \mathcal{R}_{,4})e^1 + (\mathcal{R}_{,2} + \mathcal{Q}_{,4})e^3 \end{aligned}$$

and

(3.3) $$\begin{aligned} \Gamma_{41} &= -\mathcal{P}_{,4}e^1 - \mathcal{R}_{,4}e^3, \quad \Gamma_{32} = \mathcal{R}_{,2}e^1 + \mathcal{Q}_{,2}e^3 \\ -\Gamma_{12} + \Gamma_{34} &= (-\mathcal{P}_{,2} + \mathcal{R}_{,4})e^1 + (-\mathcal{R}_{,2} + \mathcal{Q}_{,4})e^3. \end{aligned}$$

The curvature form $\mathcal{A}$ from (2.9) identically vanishes. Because:

(3.4) $$d(\Gamma_{12} + \Gamma_{34}) = (\mathcal{P}_{,2} + \mathcal{R}_{,4})_{,a}e^a \wedge e^1 + (\mathcal{R}_{,2} + \mathcal{Q}_{,4})_{,a}e^a \wedge e^3$$

we can now easily evaluate the curvature form:

(3.5) $$\begin{aligned} \mathcal{B} \equiv d(\Gamma_{12} + \Gamma_{34}) + 2\Gamma_{42} \wedge \Gamma_{31} &= C^{(4)}e^4 \wedge e^2 - R_{41}e^4 \wedge e^1 + C^{(2)}e^3 \wedge e^4 \\ &+ R_{32}e^3 \wedge e^2 + \left[C^{(3)} + \frac{R}{12} + \frac{1}{2}(R_{12} - R_{34})\right]e^1 \wedge e^2 + \left[C^{(3)} + \frac{R}{12} - \frac{1}{2}(R_{12} - R_{34})\right]e^3 \wedge e^4. \end{aligned}$$

Equations (2.12) are already fulfilled with $C^{(4)} = 0$; additionally, we have in the present case $C^{(3)} = 0$; therefore, with $R_{ab} = 0$, the coefficients at $e^3 \wedge e^4$, $e^1 \wedge e^2$, $e^4 \wedge e^1$ and $e^3 \wedge e^2$ in (3.5) must all vanish; this gives:

(3.6) $e^1 \wedge e^2 : \quad f_x = 0 \qquad e^4 \wedge e^1 : \quad f_y = 0 ,$

$e^3 \wedge e^2 : \quad g_x = 0 \qquad e^3 \wedge e^4 : \quad g_y = 0 ,$

where we denoted

(3.7) $f := P_x + R_y \qquad g := R_x + Q_y$

[Notice that we have then:

(3.8) $\Gamma_{12} + \Gamma_{34} = f du + g dv .$]

Consequently, these conditions imply that f and g are some arbitrary functions of u and v only, $f = f(uv)$, $g = g(uv)$. Knowing this and reading off in (3.5) the coefficient at $e^3 \wedge e^1$ we obtain:

(3.9) $$C^{(2)} = f_{,3} - g_{,1} = f_v - g_u$$

so that also $C^{(2)}$ depends on u and v only.

At this point, we should like to observe that the functions f and g can be affected by transformations which conserve the form of metrics of type I: these constitute the group $\mathcal{G}_I$, described in some detail in the Appendix A_3, formulae (A_3 .29) up to (A_3 .36). In particular, (A_3 .36b) asserts that by using the freedom of the functions $u' = u'(uv)$, $v' = v'(uv)$, we can always replace f and g by f' and g' such that:

(3.10) $$(f' du' + g' dv') - (f du + g dv) = d \ln \frac{\partial(u'v')}{\partial(uv)} .$$

Note that by acting with d on this equation one obtains:

(3.11) $$C^{(2)} \equiv f_v - g_u = \frac{\partial(u'v')}{\partial(uv)} [f'_{v'} - g'_{u'}] \equiv \Delta C^{(3)} ,$$

which is consistent with (A_3 .35b).

Therefore, although -- consistently with the rules of the game -- the property of $C^{(2)}$ being equal or different from zero cannot be affected by $\mathcal{G}_I$ gauge, nevertheless, by using its freedom f and g can be selected as convenient to a considerable extent. In particular, if $C^{(2)} = 0 \rightarrow f = \partial_u \chi(uv)$, $g = \partial_v \chi(uv)$, we can always arrange such a $\mathcal{G}_I$ transformation that $\Delta = \frac{\partial(u'v')}{\partial(uv)} = \chi$, and consequently to gauge out f and g so that $f' = 0$, $g' = 0$ Therefore, if $C^{(2)} = 0$, we can always assume not losing generality that $f = 0 = g$. If $C^{(2)} \neq 0$, we find it conveient to maintain the freedom of $\mathcal{G}_I$ gauge, working with f and g as general functions of u and v ; later on, we will examine which sort of simplification can be achieved by employing the freedom of this gauge.

In the next step, we seek $\mathcal{R}$ in the form of:

$$\mathcal{R} = \textcircled{H}_{xy} + \tfrac{1}{2}(f\cdot y + g\cdot x)\,. \tag{3.12}$$

Then equations (3.7) take the form of:

$$\partial_x\left[\mathcal{P} + \textcircled{H}_{yy} - \tfrac{1}{2}(f\cdot x + g\cdot y)\right] = 0\,,\quad \partial_y\left[\mathcal{Q} + \textcircled{H}_{xx} - \tfrac{1}{2}(f\cdot x + g\cdot y)\right] = 0, \tag{3.13}$$

and can be integrated in the form of: $\mathcal{Q} = -\textcircled{H}_{xx} - \tfrac{1}{2}(f\cdot x + g\cdot y) + G_{xx}$ $\mathcal{P} = -\textcircled{H}_{yy} + \tfrac{1}{2}(f\cdot x + g\cdot y) + F_{yy}$, where $F(yuv)$ and $G(xuv)$ are arbitrary. Therefore, by gauging $\textcircled{H}$ according to $\textcircled{H} \to \textcircled{H} + F + G$ we conclude that the three structural functions of the case I assume the form of:

$$\mathcal{P} = -\textcircled{H}_{yy} + \tfrac{1}{2}(f\cdot x + g\cdot y)\,,\quad \mathcal{Q} = -\textcircled{H}_{xx} + \tfrac{1}{2}(f\cdot x + g\cdot y)$$

$$\mathcal{R} = \textcircled{H}_{xy} + \tfrac{1}{2}(g\cdot x + f\cdot y)\,. \tag{3.14}$$

In the next step, by using (3.16) and the explicit expressions for ∂_1 and ∂_3, we find after some work that:

$$\mathcal{R}_{,1} - \mathcal{P}_{,3} = \partial_y\left\{\Xi - \tfrac{1}{2}xy\,C^{(2)}\right\}\,,\quad \mathcal{R}_{,3} - \mathcal{Q}_{,1} = \partial_x\left\{\Xi + \tfrac{1}{2}xy\,C^{(2)}\right\}, \tag{3.15}$$

where we denoted:

$$\Xi := \textcircled{H}_{xx}\textcircled{H}_{yy} - \textcircled{H}_{xy}\textcircled{H}_{xy} + \textcircled{H}_{xu} + \textcircled{H}_{yv} +$$
$$-\tfrac{1}{4}(f+g)(x+y)^2(\partial_x+\partial_y)(x+y)^{-1}(\partial_x+\partial_y)\textcircled{H} - \tfrac{1}{4}(f-g)(x-y)^2(\partial_x-\partial_y)(x-y)^{-1}(\partial_x-\partial_y)\textcircled{H} +$$
$$-\tfrac{1}{4}(x-y)(x+y)(f_u - g_v)\,. \tag{3.16}$$

Therefore, we have for connections:

$$\Gamma_{12} + \Gamma_{34} = f e^1 + g e^3\,,\quad \Gamma_{31} = \left[\Xi - \tfrac{1}{2}xy\,C^{(2)}\right]_y e^1 - \left[\Xi + \tfrac{1}{2}xy\,C^{(2)}\right]_x e^3, \tag{3.17}$$

and the curvature form:

(3.18) $$\mathcal{C} \equiv d\Gamma_{31} + (\Gamma_{12}+\Gamma_{34})\wedge\Gamma_{31} = [\tfrac{1}{2}C^{(3)} - \tfrac{R}{12}]e^4\wedge e^2 - \tfrac{1}{2}R_{11}\, e^4\wedge e^1 + \\ + \tfrac{1}{2}C^{(1)}e^3\wedge e^1 - \tfrac{1}{2}R_{33}\, e^3\wedge e^2 + \tfrac{1}{2}[C^{(2)} - R_{13}]e^1\wedge e^2 + \tfrac{1}{2}[C^{(2)} + R_{13}]e^3\wedge e^4,$$

can be easily computed as:

(3.19) $$\mathcal{C} = [\Xi - \tfrac{1}{2}xy\,C^{(2)}]_{ya}\, e^a\wedge e^1 + [\Xi + \tfrac{1}{2}C^{(2)}.xy]_{xa}\, e^3\wedge e^a + \\ + e^3\wedge e^1\{ f[\Xi + \tfrac{1}{2}xy\,C^{(2)}]_x + g[\Xi - \tfrac{1}{2}xy\,C^{(2)}]_y \}.$$

The term $e^4\wedge e^2$ is duly absent in $\mathcal{C}$ $[C^{(3)} = 0 = R]$. The vanishing of R_{11}, R_{33} and R_{13} implies the three conditions:

(3.20) $$-\tfrac{1}{2}R_{11} \equiv \Xi_{yy} = 0 \quad , \quad \tfrac{1}{2}R_{13} \equiv \Xi_{xy} = 0 \;, \; -\tfrac{1}{2}R_{33} \equiv \Xi_{xx} = 0,$$

and $C^{(1)}$ is given by:

(3.21) $$\tfrac{1}{2}C^{(1)} = (\partial_3 + g)(\Xi - \tfrac{1}{2}xy\,C^{(2)})_y + (\partial_1 + f)(\Xi + \tfrac{1}{2}xy\,C^{(2)})_x .$$

Therefore, integrating the three last field equations, (3.20), we conclude that necessarily:

(3.22) $$\Xi = \alpha.x + \beta.y + \gamma$$

where α, β and γ are arbitrary functions of u and v only. This information used in (3.21) gives:

(3.23) $$\tfrac{1}{2}C^{(1)} = (\partial_u + f)(\alpha + \tfrac{1}{2}y\,C^{(2)}) + (\partial_v + g)(\beta - \tfrac{1}{2}x\,C^{(2)})$$

Next comes the important point: what we can say about the transformational properties of the objects α and β from the point of view of the gauge group $\mathcal{G}_I$?
[See A_3, formulae (A_3 .29) up to (A_3 .36)]

Now, according to (3.24) and (3.17), by using (3.2) we have in the present case:

(3.24) $\Gamma_{42} = 0 \quad , \quad \Gamma_{31} = (\beta - \frac{1}{2} x C^{(2)}) e^1 - (\alpha + \frac{1}{2} y C^{(2)}) e^3$

$$\Gamma_{12} + \Gamma_{34} = f \cdot e^1 + g \cdot e^3$$

Of course, parallel formulae, with all objects primed, must hold in a new gauge frame. From this, by using (A₃ .36) we easily infer that while $C^{(a)}$'s transform according to (A₃ .25) [with $\varkappa = \frac{1}{2}(\varrho_{v'} - \sigma_{u'})$], the functions α and β transform according to:

(3.25)

$$\alpha' = \Delta^{-2}(u'_u \alpha + u'_v \beta) - \tfrac{1}{2}\sigma C'^{(2)} - (\partial_{v'} + g')\varkappa ,$$

$$\beta' = \Delta^{-2}(v'_u \alpha + v'_v \beta) + \tfrac{1}{2}\varrho C'^{(2)} + (\partial_{u'} + f')\varkappa .$$

[It can be shown that these rules are consistent with the structure and the transformation rules of $C^{(1)}$ and $C^{(2)}$.]

We can now observe that (3.14) substituted into (3.3) gives for the corresponding connections:

(3.26)
$$\Gamma_{41} = (\Theta_{yyy} - \tfrac{1}{2} g) e^1 - (\Theta_{xyy} + \tfrac{1}{2} f) e^3 ,$$
$$-\Gamma_{12} + \Gamma_{34} = 2\Theta_{xyy} e^1 - 2\Theta_{xxy} e^3 ,$$
$$\Gamma_{32} = (\Theta_{xxy} + \tfrac{1}{2} g) e^1 - (\Theta_{xxx} - \tfrac{1}{2} f) e^3 .$$

Having these, we can now compute the "dotted" (hellish) components of the Cartan structure equations, i.e.:

(3.27) $\bar{\mathcal{A}} \equiv d\Gamma_{41} + \Gamma_{41} \wedge (-\Gamma_{12} + \Gamma_{34}) = \frac{1}{2}\bar{C}^{(5)} e^4 \wedge e^1 - \frac{1}{2} R_{44} e^4 \wedge e^2 +$
$+ [\frac{1}{2}\bar{C}^{(3)} - \frac{R}{12}] e^3 \wedge e^2 - \frac{1}{2} R_{11} e^3 \wedge e^1 - \frac{1}{2}[\bar{C}^{(4)} + R_{14}] e^1 \wedge e^2 + \frac{1}{2}[\bar{C}^{(4)} - R_{14}] e^3 \wedge e^4$

$$\bar{\mathcal{D}} \equiv d(-\Gamma_{12} + \Gamma_{34}) + 2\Gamma_{41} \wedge \Gamma_{32} = \bar{C}^{(4)} e^4 \wedge e^1 - R_{24} e^4 \wedge e^2 +$$
$$+ \bar{C}^{(2)} e^3 \wedge e^2 + R_{13} e^3 \wedge e^1 - [\bar{C}^{(3)} + \tfrac{R}{12} + \tfrac{1}{2}(R_{12} - R_{34})] e^1 \wedge e^2 + [\bar{C}^{(3)} + \tfrac{R}{12} - \tfrac{1}{2}(R_{12} - R_{34})] \cdot e^3 \wedge e^4$$

$$\bar{\mathcal{C}} \equiv d\Gamma_{32} + (-\Gamma_{12} + \Gamma_{34}) \wedge \Gamma_{32} = [\tfrac{1}{2}\bar{C}^{(3)} - \tfrac{R}{12}] e^4 \wedge e^1 - R_{22} e^4 \wedge e^2 +$$
$$+ \tfrac{1}{2}\bar{C}^{(1)} e^3 \wedge e^2 - \tfrac{1}{2} R_{33} e^3 \wedge e^1 - \tfrac{1}{2}[\bar{C}^{(2)} - R_{23}] e^1 \wedge e^2 + \tfrac{1}{2}[\bar{C}^{(2)} + R_{23}] e^3 \wedge e^4 ,$$

and determine the anti-self-dual part of the conformal curvature characterized by the quantities $\bar{C}^{(a)}$. [Of course, $R_{ab} = 0$ assuming that $\textcircled{H}$ fulfills (3.22)]. We easily find that:

$$\bar{C}^{(5)} = 2\textcircled{H}_{yyyy} \qquad \bar{C}^{(3)} = 2\textcircled{H}_{xxyy} \qquad \bar{C}^{(2)} = 2\textcircled{H}_{xxxy}$$
$$\bar{C}^{(4)} = 2\textcircled{H}_{xyyy} \qquad \bar{C}^{(1)} = 2\textcircled{H}_{xxxx}\,. \tag{3.28}$$

It is now time to re-state concisely our results for the case I. We found that the structural functions P, Q and R are given in terms of $\textcircled{H}$ and arbitrary $f(uv)$ and $g(uv)$ according to (3.14). The function $\textcircled{H}$ must fulfill (3.22) with α, β and γ being arbitrary functions of u and v only. We notice however that a gauge of $\textcircled{H}$ which leaves (3.14) invariant, $\textcircled{H} \rightarrow \textcircled{H} + a(uv)x + b(uv)y$, transforms Ξ according to: $\Xi \rightarrow \Xi + (\partial_u + \tfrac{1}{2}f)a + (\partial_v + \tfrac{1}{2}g)b$, so that selecting properly a and b we can set in (3.22) $\gamma = 0$ not affecting anything. Consequently, we have for $\textcircled{H}$:

$$\Xi \equiv \textcircled{H}_{xx}\textcircled{H}_{yy} - (\textcircled{H}_{xy})^2 + \textcircled{H}_{xu} + \textcircled{H}_{yv} +$$
$$-\tfrac{1}{4}(f+g)(x+y)^2(\partial_x+\partial_y)(x+y)^{-1}(\partial_x+\partial_y)\textcircled{H} - \tfrac{1}{4}(f-g)(x-y)^2(\partial_x-\partial_y)(x-y)^{-1}(\partial_x-\partial_y)\textcircled{H} +$$
$$-\tfrac{1}{4}(x^2-y^2)(f_u - g_v) = \alpha x + \beta y \tag{3.29}$$

The functions of u and v, α, β; f and g should be here considered as given modulo the gauge transformations (3.10) and (3.25) and can be thus selected as convenient for the three distinct possible algebraic types of the conformal curvature from the left:

(3.30)

[⅃ ||] ⊗ [Anything]	:	$C^{(2)} \neq 0$
[||||] ⊗ [Anything]	:	$C^{(2)} = 0\,,\ C^{(1)} \neq 0\,,$
[-] ⊗ [Anything]	:	$C^{(2)} = 0\,,\ C^{(1)} = 0\,,$

where $C^{(2)}$ and $C^{(1)}$ are given by (3.11) and (3.23).

When $C^{(2)} = 0$, we already saw that we can set $f = 0 = g$. If additionally $C^{(1)} = 0$, then we infer from (3.23) that $\alpha = \partial_v \chi(uv)$, $\beta = -\partial_u \chi(uv)$ and -- according to (3.25) -- we can then always arrange for such a G_I gauge that $\alpha' = 0 = \beta'$.

Therefore, for the type [-] ⊗ [A] (A ≡ Anything will be abbreviation used from now on) we can set $f = g = \alpha = \beta = 0$, reducing (3.29) simply to:

(3.31) $$\Xi \equiv \Theta_{xx}\Theta_{yy} - (\Theta_{xy})^2 + \Theta_{xu} + \Theta_{yv} = 0 ;$$

in which we recognize the second heavenly equation introduced in [3], for the key function Θ . Therefore, (3.29), the fundamental equation for the type I, generalizes the second heavenly equation.

Describing more general spaces of type I in terms of the key function Θ we leave the problem of the optimal gauges for f, g, α and β open in this paper only mentioning that when $C^{(2)} \neq 0$ the gauge where $f = g$ and $\beta = -\alpha$ has some virtues.

Closing this section, we would like to state that the spaces of type I turned out to be particularly plausible and describable basically in terms of one structural function only. The question arises: can we repeat the results of this section with respect to the general case II? Surprisingly enough, the answer to this question is positive.

4. Metrics of the Type II.

Specializing the results of the section 2 for the case II, we find it convenient to make a definite choice for the remaining freedom of the tetrad gauge so that with $\phi = x + y = \Psi$ we will also have $B = \mathcal{R} = C$, $A = \mathcal{P}$, $D = \mathcal{Q}$ and consequently:

(4.1)
$$e^1 = (x+y)^{-2}du , \qquad e^2 = dx + \mathcal{P}du + \mathcal{R}dv ,$$
$$e^3 = (x+y)^{-2}dv , \qquad e^4 = dy + \mathcal{R}du + \mathcal{Q}dv ,$$
$$\partial_2 = \partial_x , \qquad \partial_1 = (x+y)^2(\partial_u - \mathcal{P}\partial_x - \mathcal{R}\partial_y) ,$$
$$\partial_4 = \partial_y , \qquad \partial_3 = (x+y)^2(\partial_v - \mathcal{R}\partial_x - \mathcal{Q}\partial_y) .$$

With the present tetrad, $C^{(5)} = 0$ and the equations (2.12) are automatically satisfied, so that $C^{(4)} = 0$ and $R_{44} = R_{42} = R_{22} = 0$. The formula (2.13) specialized to the present case after some simple computation reduces to:

(4.2) $$\frac{1}{2} C^{(3)} = -(x+y)^3\{\partial_x (x+y)^{-2}(\mathcal{P}+\mathcal{R}) + \partial_y (x+y)^{-2}(\mathcal{R}+\mathcal{Q})\}$$

In order to obtain some further information, we specialize the previous expressions for Γ_{42} , $\Gamma_{12} + \Gamma_{34}$ and Γ_{31} to the present case. This gives:

$$(4.3)\quad \Gamma_{42} = (x+y)^{-1}(e^1 - e^3)\ ,\quad \Gamma_{12} + \Gamma_{34} = d\{\ln (x+y)^{-3}\} +$$
$$+ \{\mathcal{P}_x + \mathcal{R}_y - 2(x+y)^{-1}(\mathcal{P}+\mathcal{R})\}du + \{\mathcal{R}_x + \mathcal{Q}_y - 2(x+y)^{-1}(\mathcal{R}+\mathcal{Q})\}dv,$$
$$\Gamma_{31} = (x+y)^2(\mathcal{R}_{,1} - \mathcal{P}_{,3})e^1 + (x+y)^2(\mathcal{Q}_{,1} - \mathcal{R}_{,3})e^3 +$$
$$- (x+y)(\mathcal{R}+\mathcal{Q})e^2 + (x+y)(\mathcal{P}+\mathcal{R})e^4 .$$

The present expressions for these 1 -- forms are so arranged that it is very easy to determine:

$$(4.4)\quad \mathcal{B} \equiv d(\Gamma_{12} + \Gamma_{34}) + 2\Gamma_{42}\wedge\Gamma_{31} =$$
$$= (x+y)^2\{\mathcal{P}_x + \mathcal{R}_y - 2(x+y)^{-1}(\mathcal{P}+\mathcal{R})\}_{,a}\, e^a\wedge e^1$$
$$(x+y)^2\{\mathcal{R}_x + \mathcal{Q}_y - 2(x+y)^{-1}(\mathcal{R}+\mathcal{Q})\}_{,a}\, e^a\wedge e^3$$
$$+ 2(x+y)\{(\mathcal{P}+\mathcal{R})_{,3} - (\mathcal{Q}+\mathcal{R})_{,1}\}e^3\wedge e^1 +$$
$$- 2(\mathcal{R}+\mathcal{Q})(e^1 - e^3)\wedge e^2 + 2(\mathcal{P}+\mathcal{R})(e^1 - e^3)\wedge e^4$$

We can now compare it with the right hand members of (3.5); $C^{(4)}$ duly vanishes. The vanishing of R_{14} and of R_{23} gives the conditions:

$$(4.5)\quad e^4\wedge e^1:\quad (x+y)^2\{\mathcal{P}_x + \mathcal{R}_y - 2(x+y)^{-1}(\mathcal{P}+\mathcal{R})\}_y - 2(\mathcal{P}+\mathcal{R}) = 0\ ,$$

$$(4.6)\quad e^3\wedge e^2:\quad (x+y)^2\{\mathcal{R}_x + \mathcal{Q}_y - 2(x+y)^{-1}(\mathcal{R}+\mathcal{Q})\}_x - 2(\mathcal{R}+\mathcal{Q}) = 0\ .$$

Because $R_{12} = 0 = R_{34}$ the coefficients at $e^1\wedge e^2$ and $e^3\wedge e^4$ after subtracting from them $C^{(3)}$ [known in the form of (4.2)] should vanish; this gives the further two conditions:

$$(4.7)\quad e^1\wedge e^2:\quad -(x+y)^2\{\mathcal{P}_x + \mathcal{R}_y - 2(x+y)^{-1}(\mathcal{P}+\mathcal{R})\}_x - 2(\mathcal{R}+\mathcal{Q}) +$$
$$+ 2(x+y)^3\{\partial_x (x+y)^{-2}(\mathcal{P}+\mathcal{R}) + \partial_y (x+y)^{-2}(\mathcal{R}+\mathcal{Q})\} = 0\ ,$$

$$(4.8)\quad e^3\wedge e^4:\quad -(x+y)^2\{\mathcal{R}_x + \mathcal{Q}_y - 2(x+y)^{-1}(\mathcal{R}+\mathcal{Q})\}_y - 2(\mathcal{P}+\mathcal{R}) +$$
$$+ 2(x+y)^3\{\partial_x (x+y)^{-2}(\mathcal{P}+\mathcal{R}) + \partial_y (x+y)^{-2}(\mathcal{R}+\mathcal{Q})\} = 0$$

The conditions (4.10-11) after obvious cancellations can be rewritten in the form of:

(4.9) $$\partial_x (x+y)^{-2} \mathcal{P}_y + \partial_y (x+y)^{-2} \mathcal{R}_y = 0$$

and

(4.10) $$\partial_x (x+y)^{-2} \mathcal{R}_x + \partial_y (x+y)^{-2} \mathcal{Q}_x = 0 \, .$$

It is much more difficult to bring the conditions (4.7) and (4.8) to a plausible form. First we execute differentiations and we order these relations in the form of:

(4.11) $$-(x+y)^2 (\mathcal{P}_{xx} + \mathcal{R}_{xy}) + 4(x+y)(\mathcal{P}_x + \mathcal{R}_x) + 2(x+y)(\mathcal{R}_y + \mathcal{Q}_y) + - 6\,[(\mathcal{P}+\mathcal{R}) + (\mathcal{R}+\mathcal{Q})] = 0$$

(4.12) $$-(x+y)^2 (\mathcal{R}_{xy} + \mathcal{Q}_{yy}) + 4(x+y)(\mathcal{R}_y + \mathcal{Q}_y) + 2(x+y)(\mathcal{P}_x + \mathcal{R}_x) + - 6\,[(\mathcal{P}+\mathcal{R}) + (\mathcal{R}+\mathcal{Q})] = 0$$

Now, subtracting these relations and dividing by (x + y), we obtain:

(4.13) $$(x+y)(\mathcal{Q}_{yy} - \mathcal{P}_{xx}) + 2(\mathcal{P}_x + \mathcal{R}_x) - 2(\mathcal{R}_y + \mathcal{Q}_y) = 0 \, ,$$

which is of course equivalent to the plausible:

(4.14) $$\partial_y \left[(x+y)\mathcal{Q}_y - 3\mathcal{Q} - 2\mathcal{R} \right] - \partial_x \left[(x+y)\mathcal{P}_x - 3\mathcal{P} - 2\mathcal{R} \right] = 0.$$

On the other hand, by adding (4.11) and (4.12) and multiplying the output by the factor $(x + y)^{\sigma}$, exponent σ to be determined, we obtain:

(4.15) $$-(x+y)^{\sigma+2} (\mathcal{P}_{xx} + 2\mathcal{R}_{xy} + \mathcal{Q}_{yy}) + 6(x+y)^{\sigma+1} (\mathcal{P}+\mathcal{R})_x + + 6(x+y)^{\sigma+1} (\mathcal{R}+\mathcal{Q})_y - 12(x+y)^{\sigma} [(\mathcal{P}+\mathcal{R}) + (\mathcal{R}+\mathcal{Q})] = 0 .$$

Completing now the natural divergences implicit in this equation, we find it equivalent to:

(4.16) $$\partial_x\{-(x+y)^{\sigma+2}(\mathcal{P}_x+\mathcal{R}_y)+(\sigma+8)(x+y)^{\sigma+1}(\mathcal{P}+\mathcal{R})\}+$$
$$\partial_y\{-(x+y)^{\sigma+2}(\mathcal{R}_x+\mathcal{Q}_y)+(\sigma+8)(x+y)^{\sigma+1}(\mathcal{R}+\mathcal{Q})\}+$$
$$-(\sigma+4)(\sigma+5)\{(\mathcal{P}+\mathcal{R})+(\mathcal{R}+\mathcal{P})\}(x+y)^{\sigma}=0$$

Therefore, with $\sigma=-4$ and $\sigma=-5$ equation (4.21) reduces to a divergence:

(4.17) $$\sigma=-4:\quad \partial_y\{(x+y)^{-2}(\mathcal{R}_x+\mathcal{Q}_y)-4(x+y)^{-3}(\mathcal{R}+\mathcal{Q})\}+$$
$$\partial_x\{(x+y)^{-2}(\mathcal{P}_x+\mathcal{R}_y)-4(x+y)^{-3}(\mathcal{P}+\mathcal{R})\}=0,$$

(4.18) $$\sigma=-5:\quad \partial_x\{(x+y)^{-3}(\mathcal{P}_x+\mathcal{R}_y)-3(x+y)^{-4}(\mathcal{P}+\mathcal{R})\}+$$
$$\partial_y\{(x+y)^{-3}(\mathcal{R}_x+\mathcal{Q}_y)-3(x+y)^{-4}(\mathcal{R}+\mathcal{Q})\}=0.$$

Both forms of this relation, with $\sigma=-4$ and $\sigma=-5$ are useful. In particular, we observe that (4.18) can be also rewritten in the form of:

(4.19) $$\partial_x\partial_x(x+y)^{-3}\mathcal{P}+2\partial_x\partial_y(x+y)^{-3}\mathcal{R}+\partial_y\partial_y(x+y)^{-3}\mathcal{Q}=0$$

Surprisingly enough, the four differential conditions obtained [i.e. (4.9), (4.10), and (4.14), (4.19)] can be easily integrated in all generality. Indeed, represent $\mathcal{P}$ and $\mathcal{Q}$ according to:

(4.20) $$(x+y)^{-3}\mathcal{P}=F_{yy}\qquad (x+y)^{-3}\mathcal{Q}=G_{xx}$$

According to (4.19) we have then:

(4.21) $$\partial_x\partial_y\{2(x+y)^{-3}\mathcal{R}+\partial_x\partial_y(F+G)\}=0.$$

Therefore, $2(x+y)^{-3}\mathcal{R}+\partial_x\partial_y(F+G)=f_x(x)+g_y(y)$ [the arguments u and v which are inessential in the present discussion will be suppressed in the symbols of functions]. But executing the gauge -- $F\to F+yf(x)$, $G\to G+xg(y)$ -- this gauge conserves the formulae (4.20) -- we can get rid of f and g. Therefore, without losing generality, we have as the consequence of (4.21):

(4.22) $$\mathcal{R}=-\tfrac{1}{2}(x+y)^3\partial_x\partial_y(F+G).$$

In the next step, we rewrite (4.19) in the form of:

(4.23)
$$\partial_y[(x+y)^4\partial_y(x+y)^{-3}\mathcal{Q}-2\mathcal{R}]-\partial_x[(x+y)^4\partial_x(x+y)^{-3}\mathcal{P}-2\mathcal{R}]=0.$$

Using here (4.20) and (4.22), executing the differentiations and cancelling the result by $-\frac{1}{2}(x+y)^3$ we obtain:

(4.24) $$\partial_x \partial_y \left\{ \tfrac{i}{2}(x+y)(F-G)_{xy} + F_y - G_x \right\} = 0 .$$

A very satisfactory result is then obtained if we feed (4.20) and (4.22) into the equations (4.9) and (4.10); after executing the differentiations, the resulting equations can be ordered respectively in the form of:

(4.25) $$\partial_y \partial_y \left\{ \tfrac{i}{2}(x+y)(F-G)_{xy} + F_y - G_x \right\} = 0 ,$$

and

(4.26) $$\partial_x \partial_x \left\{ \tfrac{i}{2}(x+y)(F-G)_{xy} + F_y - G_x \right\} = 0 .$$

It thus follows that F and G must be such that:

(4.27) $$\tfrac{i}{2}(x+y)(F-G)_{xy} + F_y - G_x = \alpha x + \beta y + \gamma$$

where α, β and γ can be functions of u and v only.

This remarkable result has an interesting consequence: indeed, feeding (4.20) and (4.22) into (4.2) one obtains after some ordering that:

(4.28) $$\tfrac{i}{2} C^{(3)} = -(x+y)^3 (\partial_y - \partial_x) \left\{ \tfrac{i}{2}(x+y)(F-G)_{xy} + F_y - G_x \right\} = 2(\alpha - \beta)(x+y)^3$$

This implies that $\alpha - \beta$ has a geometrical significance. The function $\gamma = \gamma(u,v)$ has no geometrical significance and can be set equal to zero without the loss of the generality [The gauge $F \to F + y\gamma$ -- it maintains (4.20), (4.22) -- removes γ in (4.27).] We observe additionally that (4.20) and (4.22) are also conserved by the gauge $F \to F + \rho\, xy$, G G - $\rho\, xy$ with $\rho = \rho(u,v)$. Consequently, one easily sees that this gauge changes (4.27) into:

(4.29) $$\tfrac{i}{2}(x+y)(F-G)_{xy} + F_y - G_x = (\alpha - 2\rho)x + (\beta - 2\rho)y .$$

Therefore, selecting $\rho = \frac{i}{4}(\alpha + \beta)$, we reduce this equation to:

(4.30) $$\tfrac{i}{2}(x+y)(F-G)_{xy} + F_y - G_x = -\tfrac{i}{2}\mu(x-y)$$

[with $\mu = \beta - \alpha$] so that:

(4.31) $$C^{(3)} = -2\mu(x+y)^3 .$$

Now, the constraint on the structural functions, (4.30), can be integrated. Indeed, let $A = \frac{1}{2}(F + G)$ and $B = \frac{1}{2}(F - G)$. Then one easily sees that the equation (4.30) takes the form:

(4.32) $$\partial_x \left(x B_y - A - \tfrac{i}{2}\mu xy \right) + \partial_y \left(y B_x + A + \tfrac{i}{2}\mu xy \right) = 0 .$$

Therefore, there exists such a function χ that:

$$x B_y - A - \tfrac{1}{2}\mu xy = \chi_y \quad , \quad y B_x + A + \tfrac{1}{2}\mu xy = -\chi_x \tag{4.33}$$

Adding these two equations we obtain:

$$\partial_y (xB - \chi) + \partial_x (yB + \chi) = 0. \tag{4.34}$$

This, however, implies that there exists a function Λ such that:

$$\begin{aligned} xB - \chi &= \Lambda_x \\ yB + \chi &= -\Lambda_y \ . \end{aligned} \tag{4.35}$$

It follows that:

$$\begin{aligned} B &= (x+y)^{-1}(\Lambda_x - \Lambda_y) \ , \\ \chi &= -(x+y)^{-1}(y\Lambda_x + x\Lambda_y) \ . \end{aligned} \tag{4.36}$$

Substituting it into any of the two relations (4.33) one obtains:

$$A = \Lambda_{xy} - \tfrac{1}{2}\mu xy \tag{4.37}$$

From this we have:

$$\begin{aligned} F &= A + B = \Lambda_{xy} + (x+y)^{-1}(\Lambda_x - \Lambda_y) - \tfrac{1}{2}\mu xy \ , \\ G &= A - B = \Lambda_{xy} - (x+y)^{-1}(\Lambda_x - \Lambda_y) - \tfrac{1}{2}\mu xy \ . \end{aligned} \tag{4.38}$$

Using this inside of our structural functions, we finally obtain:

$$\begin{aligned} \mathcal{P} &= (x+y)^3 \partial_y \partial_y \{\Lambda_{xy} + (x+y)^{-1}(\Lambda_x - \Lambda_y)\} \ , \\ \mathcal{Q} &= (x+y)^3 \partial_x \partial_x \{\Lambda_{xy} - (x+y)^{-1}(\Lambda_x - \Lambda_y)\} \\ \mathcal{R} &= -(x+y)^3 \partial_x \partial_y \{\Lambda_{xy} - \tfrac{1}{2}\mu xy\} \ , \end{aligned} \tag{4.39}$$

where $\Lambda = \Lambda(xyuv)$ and $\mu = \mu(uv)$ are arbitrary functions.

With this shape of the structural functions, we are already sure that $C^{(5)} = C^{(4)} = 0$ and $R_{44} = R_{42} = R_{22} = 0$ and also $R_{12} = R_{34} = R_{14} = R_{23} = 0$; for $C^{(3)}$ we have:

$$C^{(3)} = -2\mu (x+y)^3 \tag{4.40}$$

In the next step, we read off from (4.4) the coefficient at $e^3 \wedge e^1$ determining in this manner the conformal curvature component $C^{(2)}$:

(4.41) $C^{(2)} = (x+y)^2\{[\mathcal{P}_x + \mathcal{R}_y - 2(x+y)^{-1}(\mathcal{P}+\mathcal{R})]_{,3} - [\mathcal{R}_x + \mathcal{Q}_y - 2(x+y)^{-1}(\mathcal{R}+\mathcal{Q})]_{,1}\} +$
$+ 2(x+y)\{(\mathcal{P}+\mathcal{R})_{,3} - (\mathcal{Q}+\mathcal{R})_{,1}\}.$

After the obvious cancellations, we find this expression to be equal to:

(4.42) $C^{(2)} = (x+y)^4\{(\partial_v - \mathcal{R}\partial_x - \mathcal{Q}\partial_y)(\mathcal{P}_x + \mathcal{R}_y) - (\partial_u - \mathcal{P}\partial_x - \mathcal{R}\partial_y)(\mathcal{R}_x + \mathcal{Q}_y)\}.$

We must now determine the differential form C defined by (3.18). In order to be able to obtain it easily, we rewrite the connections (4.3) in the form of:

(4.43) $\Gamma_{12} + \Gamma_{34} = [(x+y)(\mathcal{P}+\mathcal{R}) + (x+y)^2(\mathcal{P}_x + \mathcal{R}_y)]e^1 +$
$+ [(x+y)(\mathcal{R}+\mathcal{Q}) + (x+y)^2(\mathcal{R}_x + \mathcal{Q}_y)]e^3 - 3(x+y)^{-1}(e^2+e^4)$
$\Gamma_{31} = (\mathcal{R}_{,1} - \mathcal{P}_{,3})du + (\mathcal{Q}_{,1} - \mathcal{R}_{,3})dv +$
$- (x+y)(\mathcal{R}+\mathcal{Q})e^2 + (x+y)(\mathcal{P}+\mathcal{R})e^4.$

We then easily find:

(4.44) $(\Gamma_{12} + \Gamma_{34}) \wedge \Gamma_{31} =$
$e^4 \wedge e^1\{-[(x+y)^2(\mathcal{P}+\mathcal{R}) + (x+y)^3(\mathcal{P}_x + \mathcal{R}_y)](\mathcal{P}+\mathcal{R}) + 3(x+y)(\mathcal{P}_{,3} - \mathcal{R}_{,1})\} +$
$e^3 \wedge e^2\{-[(x+y)^2(\mathcal{R}+\mathcal{Q}) + (x+y)^3(\mathcal{R}_x + \mathcal{Q}_y)](\mathcal{R}+\mathcal{Q}) + 3(x+y)(\mathcal{Q}_{,1} - \mathcal{R}_{,3})\} +$
$e^1 \wedge e^2\{-[(x+y)^2(\mathcal{P}+\mathcal{R}) + (x+y)^3(\mathcal{P}_x + \mathcal{R}_y)](\mathcal{R}+\mathcal{Q}) - 3(x+y)(\mathcal{P}_{,3} - \mathcal{R}_{,1})\} +$
$e^3 \wedge e^4\{+[(x+y)^2(\mathcal{R}+\mathcal{Q}) + (x+y)^3(\mathcal{R}_x + \mathcal{Q}_y)](\mathcal{P}+\mathcal{R}) + 3(x+y)(\mathcal{Q}_{,1} - \mathcal{R}_{,3})\} +$
$e^3 \wedge e^1\{[\mathcal{P}+\mathcal{R} + (x+y)(\mathcal{P}_x + \mathcal{R}_y)](\mathcal{R}_{,3} - \mathcal{Q}_{,1}) + [\mathcal{R}+\mathcal{Q} + (x+y)(\mathcal{R}_x + \mathcal{Q}_y)](\mathcal{R}_{,1} - \mathcal{P}_{,3})\}$
$+ e^4 \wedge e^2\{\ldots\}$

and

(4.45) $d\Gamma_{31} = (x+y)^2(\mathcal{R}_{,1} - \mathcal{P}_{,3})_{,a} e^a \wedge e^1 + (x+y)^2(\mathcal{Q}_{,1} - \mathcal{R}_{,3})_{,a} e^a \wedge e^3 +$
$- [(x+y)(\mathcal{R}+\mathcal{Q})]_{,a} e^a \wedge e^2 + [(x+y)(\mathcal{P}+\mathcal{R})]_{,a} e^a \wedge e^4 +$
$- (x+y)(\mathcal{R}+\mathcal{Q})\{(x+y)^2 \mathcal{P}_{,a} e^a \wedge e^1 + (x+y)^2 \mathcal{R}_{,a} e^a \wedge e^3\} +$
$+ (x+y)(\mathcal{P}+\mathcal{R})[(x+y)^2 \mathcal{R}_{,a} e^a \wedge e^1 + (x+y)^2 \mathcal{Q}_{,a} e^a \wedge e^3\}.$

Now, comparing with (3.18) and remembering the field equations $R_{11} = 0 = R_{33}$ we obtain the conditions:

$$(4.46)\quad -\tfrac{1}{2}R_{11} = (x+y)^2(\mathcal{R}_{,1}-\mathcal{P}_{,3})_y - [(x+y)(\mathcal{P}+\mathcal{R})]_{,1} - (x+y)^3(\mathcal{R}+\mathcal{Q})\mathcal{P}_y + (x+y)^3(\mathcal{P}+\mathcal{R})\mathcal{R}_y + 3(x+y)(\mathcal{P}_{,3}-\mathcal{R}_{,1}) - (x+y)^2(\mathcal{P}+\mathcal{R})^2 + - (x+y)^3(\mathcal{P}+\mathcal{R})(\mathcal{P}_x+\mathcal{R}_y) = 0,$$

$$(4.47)\quad -\tfrac{1}{2}R_{33} = -(x+y)^2(\mathcal{Q}_{,1}-\mathcal{R}_{,3})_x - [(x+y)(\mathcal{R}+\mathcal{Q})]_{,3} + (x+y)^3(\mathcal{R}+\mathcal{Q})\mathcal{R}_x + - (x+y)^3(\mathcal{P}+\mathcal{R})\mathcal{Q}_x + 3(x+y)(\mathcal{Q}_{,1}-\mathcal{R}_{,3}) - (x+y)^2(\mathcal{R}+\mathcal{Q})^2 + - (x+y)^3(\mathcal{R}+\mathcal{Q})(\mathcal{R}_x+\mathcal{Q}_y) = 0.$$

The last field equation is obtained by subtracting from the coefficient at $e^3 \wedge e^4$ in $\mathcal{C}$ the coefficient at $e^1 \wedge e^2$:

$$(4.48)\quad R_{13} = -(x+y)^2(\mathcal{Q}_{,1}-\mathcal{R}_{,3})_y + (x+y)^2(\mathcal{R}_{,1}-\mathcal{P}_{,3})_x + [\ +y)(\mathcal{P}+\mathcal{R})]_{,3} + [(x+y)(\mathcal{R}+\mathcal{Q})]_{,1} + (x+y)^3(\mathcal{R}+\mathcal{Q})\mathcal{R}_y - (x+y)^3(\mathcal{P}+\mathcal{R})\mathcal{Q}_y - (x+y)^3(\mathcal{R}+\mathcal{Q})\mathcal{P}_x + (x+y)^3(\mathcal{P}+\mathcal{R})\mathcal{R}_x + + \{(x+y)^2(\mathcal{R}+\mathcal{Q}) + (x+y)^3(\mathcal{R}_x+\mathcal{Q}_y)\}(\mathcal{P}+\mathcal{R}) + 3(x+y)(\mathcal{Q}_{,1}-\mathcal{R}_{,3}) + + \{(x+y)^2(\mathcal{P}+\mathcal{R}) + (x+y)^3(\mathcal{P}_x+\mathcal{R}_y)\}(\mathcal{R}+\mathcal{Q}) + 3(x+y)(\mathcal{P}_{,3}-\mathcal{R}_{,1}) = 0$$

By executing the obvious cancellations, we obtain:

$$(4.49)\quad -\tfrac{1}{2}R_{11} = (x+y)[3\mathcal{P}_{,3}-\mathcal{P}_{,1}-4\mathcal{R}_{,1}] + (x+y)^2(\mathcal{R}_{,1}-\mathcal{P}_{,3})_y + -(x+y)^3[(\mathcal{R}+\mathcal{Q})\mathcal{P}_y + (\mathcal{P}+\mathcal{R})\mathcal{P}_x] = 0,$$

$$(4.50)\quad -\tfrac{1}{2}R_{33} = (x+y)[3\mathcal{Q}_{,1}-\mathcal{Q}_{,3}-4\mathcal{R}_{,3}] + (x+y)^2(\mathcal{R}_{,3}-\mathcal{Q}_{,1})_x + - (x+y)^3[(\mathcal{R}+\mathcal{Q})\mathcal{Q}_y + (\mathcal{P}+\mathcal{R})\mathcal{Q}_x] = 0,$$

$$(4.51)\quad R_{13} = (x+y)[4\mathcal{P}_{,3}+4\mathcal{Q}_{,1}-2\mathcal{R}_{,3}-2\mathcal{R}_{,1}] + +(x+y)^2[(\mathcal{R}_{,3}-\mathcal{Q}_{,1})_y+(\mathcal{R}_{,1}-\mathcal{P}_{,3})_x] + (x+y)^3[2(\mathcal{R}+\mathcal{Q})\mathcal{R}_y+2(\mathcal{P}+\mathcal{R})\mathcal{R}_x]=0.$$

The field equations obtained above written in the present form appear to be very involved. We will be able, however, to bring them to a more plausible form and then to execute the integration process up to the very end, in the complete analogy with the case I studied in the previous section.

First, using the explicit form of ∂_1 and ∂_3 and the fact that $\mathcal{P} = (x + y)^3 .\ Fyy$, $\mathcal{Q} = (x + y)^3 .\ Gxx$, $\mathcal{R} = -(x + y)^3 .\ Axy$, with $F = A + B$ and $G = A - B$ (but at this time still not using the specific form of A and B) we can organize the last three field equations into the plausible form of:

$$(4.52) \quad (a)\ R_{11} = 2(x+y)^6 \partial_y \mathcal{Y} = 0 \ , \qquad (b)\ R_{33} = 2(x+y)^6 \partial_x \mathcal{X} = 0 ,$$
$$(c)\ -R_{13} = (x+y)^6 [\partial_x \mathcal{Y} + \partial_y \mathcal{X}] = 0 ,$$

where we introduced the abbreviations:

$$(4.53) \quad \mathcal{X} := (\partial_u + \partial_v) G_x + (x+y)^{-4} (\mathcal{Q}_{,1} - \mathcal{R}_{,3}) - \tfrac{1}{4} (\mu_u - \mu_v) y \ ,$$
$$\mathcal{Y} := (\partial_u + \partial_v) F_y + (x+y)^{-4} (\mathcal{P}_{,3} - \mathcal{R}_{,1}) + \tfrac{1}{4} (\mu_u - \mu_v) x \ .$$

In the next step, we order inside of $\mathcal{X}$ and $\mathcal{Y}$ the terms which contain ∂_u and ∂_v obtaining:

$$(4.54) \quad \mathcal{X} = \partial_x \{ (x+y)(G_{xu} + F_{yv}) + \tfrac{1}{4}(x-y)(\mu_v x - \mu_u y) \} +$$
$$+ (x+y)^{-2} \{ \mathcal{R}\mathcal{R}_x + \mathcal{Q}\mathcal{R}_y - \mathcal{P}\mathcal{Q}_x - \mathcal{R}\mathcal{Q}_y \} ,$$
$$\mathcal{Y} = \partial_y \{ (x+y)(G_{xu} + F_{yv}) + \tfrac{1}{4}(x-y)(\mu_v x - \mu_u y) \} +$$
$$+ (x+y)^{-2} \{ \mathcal{P}\mathcal{R}_x + \mathcal{R}\mathcal{R}_y - \mathcal{R}\mathcal{P}_x - \mathcal{Q}\mathcal{P}_y \} \qquad {}^{*)}$$

Now, (with the here inessential variables u and v suppressed) we would like to demonstrate that the form:

$$(4.55) \qquad \omega := \mathcal{X} dx + \mathcal{Y} dy$$

is closed, i.e., that there exists a function Ξ such that $\mathcal{X} = \Xi_x$, $\mathcal{Y} = \Xi_y$.

Using (4.54) we have:

$$(4.56) \quad \omega' := \omega - d\{ (x+y)(G_{xu} + F_{yv}) + \tfrac{1}{4}(x-y)(\mu_v x - \mu_u y) \} =$$
$$= (x+y)^{-2} \{ \mathcal{R} d\mathcal{R} + [\mathcal{Q}\mathcal{R}_y - \mathcal{R}\mathcal{Q}_y - \mathcal{P}\mathcal{Q}_x] dx + [\mathcal{P}\mathcal{R}_x - \mathcal{R}\mathcal{P}_x - \mathcal{Q}\mathcal{P}_y] dy \} =$$
$$= \tfrac{1}{2}(x+y)^{-2} d(\mathcal{R}^2 - \mathcal{P}\mathcal{Q}) + (x+y)^{-2} \{ [\mathcal{Q}\mathcal{R}_y - \mathcal{R}\mathcal{Q}_y + \tfrac{1}{2}(\mathcal{Q}\mathcal{P}_x - \mathcal{P}\mathcal{Q}_x)] dx +$$
$$+ [\mathcal{P}\mathcal{R}_x - \mathcal{R}\mathcal{P}_x + \tfrac{1}{2}(\mathcal{P}\mathcal{Q}_y - \mathcal{Q}\mathcal{P}_y)] dy \} .$$

In the next step, we can conveniently substitute here $\mathcal{P}, \mathcal{Q}$ and $\mathcal{R}$ as expressed through A and B:

---

*We selected so the linear terms in x and y in (4.53) for the purpose of obtaining then all the ∂_u and ∂_v terms in $\mathcal{X}$ and $\mathcal{Y}$ in the form of a simple gradient; of course, equations (4.52) stay invariant with respect to the gauge: $\mathcal{X} \to \mathcal{X} - \alpha \cdot y$, $\mathcal{Y} \to \mathcal{Y} + \alpha \cdot x$, $\alpha(uv)$ being arbitrary.

(4.57) $$\omega' = \tfrac{1}{2}(x+y)^{-2}\, d\{(x+y)^6 (A_{xy}A_{xy} - F_{yy}G_{xx})\} + (x+y)^4\{[A_{xy}G_{xxy} - G_{xx}A_{xyy} + \tfrac{1}{2}(G_{xx}F_{yyx} - F_{yy}G_{xxx})]dx + [A_{xy}F_{yyx} - F_{yy}A_{yxx} + \tfrac{1}{2}(F_{yy}G_{xxy} - G_{xx}F_{yyy})]dy\}.$$

Then adding to the first term $\frac{1}{2}(x+y)^4\, d(A_{xy}A_{xy} - F_{yy}G_{xx})$ and subtracting it from the second term, after executing cancellations, we obtain:

(4.58) $$\omega' = (x+y)^4 d(A_{xy}A_{xy} - F_{yy}G_{xx}) + 3(x+y)^3(A_{xy}A_{xy} - F_{yy}G_{xx})d(x+y) + (x+y)^4\{[-A_{xy}B_{xxy} + G_{xx}B_{yyx}]dx + [A_{xy}B_{yyx} - F_{yy}B_{xxy}]dy\}.$$

Now, it is time to use the information that $A = \Lambda_{xy} - \frac{1}{2}\mu xy$, $B = (\Lambda_x - \Lambda_y)/(x+y)$. Differentiating, we have:

(4.59) $$B_x = (x+y)^{-1}(\Lambda_{xx} - \Lambda_{xy} - B), \quad B_y = (x+y)^{-1}(\Lambda_{xy} - \Lambda_{yy} - B),$$

and

(4.60) $$B_{xy} = (x+y)^{-1}(\Lambda_{xyx} - \Lambda_{yxy} - B_x - B_y) = (x+y)^{-1}[A_x - A_y - B_x - B_y - \tfrac{1}{2}\mu(x-y)].$$

This implies:

(4.61) $$B_{xxy} = (x+y)^{-1}(G_{xx} - A_{xy} - 2B_{xy} - \tfrac{1}{2}\mu),$$
$$B_{yyx} = (x+y)^{-1}(A_{xy} - F_{yy} - 2B_{xy} + \tfrac{1}{2}\mu).$$

By substituting this into (4.58), after cancellations, we can order the result in the form of:

(4.62) $$\omega' = (x+y)^4 d(A_{xy}A_{xy} - G_{xx}F_{yy}) + 4(x+y)^3(A_{xy}A_{xy} - G_{xx}F_{yy})d(x+y) + 2(x+y)^3 B_{xy}\{[A_y - A_x + B_x]_x dx + [A_y - A_x + B_y]_y dy\} + \tfrac{1}{2}\mu(x+y)^3\{(A_x + A_y - B_x)_x dx + (A_x + A_y + B_y)_y dy\}.$$

The first and the second lines of (4.62) if we eliminate $A_x - A_y$ by using (4.60) amount to:

(4.63) $$\mathrm{I} + \mathrm{II} = d\{(x+y)^4(A_{xy}A_{xy} - F_{yy}G_{xx})\} + 2(x+y)^3\{[-2B_{xy} - \tfrac{1}{2}\mu - (x+y)B_{xxy}]dx + [-2B_{xy} + \tfrac{1}{2}\mu - (x+y)B_{yyx}]dy\} = d\{(x+y)^4(A_{xy}A_{xy} - B_{xy}B_{xy} - F_{yy}G_{xx})\} - \mu(x+y)^3 B_{xy}\, d(x-y)$$

Therefore, remembering that $(A_{xy})^2 - (B_{xy})^2 = F_{yx}G_{xy}$, we conclude that:

(4.64)
$$\omega'' := \omega' - d\{(x+y)^4(F_{xy}G_{xy} - F_{yy}G_{xx})\} =$$
$$= +\tfrac{1}{2}\mu(x+y)^3 d(A_x + A_y) - \tfrac{1}{2}\mu(x+y)^3\{[B_{xx}+2B_{xy}]dx - [B_{yy}+2B_{xy}]dy\}.$$

It remains to bring this form -- which is already linear in the structural functions -- to the shape of a differential. For this purpose, we notice that:

(4.65)
$$B_{xx} = (x+y)^{-1}(\Lambda_{xxx} - \Lambda_{xxy} - 2B_x),$$
$$B_{yy} = (x+y)^{-1}(\Lambda_{xyy} - \Lambda_{yyy} - 2B_y).$$

We can now substitute this -- and the first line of (4.60) -- into our expression for ω'' ; additionally, because $A_x + A_y = \Lambda_{xyy} + \Lambda_{yxx} - \tfrac{1}{2}\mu(x+y)$, differentiating "by parts" the first term for ω'' and ordering we obtain:

(4.66)
$$\omega''' := \omega'' + d\{\tfrac{1}{16}\mu^2(x+y)^4 - \tfrac{1}{2}\mu(x+y)^3\partial_x\partial_y(\Lambda_x+\Lambda_y)\} =$$
$$= -\tfrac{1}{2}\mu(x+y)^2 d(\Lambda_{xx}+4\Lambda_{xy}+\Lambda_{yy}) - \mu(x+y)^2\{-(2B_x+B_y)dx + (B_x+2B_y)dy\}.$$

The first term of the second line we differentiate "by parts" and for B_x and B_y we use (4.59). This after ordering gives:

(4.67)
$$\omega^{IV} := \omega''' + d\{\tfrac{1}{2}\mu(x+y)^2(\Lambda_{xx}+4\Lambda_{xy}+\Lambda_{yy})\} =$$
$$= 3\mu(x+y)d(\Lambda_x+\Lambda_y) - 3\mu(x+y)B\, d(x-y).$$

This again we differentiate "by parts" and we substitute for B its value, $B = (\Lambda_x - \Lambda_y)/(x+y)$. This gives:

(4.68)
$$\omega^{V} := \omega^{IV} + d\{-3\mu(x+y)(\Lambda_x+\Lambda_y)\} =$$
$$= -3\mu(\Lambda_x+\Lambda_y)d(x+y) - 3\mu(\Lambda_x-\Lambda_y)d(x-y) =$$
$$= -6\mu(\Lambda_x dx + \Lambda_y dy) = d\{-6\mu\Lambda\}.$$

Therefore, we demonstrated that:

(4.69)
$$\omega = X dx + Y dy = d\{\tfrac{1}{4}(x-y)(\mu_v x - \mu_u y) + (x+y)(G_{xu}+F_{yv}) +$$
$$+ (x+y)^4(F_{xy}G_{xy} - F_{yy}G_{xx} - \tfrac{1}{16}\mu^2) + \tfrac{1}{2}\mu(x+y)^3\partial_x\partial_y(\Lambda_x+\Lambda_y) +$$
$$- \tfrac{1}{2}\mu(x+y)^2(\Lambda_{xx}+4\Lambda_{xy}+\Lambda_{yy}) + 3\mu(x+y)(\Lambda_x+\Lambda_y) - 6\mu\Lambda\}.$$

[Observe that u and v are treated here as parameters!] The same result, however, can be written more compactly. One easily shows that:

(4.70) $$-\tfrac{1}{2}\mu(x+y)^3\partial_x\partial_y(\Lambda_x+\Lambda_y)+\tfrac{1}{2}\mu(x+y)^2(\Lambda_{xx}+4\Lambda_{xy}+\Lambda_{yy})+ -3\mu(x+y)(\Lambda_x+\Lambda_y)+6\mu\Lambda =$$
$$= \mu(x+y)^4\partial_x\partial_y\{(x+y)^{-2}[\Lambda-\tfrac{1}{2}(x+y)(\Lambda_x+\Lambda_y)]\}.$$

Therefore, introducing the function:

(4.71) $$\Xi := \tfrac{1}{4}(x-y)(\mu_v x-\mu_u y)+(x+y)(G_{xu}+F_{yv})+ + (x+y)^4\{F_{yx}G_{xy}-F_{yy}G_{xx}-\mu\,\partial_x\partial_y(x+y)^{-2}[\Lambda-\tfrac{1}{2}(x+y)(\Lambda_x+\Lambda_y)]-\tfrac{1}{16}\mu^2\}$$

we have the result that:

(4.72) $$X = \partial_x\Xi\ , \qquad Y = \partial_y\Xi\ .$$

The form of Ξ given in (4.71) can be still improved; first we replace the structural function Λ by an equivalent function Θ defined by:

(4.73) $$\Lambda := \Theta + \tfrac{1}{4}\mu\cdot\tfrac{1}{4!}(x+y)^4.$$

Then (4.71) assumes the form of:

$$\Xi := \tfrac{1}{4}(x-y)(\mu_v x-\mu_u y)+(x+y)(G_{xu}+F_{yv})+$$

(4.74)
$$+(x+y)^4\{F_{yx}G_{xy}-F_{yy}G_{xx}-\mu\,\partial_x\partial_y(x+y)^{-2}[\Theta-\tfrac{1}{2}(x+y)(\Theta_x+\Theta_y)]\},$$

with F and G being now the abbreviations for:

(4.75) $$F = \Theta_{xy} + \frac{\Theta_x-\Theta_y}{x+y} + \tfrac{1}{8}\mu(x-y)^2,$$
$$G = \Theta_{xy} - \frac{\Theta_x-\Theta_y}{x+y} + \tfrac{1}{8}\mu(x-y)^2.$$

The functions which describe the metric are then:

(4.76) $$\mathcal{P} = (x+y)^3F_{yy} = (x+y)^3\{(\Theta_{xy}+\frac{\Theta_x-\Theta_y}{x+y})_{yy}+\tfrac{1}{4}\mu\},$$
$$\mathcal{Q} = (x+y)^3G_{yy} = (x+y)^3\{(\Theta_{xy}-\frac{\Theta_x-\Theta_y}{x+y})_{xx}+\tfrac{1}{4}\mu\},$$
$$\mathcal{R} = -\tfrac{1}{2}(x+y)^3(F+G)_{xy} = (x+y)^3\{(-\Theta_{xy})_{xy} \quad +\tfrac{1}{4}\mu\}.$$

The representation of Ξ given above has the virtue that with Θ considered as a structural function, Ξ is now just linear in the structural function $\mu = \mu(uv)$.

[The ∂_u and ∂_v derivatives of μ in <u>all</u> terms of Ξ amount to $\tfrac{1}{4}(x - y)(\mu_u x - \mu_v y)$.]

The field equations (4.52) can be now written simply in the form of:

$$R_{11} = 2(x+y)^6 \Xi_{yy} = 0 \qquad R_{33} = 2(x+y)^6 \Xi_{xx} = 0 \tag{4.77}$$
$$-R_{13} = 2(x+y)^6 \Xi_{xy} = 0 .$$

If--- with Θ and μ arbitrary -- we maintain for a moment R_{11}, R_{33} and R_{13} in general different from zero, then because $e^1 = (x + y)^{-2} u_{,\mu} dx^\mu$, $e^3 = (x + y)^{-2} v_{,\mu} dx^\mu$, we have in the arbitrary coordinates for the Einstein tensor:

$$G_{\mu\nu} = R_{\mu\nu} = R_{ab}\, e^a{}_\mu\, e^b{}_\nu = \tag{4.78}$$

$$= 2(x+y)^2 \{ \Xi_{yy}\, u_{,\mu} u_{,\nu} - 2\Xi_{xy}\, u_{,(\mu} v_{,\nu)} + \Xi_{xx}\, v_{,(\mu} v_{,\nu)} \} .$$

Therefore, if only the hessian of Ξ :

$$h(\Xi) := \begin{vmatrix} \Xi_{xx} & \Xi_{xy} \\ \Xi_{yx} & \Xi_{yy} \end{vmatrix} \tag{4.79}$$

is $\neq 0$, then our Einstein tensor has the algebraic structure of $G_{\mu\nu} = k_{(\mu} \ell_{\nu)}$ where k_μ and ℓ_μ are the linearly independent mutually orthogonal null vectors. If however $h = 0$ then $G_{\mu\nu} = k_\mu k_\nu$ where k_μ is some null vector. [This is of course the case when Ξ is a solution of the differential equation of the type $H(\Xi_x \Xi_y\, uv) = 0$ i.e., the case when (with u and v suppressed) $\Xi = \Xi(xy)$ is a developable surface].

The most interesting for our purposes is of course the case when all field equations are fulfilled, $R_{ab} = 0$. According to (4.77) we have here:

$$\Xi \equiv \tfrac{1}{4}(x-y)(\mu_v x - \mu_u y) + (x+y)(G_{xu} + F_{yv}) + \tag{4.80}$$

$$+ (x+y)^4 \{ F_{yx} G_{xy} - F_{yy} G_{xx} - \mu\, \partial_x \partial_y (x+y)^{-2} [\Theta - \tfrac{1}{2}(x+y)(\Theta_x + \Theta_y)] \} =$$
$$= \tfrac{1}{2} \gamma(uv)(x-y) + \varkappa(uv)(x+y) + \lambda(uv)$$

where $\gamma(uv)$, $\varkappa(uv)$ and $\lambda(uv)$ are the arbitrary functions of the two variables u and v. It is clear that without losing generality, we can set $\varkappa = 0$; indeed, the metric is defined entirely by the functions (4.76) which stay invariant when we gauge Θ according to: $\Theta \to \Theta + \alpha + \beta(x + y) + \tfrac{1}{2}\gamma(x + y)^2 + \tfrac{1}{3!}\delta(x + y)^3$ with α, β, γ and δ being functions of u and v only. Under this gauge Ξ transforms according to: $\Xi \to \Xi - 6\mu\alpha + (x + y)(\delta_u + \delta_v)$. Thus, selecting properly δ we can always make $\varkappa = 0$ in the equation (4.80).

If $\mu \neq 0$ then by selecting properly α we can also make $\lambda(uv) = 0$.

Therefore, it is enough to consider the equation (4.80) in the two sub-cases: the general sub-case with $\mu \neq 0$ and the special sub-case $\mu = 0$ i.e.,

(4.81) $$G: \mu \neq 0, \quad \tfrac{1}{4}(x-y)(\mu_v x - \mu_u y) + (x+y)(G_{xu} + F_{yv}) + \\ + (x+y)^4\{F_{yx}G_{xy} - F_{yy}G_{xx} - \mu\,\partial_x\partial_y\,(x+y)^{-2}[\Theta - \tfrac{1}{2}(x+y)(\Theta_x + \Theta_y)]\} = \\ = \tfrac{1}{2}\nu(x-y),$$

(4.82) $$S: \mu = 0 \\ (x+y)(G_{xu} + F_{yv}) + (x+y)^4\{F_{yx}G_{xy} - G_{xx}F_{yy}\} = \tfrac{1}{2}\nu(x-y) + \lambda,$$

We can mention that by using the explicit form of the commutation relation from the appendix A_1, we can represent $C^{(2)}$ given according to (4.42) in the form of:

(4.83) $$C^{(2)} = (x+y)^2[(\mathcal{P}_{,2} + \mathcal{R}_{,4})_{,3} - (\mathcal{R}_{,2} - \mathcal{Q}_{,4})_{,1}]$$

into the more plausible:

(4.84) $$C^{(2)} = (x+y)^4\{\partial_x (x+y)^{-2}(\mathcal{P}_{,3} - \mathcal{R}_{,1}) - \partial_y (x+y)^{-2}(\mathcal{Q}_{,1} - \mathcal{R}_{,3})\}.$$

Substituting here for $\mathcal{P}_{,3} - \mathcal{R}_{,1}$ and $\mathcal{Q}_{,1} - \mathcal{R}_{,3}$ from (4.53) with $X = \Xi_x = \tfrac{1}{2}\nu$, $Y = \Xi_y = -\tfrac{1}{2}\nu$ and F and G given in the form of (4.75), we obtain:

(4.85) $$C^{(2)} = -2\nu(x+y)^5 + 2(\mu_v x - \mu_u y)(x+y)^5.$$

This exhibits explicitly the mechanism according to which ν enters into the geometrical quantities.

Now, the equation:

(4.86) $$\Xi \equiv \tfrac{1}{4}(x-y)(\mu_v x - \mu_u y) + (x+y)(G_{xu} + F_{yv}) + \\ + (x+y)^4\{F_{yx}G_{xy} - F_{yy}G_{xx} - \mu\,\partial_x\partial_y\,(x+y)^{-2}[\Theta - \tfrac{1}{2}(x+y)(\Theta_x + \Theta_y)]\} = \\ = \tfrac{1}{2}\nu(uv)(x-y) + \lambda(uv),$$

$$F := \Theta_{xy} + \frac{\Theta_x - \Theta_y}{x+y} + \tfrac{1}{8}\mu(x-y)^2$$

$$G := \Theta_{xy} - \frac{\Theta_x - \Theta_y}{x+y} + \tfrac{1}{8}\mu(x-y)^2$$

with $\mu(uv)$, $\nu(uv)$ and $\lambda(uv)$ considered as arbitrarily given functions of u and v only, can be considered as a differential equation of the fourth order for the unknown "key function" Θ. The non-linear terms which appear in this equation, are quadratic. It is now clear that the integral variety of this equation determines all

Einstein flat minimally degenerated complex spaces V_4 *). It turns out that, however, we can state our result in a still better form, in particular lowering the order of the equation to the second order only.

Indeed, in the place of working with the key function Θ, we can use the equivalent key function Υ defined by:

(4.87) $$\Theta := (x+y)\Upsilon$$

Then the formulae (4.86) assume the form:

(4.88) $$\Xi = \tfrac{1}{4}(x-y)(\mu_v x - \mu_u y) + (x+y)(G_{xu} + F_{yv}) +$$
$$+ (x+y)^4\{F_{yx}G_{xy} - F_{yy}G_{xx} + \tfrac{1}{2}\mu(\Upsilon_x + \Upsilon_y)_{xy}\} = \tfrac{1}{2}\nu(uv)(x-y) + \lambda(uv),$$
$$F := (x+y)\Upsilon_{xy} + 2\Upsilon_x + \tfrac{1}{8}(x-y)^2,$$
$$G := (x+y)\Upsilon_{xy} + 2\Upsilon_y + \tfrac{1}{8}(x-y)^2.$$

where $\mu \neq 0 \rightarrow \lambda = 0$. One easily finds then that with the present structural functions we have:

(4.89) $$\mathcal{P} = (x+y)^3 F_{yy} = \partial_y (x+y)^4 \Upsilon_{xyy} + \tfrac{1}{4}\mu(x+y)^3$$
$$\mathcal{Q} = (x+y)^3 G_{xx} = \partial_x (x+y)^4 \Upsilon_{yxx} + \tfrac{1}{4}\mu(x+y)^3$$
$$\mathcal{R} = -\tfrac{1}{2}(x+y)^3(F+G)_{xy} = -(x+y)^3\partial_x\partial_x\partial_y\partial_y(x+y)\Upsilon + \tfrac{1}{4}\mu(x+y)^3$$

At the same time we have for $C^{(3)}$ and $C^{(2)}$:

(4.90) $$C^{(3)} = -2\mu(x+y)^3,$$
$$C^{(2)} = -2\nu(x+y)^5 + 2(\mu_v x - \mu_u y)(x+y)^5.$$

This, however, is just only one step from our result in its final form: we introduce the fundamental key function, Π defined by:

(4.91) $$\Pi := \Upsilon_{xy}$$

Then, in terms of this object we express the structural functions which describe the shape of the metric according to:

*All modulo the limiting transitions (contractions) to the case I which was investigated in some details in section 3. We will elucidate this point more amply subsequently.

$$\begin{aligned}
\mathcal{P} &= \partial_y (x+y)^4 \partial_y \Pi + \tfrac{1}{4}\mu (x+y)^3 \\
\mathcal{Q} &= \partial_x (x+y)^4 \partial_x \Pi + \tfrac{1}{4}\mu (x+y)^3 \\
\mathcal{R} &= -\tfrac{1}{2}\left[\partial_x (x+y)^4 \partial_y + \partial_y (x+y)^4 \partial_x\right]\Pi + \tfrac{1}{4}\mu (x+y)^3
\end{aligned} \tag{4.92}$$

Moreover, defining:

$$\begin{aligned}
\mathcal{K} &:= F_y = (x+y)\partial_y \Pi + 3\Pi - \tfrac{1}{4}\mu(x-y) \\
\mathcal{L} &:= G_x = (x+y)\partial_x \Pi + 3\Pi + \tfrac{1}{4}\mu(x-y)
\end{aligned} \tag{4.93}$$

we find that our equation assumes the form:

$$\begin{aligned}
\Xi \equiv {} & \tfrac{1}{4}(x-y)(\mu_v x - \mu_u y) + (x+y)(\mathcal{L}_u + \mathcal{K}_v) + \\
& + (x+y)^4\{\mathcal{K}_x \mathcal{L}_y - \mathcal{K}_y \mathcal{L}_x + \tfrac{1}{2}\mu(\partial_x + \partial_y)\Pi\} = \tfrac{1}{2}\nu(x-y) + \lambda
\end{aligned} \tag{4.94}$$

The same, distinguishing explicitly all terms which contain μ and ordering, we can state as follows: Let:

$$\mathcal{F} := (x+y)\partial_y \Pi + 3\Pi \ , \qquad \mathcal{G} := (x+y)\partial_x \Pi + 3\Pi . \tag{4.95}$$

Then our equation requires that:

$$\begin{aligned}
\Xi \equiv {} & \tfrac{1}{4}(x-y)(x\mu_u - y\mu_v) + \tfrac{1}{4}\mu(\partial_x + \partial_y)(x+y)^5(\partial_x + \partial_y)\Pi + \\
& + (x+y)(\mathcal{G}_u + \mathcal{F}_v) + (x+y)^4(\mathcal{F}_x \mathcal{G}_y - \mathcal{F}_y \mathcal{G}_x) = \tfrac{1}{2}\nu(x-y) + \lambda .
\end{aligned} \tag{4.96}$$

Therefore, considering μ, ν and λ as arbitrary given functions of u and v only, we can now interpret (4.96) as an equation of the second order -- and with the quadratic non-linearity -- for our fundamental key function Π.

It remains now to work out the remaining components of the conformal curvature besides $C^{(3)}$ and $C^{(2)}$ already known in the form of (4.90).

Reading off $e^3 \wedge e^1$ component in (3.18) we obtain by using (4.44) and (4.45) after some preliminary ordering that:

$$\begin{aligned}
\tfrac{1}{2}C^{(1)} = {} & -(x+y)^4\{\partial_3 (x+y)^{-2}(\mathcal{P}_{,3} - \mathcal{R}_{,1}) + (\mathcal{R}_x + \mathcal{Q}_y)(\mathcal{P}_{,3} - \mathcal{R}_{,1}) + \\
& + \partial_1 (x+y)^{-2}(\mathcal{Q}_{,1} - \mathcal{R}_{,3}) + (\mathcal{P}_x + \mathcal{R}_y)(\mathcal{Q}_{,1} - \mathcal{R}_{,3})\} .
\end{aligned} \tag{4.97}$$

On the other hand, from (4.53) we easily find that in the present notation:

$$\begin{aligned}
\mathcal{P}_{,3} - \mathcal{R}_{,1} &= (x+y)^4\{-\tfrac{1}{2}\nu - (\partial_u + \partial_v)\mathcal{F} - \tfrac{1}{4}y\mu_u + \tfrac{1}{4}(2x-y)\mu_v\} \\
\mathcal{Q}_{,1} - \mathcal{R}_{,3} &= (x+y)^4\{\tfrac{1}{2}\nu - (\partial_u + \partial_v)\mathcal{G} + \tfrac{1}{4}(2y-x)\mu_u - \tfrac{1}{4}x\mu_v\}
\end{aligned} \tag{4.98}$$

By using (4.92) one also easily finds:

$$\mathcal{R}_x + \mathcal{Q}_y = \tfrac{3}{2}\mu(x+y)^2 + 2\partial_x\left[(x+y)^3(\Pi_x - \Pi_y)\right], \tag{4.99}$$
$$\mathcal{P}_x + \mathcal{R}_y = \tfrac{3}{2}\mu(x+y)^2 - 2\partial_y\left[(x+y)^3(\Pi_x - \Pi_y)\right].$$

Although one could proceed with the computation of $C^{(1)}$ maintaining its present generality, it is convenient to restrict it at this point by using the remaining freedom of choice for coordinates. [See A_3, beginning from (A_3.37).] Indeed, without loss of generality we can make:

$$\mu = \text{const} \tag{4.100}$$

This is so because according to (A_3.25) $C^{(3)} = C'^{(3)}$ and by using (A_3.38) in the new coordinates $C'^{(3)} = -2\mu'(x' + y')^3$, $\mu' = \mu(\xi_{\xi'})^3$, with ξ being disposable function of the 2 variables equivalent to u and v.

Then, with μ = const, the computation of $C^{(1)}$ -- still formidable because it involves the iterated use of our basic equation (4.96) -- can be executed yielding the result that:

$$C^{(1)} = 2\Phi^7\left[x\nu_v - y\nu_u + 2\nu(\partial_x - \partial_y)\Phi^3\Pi + (\partial_u + \partial_v)\{\lambda - \mu\Phi^{-1/2}(\partial_x + \partial_y)\Phi^{3/2}\Pi\}\right], \tag{4.101}$$

where of course $\Phi = x + y$. We would like to express our gratitude to Drs. J. D. Finley III and A. Garcia for deriving (independently) for us the expression for $C^{(1)}$.

Now, as far as the remaining components of the anti-self-dual conformal curvature are concerned, they can be computed in the form of:

$$\bar{C}^{(5)} = -2\Phi^2\mathcal{P}_{yy}, \qquad \bar{C}^{(1)} = -2\Phi^2\mathcal{Q}_{xx},$$
$$3\bar{C}^{(3)} = \Phi^2[4\mathcal{R}_{xy} - (\mathcal{Q}_{yy} + \mathcal{P}_{xx})], \tag{4.102}$$
$$\bar{C}^{(4)} = \Phi^2(\mathcal{R}_{yy} - \mathcal{P}_{xy}), \qquad \bar{C}^{(2)} = \Phi^2[\mathcal{R}_{xx} - \mathcal{Q}_{yx}].$$

where $\mathcal{P}$, $\mathcal{Q}$ and $\mathcal{R}$ have to be interpreted as the right hand members of (4.92).

5. Concluding Remarks.

We have obtained the result that all one-sidedly degenerate metrics which fulfil Einstein's vacuum equations can be represented as:

(5.1) $$ds^2 = 2e^1 \otimes e^2 + 2e^3 \otimes e^4$$

$$e^1 = \Phi^{-2} du \qquad e^3 = dx + \mathcal{P} du + \mathcal{R} dv$$

$$e^3 = \Phi^{-2} dv \qquad e^4 = dy + \mathcal{R} du + \mathcal{Q} dv ,$$

where

(5.2) $$\mathrm{I}: \quad \Phi = 1 \qquad\qquad \mathrm{II}: \quad \Phi = x + y$$

respectively in the non-diverging and diverging cases, in the sense introduced in the section 2. In the case II then there exist such a constant μ and a function $\Pi = \Pi(xyuv)$ that:

(5.3) $$\mathcal{P} = \partial_y \Phi^4 \partial_y \Pi + \tfrac{1}{4}\mu\Phi^3 , \qquad \mathcal{Q} = \partial_x \Phi^4 \partial_x \Pi + \tfrac{1}{4}\mu\Phi^3$$
$$\mathcal{R} = -\tfrac{1}{2}(\partial_x \Phi^4 \partial_y + \partial_y \Phi^4 \partial_x)\Pi + \tfrac{1}{4}\mu\Phi^3$$

This function must fulfil the equation:

(5.4) $$\Phi^4(\mathcal{F}_x \mathcal{G}_y - \mathcal{F}_y \mathcal{G}_x) + \Phi(\mathcal{G}_u + \mathcal{F}_v) +$$
$$\tfrac{1}{4}\mu(\partial_x + \partial_y)\Phi^5(\partial_x + \partial_y)\Pi = \tfrac{1}{2}\nu(x-y) + \lambda ,$$

where ν and λ are arbitrary functions of u and v only (if $\mu \neq 0$ then $\lambda = 0$) and:

(5.5) $$\mathcal{F} := \Phi^{-2}\partial_y \Phi^3 \Pi , \qquad \mathcal{G} := \Phi^{-2}\partial_x \Phi^3 \Pi .$$

The (degenerate) left conformal curvature is then characterized by the quantities:

(5.6) $$C^{(5)} = C^{(4)} = 0 , \qquad C^{(3)} = -2\mu\Phi^3 , \qquad C^{(2)} = -2\nu\Phi^5$$

$$C^{(1)} = 2\Phi^7 \big[x\nu_v - y\nu_u + 2\nu(\partial_x - \partial_y)\Phi^3 \Pi +$$
$$(\partial_u + \partial_v)\{\lambda - \mu\Phi^{-1/2}(\partial_x + \partial_y)\Phi^{3/2}\Pi\} \big]$$

One can show that case I studied in section 3, can be then obtained from the case II by a contraction, i.e., the corresponding limiting transition consisting in the coordinates transformation $x \to \tfrac{1}{2} + \varepsilon x$, $y \to \tfrac{1}{2} + \varepsilon y$, $u \to \varepsilon^{-1} u$, $v \to \varepsilon^{-1} v$ accompanied by the suitable tetrad transformation, and then in taking the limit $\varepsilon \to 0$.

It will be perhaps of some interest to enumerate the different types of solutions of Einstein's vacuum equations reduced to our single equation (5.4) or to its limiting form for the type I, (3.29).

In the obvious symbolism, we have basically seven distinct spaces:

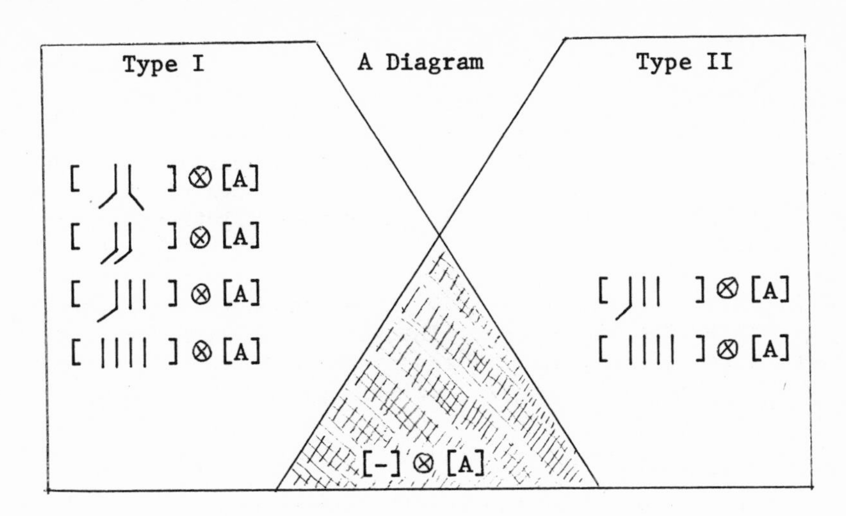

where (A $\equiv$ Anything). This diagram is to be interpreted as follows: besides the four obvious sub-cases of the type II

(5.7)

$$[\ \lambda\!\mid\ ]\otimes[A]\ :\quad C^{(3)}\neq 0,\quad \delta := 2[C^{(2)}]^2-3C^{(3)}C^{(1)}\neq 0$$

$$[\ /\!/\ ]\otimes[A]\ :\quad C^{(3)}\neq 0,\quad \delta = 0$$

$$[\ /\!\mid\mid\ ]\otimes[A]\ :\quad C^{(3)}=0,\quad C^{(2)}\neq 0$$

$$[\ \mid\mid\mid\mid\ ]\otimes[A]\ :\quad C^{(3)}=0=C^{(2)};\quad C^{(1)}\neq 0.$$

we have the $[\ /\!\mid\mid\ ]\otimes[A]$ and $[\ \mid\mid\mid\mid\ ]\otimes[A]$ sub-cases of the type I. Finally, within type I, the case [-] ⊗ [A] because of the results of [3] must overlap with the case [-] ⊗ [A] of type II. [One should notice, however, that the intersection of the type I and II solutions also includes solutions which are degenerate on both sides with the property that the expansion vector vanishes from one side only.]

These particular spaces, [-] ⊗ [A] Newman called "heavens" (H - spaces) at GR7 Conference, Tel Aviv, 1974, and one of us described in terms of the first and second heavenly equations for the key functions Ω and Θ, [3]. The basic results of the present work contain, of course, the equation for the key function of heavens, Θ, as the most special case. For that reason, if one were tempted to follow the previous terminology one could perhaps call the group of complex space-times from our diagram, HH metrics, the second H standing may be, for "higher" or "hyper" heavens. Equation (5.4) and its form for the type I, (3.29) could be then referred to as HH equations.

It is to be pointed out that although $[-]\otimes[A]$ spaces are highly interesting -- out of their nature -- they certainly do not permit any real "earthly" cross-section. The complex metrics described in this paper -- out of their very nature -- do contain all results of the theory of real algebraically degenerate solutions as earthly cross-sections.*

We remark, incidentally, that Mr. K. Rózga has developed recently (Ph.d. thesis at the Institute of Theoretical Physics, University of Warsaw, March 1976) a general theory of real cross-sections with the signature (+ + + -) of complex V_4's. This should provide a useful tool in the study of real cross-sections of HH spaces.

We will close this paper by the remark that in order to see whether our reduction of the complex "degenerated gravity" to one single equation is consistent, it is of interest to investigate the characteristic surfaces of (5.4), i.e. the surfaces where the second derivatives of Π are permitted to possess a jump. One easily finds that these jumps are permitted only on the eiconal surfaces of the complex space-time, i.e., surfaces S = 0 such that $g^{\mu\nu} S_{,\mu} S_{,\nu} = 0$, where $g^{\mu\nu}$ is just the riemannian metric defined by our ds^2.

APPENDIX A_1

The commutator (2.3) specialized for ab = 21 gives:

$$(A_1.1)\quad (\ln \Phi^{-2})_{,2}\partial_1 + \Phi^2 A_{,2}\partial_2 + \Phi^2 C_{,2}\partial_4 = (\Gamma_{221}-\Gamma_{212})\partial_1 + (\Gamma_{121}-\Gamma_{112})\partial_2 + (\Gamma_{421}-\Gamma_{412})\partial_3 + (\Gamma_{321}-\Gamma_{312})\partial_4$$

so that consequently:

$$(A_1.2)\qquad (a)\ \Gamma_{122} = (\ln\Phi^{-2})_{,2} \qquad (b)\ \Gamma_{121} = \Phi^2 A_{,2}$$

$$(c)\ \Gamma_{421} - \Gamma_{412} = 0 \qquad (d)\ \Gamma_{321} - \Gamma_{312} = \Phi^2 C_{,2}$$

Specializing (2.3) for ab = 23 we obtain:

$$(A_1.3)\quad (\ln \Psi^{-2})_{,2}\partial_3 + \Psi^2 B_{,2}\partial_2 + \Psi^2 D_{,2}\partial_4 = (\Gamma_{223}-\Gamma_{232})\partial_1 + (\Gamma_{123}-\Gamma_{132})\partial_2 + (\Gamma_{423}-\Gamma_{432})\partial_3 + (\Gamma_{323}-\Gamma_{332})\partial_4$$

---

*Then is there mirth in heaven -- when earthly things made even -- atone together. (W. Shakespeare, As You Like It, V, IV. 115-117) so that consequently:

$(A_1.4)$ (a) $\Gamma_{322} = 0$ (b) $\Gamma_{323} = \Psi^2 D_{,2}$

(c) $\Gamma_{123} + \Gamma_{312} = \Psi^2 B_{,2}$ (d) $\Gamma_{423} + \Gamma_{342} = (\ln \Psi^{-2})_{,2}$.

The next specialization of (2.3) is ab = 41 and gives:

$$(A_1.5)\quad (\ln \Phi^{-2})_{,4}\partial_1 + \Phi^2 A_{,4}\partial_2 + \Phi^2 C_{,4}\partial_4 = (\Gamma_{241} - \Gamma_{214})\partial_1 + (\Gamma_{141} - \Gamma_{114})\partial_2 + (\Gamma_{441} - \Gamma_{414})\partial_3 + (\Gamma_{341} - \Gamma_{314})\partial_4$$

so that consequently:

$(A_1.6)$ (a) $\Gamma_{411} = -\Phi^2 A_{,4}$ (b) $\Gamma_{414} = 0$

(c) $\Gamma_{124} - \Gamma_{421} = (\ln \Phi^{-2})_{,4}$ (d) $\Gamma_{341} - \Gamma_{314} = \Phi^2 C_{,4}$

Then we set in (2.3) ab = 43 obtaining:

$$(A_1.7)\quad (\ln \Psi^{-2})_{,4}\partial_3 + \Psi^2 B_{,4}\partial_2 + \Psi^2 D_{,4}\partial_4 = (\Gamma_{243} - \Gamma_{234})\partial_1 + (\Gamma_{143} - \Gamma_{134})\partial_2 + (\Gamma_{443} - \Gamma_{434})\partial_3 + (\Gamma_{343} - \Gamma_{334})\partial_4$$

so that:

$(A_1.8)$ (a) $\Gamma_{344} = (\ln \Psi^{-2})_{,4}$ (b) $\Gamma_{343} = \Psi^2 D_{,4}$

(c) $\Gamma_{324} - \Gamma_{423} = 0$ (d) $\Gamma_{314} - \Gamma_{413} = \Psi^2 B_{,4}$

Now, the case of ab = 42 gives:

$$(A_1.9)\quad 0 = (\Gamma_{242} - \Gamma_{224})\partial_1 + (\Gamma_{142} - \Gamma_{124})\partial_2 + (\Gamma_{442} - \Gamma_{424})\partial_3 + (\Gamma_{342} - \Gamma_{324})\partial_4$$

so that it follows that:

$(A_1.10)$ (a) $\Gamma_{422} = 0$ (b) $\Gamma_{424} = 0$

(c) $\Gamma_{412} + \Gamma_{124} = 0$ (d) $\Gamma_{342} - \Gamma_{324} = 0$

It remains to specialize (2.3) for ab = 13. This gives:

$$(A_1.11)\quad -(\ln \Phi^{-2})_{,3}\partial_1 - \Phi^2 A_{,3}\partial_2 - \Phi^2 C_{,3}\partial_4 + (\ln \Psi^{-2})_{,1}\partial_3 + \Psi^2 B_{,1}\partial_2 + \Psi^2 D_{,1}\partial_4 = (\Gamma_{213} - \Gamma_{231})\partial_1 + (\Gamma_{113} - \Gamma_{131})\partial_2 + (\Gamma_{413} - \Gamma_{431})\partial_3 + (\Gamma_{313} - \Gamma_{331})\partial_4 .$$

Therefore, we have:

(A$_1$.12) (a) $\Gamma_{311} = -\Phi^2 A_{,3} + \Psi^2 B_{,1}$ (b) $\Gamma_{313} = -\Phi^2 C_{,3} + \Psi^2 D_{,1}$

(c) $\Gamma_{321} - \Gamma_{123} = -(\ln \Phi^{-2})_{,3}$ (d) $\Gamma_{413} + \Gamma_{341} = (\ln \Psi^{-1})_{,1}$.

In the next step, we gather the (c - d) relations which involve two Γ's into four triplets:

(A$_1$.13)

$$\Gamma_{412} + \Gamma_{124} = 0$$
$$\Gamma_{421} - \Gamma_{412} = 0$$
$$\Gamma_{124} - \Gamma_{421} = (\ln \Phi^{-2})_{,4}$$

(A$_1$.14)

$$\Gamma_{342} - \Gamma_{324} = 0$$
$$\Gamma_{324} - \Gamma_{423} = 0$$
$$\Gamma_{423} + \Gamma_{342} = (\ln \Psi^{-2})_{,2}$$

(A$_1$.15)

$$\Gamma_{321} - \Gamma_{312} = \Phi^2 C_{,2}$$
$$\Gamma_{123} + \Gamma_{312} = \Psi^2 B_{,2}$$
$$\Gamma_{123} - \Gamma_{321} = (\ln \Phi^{-2})_{,3}$$

(A$_1$.16)

$$\Gamma_{341} - \Gamma_{314} = \Phi^2 C_{,4}$$
$$\Gamma_{314} - \Gamma_{413} = \Psi^2 B_{,4}$$
$$\Gamma_{413} + \Gamma_{341} = (\ln \Psi^{-2})_{,1}$$

We solve now these simple linear equations obtaining:

(A$_1$.17) $\Gamma_{124} = \frac{1}{2}(\ln \Phi^{-2})_{,4}$, $\Gamma_{421} = -\frac{1}{2}(\ln \Phi^{-2})_{,4}$, $\Gamma_{412} = -\frac{1}{2}(\ln \Phi^{-2})_{,4}$,

(A$_1$.18) $\Gamma_{342} = \frac{1}{2}(\ln \Psi^{-2})_{,2}$, $\Gamma_{324} = \frac{1}{2}(\ln \Psi^{-2})_{,2}$, $\Gamma_{423} = \frac{1}{2}(\ln \Psi^{-2})_{,2}$

and

(A$_1$.19)

$$\Gamma_{123} = \frac{1}{2}[\Phi^2 C_{,2} + \Psi^2 B_{,2} + (\ln \Phi^{-2})_{,3}]$$
$$\Gamma_{321} = \frac{1}{2}[\Phi^2 C_{,2} + \Psi^2 B_{,2} - (\ln \Phi^{-2})_{,3}]$$
$$\Gamma_{312} = \frac{1}{2}[-\Phi^2 C_{,2} + \Psi^2 B_{,2} - (\ln \Phi^{-2})_{,3}] ,$$

and finally:

(A$_1$.20) $$\Gamma_{341} = \tfrac{1}{2}\left[\Phi^2 C_{,4} + \Psi^2 B_{,4} + (\ln \Psi^{-2})_{,1}\right]$$
$$\Gamma_{344} = \tfrac{1}{2}\left[-\Phi^2 C_{,4} + \Psi^2 B_{,4} + (\ln \Psi^{-2})_{,1}\right]$$
$$\Gamma_{413} = \tfrac{1}{2}\left[-\Phi^2 C_{,4} - \Psi^2 B_{,4} + (\ln \Psi^{-2})_{,1}\right].$$

The Γ_{abc}'s computed here give the formulae for $\Gamma_{ab} = \Gamma_{abc}\, e^c$ listed in the text as (2.4) up to (2.8).

APPENDIX A$_2$

In the work which uses the notion of the minimal (one-sided) degeneration of the conformal curvature it is proper to provide a lucid description of this concept.

It was explained in [3] that the conformal curvature tensor of a complex V_4, $C_{\alpha\beta\gamma\delta}$, is algebraically equivalent to a pair (in general of autonomic and independent) spinorial objects $C_{ABCD} = C_{(ABCD)}$ and $\bar{C}_{\dot{A}\dot{B}\dot{C}\dot{D}} = \bar{C}_{(\dot{A}\dot{B}\dot{C}\dot{D})}$. These objects, according to Penrose [14] can be always represented as $C_{ABCD} = \alpha_{(A}\beta_B\gamma_C\delta_{D)}$ and $\bar{C}_{\dot{A}\dot{B}\dot{C}\dot{D}} = \bar{C}_{(\dot{A}\dot{B}\dot{C}\dot{D})}$. The coincidences (up to proportionality) among the un-dotted and dotted Penrose spinors give the distinct algebraic types, described as e.g. [//]⊗[||||], etc. (For the notion of the tensor product see [3], footnote 5.) It is quite easy to provide a tensorial transcription of these facts. Indeed, let:

(A$_2$.1) $$C_{(\pm)}^{\alpha\beta\gamma\delta} := \tfrac{1}{2}\left(C^{\alpha\beta\gamma\delta} \pm \frac{i}{2\sqrt{-g}}\,\epsilon^{\alpha\beta\bullet\bullet}{}_{\mu\nu}\, C^{\mu\nu\gamma\delta}\right)$$

be the self dual and anti-self dual parts of the conformal curvature. Then, from the point of view of the spinorial images:

(A$_2$.2)
$$C_{(+)}^{\alpha\beta\gamma\delta} \rightarrow \begin{cases} C^{(+)}_{ABCD} = C_{ABCD} \\ \bar{C}^{(+)}_{\dot{A}\dot{B}\dot{C}\dot{D}} = 0 \end{cases} \qquad C_{(-)}^{\alpha\beta\gamma\delta} \rightarrow \begin{cases} C^{(-)}_{ABCD} = 0 \\ \bar{C}^{(-)}_{\dot{A}\dot{B}\dot{C}\dot{D}} = \bar{C}_{\dot{A}\dot{B}\dot{C}\dot{D}} \end{cases}$$

Therefore, with $K_\mu = \tfrac{1}{2} g_{\mu A\dot{B}}\, k^A \bar{k}^{\dot{B}}$, the linear operation $\mathcal{D}(K_\mu)$ defined in [3] as applied to $C_{(\pm)}^{\alpha\beta\gamma\delta}$ gives:

(A$_2$.3) $$\mathcal{D}(K_\mu) C_{(\pm)}^{\alpha\beta\gamma\delta} := K_\mu K^{[\alpha} C_{(\pm)}^{\beta]\mu\nu[\gamma} K^{\delta]} K_\nu \equiv K_{(\mp)}^{\alpha\beta\gamma\delta}$$

and produces the two objects which 1o. are endowed with all symmetries of the conformal curvature, and, according to the work of [3], 2o. have the spinorial images given by:

$$(A_2.4)\quad K^{(-)}_{ABCD}=0\ ,\ \bar{K}^{(-)}_{\dot A\dot B\dot C\dot D}=\tfrac{1}{16}\bar{k}_{\dot A}\bar{k}_{\dot B}\bar{k}_{\dot C}\bar{k}_{\dot D}\ C_{ABCD}k^A k^B k^C k^D\ ,$$

$$K^{(+)}_{ABCD}=\tfrac{1}{16}k_A k_B k_C k_D\ \bar{C}_{\dot A\dot B\dot C\dot D}\bar{k}^{\dot A}\bar{k}^{\dot B}\bar{k}^{\dot C}\bar{k}^{\dot D}\ ,\ \bar{K}^{(+)}_{\dot A\dot B\dot C\dot D}=0.$$

We can now define the "left" generalized Debever - Penrose vectors K_μ as the objects which 1o. All are linearly independent in pairs 2o. All are mutually orthogonal null vectors 3o. All have the property that $\mathcal{D}(K_\mu)C^{\alpha\beta\gamma\delta}_{(+)} = 0$. If the number of these vectors is ≤ 3, we say that the curvature is degenerated from the left. The parallel definition applies with respect to the right-sided degeneration. The minimally degenerate curvature must be degenerate at least from one side.

APPENDIX A_3

The canonical tetrad is given for our two cases in the form of

$$(A_3.1)\quad e^1=\Phi^{-2}du\ ,\ e^2=dx+\mathcal{P}du+\mathcal{R}dv\ ,$$

$$e^3=\Phi^{-2}dv\ ,\ e^4=dy+\mathcal{R}du+\mathcal{Q}dv\ ,$$

where, respectively in the cases I and II, the factor is given by

$$(A_3.2)\qquad I:\quad \Phi=1 \qquad\qquad II:\quad \Phi=x+y$$

The transformation of the variables u and v,

$$(A_3.3)\qquad u'=u'(uv)\qquad v'=v'(uv)$$

$$\Delta:=\frac{\partial(u'v')}{\partial(uv)}\neq 0$$

changes the parametrization of our congruence of the null strings only. Therefore, it is to be expected that when working with u' and v' in the place of u and v , by re-orienting properly the tetrad and by re-defining suitably the coordinates x' and y' , one should be able to maintain our canonical tetrad form-invariant, with the transformation functions $u'=u'(uv)$ and $v'=v'(uv)$ being arbitrary in the case I and perhaps somehow more specific in the case II.

This argument gives the intuitive motivations in the process of the construction of the group of automorphisms of the metrics under study. Indeed, it is quite clear that a GL(2,$\mathbb{C}$) transformations of e^1 and e^3 with coefficients dependent on u and v only can be always expressed in the form of

$$(A_3.4)\quad e^{1'}=\lambda(u'_u e^1+u'_v e^3)=\lambda\Phi^{-2}du'\ ,\ e^{3'}=\mu(v'_u e^1+v'_v e^3)=\mu\Phi^{-2}dv'$$

where λ and μ are non-zero functions of u and v only. Now with (A_3.4) postulated, one easily deduces from the invariance condition,

(A_3.5) $$ds^2 = 2e^1\otimes e^2 + 2e^3\otimes e^4 = 2e^{1'}\otimes e^{2'} + 2e^{3'}\otimes e^{4'}$$

that the corresponding $e^{2'}$ and $e^{4'}$ must possess the form of

(A_3.6) $$e^{2'} = \lambda^{-1}(u_{u'}e^2 + v_{u'}e^4 + \varkappa dv') \qquad e^{4'} = \mu^{-1}(u_{v'}e^2 + v_{v'}e^4 - \varkappa du')$$

where $\varkappa$ can depend on all four variables. [We are concerned here with the orientation-preserving tetrad transformation such that $e^1\wedge e^2\wedge e^3\wedge e^4 = e^{1'}\wedge e^{2'}\wedge e^{3'}\wedge e^{4'}$.] It is of some interest to observe that

(A_3.7) $$e^{3'}\wedge e^{1'} = (\lambda\mu\Delta)e^3\wedge e^1, \quad e^{1'}\wedge e^{2'} + e^{3'}\wedge e^{4'} = e^1\wedge e^2 + e^3\wedge e^4 - 2\varkappa\Delta\Phi^2 e^3\wedge e^1$$
$$e^{4'}\wedge e^{2'} = (\lambda\mu\Delta)^{-1}\{e^4\wedge e^2 - \varkappa\Delta\Phi^2(e^1\wedge e^2 + e^3\wedge e^4) + \varkappa^2\Delta^2\Phi^4 e^3\wedge e^1\}$$

Now, it is clear that with $e^{2'} = \lambda^{-1}(u_{u'}dx + v_{u'}dy) + \ldots$ and $e^{4'} = \mu^{-1}(u_{v'}dx + v_{v'}dy) + \ldots$, where $(\ldots)$ denote terms linear in du and dv, if we intend to keep $e^{2'}$ and $e^{4'}$ form-invariant with respect to (A_3.1), then we should define new coordinates x and y such that

(A_3.8) $$x' := \lambda^{-1}(u_{u'}x + v_{u'}y) + \rho \qquad y' := \mu^{-1}(u_{v'}x + v_{v'}y) + \sigma\,,$$

where ρ and σ are some functions of u and v only. Observe that if we consider (A_3.3) and (A_3.8) as a coordinate transformation we have $\partial(uvxy)/\partial(u'v'x'y') = 1$.

We must now distinguish the cases I and II. In the case I the form-invariance of $e^{1'}$ and $e^{2'}$ requires that

(A_3.9) I : $$\lambda = 1 = \mu$$

In the case II the form-invariance of $e^{1'}$ and $e^{3'}$ requires that $\lambda = \mu$ and $x' + y' = \frac{1}{\sqrt{\lambda}}(x + y)$. By using in the last condition (A_3.8) we find it equivalent to

(A_3.10) II :
(a) $\rho + \sigma = 0$ (b) $\frac{1}{2}(\partial_{u'} + \partial_{v'})(u - v) = 0$

(c) $\sqrt{\lambda} = \frac{1}{2}(\partial_{u'} + \partial_{v'})(u + v) = \sqrt{\mu}$.

For the purposes of the discussion of case II it is thus convenient to replace u,v and u', v' by the variables

(A_3.11) $$\xi := \frac{1}{\sqrt{2}}(u+v) \qquad \eta := \frac{1}{\sqrt{2}}(u-v)$$
$$\xi' := \frac{1}{\sqrt{2}}(u'+v') \qquad \eta' := \frac{1}{\sqrt{2}}(u'-v')$$

in terms of which (A .10b-c) assume the simple form of $\partial_{\xi'}\eta = 0$, $\sqrt{\lambda} = \partial_{\xi'}\xi$. Therefore, in case II the (still permitted) transformation of the u,v variables amounts to

$$(A_3.12) \qquad \xi = \xi(\xi'\eta') \quad , \quad \eta = \eta(\eta')$$

where the last two functions are arbitrary, subject only to the condition that

$$(A_3.13) \qquad \Delta^{-1} = \frac{\partial(uv)}{\partial(u'v')} = \frac{\partial(\xi\eta)}{\partial(\xi'\eta')} = \xi_{\xi'}\eta_{\eta'} \neq 0$$

while at the same time for λ and μ we have that

$$(A_3.14) \qquad \lambda = (\xi_{\xi'})^2 = \mu$$

In the next step -- remembering that for both cases we can set $\lambda = \mu$ we can write the complete expressions for $e^{2'}$ and $e^{4'}$:

$$(A_3.15) \qquad \begin{aligned} e^{2'} &= dx' - x\,d(\lambda^{-1}u_{u'}) - y(\lambda^{-1}v_{u'}) - d\rho + \lambda^{-1}[u_{u'}(P du + R dv) + v_{u'}(R du + Q dv) + \varkappa dv'], \\ e^{4'} &= dy' - x\,d(\lambda^{-1}u_{v'}) - y(\lambda^{-1}v_{v'}) - d\sigma + \lambda^{-1}[u_{v'}(P du + R dv) + v_{v'}(R du + Q dv) - \varkappa du']. \end{aligned}$$

Because however the coefficients of dv in $e^{2'}$ and du in $e^{4'}$ should be equal to $\mathcal{R}'$, we obtain as the condition of our form-invariance:

$$(A_3.16) \qquad \begin{aligned} \mathcal{R}' &\equiv -x(\lambda^{-1}u_{u'})_{v'} - y(\lambda^{-1}v_{u'})_{v'} - \rho_{v'} + \lambda^{-1}[u_{u'}(P u_{v'} + R v_{v'}) + v_{u'}(R u_{v'} + Q v_{v'}) + \varkappa] = \\ &= -x(\lambda^{-1}u_{v'})_{u'} - y(\lambda^{-1}v_{v'})_{u'} - \sigma_{u'} + \lambda^{-1}[u_{v'}(P u_{u'} + R v_{u'}) + v_{v'}(R u_{u'} + Q v_{u'}) - \varkappa]. \end{aligned}$$

By executing the obvious cancellations and ordering we obtain thus a condition which fixes the value of $\varkappa$:

$$(A_3.17) \qquad \varkappa = \tfrac{1}{2}\lambda(\rho_{v'} - \sigma_{u'}) + x\,\frac{\partial(\ln\lambda^{1/2}, u)}{\partial(u', v')} + y\,\frac{\partial(\ln\lambda^{1/2}, v)}{\partial(u', v')} .$$

In the case I this amounts simply to

$$(A_3.18) \qquad \mathrm{I}: \qquad \varkappa = \tfrac{1}{2}(\rho_{v'} - \sigma_{u'})$$

and in the case II we conclude that

$$(A_3.19) \qquad \mathrm{II}: \quad \varkappa = \tfrac{1}{2}(\xi_{\xi'})^2(\rho_{v'} - \sigma_{u'}) + \frac{1}{\sqrt{2}}\,\frac{x+y}{\xi_{\xi'}}(\xi_{\xi'}\xi_{\xi'\eta'} - \xi_{\eta'}\xi_{\xi'\xi'}) - \frac{1}{\sqrt{2}}\,\frac{x-y}{\xi_{\xi'}}\eta_{\eta'}\xi_{\xi'\xi'}$$

By using $(A_3.17)$ in $(A_3.16)$ we find $\mathcal{R}'$ such that

$$(A_3.20) \qquad \begin{aligned} \mathcal{R}' &= \lambda^{-1}[u_{u'}u_{v'}P + (u_{u'}v_{v'} + u_{v'}v_{u'})R + v_{u'}v_{v'}Q] + \\ &- \tfrac{1}{2}x[(\lambda^{-1}u_{u'})_{v'} + (\lambda^{-1}u_{v'})_{u'}] - \tfrac{1}{2}y[(\lambda^{-1}v_{u'})_{v'} + (\lambda^{-1}v_{v'})_{u'}] - \tfrac{1}{2}(\rho_{v'} + \sigma_{u'}) , \end{aligned}$$

and we easily read off from (A_3.15) the values of $\mathcal{P}'$ and $\mathcal{Q}'$

(A_3.21) $\mathcal{P}' = \lambda^{-1}[u_{u'}u_{u'}\mathcal{P} + 2u_{u'}v_{u'}\mathcal{R} + v_{u'}v_{u'}\mathcal{Q}] - x(\lambda^{-1}u_{u'})_{u'} - y(\lambda^{-1}v_{u'})_{u'} - \rho_{u'}$,

$$\mathcal{Q}' = \lambda^{-1}[u_{v'}u_{v'}\mathcal{P} + 2u_{v'}v_{v'}\mathcal{R} + v_{v'}v_{v'}\mathcal{Q}] - x(\lambda^{-1}u_{v'})_{v'} - y(\lambda^{-1}v_{v'})_{v'} - \sigma_{v'} .$$

One can add that with $\lambda = \mu$ the formulae inverse to (A_3.8) are:

(A_3.22) $x = \lambda\{u'_u(x'-\rho) + v'_u(y'-\sigma)\}$, $y = \lambda\{u'_v(x'-\rho) + v'_v(y'-\sigma)\}$

and can be used in (A_3.20-21) if one wishes to work with x', y' in place of x,y .

Now, as a general comment, we would like to observe that our tetrad transformation, $e^{a'} = T^{a'}{}_b\, e^b$, described by (A_3.4) and (A_3.6) is characterized by the matrix

(A_3.23)
$$(T^{a'}{}_b) = \begin{pmatrix} \lambda u'_u & 0 & \lambda u'_v & 0 \\ \varkappa\Phi^2 v'_u & \lambda^{-1}u_{u'} & \varkappa\Phi^2 v'_v & \lambda^{-1}v_{u'} \\ \lambda v'_u & 0 & \lambda v'_v & 0 \\ -\varkappa\Phi^2 u'_u & \lambda^{-1}u_{v'} & -\varkappa\Phi^2 u'_v & \lambda^{-1}v_{v'} \end{pmatrix} ,$$

which can be interpreted as induced according to the standard rules by the more elementary spinorial transformations:

(A_3.24)
(a) $$(\ell^{A'}{}_A) = \begin{pmatrix} 1/\lambda\sqrt{\Delta} & , -\varkappa\sqrt{\Delta}\Phi^2/\lambda \\ 0 & , \lambda\sqrt{\Delta} \end{pmatrix} \to (\ell^{-1}{}^A{}_{A'}) = \begin{pmatrix} \lambda\sqrt{\Delta} & , \varkappa\sqrt{\Delta}\Phi^2/\lambda \\ 0 & , 1/\lambda\sqrt{\Delta} \end{pmatrix}$$

(b) $$(\bar{\ell}^{\dot{B}'}{}_{\dot{B}}) = \begin{pmatrix} v_{v'} & , u_{v'} \\ v_{u'} & , u_{u'} \end{pmatrix}\Delta^{1/2} \to (\bar{\ell}^{-1}{}^{\dot{B}}{}_{\dot{B}'}) = \begin{pmatrix} v'_v & , u'_v \\ v'_u & , u'_u \end{pmatrix}.\Delta^{-1/2}$$

[That is, with $(g^{A\dot{B}}) = \begin{pmatrix} e^4 & ; e^2 \\ e^1 & ; -e^3 \end{pmatrix}\sqrt{2}$ the new tetrad is induced by $(g^{A\dot{B}}) = (\ell^{A'}{}_A\bar{\ell}^{\dot{B}'}{}_{\dot{B}}\, g^{A\dot{B}}) = \begin{pmatrix} e^{4'} & , e^{2'} \\ e^{1'} & , -e^{3'} \end{pmatrix}\sqrt{2}$.]

Knowing this, one easily shows that the quantities $C^{(a)}$ -- equivalent to $2C_{ABCD}$ -- with $C^{(5)} = 0 = C^{(4)}$ -- transform according to the rules:

(A_3.25) $C'^{(3)} = C^{(3)}$, $C'^{(2)} = \lambda^{-2}[\Delta^{-1}C^{(2)} + 3\varkappa\Phi^2C^{(3)}]$

$$C'^{(1)} = \lambda^{-4}\{\Delta^{-2}C^{(1)} + 4\varkappa\Phi^2\Delta^{-1}C^{(2)} + 6\varkappa^2\Phi^4C^{(3)}\} .$$

Similarly, quite easily, remembering that the tetradial and the spinorial connection forms are related by

(A$_3$.26) $$\Gamma_{AB} = -\begin{pmatrix} \Gamma_{42} , & \frac{1}{2}[\Gamma_{12}+\Gamma_{34}] \\ \frac{1}{2}[\Gamma_{12}+\Gamma_{34}] , & \Gamma_{31} \end{pmatrix}$$

we easily find by applying the transformation law,

(A$_3$.27) $$\Gamma^{A'}{}_{B'} = \ell^{A'}{}_{A}\,\ell^{-1\,B}{}_{B'}\,\Gamma^{A}{}_{B} + \ell^{A'}{}_{S}\,d\ell^{-1\,S}{}_{B'}$$

that the "heavenly" connection forms transform under our tetrad gauge according to

(A$_3$.28) $$\Gamma_{4'2'} = \lambda^2\Delta\,\Gamma_{42} \;,\quad \Gamma_{1'2'}+\Gamma_{3'4'} = \Gamma_{12}+\Gamma_{34}+2\varkappa\Phi^2\Gamma_{42}+d\ln(\lambda^2\Delta),$$
$$\Gamma_{3'1'} = \Delta^{-1}\lambda^{-2}\Gamma_{31}+\varkappa\lambda^{-2}\Phi^2(\Gamma_{12}+\Gamma_{34})+\varkappa^2\lambda^{-2}\Delta\Phi^4\Gamma_{42}+\varkappa\Phi^2\lambda^{-2}d\ln(\Delta\Phi^2\varkappa).$$

At this point, we find it convenient to bifurcate these considerations of the specific cases I and II and to outline the basic facts concerning the corresponding groups of automorphisms separately for these cases.

Thus, in <u>case I</u>, the coordinate transformations

(A$_3$.29) $$u' = u'(uv) \qquad\qquad v' = v'(uv)$$
$$\mathcal{G}_I:\quad x' = u_{u'}x + v_{u'}y + \rho(uv) \qquad\qquad y' = u_{v'}x + v_{v'}y + \sigma(uv)$$
$$\Delta := \frac{\partial(u'v')}{\partial(uv)} \neq 0 \;,$$

with the inverse transformations

(A$_3$.30) $$u = u(u'v') \qquad\qquad v = v(u'v')$$
$$x = u'_u(x'-\rho)+v'_u(y'-\sigma) \qquad\qquad y = u'_v(x'-\rho)+v'_v(y'-\sigma) \;,$$

leave ds^2 <u>form-invariant</u>, with $u'=u'(uv)$, $v'=v'(uv)$, $\rho(uv)$ and $\sigma(uv)$ being four arbitrary functions. The canonical tetrad (A$_3$.1) [with $\Phi = 1$] stays then also <u>form-invariant</u> if (A$_3$.29) is accompanied by the associated tetrad transformation

(A$_3$.31) $$e^{1'} = u'_u e^1 + u'_v e^3 \;,\; e^{3'} = v'_u e^1 + v'_v e^3$$
$$\mathcal{G}_I \qquad e^{2'} = u_{u'}e^2 + v_{u'}e^4 + \varkappa(v'_u e^1 + v'_v e^3) \qquad \varkappa := \tfrac{1}{2}(\rho_{v'}-\sigma_{u'})$$
$$e^{4'} = u_{v'}e^2 + v_{v'}e^4 - \varkappa(u'_u e^1 + u'_v e^3)$$

The structural functions are then transformed according to the rules

(A$_3$.32) $$P' = u_{u'}u_{u'}P + 2u_{u'}v_{u'}R + v_{u'}v_{u'}Q - xu_{u'u'} - yv_{u'u'} - \rho_{u'}$$
$$Q' = u_{v'}u_{v'}P + 2u_{v'}v_{v'}R + v_{v'}v_{v'}Q - xu_{v'v'} - yv_{v'v'} - \sigma_{v'}$$
$$R' = u_{u'}u_{v'}P + (u_{u'}v_{v'} + u_{v'}v_{u'})R + v_{u'}v_{v'}Q - xu_{u'v'} - yv_{u'v'} - \tfrac{1}{2}(\rho_{v'} + \sigma_{u'}).$$

It is useful to notice that

(A$_3$.33) $$\partial_x = u_{u'}\partial_{x'} + u_{v'}\partial_{y'} \qquad \partial_{x'} = u'_u\partial_x + u'_v\partial_y$$
$$\partial_y = v_{u'}\partial_{x'} + v_{v'}\partial_{y'} \qquad \partial_{y'} = v'_u\partial_x + v'_v\partial_y$$

In the present case (A$_3$.24a) reduces to

(A$_3$.34) $$(\ell^{A'}{}_A) = \begin{pmatrix} \Delta^{-1/2}, & -\varkappa\Delta^{1/2} \\ 0, & \Delta^{1/2} \end{pmatrix}, \quad (\ell^{-1\,A}{}_{A'}) = \begin{pmatrix} \Delta^{1/2}, & \varkappa\Delta^{1/2} \\ 0, & \Delta^{-1/2} \end{pmatrix}$$

and, with $C^{(3)} = 0$, (A$_3$.25) gives that

(A$_3$.35) (a) $C'^{(3)} = 0$, (b) $C'^{(2)} = \Delta^{-1}C^{(2)}$

(c) $C'^{(1)} = \Delta^{-2}C^{(1)} + 4x\Delta^{-1}C^{(2)}$

At the same time, with $\Gamma_{42} = 0 = \ln\lambda$ for the case I, (A$_3$.28) gives:

(A$_3$.36) (a) $\Gamma_{4'2'} = 0$ (b) $\Gamma_{1'2'} + \Gamma_{3'4'} = \Gamma_{12} + \Gamma_{34} + d\ln\Delta$

(c) $\Gamma_{3'1'} = \Delta^{-1}\Gamma_{31} + \varkappa(\Gamma_{12} + \Gamma_{34}) + \varkappa\, d\ln(\varkappa\Delta)$.

Now, discussing the <u>case II</u>, it is convenient -- parallel with ξ and η which replace u and v according to (A$_3$.11) -- to apply in place of x and y the variables

(A$_3$.37) $$w := \tfrac{1}{\sqrt{2}}(x+y), \quad z := \tfrac{1}{\sqrt{2}}(x-y)$$

Then, one easily sees that the form-invariance of ds^2 for the case II permits the group of coordinate transformations

(A$_3$.38) $$\xi' = \xi'(\xi\eta), \qquad \eta' = \eta'(\eta)$$

G_{II} $$w' = \xi'_\xi w, \quad z' = \xi'_\xi\eta_{\eta'}(\xi'_\xi z - \xi'_\eta w) + \tau$$
$$\Delta = \frac{\partial(u'v')}{\partial(uv)} = \frac{\partial(\xi'\eta')}{\partial(\xi\eta)} = \xi'_\xi\eta'_\eta \neq 0,$$

which is characterized by the two arbitrary functions of two variables [$\xi'(\xi\eta)$ and $\tau(\xi\eta)$] and one arbitrary function of one variable [$\eta'(\eta)$]. (Of course $\tau := \frac{1}{\sqrt{2}}(\rho - \sigma)$ while $\rho + \sigma = 0$.) Notice that the inverse transformation to (A$_3$.38) is given by:

$$(A_3.39) \qquad \xi = \xi(\xi'\eta') \qquad \eta = \eta(\eta')$$

$$w = \xi_{\xi'} w' , \quad z = \xi_{\eta'}\eta'_{\eta}\left[\xi_{\eta'}(z'-\tau) - \xi_{\eta'} w'\right] .$$

Of course, the canonical tetrad of case II stays the also <u>form-invariant</u> if (A_3.38) is accompanied by an associated tetrad transformation $e^{a'} = T^{a'}{}_a e^a$ with $(T^{a'}{}_a)$ given in the form of (A_3.23), where $\varkappa$ from (A_3.19) in the present notation assumes the form,

$$(A_3.40) \qquad \varkappa = \tfrac{1}{2}(\xi_{\eta'})^2 \tau_{\xi'} + \eta_{\eta'}\left[w(\xi_{\xi'})_{\eta} - z(\xi_{\eta'})_{\xi}\right] .$$

The formulae (A_3.23) up to (A_3.28) maintain the validity in the case under discussion with the understanding that (A_3.40) defines the values of $\varkappa$, $\Phi = \sqrt{2}\,w$ and $\Delta^{-1} = \xi_{\eta'}\eta_{\eta'}$ while for λ we take $\lambda = (\xi_{\eta'})^2$.

REFERENCES

[1] G. C. Debney, R. P. Kerr, and A. Schild, J. Math. Phys 10 (1969), 1842.

[2] J. F. Plebański, "Spinors, Tetrads and Forms"; a monograph of the Centro de Investigacion, 297 pages, (1974), unpublished.

[3] J. F. Plebański, J. Math. Phys. 16 (1975), 2395.

[4] J. Goldberg and R. Sachs, Acta Phys. Polon. Suppl. 22, 13 (1962), see also I. Robinson and A. Schild, J. Math. Phys. <u>4</u>, 484 (1963).

[5] I. Robinson and A. Trautman, Phys. Rev. Lett. <u>4</u>, 431 (1960), and I. Robinson and A. Trautman in Conference Int. sur les Theories Relativistes, July 25-31, 1962, Gauthier-Villars , and PWN, Paris-Warszawa, 1964, p. 107-114.

[6] J. F. Plebański and S. Hacyan, J. Math. Phys. 16 (1975), 2403.

[7] J. F. Plebański, Ann. Phys. 90, 196 (1975).

[8] B. Carter, Commun. Math. Phys. 10, 280 (1968).

[9] J. F. Plebański and M. Demianski, Rotating, Charged and Uniformly Accelerating Mass in General Relativity, Orange Aid Preprint 401 (1975), to be published in Ann. Phys. (1976). See also J. F. Plebański, Annals of N. Y. Ac. of Sciences, 262 (1975) 246.

[10] J. F. Plebański and A. Schild, Complex Relativity and Double KS Metric, Proceeding of the International Symposium on Mathematical Physics, Mexico City, January 5 to 8, 1976, pp. 765-785; see also Nuevo Cimento 1976, to be published.

[11] W. Slebodziński, Exterior Forms and Their Applications, Monografie Matematyczne, Warszawa, 1970, PWN.

[12] H. Flanders, Differential Forms with Applications to the Physical Sciences, Ac. Press, New York-London 1963.

[13] J.D. Finley and J.F. Plebański, J. Math. Phys. <u>17</u>, 585 (1976).

[14] R. Penrose, Ann. Phys. (N.Y.), 10, 171 (1960).

Local Supersymmetry and Gravitation

P. van Nieuwenhuizen*

Institute for Theoretical Physics
State University of New York
Stony Brook, Long Island, New York 11794

ABSTRACT

We discuss a recently[1,2,3] discovered Lagrangian field theory for gravitation, which is locally supersymmetric ("supergravity") and which uses normal four dimensional spacetime instead of superspace. The action contains the Einstein action and the minimally coupled spin 3/2 action together with a non-derivative four-fermion interaction of gravitational strength. When recast in a first order formalism with torsion, the action contains no four-fermion term but the spin 3/2 field is coupled in a non-minimal way.

The commutator of two local supersymmetry transformations produces a general coordinate and local Lorentz transformation. This deeper symmetry which connects the structure of spacetime with transformations between bosons and fermions, seems interesting. Applications to quantum gravity are discussed.

*Supported in part by the NSF Grant #MPS-74-13208-A01.

I. INTRODUCTION

Supersymmetry was discovered two years ago by Wess and Zumino as a symmetry which transforms bosons into fermions and fermions into bosons with parameters which are constant (i.e., spacetime independent) spinors. There are now a certain number of Lagrangian field theories known, whose action is invariant under global supersymmetry transformations. When one wants to extend this global symmetry to a local symmetry, one obtains a gauge theory, and experience with other gauge theories leads one to expect that one needs the (?) gauge field of supersymmetry to construct gauge-invariant actions. As we will see, simple arguments indicate that this gauge field is in fact a gauge multiplet which consists of (at least) the graviton and the spin 3/2 field.* Therefore the name "supergravity" is a synonym for local supersymmetry. There exists now one locally supersymmetric action; it is the action for the gauge multiplet of supergravity and is thus the supersymmetric extension of the Hilbert action for general relativity. Work is in progress to couple this gauge multiplet to matter multiplets in a locally supersymmetric way.

As will be discussed below, general coordinate transformations and local Lorentz rotations can be obtained by applying twice a supersymmetry transformation. Local supersymmetry is thus in this sense a symmetry which underlies general relativity and this deeper structure of general relativity seems very interesting. Also, on the quantum level, supersymmetry has interesting properties. Certain globally supersymmetric field theories have less divergent quantum corrections than one might expect from naive power counting due to cancellations associated with the higher symmetry. Perhaps similar cancellations of divergences occur in locally supersymmetric models, and maybe there is a surprise waiting with respect to the quantization of "hitherto unrenormalizable field theories" (by which one, of course, means gravitation.)

There exist two approaches to supergravity: one which uses superspace and one which uses only normal four-dimensional space. The superspace approach is due to Arnowitt, Nath and Zumino. It is a very elegant formalism, but due to space limitations I cannot discuss it here in detail. The basic idea is to define superspace, which consists of the four usual coordinates x and four anti-commuting coordinates θ^α (hence $(\theta^\alpha)^2=0$) and to define a metric $g_{\mu\nu}(x,\theta)$, where $\mu,\nu = 1,8$, which satisfies the matter-free Einstein

*The classical tests of Einstein theory remain of course unaffected in super-gravity because a (massless or massive) fermion can never give a long-range force.

equation $R_{\mu\nu} = 0$. Expanding $g_{\mu\nu}(x,\theta)$ in θ yields a finite number of terms (because $\theta_\alpha^2 = 0$), which are supposed to represent the gravitational field, the electromagnetic field, the Yang-Mills field, the Dirac field and many other fields. Matter is thus determined by gravitation in this approach, in agreement with Einstein's later ideas. Zumino has started a different theory which still uses superspace, but where matter is no longer contained in $g_{\mu\nu}(x,\theta)$ but is introduced independently from gravitation. Due to algebraic complexities, the true physical content of these superspace theories is at present unknown, but it is possible that the superspace approach and the normal space approach will eventually turn out to be equivalent.

The normal space approach to supergravity uses, as the work suggests, only the usual four-dimensional spacetime. This brings supergravity to the level of other gauge theories and allows one to use methods and concepts developed for gauge theories in general. It is this approach which will be discussed here.

This account starts,for pedagogical reasons, with a very simple globally supersymmetric Lagrangian field theory model which contains, however, the essential ingredients we will need. Then it will be argued heuristically why there is an intimate relation between gravitation and local supersymmetry. As we will see, the gauge particles of supersymmetry are expected to be the graviton and the real, massless spin 3/2 field. Some elements of spin 3/2 field theory are therefore presented separately, and finally the classical theory of supergravity will be constructed.

II. THE SIMPLEST SUPERSYMMETRIC MODEL IN FLAT SPACE

Consider the Lagrangian for a massless real scalar field A, a massless real pseudoscalar field B and a massless real (Majorana) fermion field ψ.

$$\mathcal{L} = -\frac{1}{2}(\partial_\mu A \partial_\nu A + \partial_\mu B \partial_\nu B)\eta^{\mu\nu} - \frac{1}{2}\bar{\psi}\gamma^\mu \partial_\mu \psi \tag{1}$$

This is essentially the first supersymmetric model, discovered by Wess and Zumino.[4] Although one might not have expected it at first, the action $\int \mathcal{L}\, d^4x$ is globally supersymmetric under the transformatio

$$\delta A = \bar{\epsilon}\psi \qquad \delta B = \qquad \delta\psi = \partial_\mu(A-$$

The parameters ε are anticommuting constant spinors, satisfying $[\varepsilon,A] = [\varepsilon,B] = \{\varepsilon,\psi\}=\{\varepsilon_1,\varepsilon_2\} = 0$. One can view them as classical fermion fields which anticommute since $\hbar=0$. Both ε and ψ are real (Majorana) spinors (the fermion is its own antiparticle), satisfying

$$\bar{\psi} = \psi^T (C^{-1})^T \qquad \bar{\epsilon} = \epsilon^T (C^{-1})^T \tag{3}$$

where the charge conjugation matrix C satisfies[5]

$$C\gamma^\mu C^{-1} = -(\gamma^\mu)^T \qquad C^{-1} = -C \tag{4}$$

With these rules one easily verifies that

$$\delta\mathcal{L} = \partial_\mu K^\mu \,, \qquad K^\mu = -\frac{1}{2}\bar{\epsilon}\gamma^\mu(\partial\!\!\!/ A - \tag{5}$$

where $\partial\!\!\!/ = \gamma^\mu \partial_\mu$.

We make four observations:

(1) The bosons are transformed into the fermion and vice-versa. Since one of the fields is massless, so must be the others (due to symmetry). One can introduce mass terms and couplings, invariant under (2), but since we are heading for the supersymmetric gauge fields, we consider only massless fields in this example.

(2) The spins of the bosons and fermions are J and J + ½ with J=0. Also the combination J and J - ½ is possible; this will actually be the case in supersymmetry.

(3) There should be as many physical boson states as fermion states. In our example there are two boson states and two fermion states with helicities (0,0,½,-½). For non-zero boson spin, one needs one real boson and real fermion because now each has two helicity states.

(4) The action (but not the Lagrangian) is invariant under the global supersymmetry transformations in (2).

(Actually, one can even put B = 0 in (1) and (2); one still has, in this case, an invariant action, but the algebra (see below) is not what one wants).

Since the action has a global invariance, there is a current which is conserved when the field equations are satisfied (the Noether current). It is generally given by (denoting all fields in the action by χ_i)

$$j^\mu_N = \sum_i \frac{\partial \mathcal{L}}{\partial(\partial_\mu \chi_i)} \delta\chi^i - K^\mu \quad , \quad \partial_\mu j^\mu_N = 0 \tag{6}$$

where $\partial_\mu K^\mu$ is the variation of the action (see (5)). In our example it is equal to (omitting ε)

$$j^\mu_N = - \quad (\gamma A - \gamma B \gamma^5)\gamma^\mu \psi \tag{7}$$

One easily verifies that it is conserved when $\Box A = \Box B = \gamma^\mu \partial_\mu \psi = 0$. Omitting (as is usual) the transformation parameter $\bar{\varepsilon}$ in the current, one obtains a <u>spinorial current</u>, which is not an observable.

One can now construct the four charges

$$Q_a = \int [j^0_N(\vec{x},t)]_a d^3x \quad , \quad a = 1,4 \tag{8}$$

which are constant (in time) and again spinors (with index a). Straightforward algebra shows that they satisfy a so-called graded (because there are anticommutation as well as commutation relations) Lie algebra

$$\{Q_a, \bar{Q}_b\} = (\gamma^\alpha)_{ab} P_\alpha \; , \; [Q_a, P_\alpha] = [P_\alpha, P_\beta] = 0 \tag{9}$$

where P_α are the Lorentz generators. The Dirac matrices are thus the structure constants. The graded Lie algebra in (9) was here derived from the model in (1), but it holds for any globally supersymmetric Lagrangian field theory and is thus model independent.

III. RELATION BETWEEN LOCAL SUPERSYMMETRY AND GRAVITATION

Since the parameters ε are anticommuting, one can rewrite (9) in terms of scalar charges $(\bar{\varepsilon}Q)$ as

$$[\bar{\epsilon}_1 Q, \bar{\epsilon}_2 Q] = (\bar{\epsilon}_1 \gamma^\alpha \epsilon_2) P_\alpha \tag{10}$$

together with $[\bar{\varepsilon}Q, P_\alpha] = [P_\alpha, P_\beta] = 0$. The Lorentz generators P_α describe translations in spacetime, everywhere over the same distance $(\bar{\varepsilon}_2\gamma^\alpha\varepsilon_1)$. If, however, ε becomes spacetime dependent (as in a gauge theory of supersymmetry), then we have spacetime dependent translations which (at least for scalars) describe a general coordinate transformation as given by general relativity. According to the equivalence principle, general relativity implies gravitation, hence the right hand side of (10) has, in an admittedly loose sense, something to do with gravitation when we have a local gauge invariance $(\varepsilon=\varepsilon(x))$. Conversely, suppose that we have gravitation and hence curved space. Then the concept of a constant spinor is untenable, because if ε is constant in one frame, then in general, it will pick up a spacetime dependence after a local Lorentz rotation. Hence, the algebraic relation in (10) suggests an intimate relation between gravitation and local supersymmetry, reason why one sometimes speaks of supergravity instead of local supersymmetry.

There exists a theorem[6] which states the existence of irreducible representations of the Lie algebra in (9) in the Hilbert space spanned by the helicity states of two massless particles: one real boson with spin $J>0$ and one real (Majorana) fermion with spin $J\pm\frac{1}{2}$. The theorem says absolutely nothing about the existence of a locally supersymmetric field theory, but let us assume that such a theory exists. As argued previously, the real massless boson is expected to be the graviton, hence $J=2$. For the fermion we can then have either spin 3/2 or 5/2. It is simplest to choose 3/2; this also follows from the rule of thumb that a gauge field has always one more Lorentz index than its gauge parameter (electromagnetism $\Lambda(x)$ and $A_\mu(x)$, gravitation $\varepsilon^\alpha(x)$ and $g_{\mu\nu}(x)$, supersymmetry $\varepsilon(x)$ and $\psi_\mu(x)$ where the vectorial spinor ψ_μ contains indeed spin 3/2 as shown below).[7] It has in fact been shown that there exist no globally supersymmetric interactions between spin 2 and spin 5/2: kinematical singularities constrain the tree graphs so severely that the coupling constant has to vanish.[8] Also, no gauge action exists for spin 5/2.

Thus we arrive at the following task: construct a locally supersymmetric field theory containing (at least) spin 2 and spin 3/2 fields (since spin 2 is present, this theory must also be generally covariant.) Our philosophy will be to start from only what is absolutely necessary: the Einstein and (generally covariant) spin 3/2 action. Then we will try to find criteria which will enable us to postulate (trial) transformation laws for the fields in this action, and finally we will see how far one gets in this approach. First we will now review some spin 3/2 theory.

IV. THE SPIN 3/2 LAGRANGIAN

The spin 3/2 Lagrangian, as essentially given by Fierz and Pauli, and Rarita and Schwinger, reads in flat space[7]

$$\mathcal{L}_{3/2} = -\frac{1}{2} \varepsilon^{\mu\nu\rho\sigma} \bar{\psi}_\mu \gamma_5 \gamma_\nu \partial_\rho \psi_\sigma \tag{11}$$

Since ψ_μ is its own antiparticle, it satisfies (3) and there is a factor $\frac{1}{2}$ in (11). One can actually find an equivalent one parameter set of Lagrangians by substitution of $\psi_\mu = \hat{\psi}_\mu + a\gamma_\mu \gamma \cdot \hat{\psi}$ but there are also other, inequivalent Lagrangians for spin 3/2. The spin content of the vectorial spinor (or spinorial vector) ψ_μ is

$$(\text{spin } \psi) \otimes (\text{spin } \mu) = \frac{1}{2} \otimes (1+0) = \frac{3}{2} \oplus \frac{1}{2} \oplus \frac{1}{2} \tag{12}$$

and ψ_μ contains the required spin 3/2 plus lower spins (gauge components), analogous to the electromagnetic or gravitational fields. The flat-space Lagrangian has an electromagnetic type of <u>gauge</u> invariance with a <u>spinorial parameter</u>

$$\delta \psi_\sigma(x) = \partial_\sigma \epsilon(x) \tag{13}$$

This is exactly the kind of transformation we are looking for in curved space. The philosophy of ref. (1) was now to consider this invariance as the flat space counterpart of the full supersymmetric gauge transformation of ψ_σ in curved space. To go the way back into curved space, we simply covariantize (13) to

$$\delta \psi_\mu^{(1)} \sim D_\mu \epsilon = \partial_\mu \epsilon + \frac{1}{2} \omega_{\mu ab}(V) \sigma^{ab} \epsilon$$

$$\sigma^{ab} = (\gamma^a \gamma^b - \gamma^b \gamma^a)/4 \quad , \quad \{\gamma^a, \gamma^b\} = 2\eta^{ab} \tag{14}$$

where D is the usual covariant derivative on a spinor with spinor connection $\omega_{\mu ab}(V)$ in terms of the Vierbein fields $V_{a\mu}$. Since (14) will turn out to be only half of the correct transformation law, we have given it the superfix (1).* The covariantized action for spin 3/2 is simply obtained from (11) by replacing ∂ by D since the ε-symbol is already a density.

$$\mathcal{L}_{3/2}(\text{curved space}) = -\frac{1}{2}\varepsilon^{\mu\nu\rho\sigma}\bar{\psi}_\mu\gamma_5\gamma_\nu D_\rho\psi_\sigma \tag{15}$$

where, as usual

$$D_\rho\psi_\sigma = \partial_\rho\psi_\sigma + \frac{1}{2}\omega_{\rho ab}(V)\sigma^{ab}\psi_\sigma - \Gamma^\alpha_{\rho\sigma}\psi_\alpha \tag{16}$$

Actually, the last term in (16) can be and will be omitted due to the ε-symbol in (15). We note that then the only dependence of (15) on the Vierbein field is through $\gamma_\nu = \gamma^a V_{a\nu}$ and $\omega_{\rho ab}(V)$, because γ^a and γ_5 are constant Dirac matrices.

---

*As always, one needs a square root when dealing with fermions. In general relativity one cannot couple fermions directly to $g_{\mu\nu}$ but only to its square root, the Vierbein field $V_{a\mu}$ which satisfies $g_{\mu\nu} = V_{a\mu}V_{b\nu}\,\delta^{ab}$.

V. FIRST STEP TO CONSTRUCT A THEORY OF SUPERGRAVITY

As discussed before, we start from the minimally coupled Einstein and Rarita-Schwinger actions

$$\mathcal{L}_2+\mathcal{L}_{3/2} = -2\kappa^{-2}(-g)^{1/2}R(g) - \frac{1}{2}\varepsilon^{\mu\nu\rho\sigma}\bar{\psi}_\mu\gamma_5\gamma_\nu D_\rho\psi_\sigma \tag{17}$$

where $\kappa^2=32\pi G$ with G=Newton's constant. We must now complete (14) with the transformation law for $V_{a\mu}$. From (2) we expect that it involves $\bar{\varepsilon}$ and ψ_μ, but that still leaves the index a. The simplest choice is $\delta V^a{}_\mu \sim \bar{\varepsilon}\gamma^a\psi_\mu$. (Another choice would be $\delta V^a{}_\mu \sim \bar{\varepsilon}\gamma_\mu\psi^a$ with $\psi^a=V^{a\nu}\psi_\nu$ but this transformation contains no term linear in the fermion field). We try thus the set of transformation laws (with the correct constants already inserted and $\varepsilon=\varepsilon(x)$ in what follows)

$$\delta\psi_\mu = 8\kappa^{-1}D_\mu\epsilon \quad , \quad \delta V^a_\mu = \bar{\epsilon}\gamma^a\psi_\mu \quad , \quad \delta g_{\mu\nu} = \kappa(\bar{\epsilon}\gamma_\mu\psi_\nu + \bar{\epsilon}\gamma_\nu\psi_\mu) . \tag{18}$$

The last result follows easily from $g_{\mu\nu}=V_{a\mu}V^a{}_\nu$.

Under the local supersymmetry transformations in (18), the action defined by (17) transforms schematically as

$$\int\left(\frac{\delta\mathcal{L}_2}{\delta g_{\mu\nu}}\delta g_{\mu\nu} + \frac{\delta\mathcal{L}_{3/2}}{\delta\psi_\mu}\delta\psi^{(1)}_\mu + \frac{\delta\mathcal{L}_{3/2}}{\delta V^a_\mu}\delta V^a_\mu\right)d^4x . \tag{19}$$

Cleary the first two terms are linear in ψ_μ where as the last is cubic in ψ_μ. Hence both kinds of terms must vanish separately. The first term is of course proportional to the Einstein tensor. But so is the second; and their sum cancels (!)

$$\int\left(\frac{\delta\mathcal{L}_2}{\delta g_{\mu\nu}}\delta g_{\mu\nu} + \frac{\delta\mathcal{L}_{3/2}}{\delta\psi_\mu}\delta\psi^{(1)}_\mu\right)d^4x = 0 \tag{20}$$

To prove this result, note that replacing ψ_σ by $D_\sigma\varepsilon$ in (17) yields due to the ε- symbol, a curvature term

$$[D_\rho, D_\sigma]\epsilon = \frac{1}{2} R_{\rho\sigma ab}\sigma^{ab}\epsilon \tag{21}$$

where

$$R_{\rho\sigma ab} = (\partial_\rho \omega_{\sigma ab} + \omega_{\rho a}{}^{c}\omega_{\sigma cb}) - (\rho \leftrightarrow \sigma) . \tag{22}$$

(Also $R=R_{\rho\sigma ab}V^{b\rho}V^{a\sigma}$ in (17)). World and local Lorentz indices are raised and lowered, as usual by $g_{\mu\nu}$ and η_{ab} respectively. Manipulation with Dirac matrices yields

$$\gamma_\nu \sigma^{ab} = \frac{1}{2}\left(V^a{}_\nu \gamma^b - V^b{}_\nu \gamma^a + V_{c\nu}\,\varepsilon^{abcd}\gamma_5\gamma_d\right) . \tag{23}$$

Only the last (axial vector) term contributes, because $\varepsilon^{\mu\nu\rho\sigma} R_{\rho\sigma ab}V^a_\nu$ vanishes, according to the cyclic identity

$$R_{[\rho\sigma\nu]b} = 0 . \tag{24}$$

Evaluating the product of the two ε-symbols $\varepsilon^{\mu\nu\rho\sigma}$ ε^{abcd} in terms of products of Vierbein fields leads to (20). Hence, the group property (21), details of spinor spinor algebra (23) and a special property of the Riemann tensor in (24) lead to the non-trival cancellation of the order ψ terms in (19)! (variation of $\bar{\psi}_\mu$ in (17) gives the same result as the variation ψ_σ which we discussed. To prove this, integrate partially and transpose, using the Majorana properties in (3) and (4)).

Unfortunately, the last term in (19) does not vanish by itself as for example trial functions show. The previous result was however too nice to abandon, so we will add terms of higher order in κ to the action and transformation rules in order to cancel the cubic term in (19) as well.

VI. ADDING TERMS TO ACTION AND TRANSFORMATION LAWS

Since we want to cancel the last term in (19), let us have a closer look at it. As discussed below (16), it contains only two terms, obtained by varying γ_ν and $\omega_{\rho ab}(V)$ in (15). From the metric postulate $D_\mu V_{a\nu} = \partial_\mu V_{a\nu} + \omega_{\mu ab} V^b_\nu - \{{\alpha \atop \mu\nu}\} V_{a\alpha} = 0$ one can obtain ω in terms of $V_{a\mu}$. The variation $\delta\omega_{\rho ab}$ is a tensor and can be determined by taking its linearized limit first and covariantizing afterwards. One easily finds

$$\delta\omega_{\rho ab}\sigma^{ab} = \kappa\left[-D_\rho(\bar\epsilon\gamma_a\psi_b) + D_b(\bar\epsilon\gamma_a\psi_\rho) + D_b(\bar\epsilon\gamma_\rho\psi_a)\right]\sigma^{ab} \tag{25}$$

One finds thus

$$\frac{\delta\mathcal{L}_{3/2}}{\delta V^a_\mu}\delta V^a_\mu = -\frac{1}{2}\,\epsilon^{\mu\nu\rho\sigma}\Big[(\bar\psi_\mu\gamma_5\gamma_a D_\rho\psi_\sigma)(\bar\epsilon\gamma^a\psi_\nu) + (\bar\psi_\mu\gamma_5\gamma_\nu\sigma^{ab}\psi_\sigma)(\tfrac{1}{2}\delta\omega_{\rho ab})\Big] \tag{26}$$

We must now make a small computation. Interchanging $(D_\rho\psi_\sigma)$ and $(\gamma^a\psi_\nu)$ by means of a so-called Fierz rearrangement*, one obtains for the first term between the square brackets $\frac{1}{4}(\bar\psi_\mu\gamma^d\psi_\nu)(\bar\epsilon\gamma_d\gamma_5 D_\rho\psi_\sigma)$ where we used that for a Majorana field

$$\bar\psi_\mu\gamma_a\cdots\gamma_q\psi_\nu = \bar\psi_\nu(-\gamma_q)\cdots(-\gamma_a)\psi_\mu \tag{27}$$

(which follows from (4) after transposing) and the antisymmetry in $(\mu\nu)$. Using (20), also the second term in (24) can be cast in a similar form (using antisymmetry in μ and σ) and one obtains**

---

*According to a Fierz transformation on Majorana fields. $(\bar\phi\chi)(\bar\eta\zeta) = -\frac{1}{4}\sum_A(\bar\phi O^A\zeta)(\bar\eta O^A\chi)$ where $O^A = 1, \gamma_5, \gamma_\mu, i\gamma_\mu\gamma_5$ and $(\gamma_\mu\gamma_\nu - \gamma_\nu\gamma_\mu)/2i$ are the sixteen Dirac matrices. Choose $\bar\phi = \bar\psi\gamma_5\gamma_a$, $\chi = D_\rho\psi_\sigma$, $\bar\eta = \bar\epsilon$, $\zeta = \gamma^a\psi_\nu$

**For a Majorana field $(\bar\psi_\mu O^F\psi_\nu) = (\bar\psi_\nu O^F\psi_\mu)(-)^{\sigma(F)}$, where $\sigma(F) = +1$ for $F = 1, \gamma_5, \gamma_\mu\gamma_5$ and $\sigma(F) = -1$ for $F = \gamma_\mu, \sigma_{\mu\nu}$. See for example ref (5).

$$\frac{\delta \mathcal{L}_{3/2}}{\delta V^{2}_{\mu}} \delta V^{2}_{\mu} = -\frac{1}{4}\varepsilon^{\mu\nu\rho\sigma}(\bar{\psi}_{\mu}\gamma^{d}\psi_{\nu}) \left[\bar{\varepsilon}\gamma^{d}\gamma_{5}D_{\rho}\psi_{\sigma} - \frac{1}{2}\varepsilon_{cabd}V^{c}_{\nu}\,\delta\omega_{\rho ab}\right]. \tag{28}$$

This is the end of the small computation.

A moment of inspection of (28) and (25) shows an interesting structure: There are two kinds of terms

(1) curl terms with $D_{\rho}\psi_{\sigma} - D_{\sigma}\psi_{\rho}$

(2) gradiant terms with $D_{\mu}\varepsilon$.

Suppose now that we would add to $\delta\bar{\psi}_{\mu}$ an extra term of the form $\delta\bar{\psi}_{\mu}^{(2)} \sim (\bar{\psi}\gamma\psi)(\bar{\varepsilon}\ N)$. Substituting this into (17) would indeed lead to curl terms of the right form $(\bar{\psi}\gamma\psi)(\bar{\varepsilon}\cdots D_{\rho}\psi_{\sigma})$. Of course all the gamma matrices have to work out right, but it works and the precise result is given in (35). (As before, varying ψ_{σ} in (17) gives the same result). This leaves us with gradiant terms. They are of the form $(\bar{\psi}\gamma\psi)(D_{\mu}\bar{\varepsilon}\cdots\psi)$ and can in principle be cancelled by adding to the action of a non-derivative four-fermion coupling $\mathcal{L}_{4}=(\bar{\psi}P\psi)(\bar{\psi}D\psi)$ of gravitational strength since its variation $(\delta\mathcal{L}_{4}/\delta\bar{\psi}_{\rho})\ D_{\rho}\bar{\varepsilon}$ is of the correct form. Again the gamma matrices work out, and this precise result is given in (33).

We have thus achieved that

$$\int\left(\frac{\delta\mathcal{L}_{3/2}}{\delta V^{a}_{\mu}}\delta V^{a}_{\mu} + \frac{\delta\mathcal{L}_{3/2}}{\delta\psi_{\mu}}\delta\psi^{(2)}_{\mu} + \frac{\delta\mathcal{L}_{4}}{\delta\psi_{\mu}}\delta\psi^{(1)}_{\mu}\right)d^{4}x = 0 \tag{29}$$

but all we have really done is push the program one step further, from terms cubic in ψ_{μ} to quintic terms. Indeed, the new term $\mathcal{L}_{4}$ in the action gives two new contributions to the variation of the action

$$\int\left(\frac{\delta\mathcal{L}_{4}}{\delta\psi_{\mu}}\delta\psi^{(2)}_{\mu} + \frac{\delta\mathcal{L}_{4}}{\delta V^{a}_{\mu}}\delta V^{a}_{\mu}\right)d^{4}x. \tag{30}$$

which are both of order ψ^{5}, as follows from (33) and (35). One might try once more to add new terms to action and/or transformation laws, but since (30) contains no derivatives (unlike (28)), this program cannot work. So the crucial question is whether (30) vanishes by itself due to over-antisymmetrization, without outside help. This is a difficult problem because the five Majorana fields ψ_{μ} can be recoupled in all possible ways so that many hidden symmetries are expected to be present.

Actually, the question whether (30) vanishes by itself was solved in a rather unorthodox way. The general structure of (30) is

$$\sum_{m,\mu} C(m_0, \cdots m_5, \mu_1, \cdots \mu_5)\bar{\epsilon}^{m_0} \bar{\psi}_{\mu_1}^{m_1} \psi_{\mu_2}^{m_2} \bar{\psi}_{\mu_3}^{m_3} \psi_{\mu_4}^{m_4} \psi_{\mu_5}^{m_5} \tag{31}$$

where $m_0, \cdots, m_5$ are the Dirac indices and the coefficients C are built from ε- symbols, Kronecker deltas and Dirac matrices. For any independent combination of five spin 3/2 fields the coefficient C was calculated by means of a simple Fortran program (since the C's have absolute value one, no great precision was required). It was a memorable moment when (after a few seconds) the computer displayed all 3000 coefficients: all were zero!

VII. A THEORY OF SUPERGRAVITY

The final result is the action

$$\mathcal{L} = \mathcal{L}_2 + \mathcal{L}_{3/2} + \mathcal{L}_4 \tag{32}$$

with $\mathcal{L}_2$ and $\mathcal{L}_{3/2}$ given in (17) and (22) and $\mathcal{L}_4$ by

$$\mathcal{L}_4 = \frac{\kappa^2}{32}(\det V^a{}_\mu)\left[(\bar{\psi}^b \gamma^a \psi^c)(\bar{\psi}_b \gamma_a \psi_c + 2\bar{\psi}_a \gamma_b \psi_c) - 4(\bar{\psi}_a \gamma^b \psi_b)^2\right] \tag{33}$$

This action is invariant under the local supersymmetry transformations

$$\delta V^a{}_\mu = \kappa(\bar{\epsilon}\gamma^a \psi_\mu) \tag{34}$$

$$\delta \psi_\mu = 2\kappa^{-1}\left[D_\mu \epsilon + \frac{\kappa^2}{8}(2\bar{\psi}_\mu \gamma_a \psi_b + \bar{\psi}_a \gamma_\mu \psi_b)(\sigma^{ab}\epsilon)\right] \tag{35}$$

This result was first obtained by D.Z.Freedman, S. Ferrara and myself.[1]

The presence of a four fermion term seems to indicate that torsion is present. This led Deser and Zumino[2] to an alternative much shorter derivation which does not need a computer calculation. Consider first order formalism in which the spinor connection $\omega_{\mu ab}$ is an independent field, on the same footing as ψ_μ and V^a_μ. Consider the action

$$\mathcal{L}(V, \psi_\mu, \omega) = \mathcal{L}_2(V, \omega) + \mathcal{L}_{3/2}(V, \psi_\mu, \omega) \tag{36}$$

$$\mathcal{L}_2 = -2\kappa^{-2}(\det V^a{}_\mu)\, V^{b\mu} V^{a\nu} R_{\mu\nu ab}(\omega) \tag{37}$$

where $R_{\mu\nu ab}$ was given in (22) and $\mathcal{L}_{3/2}$ is still given by (15) but $D_\rho \psi_\sigma$ replaced by $D_\rho \psi_\sigma = \partial_\rho \psi_\sigma + \frac{1}{2}\omega_{\rho ab}\sigma^{ab}\psi_\sigma$. This is a non-minimal coupling; the absence of a term $\Gamma_{\mu\nu}{}^\alpha \psi_\alpha$ in the action keeps the theory invariant under the flat space electromagnetic-type gauge invariance $\delta\psi_\alpha = \partial_\alpha \varepsilon(x)$, but why this must be so probably should follow from Lie group theory. Variation with respect $\omega_{\rho ab}$ gives the equation of motion

$$\omega_{\mu\nu\rho} = \omega_{\mu\nu\rho}(V) - \frac{\kappa^2}{16}\left(\bar{\psi}_\beta \gamma_\rho \psi_\alpha + \bar{\psi}_\rho \gamma_\beta \psi_\alpha + \bar{\psi}_\beta \gamma_\alpha \psi_\rho\right) \tag{38}$$

and substituting this back into $\mathcal{L}(V, \psi_\mu, \omega)$ yields the second order action (32), including the four fermion interaction $\mathcal{L}_4$. Also $\delta\psi_\mu$ can now be written as

$$\delta\psi_\mu = 2\kappa^{-1} D_\mu \epsilon = 2\kappa^{-1}\left(\partial_\mu \epsilon + \tfrac{1}{2}\omega_{\mu ab}\sigma^{ab}\epsilon\right) \tag{39}$$

if one substitutes again the field equation (38). Torsion is usually defined as

$$S_{\mu\nu}^{\ \ \rho} = \tfrac{1}{2}\left(\Gamma_{\mu\nu}^{\rho} - \Gamma_{\nu\mu}^{\rho}\right) \tag{40}$$

and the relation with (38) follows from the metric postulates

$$D_\mu g_{\nu\rho} = 0 \Rightarrow \Gamma_{\mu\nu}^{\rho} = \left\{ {\rho \atop \mu\nu} \right\} - K_{\mu\nu}^{\ \ \rho} \tag{41}$$

$$D_\mu V_{a\nu} = 0 \Rightarrow \omega_{\mu\nu\rho} = \omega_{\mu\nu\rho}(V) + K_{\mu\nu\rho} \tag{42}$$

where we defined

$$K_{\mu\nu\rho} = - S_{\rho\nu\mu} + S_{\nu\mu\rho} - S_{\rho\mu\nu} \tag{43}$$

Not only leads insertion of the field equation for ω from first order back to second order formalism, but also one can show that the action is invariant in first order formalism under the following local supersymmetry transformations:

$$\delta\psi_\mu = 2\kappa^{-1} D_\mu \epsilon \qquad \delta V^a_{\ \mu} = \kappa \bar{\epsilon} \gamma^a \psi_\mu \tag{44}$$

$$\delta\omega_{\mu ab} = \left(-\tfrac{1}{2}\bar{\epsilon}\gamma_5\gamma_c \mathcal{D}_\rho\psi_\sigma\right)\left(V^c{}_\mu \varepsilon^{ab\rho\sigma} + V^a{}_\mu \varepsilon^{bc\rho\sigma}\right) - (a \leftrightarrow b) \qquad (45)$$

where $\varepsilon^{ab\rho\sigma} = V_r{}^\rho V_s{}^\sigma \varepsilon^{abrs}$

It seems that this first order formalism is the natural way to describe supergravity; it is certainly much more elegant. However, why supergravity must be described in terms of torsion is a priori (to me) not clear; although, once one chooses this starting point, the result is obtained in a few lines. The derivation which uses the second order formalism has the advantage that it makes no a priori choice: one is naturally led to a result which can, in a very economical way, be rewritten in the form of torsion.

VIII. THE ALGEBRA OF SUPERGRAVITY

The commutator of two successive supersymmetry transformations in flat space is equal to a translation, see (10). It is interesting to see what happens in supergravity. After a straightforward calculation one finds for the Vierbein field for example, using the Jacobi identity and defining $\delta_1 V_\mu^a = [\bar{\epsilon}_1 Q, V_\mu^a]$ $\quad \xi^\alpha = \bar{\epsilon}_1(x)\gamma^\alpha\epsilon_2(x)$

$$\delta_1(\delta_2 V^a_\mu) - \delta_2(\delta_1 V^a_\mu) = [\bar{\epsilon}_1 Q, \bar{\epsilon}_2 Q]V^a_\mu = [\delta_1, \delta_2]V^a_\mu =$$

$$\left(\frac{\partial \xi^\alpha}{\partial x^\mu} V^a_\alpha + \xi^\alpha \partial_\alpha V^a_\mu\right) + \left(\omega_\mu{}^{ab} V_{b\alpha}\right) + \left(-\kappa\, \xi^\alpha \bar{\psi}_\alpha\right)\gamma^a \psi_\mu \tag{46}$$

+ terms which vanish when the eqs. of motions are inserted. Hence two successive local supersymmetry transformations (on V^a_μ, but the same results hold for ψ_μ as well) produce the transformations of general relativity (the first two terms in curly brackets) plus a supersymmetry transformation with field-dependent parameter $\bar{\epsilon}' = -\kappa\, \xi^\alpha \bar{\psi}_\alpha$. This is the deeper structure of general relativity to which was alluded in the introduction. In (flat space) globally supersymmetric models one obtains terms in $[\delta_1, \delta_2]$ which vanish as a consequence of the field equations, if one eliminates auxiliary fields (such as $\omega_{\mu ab}$) before evaluating the commutators. The appearance of such terms in the commutators above might indicate that there are more auxiliary fields needed than $\omega_{\mu ab}$ in order to eliminate such equation of motion terms. Usually, the theory becomes simpler when one includes all the auxiliary fields, but one can perfectly well do without them (at the cost of a little bit more labour). It is unknown at present which (if any) the extra auxiliary fields in the supersymmetric gauge multiplet are.

IX. QUANTUM GRAVITY

If one considers the sum of the spin 2 and spin 3/2 fields as a metric in extended sense, then one would expect the same results for quantum supergravity as for ordinary gravity. What are the results in ordinary quantum gravity? It is known that the gravitational self-interactions, in the absence of matter, are non-renormalizable but have an S-matrix which is finite at the one-loop level (i.e. with respect to the lowest quantum corrections). It is also known that coupling to such individual matter fields as spin 0, spin 1/2, photons, or Yang-Mills fields leads to an S-matrix which is divergent and nonrenormalizable.[10] And finally it is known in one case what happens when three things are mixed together: coupling quantum electrodynamics to gravitation leads to a nonrenormalizable S-matrix.[11]

Therefore one would expect that pure supergravity might have a one loop finite S-matrix, but that coupling to matter spoils this result. There have been speculations in the past that a first order formulation with torsion might improve the situation in Einstein gravitation (specifically, in the Dirac-Einstein system), but a recent result[12] shows this not to be true for QED coupled to gravitation. It might however be that torsion in the context of supergravity works better than in Einstein gravitation.

There exists[13] a calculation which indicates (although no proof) that the two-loop corrections of pure Einstein gravitation lead to a divergent S-matrix. In this respect supergravity might well do better: the extra symmetry (supersymmetry) might cancel enough divergences such that the S-matrix becomes finite at the two-loop level. It is known how one should quantize supergravity; the ideas of modern gauge field quantization apply here as well as in ordinary gravitation.[14] Also, no problems with causality are present.[14]

X. OUTLOOK

The action for the gauge fields of supergravity (the Vierbein and the spin 3/2 field) is now known. Coupling to matter is next required. It seems logical to consider the coupling to those matter multiplets which themselves are already invariant under global supersymmetry transformations. Once this is done, one can investigate whether spontaneous symmetry breaking can occur, and if so, whether the massless spin 3/2 gauge fermion can eat the massless Goldstone spin 1/2 fermion which, according to global supersymmetry, should exist after which banquet the spin 3/2 field becomes massive. Such a Goldstone neutrino is not seen in nature, and this has posed doubt in the past to the applicability of supersymmetry[16] - but this doubt now might disappear. It should be noted that the (universal?) coupling of a (massless or massive) spin 3/2 fermion to matter with gravitational strength leaves the classical predictions of Einstein theory unchanged, since it cannot produce long range forces. There is an intimate connection between twistor theory and supergravity, as one cannot escape feeling at this conference. Maybe cross-fertilization might be useful.

At the quantum level the one-loop divergences of supergravity should be calculated. Considering the supersymmetric gauge multiplet as a "metric in extended sense" one might expect the same results as in pure gravitation. But it is possible that the extra symmetry, provided by local supersymmetry, is so potent that supergravity is the royal road to quantization.

Note added: The matter coupling problem has been solved [see S. Ferrara, J. Scherk and the author in Phys. Rev. Lett. 37, 1035 (1976)]. The quantum corrections to supergravity are indeed finite at the one-loop level (M.T. Grisaru, J.A.M. Vermaseren and the author) and at the two-loop level the leading divergences have recently been shown to cancel (M.T. Grisaru). The Buchdahl inconsistencies for higher spin fields are circumvented because we allow not only the physical spin 3/2 but also lower spins (gauge-components). The derivative $\mathcal{D}_\rho \psi_\sigma$ might be considered covariant if ψ_σ is considered as a one-form.

ACKNOWLEDGMENT

The work presented here was done in collaboration by Dan Freedman and myself at Stony Brook and Sergio Ferrara in Paris. It is a pleasure to acknowledge the fruitful collaboration with them, as well as enlightening discussion with Peter Breitenlohner from Munich.

REFERENCES

[1]D.Z. Freedman, P. van Nieuwenhuizen and S. Ferrara, Phys. Rev. D13, 3214 (1976).

[2]S. Deser and B. Zumino, Phys. Lett. B62, 335 (1976).

[3]D.Z. Freedman and P. van Nieuwenhuizen, Phys. Rev. D14, 912 (1976).

[4]J. Wess and B. Zumino, Nucl. Phys. B70, 39 (1974).

[5]See for example G. Kallen, "Elementary Particle Physics", Addison Wesley (1964).

[6]A. Salam and J. Strathdee, Nucl. Phys. B76, 477 (1974); The extension of this theorem to massless particles is due to B. Zumino.

[7]M. Fierz and W. Pauli, Proc. Roy. Soc. (London), A173, 211 (1939). W. Rarita and J. Schwinger, Phys. Rev. 60, 61 (1941).

[8]M.T. Grisaru, H. Pendleton and P. van Nieuwenhuizen, Phys. Rev. D, to be published.

[9]G. 't Hooft and M. Veltman, Ann. Inst. H. Poincare 20, 69 (1974).

[10]P. van Nieuwenhuizen, Proc. Trieste conf. June 1976; S. Deser, Proc. Boston Gauge Conf. Sept. 1975.

[11]M.T. Grisaru, P. van Nieuwenhuizen and C.C. Wu, Phys. Rev. D12, D13 (1975).

[12]P. van Nieuwenhuizen, Proc. Caracas Conf. Dec. 1975, to be published.

[13]P. van Nieuwenhuizen, Ann. of Phys., to be published.

[14]A. Das and D.Z. Freedman, Nucl. Phys., to be published.

[15]B. Zumino, Proc. London Conf. July 1974.

[16]B. deWit and D.Z. Freedman, Phys. Rev. Letters 35, 827 (1975). W.A. Bardeen, unpublished.

CONFORMAL BUNDLE BOUNDARIES

B.G. Schmidt

Max-Planck-Institut für Physik und Astrophysik
Föhringer Ring 6, 8000 München 40, West-Germany

1. INTRODUCTION

The b-boundary construction [1], [2] attaches to any space-time a boundary, which may be used to define singularities. A generalization of this to conformal structures and projective structures was given in [3]. This way any space-time defines a conformal boundary. This boundary contains null infinity as defined by Penrose and further boundary points which may be used to define space-like and time-like infinity.

In section 2 the construction of the conformal bundle boundary is given. Section 3 contains the interpretation of the boundary points. In Section 4 some applications to asymptotically flat space time are given.

2. CONSTRUCTION OF THE CONFORMAL BOUNDARY OF A SPACE TIME

In this section it will be shown, how a conformal structure on a manifold M defines in a natural way two principle bundles over M. On one of those a preferred parallelization exists, which is used to define a positive definite metric. Via this metric and Cauchy completion the bundle acquires a boundary, which can be projected to define the conformal boundary of M.

The general mathematical background is the theory of prolongation of G-structures [4], [5]. The essential geometrical ideas, however, were already known by Cartan. Let M^4 be a manifold. A conformal structure C on M^4 is defined as an equivalence class of

Lorentz metrics defined by $g' \sim g$ if $g' = e^{2\sigma} g$, where σ is any real-valued function on M^4. Let us in the following write $g \in C$ for a metric g which is in the conformal class.

A Lorentz metric defines a reduction of $L(M^4)$, the frame bundle, to $O(M^4)$ the bundle of all orthonormal frames. Similarly a conformal structure C defines a reduction of $L(M)$, which is given by all frames, which are orthonormal for some metric $g \in C$. This bundle is denoted by $P(M)$ and its structure group is

$$CO: = \{A; A \in Gl(4,R), \eta(A\xi, A\xi') = c^2 \eta(\xi,\xi')\} \tag{2.1}$$

where η is a flat metric of Lorentz signature on the vector space R^4. Clearly CO is the direct product of the Lorentz group and the group of dilatations.

Hence $u \in P(M)$ is a frame e_a at a point $x \in M$ which is orthonormal for some $g \in C$. Any other frame at x which in $P(M)$ is then given by $\overline{e}_a = A_a{}^b e_b$ with $A_a{}^b \in CO$. In the following we will consider $P(M)$ always as a submanifold of $L(M)$.

Choose any $g \in C$. Then g determines uniquely a torsion-free connection Γ on $L(M)$. Denote by $\{H_u\}$ the collection of horizontal subspaces defined by Γ on $L(M)$, [1],[2]. Let B_i be the standard horizontal fields, which are defined by

$$(B_i)_u \in H_u, \quad \pi_*(B_i)_u = e_i \quad \text{if} \quad u = (e_i) \quad . \tag{2.2}$$

These fields are everywhere non-zero in $L(M)$, tangent to $P(M)$ and carry all information about the connection Γ.

One metric g determines uniquely a horizontal subspace at any point $u \in L(M)$. In contrast to this we get from a conformal structure a whole collection of subspaces at any point.

What is the relation between H_u, H'_u defined by g and $g' = e^{2\sigma} g$? It is determined by the relation between the two derivative operators ∇, ∇' determined by g, g' which can be expressed by the following relation between the Christoffel symbols in local coordinates

$$\Gamma'^e_{is} = \Gamma^e_{is} + \delta^e_i \sigma_{|s} + \delta^e_s \sigma_{|i} - g_{is} g^{ej} \sigma_{|j} \quad . \tag{2.3}$$

From this and the expression for B_i given in [2,6] one finds

$$B'_i = B_i - (\delta^e_i \tilde{\sigma}_{|k} + \delta^e_k \tilde{\sigma}_{|i} - \tilde{g}_{ik} \tilde{g}^{es} \tilde{\sigma}_{|s}) \overset{*}{E}{}^k_e \quad . \tag{2.4}$$

Here $\overset{*}{E}{}^k_e$ denote the standard vertical vector fields in $L(M)$ and $\tilde{\sigma}_{|i}$, $\tilde{g}_{ik}$ are functions defined on $L(M)$ in the following way: if $g(X,Y)$ is a tensor field on M^4, then $\tilde{g}_{ik}(u)$ is defined by $\tilde{g}_{ik}(u) = g(X_i, X_k)$

if $u = (X_i)$. Hence $\tilde{\sigma}_{|i}$, $\tilde{g}_{ik}(u)$ are just the components of $d\sigma$ and g in the frame u. (This convention will be used for any tensor field on M^4.) In particular $\tilde{g}_{ik}\tilde{g}^{rs}$ is constant on $P(M)$, i.e. for conformal frames; more precisely $\tilde{g}_{ik}\tilde{g}^{rs} = \eta_{ik}\eta^{rs}$ on $P(M)$ where η_{ik} is diagonal $(-1,1,1,1)$ as well as η^{ik}. Therefore we find that at $u_o \varepsilon P(M)$ the class of preferred horizontal subspaces is spanned by

$$(B'_i)_{u_o} = (B_i)_{u_o} - (\delta^l_i b_k + \delta^l_k b_i - \eta_{ik}\eta^{ls} b_s)\overset{*}{E}{}^k_l \quad . \tag{2.5}$$

The vertical fields $\overset{*}{E}_\alpha$, $\alpha = 1,\ldots 7$ are defined by

$$k < l, \; \overset{*}{M}{}^k_l := \overset{*}{E}{}^k_l - \eta_{lt}\eta^{ks}\overset{*}{E}{}^t_s, \; \overset{*}{E}_7 = \delta^i_k \overset{*}{E}{}^k_i \quad . \tag{2.6}$$

These fields are tangent to the fibres in the frame bundle and span the tangent space of the fibres of $P(M)$ since the Lie algebra of CO is spanned by the Lie algebra $E^k_e - \eta_{et}\eta^{ks}E^t_s$ of the Lorentz group and δ^i_k. (E^k_e is the matrix with 1 in the k-th row and l-th column and 0 at all other entries.) The $\overset{*}{E}_\alpha$ on the fundamental vector field defined by the group action of CO on $P(M)$.

Using this we observe that a conformal structure defines a collection of frames of $P(M)$. At a point u_o a particular frame is given by the vectors $(B_i, \overset{*}{E}_\alpha)$. All frames at u_o are given by changing B_i according to (2.5). Using the abbreviation

$$S^l_{ik}(b) := \delta^l_i b_k + \delta^l_k b_i - \eta_{ik}\eta^{ls} b_s \quad , \tag{2.7}$$

we realize that the abelian group $\overset{*}{R}{}^4$ acts on these frames according to

$$f_{c_i} : \begin{cases} B'_l \longrightarrow B'_i + S^l_{ik}(c_r)\overset{*}{E}{}^k_l \\ \overset{*}{E}_\alpha \longrightarrow \overset{*}{E}_\alpha \end{cases} \quad . \tag{2.8}$$

Clearly $S^l_{ik}(c + c') = S^l_{ik}(c) + S^l_{ik}(c')$ implies that $c_i \rightarrow f_{c_i}$ is an isomorphism of $\overset{*}{R}{}^4$ into the group of linear transformations acting on the frames $(B'_i, \overset{*}{E}_\alpha)$.

This implies that all the frames $(B'_i, \overset{*}{E}_\alpha)$ characterized by (2.5), (2.6) and labelled by $b_i \varepsilon \overset{*}{R}{}^4$ define a subset of $L(P(M))$, the frame bundle of $P(M)$. This subset is obviously a reduction of $L(P)$ to a principle subbundle with the abelian structure group $\overset{*}{R}{}^4$. We denote this bundle by $P^1(P)$ and the projection by $\pi^1: P^1(P) \longrightarrow P$.

Let us recapitulate the meaning of a point $z_o \varepsilon P^1$. It determines a point $u_o = \pi^1(z_o)$ in $P(M)$ and a frame $(B_i, \overset{*}{E}_\alpha)_{u_o}$, where the B_i are the standard horizontal sections of some connection at

u_o determined by some metric $g \in C$.

Another way of saying the same thing is this: $z_o \in P^1$ determines uniquely (and is determined by) an equivalence class of metrics $g \in C$ which coincide at x_o and have the same connection at x_o. $(x_o = (\pi \circ \pi^1)(z_o))$.

What is the advantage in passing from the space-time M^4 to $P(M)$ and to the even more complicated P^1? The basic reason is the following. The existence of a unique connection which is determined by a metric g is equivalent to the fact that g determines a complement to the fibres in the bundle of orthonormal frames. For a conformal class this is not the case; we just get a whole collection of complements, spanned by all the $(B_i)_u$ defined in (2.5). All these complements at points of $P(M)$ form the points of $P^1(P)$. We will show that we get on this space $P^1(P)$ preferred complements to the fibres in $P^1(P)$. The importance of the existence of such a preferred complement will then become obvious.

To determine the complement pick an arbitrary $z_o \in P^1(P)$ and consider all $g \in C$ which have the property that the cross section $f : P \longrightarrow P^1$, defined by the connection of g, passes through z_o. (These are just the metrics described above, which define the same connection at $x_o = (\pi \circ \pi^1)(z_o))$. What determines the tangent space of these sections? Let $f : u \longrightarrow (B_i)_u$ be one such section, then all the others can be given as

$$u \longrightarrow B_i' = B_i - S^1_{ik}(\tilde{\sigma}_{|j})\overset{*}{E}{}^k_1 \quad , \tag{2.9}$$

provided $\tilde{\sigma}_{|k}(u_o) = 0$. The tangent space at the point z_o of such a section is determined by the first order change of $\tilde{\sigma}_{|k}$ around u_o. Therefore two such sections given by $\tilde{\sigma}$, $\tilde{\sigma}'$ are tangent at z_o if and only if

$$(B_i)_{u_o}\tilde{\sigma}_{|k} = (B_i)_{u_o}\tilde{\sigma}'_{|k} \; , \quad (\overset{*}{E}_\alpha)_{u_o}\tilde{\sigma}_{|k} = (\overset{*}{E}_\alpha)_{u_o}\tilde{\sigma}'_{|k} \quad . \tag{2.10}$$

Now $\tilde{\sigma}_{|k}(u_o) = 0$ implies $(\overset{*}{E}_\alpha\tilde{\sigma}_{|k})_{u_o} = 0$ (if the components of a tensor vanishes in one frame, then in any!). Furthermore the very definition of the covariant derivative implies

$$(B_i)_{u_o}\tilde{\sigma}_{|k} = (\widetilde{\sigma_{|k\|i}})u_o \tag{2.11}$$

where the covariant derivation $\sigma_{|i\|k}$ is performed with the connection determined by $(B_i)_{u_o}$. Hence there is a 1-1 relation between $(\widetilde{\sigma_{|i\|k}})u_o$ and the tangent spaces of the sections described above.

Second derivatives of the conformal factor determine the relation between the curvature tensor of the two metrics. The part which cannot be changed is the Weyl tensor. The Ricci tensor changes according to $(n = 4)$

$$R'_{ik} = R_{ik} + 2\sigma_{|i\|k} - 2\sigma_{|i}\sigma_{|k} + g_{ik}g^{rs}(\sigma_{|r\|s} + 2\sigma_{|r}\sigma_{|s}) \qquad (2.14)$$

(where R_{ik} is the Ricci tensor of g, R'_{ik} the Ricci tensor of $g'_{ik} = e^{2\sigma}g_{ik}$ and covariant derivative is performed with respect to g).

Using (2.14) for the corresponding functions on P(M) taking into account that $\tilde{\sigma}_{|i}(u_o) = 0$, one gets, at u_o,

$$\tilde{R}'_{ik} = \tilde{R}_{ik} + \widetilde{2\sigma_{|i\|k}} + \eta_{ik}\eta^{rs}\widetilde{\sigma_{|r\|s}} \quad . \qquad (2.15)$$

This implies that there exists a unique $\sigma_{|i\|k}(x_o)$ such that $\tilde{R}'_{ik}(u_o) = 0$. Therefore a unique complement to the fibre at z_o in $P^1(P)$ is determined by the condition $R'_{ik}(x_o) = 0$.

Hence once a connection in the conformal class is given at x_o, one can uniquely determine the change of the connection to first order by the condition that its Ricci tensor vanishes at x_o. This is the geometrical implication of the existence of the unique complement in $P^1(P)$. Let us denote this complement by H_z.

From the distribution of horizontal subspaces H_z in $P^1(P)$ one gets immediately the following unique parallelization:

Choose a basis $A^m \varepsilon \overset{*}{R}^4$ and denote by $\overset{*}{A}^m$ the corresponding vector field on $P^1(P)$ (tangent to the fibres) which are determined by the fibre group $\overset{*}{R}^4$ on P^1.

For any $z \varepsilon P^1$ we define vector fields Z_i, Z_α by the conditions:

$$(Z_i)_z \varepsilon H_z, \quad \pi^1_*(Z_i)_z = (B_i)_u \quad \text{if } z = (B_i, \overset{*}{E}_\alpha)_u \quad , \qquad (2.16)$$

$$(Z_\alpha)_z \varepsilon H_z, \quad \pi^1_*(Z_\alpha)_z = (\overset{*}{E}_\alpha)_u \quad \text{if } z = (B_i, \overset{*}{E}_\alpha)_u \quad .$$

Clearly $\overset{+}{A}^{*m}$, Z_i, Z_α define a parallelization on P^1 determined intrinsically by the conformal structure C on M.

Using the vector fields of the parallelization one gets a unique positive metric g on P^1, by demanding that $\overset{*}{A}^1$, Z_i, Z_α are othonormal for g. This way (P^1, g) becomes a positive definite Riemannian space.

The distance d(z,z') defined by the positive definite metric g makes P^1 into a metric space which has a unique Cauchy completion $\bar{P}^1$ [7]. In this way we get boundary points of $\bar{P}^1$. To attach a boundary to the base manifold M^4 one proceeds as follows. In [3] it is established that P^1 can also be considered as a bundle over M, which is denoted by $P^1(M)$. The group acting on $P^1(M)$ is the semi-direct product of CO and $\overset{*}{R}{}^4$ given by the obvious action of CO on $\overset{*}{R}{}^4$. The fundamental vector fields corresponding to this action on P^1 are just $\overset{*}{A}{}^m$ and Z_α. If we consider P^1 as a bundle over M we denote by $\hat{H}_z$ the complement to the fibres in $P^1(M)$ spanned by (Z_i).

Furthermore it is proved in [3] that the action of the structure group on $P^1(M)$ is uniformly continuous with respect to $d(z,z')$. This provides us with a unique extension of the action of the structure group on P^1 onto the Cauchy completion $\bar{P}^1$.

We define the $\bar{M}^4 = \partial_c M^4 \cup M^4$ as the quotient space P^1/G, (if G is the structure group acting on $\bar{P}^1(M)/G$) on which the usual quotient topology is defined.

In this way any space-time M^4 get uniquely a boundary $\partial_c M^4$ intrinsically defined by the conformal structure of the space-time. $M^4 \cup \partial_c M^4$ is in general just a topological space.

3. INTERPRETATION OF BOUNDARY POINTS

In section 2 we attached to any space time a boundary in an unfortunately rather abstract way. The main purpose of this section is to give an interpretation of these boundary points using properties of M^4. From the consideration in section 2 it is obvious that the vector fields Z_i are the relevant part of the structure on $P^1(M)$.

Let us consider a curve $z(\lambda)$ which is always tangent to Z_1,

$$\dot{z}(\lambda) = (Z_1)_{z(\lambda)} \quad . \tag{3.1}$$

Such a curve determines the following curves in CO(M) and M:

$$u(\lambda) = \pi^1(z(\lambda)), \quad x(\lambda) = \pi(u(\lambda)) \quad . \tag{3.2}$$

Hence $z(\lambda)$ just consists of a curve $x(\lambda)$, a frame field $e_a(\lambda) = u(\lambda)$ along $x(\lambda)$ and a connection $H_{u(\lambda)}$ along $u(\lambda)$.

How are these objects now determined by (3.1)? Consider $H_u(\lambda)$ along $u(\lambda)$ in P(M). This collection of horizontal subspaces can certainly be extended in a lot of ways to define a linear connection Γ in a neighbourhood of $x(\lambda)$. This connection determines a section $f : P \longrightarrow P^1$ which contains $z(\lambda)$.

Hence the tangent space of the section contains Z_1 along $z(\lambda)$. Let R_{ik} be the Ricci tensor of Γ ; then the definition of the horizontal subspace in P^1 implies

$$R_{ik}\dot{x}^k = 0 \tag{3.3}$$

$$(\pi^1(\dot{z}) = \dot{u} \ , \quad \pi(\dot{u}) = \dot{x}) \ .$$

Let furthermore $u(\lambda)$ be given by the frame $e_a(\lambda)$ along $x(\lambda)$. Then because of (2.16)

$$\dot{u}(\lambda) = \pi^1(Z_1) = (B_1)_{u(\lambda)} \ . \tag{3.4}$$

Therefore $u(\lambda)$ is an integral curve of the standard horizontal field B_1 given by the connection Γ. Hence $x(\lambda)$ is a geodesic of Γ and e_a is parallel along $x(\lambda)$ with respect to Γ; furthermore $e_1 = \dot{x}$ is implied from $\pi(B_1) = e_1$.

Collecting all these we have the following: If $z(\lambda)$ is an integral curve of Z_1, then

$$z(\lambda) = \{x(\lambda), e_a(\lambda), \Gamma(\lambda)\} \tag{3.5}$$

satisfy the conditions: $\dot{x}(\lambda) = e_1$, e_a is parallel along $x(\lambda)$ with respect to $\Gamma(\lambda)$; if Γ is extended to a connection in the conformal class around $x(\lambda)$, the $R_{ik}\dot{x}^k = 0$ holds along $x(\lambda)$.

Considering an integral curve of $\xi^i Z_i$, ξ^i = const. one gets the corresponding statement. Instead of $\dot{x} = e_1$ one has then just $\dot{x} = \xi^a e_a$.

The interpretation of the length of an integral curve of $\xi^i Z_i$ is now straight forward. Because the Z_i are orthonormal with respect to the positive definite metric defined on P^1 and ξ^i is constant, the length is just $\Sigma(\xi^i)^2$ times the value of the parameter λ. This is however the affine parameter of the geodesic determined by $z(\lambda)$.

A vector field $\xi^i Z_i$ has a unique integral curve provided initial data $z(0)$ are given. Hence there should exist a unique system of differential equations, which determine $x(\lambda)$, $e_a(\lambda)$ and $\Gamma(\lambda)$. We can derive those equations in the following way: Choose any metric g in the conformal class and let Γ be its connection. Let x^i be local coordinates in which Γ is given by Γ^l_{ik}. Let the connection $\Gamma'(\lambda)$ we try to determine be given by

$$\Gamma'^l_{ik} = \Gamma^l_{ik} + S^l_{ik}(b_j) \ , \tag{3.6}$$

and suppose

$$(x^i(\lambda),\ e^k_a(\lambda)\ \frac{\partial}{\partial x^k}\ ,\ \Gamma'^{l}_{ik}\ (x(\lambda))) \tag{3.7}$$

defines an integral curve of $\xi^i Z_i$. Then the considerations above imply that the Ricci tensor of Γ' satisfies $R'_{ik}\dot{x}^k = 0$ which can be expressed in the following way: $(n = 4)$

$$\begin{aligned} 2(\nabla_{\dot{x}} b)_i &= -(R_{ik} - \tfrac{1}{6} Rg_{ik}) + S^l_{ik}(b)\dot{x}^k b_l \\ &= -(R_{ik} - \tfrac{1}{6} Rg_{ik}) + 2b_i b_k \dot{x}^k - g_{ik}\dot{x}^k(b_r b^r) \ . \end{aligned} \tag{3.8}$$

The curve $x^i(\lambda)$ has to be a geodesic of Γ' hence

$$\nabla'_{\dot{x}}\dot{x} = 0 \Longleftrightarrow (\nabla_{\dot{x}}\dot{x})^i = -S^i_{lk}(b)\dot{x}^k\dot{x}^l \ . \tag{3.9}$$

Finally the frame $e_a(\lambda)$ has to be parallel with respect to Γ':

$$\nabla'_{\dot{x}} e_a = 0\ ,\quad \dot{x}(0) = \xi^i e_a(0) \ . \tag{3.10}$$

The equations (3.8),(3.9),(3.10) form a system of ordinary differential equations for the functions

$$x^i(\lambda),\quad b_k(\lambda),\quad e^i_a(\lambda) \ . \tag{3.11}$$

Clearly a unique solution exists provided initial values are given. The system decouples, because (3.8),(3.9) are sufficient to determine $x(\lambda)$ and $b(\lambda)$.

The connection ∇ which is used in (3.8),(3.9) is only needed to write down an explicit differential equation for b_i. The connection

$$\nabla' \Longleftrightarrow \Gamma^i_{kl} + S^i_{kl}(b_j) \tag{3.12}$$

is the invariant object determined along $x(\lambda)$. If we would change ∇, then just b_j would change, ∇' being the same connection along $x(\lambda)$.

So far we considered curve $z(\lambda)$ which satisfied

$$\dot{z}(\lambda) = \xi^i (Z_i)_{z(\lambda)},\quad \xi^i = \text{const} \ . \tag{3.15}$$

A curve $z(\lambda)$ satisfying the above equation with $\xi^i(\lambda)$ is called horizontal. The geometrical construction in section 2 implies that locally every curve $x(\lambda)$ in M has a unique horizontal lift $z(\lambda)$, provided $z(0)$ is prescribed in $(\pi \circ \pi^1)^{-1}(x(0))$. The considerations leading to (3.8)-(3.10) where $x(\lambda)$ is a given curve in space time. The length of the curve is clearly the generalized affine length of $x(\lambda)$ with respect to the connection determined by the lift $z(\lambda)$.

In [3] a theorem is proved which establishes that for the construction of the boundary it is sufficient to consider horizontal curves in P^1.

Let us conclude this section with some examples: Consider the timelike geodesic $t(\lambda) = \lambda$, $1 \leq \lambda < \infty$, $x^\nu = 0$. We can find a conformal factor $e^{2\sigma}$ such that $e^{2\sigma}\eta_{ik}$ has vanishing Ricci tensor along the geodesic. The condition is

$$2\ddot{\sigma} = 2(\dot{\sigma})^2 - (\dot{\sigma})^2 \tag{3.17}$$

which has the general solution for $\dot{\sigma} \neq 0$

$$\sigma = -2 \int_1^t \frac{d\lambda}{\lambda + c} + \sigma(1), \quad \dot{\sigma}(1) = \frac{-2}{1 + c} \quad . \tag{3.18}$$

For $c > -1$ $\sigma(t)$ is defined for $1 \leq \lambda < \infty$ and $\sigma = \ln(t + c)^{-2}$. For the metric $e^{2\sigma}\eta$ on the curve $t(\lambda) = \lambda$, $x^\nu = 0$ is still a geodesic, whose length is

$$L = \int_1^\infty e^{\sigma(t)} dt = \frac{1}{1 + c} \quad . \tag{3.19}$$

Hence we have constructed horizontal curves which have finite length. We also realize that for $c \to -1$ the length of the lift tends to infinity. One also realizes that for $c < -1$ the conformal factor σ is only defined for $0 \leq \lambda < \lambda_o$ and that the corresponding lifts have finite length.

4. APPLICATIONS TO ASYMPTOTICALLY FLAT SPACE-TIMES

In Geroch's contribution to these proceedings the definition of asymptotically flat space time is described. The two basic notions are null infinity and space-like infinity. In both cases one passes from the space time metric $\tilde{g}$ by conformal rescaling to an unphysical metric $g = \Omega^2\tilde{g}$ which has an extension through the "points at infinity", which are given by $\Omega = 0$. The points at infinity form a null hypersurface in the case of null-infinity and one point which is added to a Cauchy surface in the case of space like infinity.

This procedure poses immediately the question whether the collection of boundary points is unique or not. Is the null hypersurface at null infinity intrinsically defined by the space time or are there possible conformal factors and extensions in which for example Ω would vanish on a regular 2-dimensional surface. This is not the case and the conformal boundary offers a particular simple way to prove this.

Consider the case of null infinity. The definition implies that the conformal structure C of space time can be extended to a conformal C' in which J is a regular null hypersurface. This implies in particular that P^1, the bundle determined by the conformal structure C, is an open submanifold of P'^1, the corresponding bundle of the extended conformal structure C'. Clearly the positive definite metric g' defined on P'^1 as described in section 2 is an extension of g defined by the conformal structure C. By the very assumption of the existence of J the Cauchy completion of g on P^1 contains all the fibres through points of J in P'^1, and this hypersurface in P'^1 is uniquely determined by P^1, hence by C, the conformal structure of space-time. Therefore the null hypersurface J, as well as its differential and conformal structure is uniquely determined by the space-time.

Minkowski space can be conformally imbedded into the Einstein universe [3]. The boundary points are $J \cup I^o \cup I^+ \cup I^-$. The same argument as above and the fact that the conformal bundle boundary of the Einstein universe is empty [3], implies that the conformal bundle boundary of Minkowski space is just $J \cup I^o \cup I^+ \cup I^-$, hence this boundary is characterized independent of the extension.

Null infinity J is always contained in the conformal bundle boundary of a weakly asymptotically simple space-time. The example of Minkowski space however indicate, that the boundary may contain more points. In fact the following is true.

Lemma: Let $(M, \tilde{g})$ be a weakly asymptotically simple space-time. Then any generator of J gets a future and past end point in $\partial_C M$, the conformal bundle boundary.

The proof is rather straight forward. We want to apply the equations (3.8)-(3.10) to calculate the length of horizontal lifts of generators of J^+. Hence we need a metric, its connection and its Ricci tensor along a generator of J^+. From Geroch's lecture or from work of Robinson [8] one gets the existence of a metric in the conformal class and coordinates near J^+ such that the following holds:

the metric at J^+, $(r = 0)$ is $(x^o,x^1,x^2,x^3) = (u,r,x^2,x^3)$

$$2dudr + \delta_{AB}dx^A dx^B, A,B = 2,3 \quad . \tag{4.1}$$

The only nonvanishing Christoffel symbols and Ricci tensor components are on J^+:

$$\Gamma^o_{AB},\ \Gamma^C_{A1},\ R_{AB},\ R_{11},\ R_{A1} \quad . \tag{4.2}$$

Furthermore the Ricci scalar vanishes. This implies for the equation for horizontal lifts of the generators: (3.8),(3.9)

$$\begin{aligned} 2\overset{o}{b}_i &= 2b_i(b_k\dot{x}^k) - g_{ik}\dot{x}^k(b_r b^r) \\ 2\ddot{x}^i &= -2\dot{x}^i(b_s\dot{x}^s) \end{aligned} \tag{4.3}$$

if $x^k(\lambda) = (u(\lambda),\ r=0,\ x^A = \text{const.}),\ b_k(\lambda)$ describe a horizontal lift which is integral curve to a certain $\xi^i Z_i$.

These are precisely the same equations as those for a metric in Minkowski space. Therefore horizontal lifts of generators of J^+ satisfy the same equations as horizontal lifts of null geodesics in Minkowski space. As any such null geodesic in Minkowski space has lifts which are finite into the future or into the past, the same holds for J^+, which completes the proof of the lemma.

During the proof we have in fact established that the space formed by all fibres through points on J^+ is isometric to the space of all fibres through a null geodesic in Minkowski space. (The isometry referring to the metric induced by the metric on P^1 on those submanifolds.)

Having established the fact that all generators of J^+ get past end points in the bundle boundary, the obvious next question is to find conditions under which these are all identified, as this in the case for Minkowski space. To find such conditions one has to calculate the bundle metric explicitly which is rather lengthy and will be done elsewhere. The results obtained so far are however:

for stationary space-times, all past endpoints of J^+ are identified. for non stationary space-times the following sufficient condition for the end points of J^+ to be identified was found

$$\lim_{u\to -\infty} u\sigma^o\ (u,\ \theta,\ \phi) = A(\theta,\ \phi) \tag{4.4}$$

exists when σ^o is the asymptotic shear of the outgoing null hypersurface of a Bondi system.

Under the conditions described above one can derive an action of the Poincaré group on J^+. Very little is known so far about the relation between the past end point of J^+ and the future end point of J^-. Furthermore one gets also boundary points for space like hypersurfaces and it is so far unknown whether these are identified with end points of J^+ and J^-.

The results obtained so far clearly indicate that the conformal bundle boundary contains useful, invariant information about space-like infinity. It is however rather difficult to extract this information, because the structure is so complicated.

REFERENCES

1. Schmidt, B.G.: G.R.G. 1, 269 (1971).

2. Hawking, S.W., Ellis, G.F.R., The large scale structure of the Universe, Cambridge, Cambridge University Press 1973.

3. Schmidt, B.G.: Commun. math. Phys. 36, 73 (1974).

4. Sternberg, S.: Lectures on Differential Geometry, Englewood Cliffs, New Jersey, Prentice-Hall, Inc. 1964.

5. Kobayashi, S.: Transformation groups in Differential Geometry. Berlin-Heidelberg-New York: Springer 1972.

6. Schmidt, B.G.: Differential Geometry in Relativity, Astrophysics and Cosmology, ed. by W. Israel, Dortrecht-Holland: D. Reidel Publ. Co. 1974.

7. Hocking, J.G., Young, G.S.: Topology. London, Addison-Wesley 1961.

8. Robinson, D.C.: J. Math. Phys. Vol. 10, 9, 1969.

INDEX